Lecture Notes in Civil Engineering

Volume 440

Lecture Notes in Civil Engineering (LNCE) publishes the latest developments in Civil Engineering—quickly, informally and in top quality. Though original research reported in proceedings and post-proceedings represents the core of LNCE, edited volumes of exceptionally high quality and interest may also be considered for publication. Volumes published in LNCE embrace all aspects and subfields of, as well as new challenges in, Civil Engineering. Topics in the series include:

- Construction and Structural Mechanics
- Building Materials
- Concrete, Steel and Timber Structures
- Geotechnical Engineering
- Earthquake Engineering
- Coastal Engineering
- Ocean and Offshore Engineering; Ships and Floating Structures
- Hydraulics, Hydrology and Water Resources Engineering
- Environmental Engineering and Sustainability
- Structural Health and Monitoring
- Surveying and Geographical Information Systems
- Indoor Environments
- Transportation and Traffic
- Risk Analysis
- Safety and Security

To submit a proposal or request further information, please contact the appropriate Springer Editor:

- Pierpaolo Riva at pierpaolo.riva@springer.com (Europe and Americas);
- Swati Meherishi at swati.meherishi@springer.com (Asia—except China, Australia, and New Zealand);
- Wayne Hu at wayne.hu@springer.com (China).

All books in the series now indexed by Scopus and EI Compendex database!

Rathish Kumar Pancharathi ·
Christopher K. Y. Leung · J. M. Chandra Kishen
Editors

Low Carbon Materials and Technologies for a Sustainable and Resilient Infrastructure

Select Proceedings of CBKR 2023

Springer

Editors
Rathish Kumar Pancharathi
Department of Civil Engineering
National Institute of Technology Warangal
Warangal, Telangana, India

J. M. Chandra Kishen
Department of Civil Engineering
Indian Institute of Science
Bengaluru, India

Christopher K. Y. Leung
Department of Civil and Environmental Engineering
Honk Kong University of Science and Technology
Kowloon, Hong Kong

ISSN 2366-2557 ISSN 2366-2565 (electronic)
Lecture Notes in Civil Engineering
ISBN 978-981-99-7463-4 ISBN 978-981-99-7464-1 (eBook)
https://doi.org/10.1007/978-981-99-7464-1

This Springer imprint is published by the registered company Springer Nature Singapore Pte Ltd.
The registered company address is: 152 Beach Road, #21-01/04 Gateway East, Singapore 189721, Singapore

Paper in this product is recyclable.

Preface

The urgent global call for sustainable and resilient infrastructure has never been more evident. As our societies continue to expand and our infrastructure needs increase, the challenge of meeting these demands while minimizing the environmental impact becomes ever more pressing. The paramount importance of low carbon materials and technologies in achieving a sustainable and resilient built environment cannot be overstated.

This book, *Low Carbon Materials and Technologies for a Sustainable and Resilient Infrastructure*, is a comprehensive compilation of research, insights, and advancements in the field. It serves as a vital resource for professionals, researchers, policymakers, and anyone invested in building a sustainable future. The chapters within this book delve into a wide range of low carbon materials and technologies that hold promise for mitigating the environmental footprint of infrastructure development. From innovative construction materials to energy-efficient technologies, each chapter provides a deep understanding of the latest advancements in the field. The contributors to this book are leading experts from academia, industry, and research institutions who have dedicated their efforts to developing sustainable solutions.

The book covers various aspects, including but not limited to low carbon concrete, sustainable building materials, green energy technologies, concrete and masonry structures, lightweight structures, MCDM methods for selection of construction, materials and technologies, green and renewable energy applications in construction and operation, repair, retrofitting, and rehabilitation, and life-cycle assessment. The chapters explore the potential of these materials and technologies to reduce greenhouse gas emissions, enhance energy efficiency, and improve the overall resilience of infrastructure systems. By sharing their research findings, the authors aim to inspire further innovation and adoption of low carbon approaches in the construction industry. Furthermore, this book highlights the importance of interdisciplinary collaboration in tackling the challenges of sustainability. The integration of knowledge from various fields, such as material science, engineering, architecture, and environmental studies, is crucial to developing holistic solutions. By bringing together experts from diverse backgrounds, this book fosters cross-disciplinary dialogue and paves the way for future collaborations.

It is worth mentioning that this book is a product of the dedication and expertise of the authors, whose research and insights are at the forefront of sustainable and resilient infrastructure development. Their tireless efforts in advancing low carbon materials and technologies have contributed to a growing body of knowledge that will shape the future of the construction industry.

We would like to express our deepest gratitude to all the authors for their invaluable contributions to this book. Their expertise and passion have made this publication possible. We would also like to thank the readers for their interest and commitment to sustainable development. It is through your engagement that the ideas and solutions presented in this book can be implemented and drive positive change. May this book serve as a catalyst for transformative action, inspiring the adoption of low carbon materials and technologies to build a more sustainable and resilient world.

Warangal, India — Rathish Kumar Pancharathi
Kowloon, Hongkong — Christopher K. Y. Leung
Bengaluru, India — J. M. Chandra Kishen

Acknowledgements

This book arose from the work on Cement and Building Koncrete for a Sustainable and Resilient Infrastructure (CBKR 2023), an International Conference conducted at the premier institute NIT Warangal, India, during 28–29 March 2023. We would like to thank all the professors, research scholars, and other delegates of various premier engineering institutions and industries, without whose efforts, this book could not have been possible. We sincerely thank all the peer review members for contributing their valuable inputs and making the reviewing process impartial. We are very much grateful to Priya Vyas, Senior Editor, Applied Sciences and Engineering, Springer Nature, for her guidance throughout the process of publication of the conference proceedings. Also, we are thankful to Mr. Suresh Dharmalingam, Project Coordinator, and staff of Springer Nature, for their constant help in the processing of the book. We are grateful to Director, NIT Warangal, for motivating us to bring out the proceedings as a book.

Warangal, India
Kowloon, Hongkong
Bengaluru, India

Rathish Kumar Pancharathi
Christopher K. Y. Leung
J. M. Chandra Kishen

Contents

About the Editors

Dr. Rathish Kumar Pancharathi is Professor of Civil Engineering at the National Institute of Technology, Warangal. He completed Doctor of Engineering in 2006 with the prestigious Japanese Govt. Monbusho Japanese Government Scholarship and was Post-Doctoral Fellow during 2009–2010 in Japan under the Japan Society for Promotion of Sciences (JSPS).

He has published 235 technical articles including 127 international and national journals and 108 international and national conferences of repute. He has contributed 17 book chapters and authored 3 books. He has three patents granted to his credit and three more patents are in advanced stages of grant. He has guided 11 doctoral students, and 13 more students are presently working with him. He has guided 73 M.Tech. students and 39 B.Tech. batches.

He is a recipient of several awards including Aftab Mufti Medal, Monbusho and JSPS Scholarships of the Japanese Government, Best Engineering Researcher Award from NITW, Heritage Scholarship to pursue research at IST Portugal, Danish Government Scholarship, Italy Government Post-Doctoral Scholarship, Heritage Scholarship under Erasmus Mundus, AIT Fellowship, ASEM-Duo India Fellowship, Slovakian Government Scholarship, Jawaharlal Nehru Memorial Fellowship, Earth Leader Award and Distinguished Alumnus Award. He received two best paper awards. Prof. Kumar has been a member of the Research Advisory Council of the National Council for Cement and Building Materials (NCCBM), Hyderabad, for the past six years.

He is an active member in several Bureau of Indian Standard (BIS) committees like CED-4, i.e. Lime and Gypsum Products of the BIS, Preparation of Handbook of lime for the BIS, Convenor for revision of all parts of Indian Standard Code IS 2542-1978 including tests for gypsum, plaster, concrete, and mortars, and Member of Revision of Indian Standard Code IS 712:1984, i.e. specification for building limes. He is on the IRC B8 Committee of Inspection, Maintenance, and Rehabilitation of Bridges. He is on the editorial board of *Journal of Facta Universitatis*, *Journal of Cement Wapno Beton* among others and is a reviewer of several international journals published by Elsevier, Thomas Telford, Taylor and Francis, etc.

Prof. Christopher K. Y. Leung is a full professor in the Department of Civil and Environmental Engineering at the Hong Kong University of Science and Technology. He served as the head of the Department from July 2009 to June 2015. His research interests include mechanics of composite materials, application of composites in civil engineering, fracture mechanics, and mechanical aspects of fibre-optic sensing. He has published over 300 journal and conference papers and delivered plenary/keynote talks at international conferences in North America, Europe, UK, China, and other countries. Prof. Leung has received a number of research-related awards, including the Best Applied Research Paper Award from ASCE Journal of Composites in Construction (2007), the First-Class Award in Natural Sciences from the China Ministry of Education (2012), and the Second-Class State Natural Science Award from the China State Department (2015).

Prof. J. M. Chandra Kishen is a professor in the Department of Civil Engineering, Indian Institute of Science, Bengaluru, India. He has been actively involved in research, teaching, and industrial consulting activities for the past twenty-seven years. His research areas include fracture and fatigue behaviour of plain and reinforced concrete, cementitious interfaces with cold joints, residual strength prediction of concrete structures, structural health assessment and monitoring, and repair and rehabilitation of damaged structures. He has set up a unique bi-axial fatigue testing facility at IISc for conducting fatigue tests in direct tension, compression, and combinations on large concrete specimens and monitoring the evolution of microcracks through acoustic emission sensing and digital imaging. He was nominated as Fellow of the International Association of Fracture Mechanics of Concrete and Concrete Structures (IAFraMCoS) in 2016. Prof. Chandra Kishen has been contributing as a principal investigator in several research projects sponsored by the Department of Science and Technology, Department of Atomic Energy, Indian Space Research Organizations, and the Indian Railways. He has carried extensive field studies on railway bridges and highway viaducts.

Prof. Chandra Kishen is an active consultant and has helped the industry solve challenging problems involving rehabilitation of metro girders damaged due to construction faults, large IT infrastructure damaged due to fire accidents, highway viaducts damaged due to loss of prestress, and others. He is providing technical support to various irrigation and dam projects, in recognition of which the Government of Karnataka awarded him the Prof. Satish Dhawan Young Engineer State award. Prof. Chandra Kishen is a popular teacher and has delivered numerous technical lectures throughout the country in short-term courses, faculty development programmes, and conferences. He has trained field engineers on good and safe construction practices as part of the quality assurance programme of the Bengaluru Metro.

Low Carbon Sustainable Cementitious Materials

An Analysis on the Material Composition in Chettinad Lime Plasters

Mechineni Shreya, V. S. Athira, Abhishek Tripathi, and Swathy Manohar

1 Introduction

Lime is derived from sedimentary rock limestones. Around 6000 years ago, Egyptians started using lime as a plaster for the pyramids of Giza, followed by the Greeks and Romans [1]. Lime became a prominent building material utilised in various capacities throughout the world until the late nineteenth century when Portland cement with its lower setting time and higher mechanical strength appeared which caused a decline in the use of lime as a binder [2, 3]. Lime is inexpensive, simple to produce, and because it makes it simple to control the flow of water in buildings, it helps with climate control and energy conservation. It has proven to be very effective both economically and environmentally [1, 4].

In India, lime has been used as a construction material from residential houses to great monuments like The Charminar, Ajanta Ellora Caves, Hampi temples, Mughal monuments, etc. India being a land of diversity has an abundance of organic materials which were incorporated in making lime plaster according to their availability [1]. Rice husk, jute fibres, egg white, sticky rice, jaggery, gum, animal glue, blood of animals, plant extracts, and other natural polymers were used as organic materials in India. These organic compounds improve the workability, boost overall binding capacity, and lessen plaster cracking [5].

M. Shreya (✉)
Department of Civil Engineering, National Institute of Technology Warangal, Warangal, Telangana, India
e-mail: Ms812012@student.nitw.ac.in

V. S. Athira · S. Manohar
Department of Civil Engineering, Indian Institute of Technology Bombay, Mumbai, India

A. Tripathi
Department of Humanities and Social Sciences, Indian Institute of Technology Bombay, Mumbai, India

R. K. Pancharathi et al. (eds.), *Low Carbon Materials and Technologies for a Sustainable and Resilient Infrastructure*, Lecture Notes in Civil Engineering 440,
https://doi.org/10.1007/978-981-99-7464-1_1

Evidence for the longevity and durability of lime is present around us in the form of magnificent art décor and monuments. It is also imperative to note that the historic sites are facing threat due to the increasing industrial and traffic activities along with heavy tourism [6]. Therefore, the structures are in need of restoration and preservation. Restoration of ancient structures with the aid of cement was found to be ineffective. This can be owed to the incompatibility of cement in terms of higher strength, low porosity, and presence of soluble salts that caused further distress to the monuments [3, 6, 7]. Even using lime mortars by themselves in restoration projects failed because lime was combined with other additions to enhance its qualities, which modern manufacture does not fully understand [3]. To revive the traditional lime plasters used, identification and characterisation of the materials are vital to produce a lime plaster that is compatible with the original one [8].

In ancient times, it was normal practise to include organic materials that were naturally available, such as dried seeds, unprocessed sugars, and extracts of locally accessible plants, into construction projects [7–9]. Identifying the compositions, proportions, and functions of the materials utilised is the main challenge in characterising historical plasters [6]. Additionally, the nature of organic compounds utilised in conjunction with lime changes with time, making it more challenging to distinguish between the two. It has also been observed that the general level of organic additions changes among archaeological periods as well as between different areas of the monument [5]. Hence, recreating compatible plasters for restoration work or for incorporation in everyday construction requires knowledge about the chemical and mineralogical composition, details on the binder type, and gradation of aggregates along with its quantitative composition [5, 10]. There is scant research on the characteristics of lime-based mortars because the ancient recipes that are preserved do not give us information about how to handle, set, and cure the mortar or plaster. As a result, the tradition and techniques used are completely lost after the Industrial Revolution in many countries [3]. Therefore, in order to develop a standard for the preparation of lime-based mortars and plaster with organic additives, an extensive and detailed analysis of the ancient structures built by them is pertinent. While, there is a lot of research and application regarding the usage of various additives such as fly-ash, ground granulated blast furnace slag (GGBS), and other pozzolanic additives for the improvement of the properties of concrete, there is limited research relating to the use of organics in construction. But there is record of the usage of various organics in archaic times which upon analysis gives us an insight to their compositions and what is the influence of adding additives to the mix. The paper also aims to make a case for the benefits of using organics in daily construction by various case studies on the structures built in the olden days incorporating organics.

1.1 Use of Organics in Lime

The authors Ravi et al. [7] state that the proteins and carbohydrates found in organic materials are what have allowed Charminar to survive. Instead of calcite, they get converted into alcohols of short chain and stabilise the formation of metastable vaterite which is reported to keep the mortar young leading to the durability of structures.

Singh et al. [8] proved through micro-analysis that the lime plaster applied over the basalt rock in Ellora caves is a mixture of aerial lime (dolomitic lime), medium-sized siliceous aggregates with the interesting addition of Cannabis sativa as an organic filler suggesting that the knowledge of hempcrete was available in ancient times. Singh and Sardesai stated that cannabis was used as an insulating agent and to provide strength and durability to the plaster in Ellora cave. As a result, the structure's longevity despite the surrounding factors can be credited to the material's properties, which include hemp's fibrousness and durability. Hemp's ability to control humidity, prevent the growth of pests, demonstrate fire resistance, and possess hygroscopic properties were also very helpful in the preservation of the Ellora caves [11].

It can be noted that proteins operate as air entraining agents in fresh mortars and boost workability, whereas fat works as a waterproof to mortar and limits the water circulation, and carbohydrates on fermentation enhance carbonation within the mortar. As they play a crucial role in the development of hydrated phases and in resisting environmental degradation, organic materials that are high in carbohydrates, proteins, and lipids are typically added to mortars [9].

1.2 *Influence of Fibres in Lime Plastering*

The addition of plant fibres in mortars and plasters was also a common practice in India. Studies showed the existence of jute fibres in the lime plasters used in the twelfth century stepwell used in New Delhi, which lowered the shrinkage of the materials during setting resulting in crack-free plaster [12]. Jute fibres were quite famously used as an additive, and it is a very common natural fibre in India as well as China and Bangladesh [13]. According to various researcher's plant additives, it provides flexural and tensile strength, shows resistance to insects, and is antifungal and antibacterial [6, 11, 14, 15].

The Vadakimnathan Temple at Thrissur, Kerala, is also constructed using laterite blocks bonded with lime mortar in which organics were a part of. In the restoration works, various herbs like jaggery (unrefined sugar), kadukai (Terminalia Chebula), and Oonjalvalli (Cisscus glauca roxb) were added to improve properties like carbonation, plasticity of mix, and enhance durability [1, 9].

Analysis of the Padmanabhapuram Palace situated in Thuckalay, Tamil Nadu, revealed that the addition of herbs like kadukkai (Terminalia Chebula), neelamari (Indigofera tinctoria), hibiscus (Rosa sinensis), palm jaggery (Borassus flabellifer), and aloe vera (Aloe barbadenis) improved the strength by forming hydrated phases and helped in resisting cracks [16].

Apart from India, the incorporation of organics into mortar and plaster has been practiced all over the world. In China, portions of the old city wall of Nanjing were constructed using sticky-rice lime mortar, which could prevent the mortar from decaying initially due to the quick lime digestion process during the preparation of the mortar and the released heat and active oxygen could help kill bacteria. In addition, Ca $(OH)_2$ is converted into $CaCO_3$ in the interior of the mortar slowly. Finally, sticky-rice lime mortar has small pores after curing, which lowers the risk of decay and loss of organics [6]. Another interesting case is the use of blood lime mortar in the making of the base layer of frescoes [6]. The study conducted by Fang et al. [17] examined the effectiveness of blood lime mortar and discovered that the blood proteins' crystalline structure and compact skin allowed them to enhance the mortar's strength, water resistance, and weather resistance. Due to their excellent water resistance, sticky rice and tung oil were used to build mortar for tombs or dams [18]. The enhanced durability and compressive strength of ottoman mortar are thought to be the result of the use of organic fibres in the mortar-making process. The inclusion of fibres also makes the structures more flexible, increasing their ability to withstand earthquakes [14].

In addition to the use of natural polymers, there were also reports of using expensive and locally unavailable materials like the usage of latex from rubber plant for water proofing on adobe masonry in Africa and South America [19].

Since it is obvious that using organic materials was beneficial, their incorporation into modern construction projects as well as restoration work should be done without a doubt. Organic materials not only improve the properties of mortar and plaster, but they are also quite sustainable and lower the total amount of embodied energy because they do nott require any energy for production or transportation because they can be found nearby. A composite use of organics containing carbohydrates, proteins, and lipids will be beneficial based on the needs and resources available because proteins give the mortar adhesive and waterproofing properties, while lipids and fats improve workability and make mortars more hydrophobic and carbohydrates improve the hydraulic component and reduce shrinkage [16]. For instance, the simulation materials advised by the researchers Santhanam and Ramadoss were to use 12.5% admixture of kadukkai and jaggery, which was found to be near to the old mortar in the restoration of the Alamparai fort located in Tamil Nadu and composed of brick masonry [20].

To better understand the nature of the lime plasters, a composite approach of chemical and mineralogical characterisation by performing analysis like X-ray diffraction (XRD), scanning electron microscope (SEM), FTIR, XRF, DTA, and TGA has to be carried out. The aim of this study is to analyse Chettinad plaster in terms of chemical and mineral composition. In addition, it is important to consider how organic elements affect the characteristics of the mortar or plaster into which they are included and to talk about potential raw materials that might be used that are cost-effective, environmentally responsible, and energy efficient.

1.3 The Exceptional Plasters—Chettinad Plasters

Chettinad, located in the southeast corner of Tamil Nadu, consists of beautiful grand residential houses often called mansions or Chettiar palaces. They were founded by the mercantile Nattukottai Chettiar Community [21]. The buildings in Chettinad are a marvel to look at and study. They were primarily made of eco-friendly materials that result in the low embodied energy of the buildings, and hence, studying them will prove useful for the construction industry. In the study [22] the authors discuss the architectural styles and eco-friendly building materials used in Asthangudi houses, which serve to keep the inside cool even when the outside temperature is high. The paper presents the preparation of tiles with sand, cement, and baby jelly using bare hands, clean glass, and a mould which has stayed intact for over 80 years without any need of repair in detail. The machu concept of providing a loft that acts as a store, below which wooden beams run through help in reducing the heat transfer within the buildings, is also discussed. But the most unique feature of the Chettinad houses would be the fascinating use of egg white in plaster with a shiny lustrous finish. In addition to the egg white, jaggery and unripe fruit of kadukkai (gallnut) is also included. The authors RadhaKrishnan and Priya describe how the Chettinad plasters are applied, which entails painting the wall in layers with a mixture of pulverised white seashells, liquid egg white, and lime base. The authors say that, "Polishing the houses with egg white gave a glossy finish that no paint can replicate." [22]

Egg white is used as an additive because it contains protein with a tough, long-chain molecule similar to those found in plastic polymers and glutinous flour, which enhances the binding property of the material in addition to giving it a smooth, reflective finish. Egg shells were once used as aggregate and egg white as an additive. Egg demonstrates that it is an useful organic material for binding, reinforcing, and finishing purposes in construction [23]. The embodied energy of a building depends greatly on the building material used. For a sustainable future, one must look at the usage of eco-friendly raw materials as they not only provide strength, but they come with additional benefits like the smooth finish provided by the addition of egg white. With the rate of degradation accelerating and our culture and values vanishing alongside it, the need to preserve and restore historic buildings is more urgent than ever. As can be seen, lime is the oldest construction material, has the most versatility, and has a structure that has withstood the test of time and significant climatic changes, demonstrating that it is a superior and safer building material than the Portland cement currently in use. Therefore, it is essential to restore lime to the construction sector with necessary modifications. The purpose of this work is to analyse the ancient Chettinad lime plasters to understand the composition and properties, in order to demonstrate the need for additional study on such plasters as well as increased initiative to use organic lime plasters and mortars in construction.

Table 1 Details of the samples collected from the two buildings and their locations in said buildings

Label	Building number	Location
S11	1	The sample was taken from the courtyards inside wall, located on the ground floor of the first building
S12	1	The sample was taken from a bedroom inside wall on the first floor of the first building
S21	2	The sample was taken from one of the pillars near the second building's main entrance
S22	2	The sample was taken from the first-floor bedroom's roof of the second building

2 Materials and Methods

Sample Collection: The lime mortar–plaster considered in this study has been collected from two residential buildings belonging to the town of Sathyamangalam, Erode in the state of Tamil Nadu. The study considered the analysis of four samples from two buildings. Two samples from each of the buildings were collected and labelled as S11, S12, S21, and S22.

Test methods: The four samples (see Table 1) utilised in the experiment are examined using scanning electron microscope (SEM) and X-ray diffraction (XRD) techniques. The XRD technique can be used to determine the mineral composition of the binder and aggregate used to make the lime mortars and plasters. Powder XRD (PXRD) patterns of the four lime plaster samples were recorded on a Rigaku Smartlab 9kw diffractometer using Cu Kα radiation at 10° to 60° 2θ angles and subsequently analysed using Topaz software. The four samples are analysed using SEM and EDS to determine the morphologies, microstructures, and chemical makeup of plasters.

3 Results and Discussion

3.1 XRD Results

XRD analysis helps us in studying the mineral composition of the samples collected. All the samples show a high percentage of quartz indicating the usage of river sand in the making of the lime plaster. Traces of calcium aluminates and magnesite were found from S11 and S12 indicating that pure limestones were not used as raw materials as observed in Figs. 1 and 2. This is also confirmed from the EDS mapping. Existence of small amount of Mg and Al indicate that hydraulic lime was used as the primary raw material. Aragonite, a polymorph of $CaCO_3$, was identified only in the first sample S11, and no trace is detected in the other samples considered, but this does not indicate that it does not exist in the samples, only that they were not

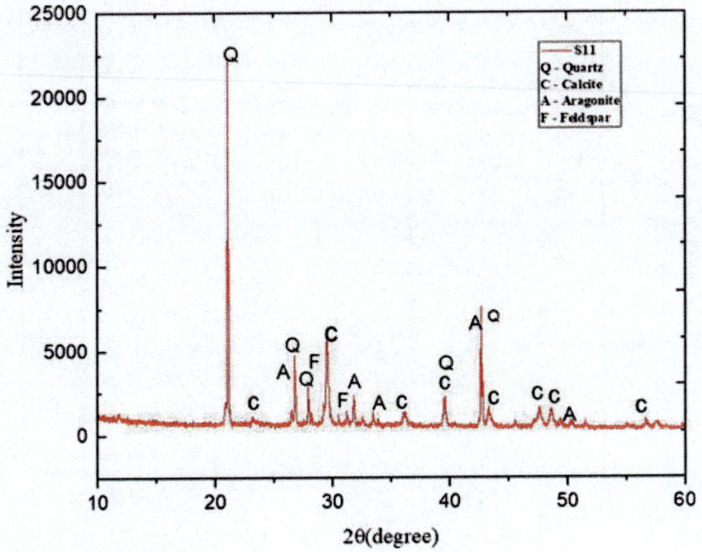

Fig. 1 XRD spectrum of sample S11

present in adequate amounts to be detected. Low percentages of Feldspar are also observed in the sample. Further trace amounts of berlinite and sodium zirconate are detected in sample S21 and S22 proving both the samples have an approximately similar composition in terms of the minerals present, as observed in Figs. 3 and 4. Calcium carbonate is a prominent binder of ancient lime plaster mostly existing in the form of calcite [24]. The study of phase and morphology of $CaCO_3$ is important for evaluation of the performance of mortars or plasters [24, 25]; it provides us with valuable information on what effect time and environment has on it.

3.2 *SEM Analysis Results*

Scanning electron microscopy (SEM) analysis coupled with EDS was done for all four plaster samples. The analysis results along with XRD correlations give better insight into the material morphology. Figure 5 shows the SEM image of plaster sample S11. The existence of rhombohedral crystals is evident that confirms the presence of calcite [26]. The EDS results show the presence of Si and Al in the sample and hence can be confirmed that the lime employed was hydraulic lime. Figure 6 shows the SEM of the sample collected from the same building at a different location (S12).

The nodule-like structures (marked in red in Fig. 6) present in the sample possibly show the presence of calcium-alumino-silicate-hydrate (CASH) [27].

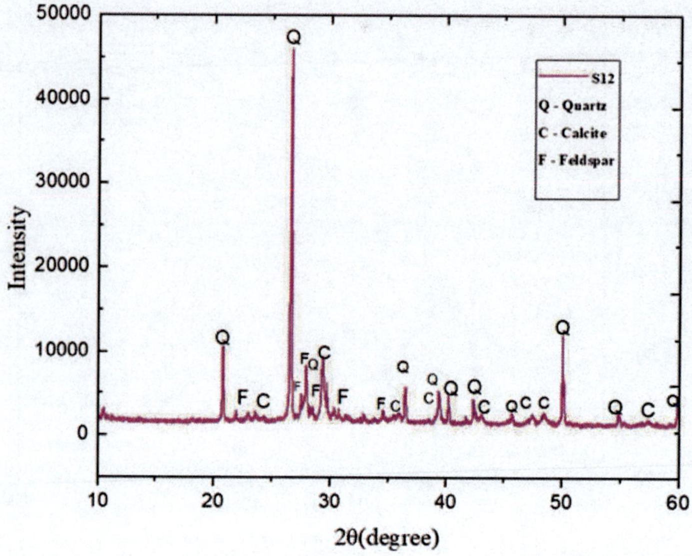

Fig. 2 XRD spectrum of sample S12

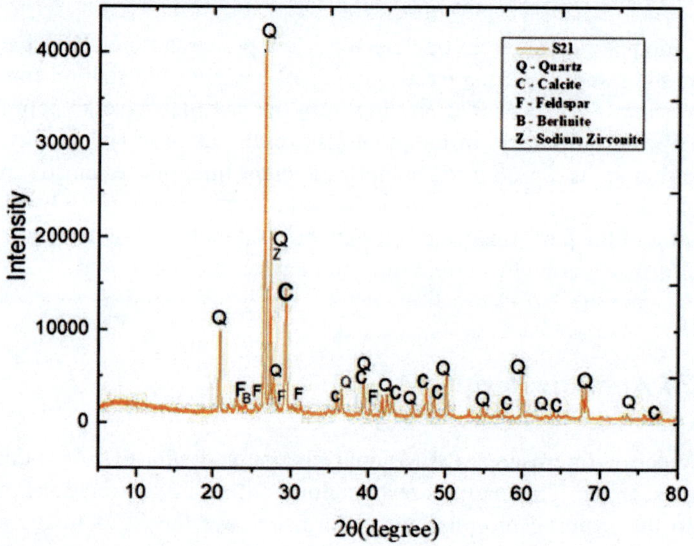

Fig. 3 XRD spectrum of sample S21

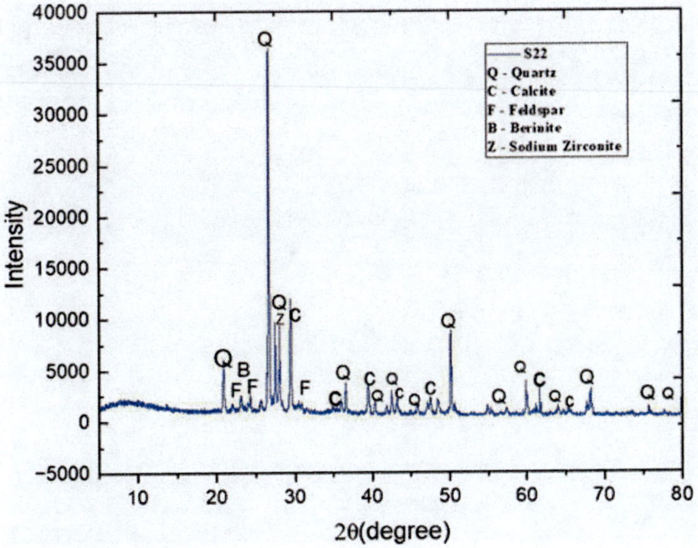

Fig. 4 XRD spectrum of sample S22

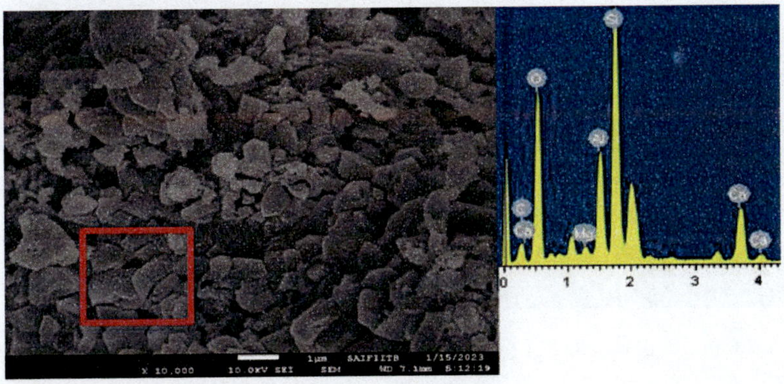

Fig. 5 SEM images of sample S11

Traces of magnesium carbonate were also found in the XRD analysis, thus indicating the use of impure limestones as raw materials.

Figure 7a, b shows the SEM images of samples S21 and S22. From EDS presence of Ca, Al and Si were observed. This indicates that the lime used was hydraulic in nature. The yellow marking on Fig. 7a indicates the presence of hexagonal portlandite crystals. The crystals present on top of portlandite can be precipitates of calcite.

The presence of unreacted portlandite can be attributed to the dense calcite precipitated at the surface, blocking the passage of CO_2. Figure 7b also shows the SEM

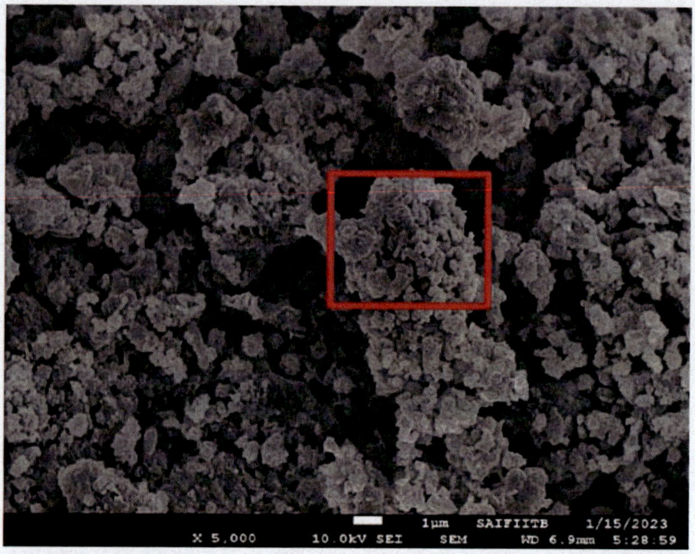

Fig. 6 SEM images of sample S12

Fig. 7 SEM images of sample S21 (**a**) and S22 (**b**)

image of building 2. The presence of dense calcite crystals as well as CASH gel can be noted. However, Mg was absent in samples S21 and S22.

4 Conclusions

1. The revival of lime is essential for the repair and conservation of heritage structures. Understanding of the materials used for the construction of ancient structures will help in their effective repair as modern materials are observed to be incompatible with the authentic traditional materials.

2. The works give an introduction on the various traditional materials used for the making of ancient structures that have been staying for long.
3. The present work gives an insight into the materials used for the smooth, glossy plasters used in Chettinad buildings. The work concludes that the plastering material was hydraulic lime added with organic additives.
4. The presence of high amount of Si and Al along with calcium, as observed from XRD and SEM–EDS, confirms the application of hydraulic lime. The presence of magnesite also confirms the presence of traces of Mg present in the raw limestone used.
5. SEM micrographs show the different forms of calcite precipitated and C–A–S–H formed.
6. The understanding on this will aid in the repair of plasters; also the same can be used for commercial purposes as they give an excellent aesthetic appearance along with thermal comfort.

References

1. Thirumalini P (2011) Study on the performance enhancement of lime mortar used in ancient temples and monuments in India. Indian J Sci Technol 4(11):1484–1487. https://doi.org/10.17485/ijst/2011/v4i11.23
2. Callebaut K, Elsen J, Van Balen K, Viaene W (2001) Nineteenth century hydraulic restoration mortars in the Saint Michael's Church (Leuven, Belgium): natural hydraulic lime or cement? Cem Concr Res 31(3):397–403. https://doi.org/10.1016/S0008-8846(00)00499-3
3. Ventolà L, Vendrell M, Giraldez P, Merino L (2011) Traditional organic additives improve lime mortars: new old materials for restoration and building natural stone fabrics. Constr Build Mater 25(8):3313–3331. https://doi.org/10.1016/j.conbuildmat.2011.03.020
4. Gireesh S (2020) Restoration toolbox the case of Chettinad lime plaster. https://portfolio.cept.ac.in/2020/M/fa/conservation-frameworks-cr4003-monsoon-2020-7450/restoration-toolbox-the-case-of-chettinad-lime-plaster-monsoon-2020-pg190949
5. Singh M, Waghmare S, Vinodh Kumar S (2014) Characterization of lime plasters used in 16th century Mughal monument. J Archaeol Sci 42:430–434. https://doi.org/10.1016/j.jas.2013.11.019
6. Fang SQ, Zhang H, Zhang BJ, Zheng Y (2014) The identification of organic additives in traditional lime mortar. J Cult Herit 15(2):144–150. https://doi.org/10.1016/j.culher.2013.04.001
7. Ravi R, Thirumalini S, Taher N (2018) Analysis of ancient lime plasters—reason behind longevity of the monument Charminar, India a study. J Build Eng 20:30–41. https://doi.org/10.1016/j.jobe.2018.04.010
8. Singh M, Vinodh Kumar S, Waghmare SA (2015) Characterization of 6–11th century AD decorative lime plasters of rock cut caves of Ellora. Constr Build Mater 98:156–170. https://doi.org/10.1016/j.conbuildmat.2015.08.039
9. Thirumalini S, Ravi R, Sekar SK, Nambirajan M (2015) Knowing from the past—ingredients and technology of ancient mortar used in Vadakumnathan temple, Tirussur, Kerala, India. J Build Eng 4:101–112. https://doi.org/10.1016/j.jobe.2015.09.004
10. Schafer J, Hilsdorf H (1993) Ancient and new lime mortars—the correlation between their composition, structure and properties
11. Singh M, Sardesai MM (2016) Cannabis sativa (Cannabaceae) in ancient clay plaster of Ellora Caves, India. Curr Sci 110(5):884–891

12. Singh SK, Dighe B, Singh MR (2020) Characterization of 12th-century brick-lime stepwell plasters from New Delhi, India. J Archaeol Sci Rep 29:102063. https://doi.org/10.1016/j.jasrep.2019.102063
13. Mohanty AK, Misra M (1995) Studies on Jute composites—a literature review. Polym Plast Technol Eng 34(5):729–792. https://doi.org/10.1080/03602559508009599
14. Binici H, Arocena J, Kapur S, Aksogan O, Kaplan H (2010) Investigation of the physico-chemical and microscopic properties of Ottoman mortars from Erzurum (Turkey). Constr Build Mater 24(10):1995–2002. https://doi.org/10.1016/j.conbuildmat.2010.03.013
15. Hejazi SM, Sheikhzadeh M, Abtahi SM, Zadhoush A (2012) A simple review of soil reinforcement by using natural and synthetic fibers. Constr Build Mater 30:100–116. https://doi.org/10.1016/j.conbuildmat.2011.11.045
16. Shivakumar M, Selvaraj T (2020) A scientific study on the role of organic lime mortars of Padmanabhapuram Palace, Thuckalay, Tamilnadu, India. Eur Phys J Plus 135(11):923. https://doi.org/10.1140/epjp/s13360-020-00896-6
17. Fang S, Zhang K, Zhang H, Zhang B (2015) A study of traditional blood lime mortar for restoration of ancient buildings. Cem Concr Res 76:232–241. https://doi.org/10.1016/j.cemconres.2015.06.006
18. Yang F, Zhang BJ, Pan C, Zeng YY (2009) Traditional mortar represented by sticky rice lime mortar—one of the great inventions in ancient China. https://doi.org/10.1007/s11431-008-0317-0
19. Clifton JR (1977) Preservation of historic adobe structures: a status report. Department of Commerce, National Bureau of Standards, Institute for Applied Technology
20. Santhanam K, Ramadoss R (2022) Conservation & restoration of historic mortars at Alamparai fort with valley conical arch, Tamilnadu, India. Constr Build Mater 339:127619. https://doi.org/10.1016/j.conbuildmat.2022.127619
21. Ślączka AA (2018) Mansions of Chettinad. Aziat Kunst 47(3):55–57. https://doi.org/10.3868/25431749-04703010
22. Radhakrishnan S, Priya RS (2014) Eco Friendly materials used in traditional buildings of Chettinadu in Tamil Nadu, India. Am J Sustain Cities Soc 1(3)
23. Gubba N, Jebakumar EK (2019) Egg as an organic building material a comparative study and understanding in Indian context. https://www.scientific.net/KEM.803.267?utm_source=researcherapp&utm_medium=referral&utm_campaign=RESR_MRKT_Researcher_inbound
24. Singh M, Vinodh Kumar S, Waghmare SA, Sabale PD (2016) Aragonite–vaterite–calcite: polymorphs of $CaCO_3$ in 7th century CE lime plasters of Alampur group of temples, India. Constr Build Mater 112:386–397. https://doi.org/10.1016/j.conbuildmat.2016.02.191
25. Hansen EF, Rodríguez-Navarro C, Balen K (2008) Lime putties and mortars. Stud Conserv 53(1):9–13. https://doi.org/10.1179/sic.2008.53.1.9
26. Cizer Ö, Rodriguez-Navarro C, Ruiz-Agudo E, Elsen J, Van Gemert D, Van Balen K (2012) Phase and morphology evolution of calcium carbonate precipitated by carbonation of hydrated lime. J Mater Sci 47(16):6151–6165. https://doi.org/10.1007/s10853-012-6535-7
27. Padmaraj D, Arnepalli DN (2021) Mechanism of carbonation in lime-stabilized silty clay from chemical and microstructure perspectives. Int J Geosynth Ground Eng 7(4):74. https://doi.org/10.1007/s40891-021-00318-2

Appraisal of Mechanical Properties of Fly Ash-Based Geopolymer Mortar Augmented with GGBS and Graphene Oxide

B. Rajmohan, Nalla Harish, R. Ramesh Nayaka, and Kim Hung Mo

1 Introduction

Cement concrete is a very popular and long-lasting construction material in practice since decades. But, the cement, a main binding matter in concrete, is inducting hazardous impact on the environment by releasing massive amounts of carbon dioxide [1, 2]. To minimize the ill effects of cement usage, several attempts are made in the past to find an alternate binder to replace the cement. Consequent results of these efforts have emerged geopolymer binders as promising future binder materials [3, 4]. The strength development in geopolymer binders occurs by the formation of aluminosilicate gel during the reaction between alumina and silica available in the precursor material and the alkaline solution [5–8]. Fly ash, an industrial waste, comprises a good amount of alumina and silica that makes it suitable to use as a precursor material for geopolymer binders. However, the previous research has indicated that the pace of geopolymerization reaction with fly ash (as a precursor) is very gradual and needs to be fastened to facilitate early high strength attainment [9, 10]. Byproduct of steelmaking facilities, ground granulated blast furnace slag (GGBS), is widely accessible and contains significant amounts of alumina and silica. Use of GGBS as a basic raw constituent for geopolymer binders makes it to set quickly and helps in achieving high early compressive strength [11]. Though, GGBS is useful in attaining high initial strengths, it creates some issues with fluidity, initial setting time, and volume stability [12]. Further, in the recent days, many researchers are looking toward to make use of benefits of nanomaterials for improving the properties

B. Rajmohan (✉) · N. Harish · R. Ramesh Nayaka
Civil Engineering Department, National Institute of Technology, Warangal, India
e-mail: rajmohan.rcr@gmail.com

K. H. Mo
Department of Civil Engineering, Faculty of Engineering, Universiti Malaya, Kaulalmpur, Malaysia

R. K. Pancharathi et al. (eds.), *Low Carbon Materials and Technologies for a Sustainable and Resilient Infrastructure*, Lecture Notes in Civil Engineering 440,
https://doi.org/10.1007/978-981-99-7464-1_2

of several construction materials. Nanoparticles help to alter the microstructure of geopolymer binders at the atomic level leading to significant improvement in its fresh and hardened characteristics. Graphene oxide is one such nanomaterial, widely chosen for use in concrete to investigate its impact on initial, hardened, and long-term attributes of concrete [13]. Therefore, the current work uses a mixture of GGBS and fly ash as a precursor to create geopolymer mortars that also contain graphene oxide as a performance enhancer to amend strength properties of geopolymer binder.

2 Materials and Methods

2.1 *Material Constituents and Mixture Design*

For making the geopolymer mortar, fly ash with GGBS mixture was employed as precursor source. Fly ash and GGBS were delivered by the Ramagundam thermal power plant and Astrra Chemicals, Chennai, respectively. Sodium hydroxide pellets (NaOH) and sodium silicate solution (Na_2SiO_3) were also acquired from Astrra Chemicals, Chennai. As a filler for the geopolymer mortar, river sand was used that had a specific gravity of 2.55 with water absorption at 1.11%. Graphene oxide (GO) was procured from Vruksha composites, Tenali, Andhra Pradesh. Superplasticizer, Conplast SP 430, was used to attain the requisite consistency of mortar along with the potable water. Table 1 lists the physical characteristics and oxide compositions of base constituents and GO. Pictures of all ingredients selected for experiment are displayed in Fig. 1.

The parametric principles listed in Table 2 were used to design the mixture proportions for geopolymer mortar.

Initial trial mixtures were designed to determine an optimum percentage of fly ash to be replaced by GGBS to attain enhanced mechanical properties in comparison

Table 1 Physical and chemical characteristics of materials

Physical characteristics of materials	Name of the material		
	Fly ash	GGBS	Graphene oxide
Specific gravity	2.17	2.91	0.48
Oxide composition (% by mass)			
SiO_2	61.35	39.85	2.06
Al_2O_3	14.23	22.83	–
Fe_2O_3	19.76	6.31	0.94
CaO	0.21	26.14	2.38
MgO	2.91	3.67	17.00
SO_3	–	–	75.41

Fig. 1 Materials used to cast the geopolymer mortar

Table 2 Parametric particulars of mixture proportion

Density of geopolymer mortar mix	2200 kg/m^3
Alkaline solution	Combination of NaOH and Na_2SiO_3
Molarity of NaOH solution	12 M
Na_2SiO_3 to NaOH ratio	2.5
Ratio of alkaline liquid to binder	0.45
Fine aggregate to binder ratio	1.45

with the geopolymer mortar made of fly ash alone. Mixture design details are as given in Table 3.

Based on the tests conducted for mechanical properties, it was noticed that the substitution of fly ash by GGBS at 30% would give better results. So, for such mixture, GO was incorporated at different weight percentages of binder to further

Table 3 Mixture types for optimizing replacement level of fly ash by GGBS

Mix type	Binder (kg/m^3)			River sand (kg)	Alkaline liquid (kg/m^3)			Water (kg/m^3)	SP (kg/m^3)
	FA	GGBS	Total		NaOH	Na_2SiO_3	Total		
M0	724.13	0	724.13	1050	93.14	232.85	325.99	72.41	28.96
M10	651.72	72.41	724.13	1050	93.14	232.85	325.99	72.41	28.96
M20	579.30	144.82	724.12	1050	93.14	232.85	325.99	72.41	28.96
M30	506.89	217.24	724.13	1050	93.14	232.85	325.99	72.41	28.96
M40	434.48	289.65	724.13	1050	93.14	232.85	325.99	72.41	28.96
M50	362.06	362.06	724.12	1050	93.14	232.85	325.99	72.41	28.96

Table 4 Mixture proportions for optimizing % of GO

Mix type	Binder (kg/m^3)			Graphene oxide (%)	River sand (kg/m^3)	Alkaline liquid (kg/m^3)			Water (kg/m^3)	SP (kg/m^3)
	FA	GGBS	Total			NaOH	Na_2SiO_3	Total		
MG1	506.89	217.24	724.13	0.05	1050	93.14	232.85	325.99	72.41	28.96
MG2	506.89	217.24	724.13	0.10	1050	93.14	232.85	325.99	72.41	28.96
MG3	506.89	217.24	724.13	0.15	1050	93.14	232.85	325.99	72.41	28.96

enhance the strength of mix. Table 4 displays the various mix proportions used to produce the GO-based geopolymer mortar.

2.2 *Preparation of Alkaline Solutions and Casting of Specimens*

NaOH solution of 12 M concentration was prepared well in advance and was cooled to the room temperature before adding requisite quantity of Na_2SiO_3 solution to it. This alkaline liquid was formed at least 1 day before it was used in geopolymer mortar mix. All the dry materials were initially poured into the pan mixer and mixed thoroughly to attain a uniform mixture. Predetermined amount of readily available alkaline solution, water, and superplasticizer were put into the homogeneous mixture followed by the mixing for another three minutes to get fresh geopolymer mortar. Such produced geopolymer mortar was poured into molds for casting test specimens. The specimens were demolded after a day and left to air dry until testing. For mixes with GO, weighed quantity of GO was dispersed into distilled water using ultrasonication, and the mixture was added to geopolymer mortar mix along with alkaline solution.

2.3 *Testing of Mortar Mix and Cast Specimens*

Initial setting time of fresh mortar mix was estimated using Vicat's apparatus as per IS-12269. Compression strength was evaluated in accordance with IS 4031-part-6 by casting and testing specimens measuring 70.6 mm × 70.6 mm × 70.6 mm at 3, 7, and 28 days [14]. For assessing flexural strength, prisms of size 40 mm × 40 mm × 160 mm were casted and tested as per IS 516-6 [15].

3 Results and Discussions

3.1 Influence of GGBS on Initial Setting Time of Fly Ash-Based Geopolymer Mortar

The consistency of geopolymer mortar, which contained solely fly ash, was sufficient at the fixed dosage of superplasticizer. But upon the addition of GGBS, to replace some portion of fly ash in the mix, produced a relatively stiffer mix that can be attributed to the faster setting of slag [16]. The increasing amount of GGBS decreased the initial setting time. Deviation in initial setting time with GGBS content is shown in Table 5. Deviation in initial setting time with change in GGBS content is graphically depicted in Fig. 2.

Table 5 Variation in initial setting time with GGBS content

Mix type	Binder content (%)		Initial setting time (min)
	FA	GGBS	
M0	100	0	235
M10	90	10	216
M20	80	20	187
M30	70	30	153
M40	60	40	134
M50	50	50	128

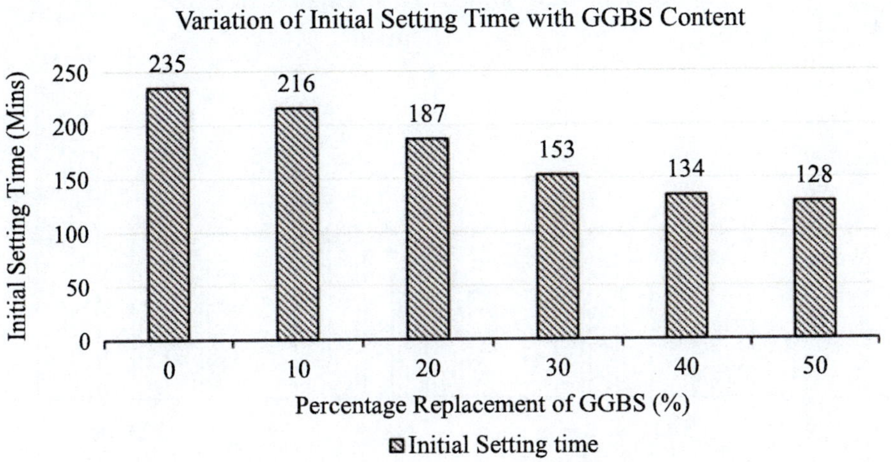

Fig. 2 Variation of initial setting time with increasing GGBS content

3.2 *Influence of GGBS Content on Compression Strength of Fly Ash Made Geopolymer Mortar*

Compressive strength of geopolymer mortar using fly ash as the only source material increased as the GGBS percentage was increased. The formation of calcium alumino silicate hydrates (C–A–S–H) gel from the complete hydration of calcium oxide contained in GGBS is primarily responsible for the increase in strength [13]. Maximum compressive strength of up to 46.80 MPa has been attained at 28 days for 50% replacement level of GGBS as shown in Table 6. However, based on the minimum strength requirement criteria of 25 MPa at 28 days, a replacement level of 30% is taken into consideration for further inquiry. Figure 3 illustrates the variance in compressive strength graphically.

Table 6 Compression test values of mortar added with GGBS

Mix type	Compressive strength (MPa)		
	3 days	7 days	28 days
M0	11.29	16.58	20.36
M10	10.43	13.03	18.19
M20	12.18	15.45	20.95
M30	16.24	21.59	25.28
M40	24.87	29.65	36.63
M50	29.09	34.75	46.80

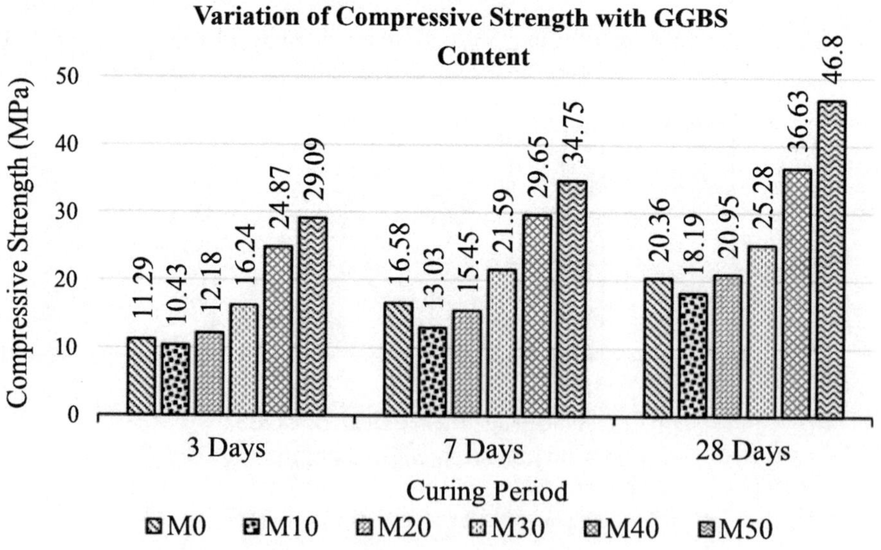

Fig. 3 Variation of compressive strength with GGBS content

3.3 Influence of GO on Initial Setting Time of Fly Ash–GGBS-Based Geopolymer Mortar

Now, for mix M30, GO was incorporated at 0.05, 0.10, and 0.15 %wt binder for ascertaining properties both at fresh and hardened conditions. It was discovered that adding GO directly affected the initial setting time of the geopolymer mixture. Because GO's higher specific surface area absorbs more water, causing a loss of consistency and a quicker setting time, the initial setting time began to decrease as GO content increased [17]. Values of initial setting time are as shown in Table 7 and are within limits as per standard requirements. Graphical presentation of results is shown in Fig. 4.

Table 7 Variation in initial setting time with GO content

Mix type	Binder content (%)		GO (%)	Initial setting time (min)
	FA	GGBS		
M30	70	30	0.00	153
MG1	70	30	0.05	107
MG2	70	30	0.10	92
MG3	70	30	0.15	79

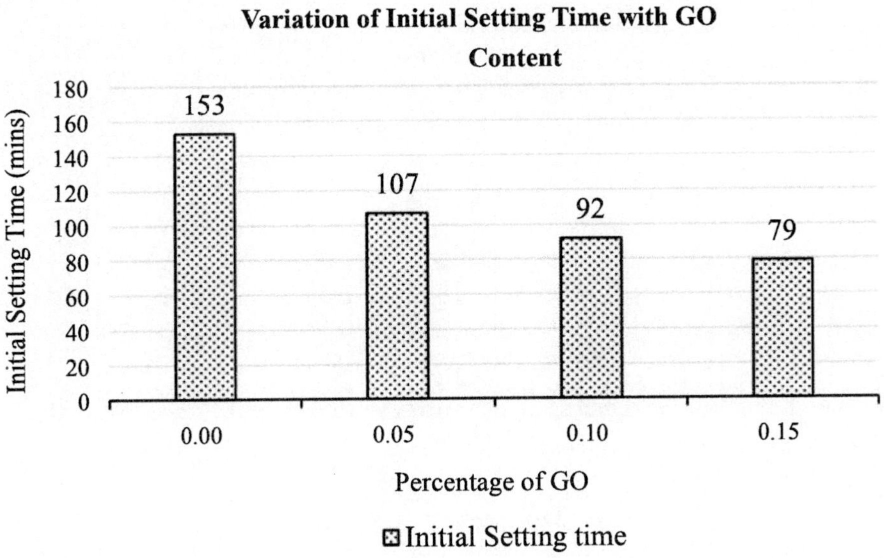

Fig. 4 Variation of initial setting time with increasing GO content

3.4 *Influence of GO Content on Compression Strength of Fly Ash–GGBS Geopolymer Mortar*

The fly ash–GGBS geopolymer mortar's compressive strength was increased to 28.73 MPa after 28 days by adding GO, up to 0.10 %wt. of binder. The very small size of the GO particles speeds up the reaction process, resulting in a denser microstructure. Denser microstructure gives a higher compressive strength to the binder material [18, 19]. However, the compressive strength started decreasing beyond 0.10 %wt of binder that could be due to ineffective dispersion of GO into the mixture. The average compressive strength figures for the geopolymer mortar with GO are shown in Table 8. Figure 5 displays the compressive strength trend over time with various GO percentages.

Table 8 Compressive strength values of mortar mix with addition of GO

Mix type	Compressive strength (MPa)		
	3 days	7 days	28 days
M30	16.24	21.59	25.28
MG1	17.90	23.48	27.90
MG2	18.43	24.96	28.73
MG3	14.14	17.70	21.36

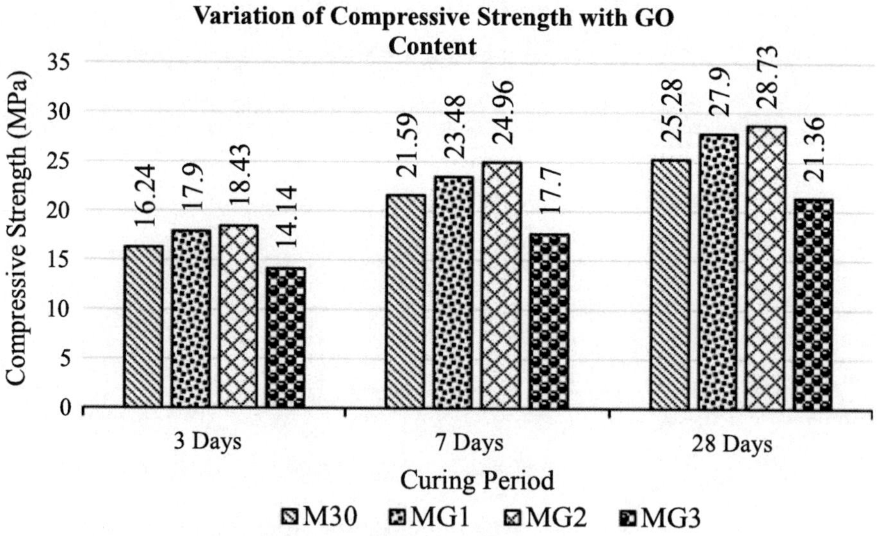

Fig. 5 Compressive strength variation with GO content

3.5 *Influence of GO Content on Flexural Strength of Fly Ash Made Geopolymer Mortar*

Finally, GO-incorporated fly ash–GGBS geopolymer mortar's flexural strength was contrasted with that of geopolymer mortar made only of fly ash. It was determined that the inclusion of GO caused the flexural strength to increase by 16.18% at 28 days. In the alkali activation reaction, GO functions as a catalyst to speed up mass and electron transfers that encourages the quicker dissolution of amorphous phases of aluminosilicate source materials resulting faster formation of geopolymer gel. So produced geopolymer gel fills the nanoscale pores with self-absorbed zeolites giving improved flexural strength [19]. On the other side, uniformly dispersed finer GO particles provide a denser microstructure through strong bonding, which enhances the geopolymer mortar's overall mechanical capabilities [20]. However, the percentage increment in flexural strength decreased as curing period increased. The findings of the tests are summarized in Table 9 and graphically displayed in Fig. 6.

Table 9 Flexural strength values of mortar mix with addition of GO

Mix type	Flexural strength (MPa)		
	3 days	7 days	28 days
M0	1.694	2.170	2.608
MG2	2.048	2.573	3.030

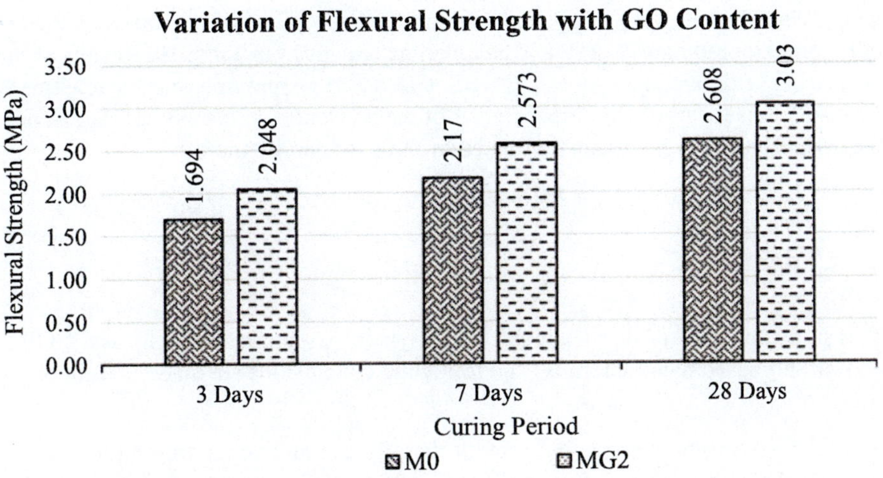

Fig. 6 Variation of flexural strength with GO content

Table 10 Carbon emission analysis for fly ash–GGBS–GO-based geopolymer mortar (per m^3)

Mortar type	Ingredients required	Carbon emission factor	Quantity of ingredients required (kg)	Quantity of carbon dioxide emission (kg)	Total quantity of carbon dioxide emission (kg)
Cement mortar	OPC	0.8800	550	484.00	577.33
	River sand	0.0139	1650	22.93	
	Water	0.3200	220	70.40	
Fly ash-based geopolymer mortar with GGBS and GO	Fly ash	0.0000	506.89	0.00	314.293
	GGBS	0.0260	217.24	5.65	
	River sand	0.0139	1050	14.59	
	Sodium hydroxide	1.1200	93.14	104.31	
	Sodium silicate	0.7100	232.85	165.32	
	Water	0.3200	72.14	23.17	
	GO	1.710	0.724	1.23	

3.6 Carbon Emission Analysis for Cement Mortar and Fly Ash–GGBS–GO Made Geopolymer Mortar

Total carbon emission in terms of kilograms per cubic meter of cement mortar and geopolymer mortar were analyzed based on the carbon emission factors of variety of required ingredients [9, 21], and calculations are given in Table 10. Results of the analysis have defined that fly ash–GGBS–GO-based geopolymer mortar releases at least 45% lesser carbon dioxide than that of the conventional cement mortar. Hence, the geopolymer can be branded as an environment-friendly mortar.

4 Conclusions

This study investigated on sustainable geopolymer mortar containing fly ash, GGBS, and carbon-based nanomaterials. The following conclusions are made in light of the study:

- GGBS having higher specific gravity than fly ash and containing high amounts of Al_2O_3 and CaO as compared to fly ash, GGBS is proven to be a potential binary binder for geopolymer mortar that could improve mechanical properties and increase the sustainable benefits.
- Adding GGBS speeds up the setting process and quickly achieves the desired compressive strength in the fly ash-geopolymer mortar.

- Incorporation of GO to fly ash-geopolymer mortar would reduce the setting period and impart higher compressive strength by forming denser microstructure
- GO fills the micropores of fly ash–GGBS geopolymer mortars and enhances flexural strength
- Fly ash–GGBS–GO made geopolymer mortars release at least 45% less carbon dioxide as compared to conventional cement mortar and hence proved to be environment-friendly mortar

The mechanical properties of fly ash-based geopolymer mortar have been significantly improved by the combined effects of GGBS and GO. However, it is important to investigate how well GO works in terms of mortar durability. Therefore, more study is necessary to determine how long GO integrated fly ash–GGBS made geopolymer mortars can last.

Acknowledgements This current research is funded by Science and Engineering research Board (SERB) "EEQ/2020/000499-Innovative Development of Sustainable White Clay and Graphene Composite Ventilation Wall Cladding Tiles to Curtail Heat Ingress". Funding by Science and Engineering Research Board (SERB), Government of India, is acknowledged.

References

1. Bhikshma V, Reddy MK, Rao TS (2012) An experimental investigation on properties of geopolymer concrete (No Cement Concrete). Asian J Civ Eng 13:841–853
2. Bondar D, Lynsdale CJ, Milestone NB (2013) Alkali-activated natural pozzolan concrete as new construction material. ACI Mater J 110:331–337
3. Hardjito D, Wallah SE, Sumajouw DMJ, Rangan BV (2005) Fly ash based geopolymer concrete. Aust J Struct Eng 6:1–10
4. Vora PR, Dave UV (2013) Parametric studies on compressive strength of geopolymer concrete. Procedia Eng 51:210–219. https://doi.org/10.1016/j.proeng.2013.01.030
5. Joshi SV, Kadu MS (2012) Role of alkaline activator in development of eco-friendly fly ash based geo polymer concrete. Int J Environ Sci Dev 3:417–421. https://doi.org/10.7763/IJESD.2012.V3.258
6. Rangan BV (2010) Fly ash-based geopolymer concrete. In: Proceedings of the international workshop on geopolymer cement and concrete, pp 68–106
7. Motorwala A, Shah V, Kammula R, Nannapaneni P, Raijiwala PDB (2013) Alkali activated fly-ash based geopolymer concrete. 3:159–166
8. Almutairi AL, Tayeh BA, Adesina A, Isleem HF, Zeyad AM (2021) Potential applications of geopolymer concrete in construction: a review. Case Stud Constr Mater 15. https://doi.org/10.1016/j.cscm.2021.e00733
9. Rajmohan B, Nayaka RR, Rajesh Kumar K, Kaleemuddin K (2022) Mechanical and durability performance evaluation of heat cured low calcium fly ash based sustainable geopolymer concrete. Mater Today Proc. https://doi.org/10.1016/j.matpr.2022.02.227
10. Silva SVA, Arachchi JNJK, Wijewardena CL, Nanayakkara SMA Development of fly ash based geopolymer concrete
11. Ismail I, Bernal SA, Provis JL, Hamdan S, Deventer JSJ (2012) Microstructural changes in alkali activated fly ash/slag geopolymers with sulfate exposure. Mater Struct 361–373. https://doi.org/10.1617/s11527-012-9906-2

12. Kamath M, Prashant S, Kumar M (2021) Micro-characterisation of alkali activated paste with fly ash-GGBS-metakaolin binder system with ambient setting characteristics. Constr Build Mater 277. https://doi.org/10.1016/j.conbuildmat.2021.122323
13. Bellum RR, Muniraj K, Indukuri CSR, Madduru SRC (2020) Investigation on performance enhancement of fly ash-GGBFS based graphene geopolymer concrete. J Build Eng 32. https://doi.org/10.1016/j.jobe.2020.101659
14. IS 4031-6 (1988) Methods of physical tests for hydraulic cement, part 6: determination of compressive strength of hydraulic cement (other than masonry cement)
15. IS 516 (2014) Method of tests for strength of concrete. New Delhi, India
16. Laskar SM, Talukdar S (2019) A study on the performance of damaged RC members repaired using ultra-fine slag based geopolymer mortar. Constr Build Mater 217:216–222. https://doi.org/10.1016/j.conbuildmat.2019.05.033
17. Sajjad U, Sheikh MN, Hadi MNS (2021) Experimental study of the effect of graphene on properties of ambient-cured slag and fly ash-based geopolymer paste and mortar. Constr Build Mater 313. https://doi.org/10.1016/j.conbuildmat.2021.125403
18. Murali M, Alaloul WS, Mohammed BS, Musarat MA, Al Salaheen M, Al-Sabaeei AM, Isyaka, A (2022) Utilizing graphene oxide in cementitious composites: a systematic review. Case Stud Constr Mater 17. https://doi.org/10.1016/j.cscm.2022.e01359
19. Liu X, Wu Y, Li M, Jiang J, Guo L, Wang W, Zhang W, Zhang Z, Duan P (2020) Effects of graphene oxide on microstructure and mechanical properties of graphene oxide-geopolymer composites. Constr Build Mater 247. https://doi.org/10.1016/j.conbuildmat.2020.118544
20. Mohammed A, Al-Saadi NTK, Sanjayan J (2018) Inclusion of graphene oxide in cementitious composites: state-of-the-art review. Aust J Civ Eng 16(2):81–95. https://doi.org/10.1080/14488353.2018.1450699
21. Serrano-Luján L, Víctor-Román S, Toledo C, Sanahuja-Parejo O, Mansour AE, Abad J, Amassian A, Benito AM, Maser WK, Urbina A (2019) Environmental impact of the production of graphene oxide and reduced graphene oxide. SN Appl Sci 1. https://doi.org/10.1007/s42452-019-0193-1

Study of Ambient Cured Fly Ash-GGBS-Metakaolin-Based Geopolymers Mortar

Banoth Gopalakrishna and Dinakar Pasla

1 Introduction

The use of waste materials or industrial by-products as a geopolymer binders in the development of GPC cement is gaining traction as a result, reduce the extreme use of natural resource materials in the manufacture of ordinary Portland cement (OPC). These industrial by-products material includes ground granulated blast furnace slag (GGBS), rice husk ash (RHA), fly ash (FA), and metakaolin (MK) as a binder in GPC cement [1]. Because there is such a huge demand for concrete, which generates a lot of carbon dioxide (CO_2) emissions, the construction industries are unable to totally replace the use of OPC with alternative cementitious materials. The utilization of natural deposits for the production of cement and concrete has led to an environmental destruction since natural resources are continually being depleted. The present global annual cement consumption may exceed 4 billion tons by 2020 [2]. According to reports, GPC cement is substantially more environmentally friendly than OPC in terms of decreased manufacturing energy requirements and reduced CO_2 emissions. The CO_2 emission from geopolymer mortar may be decreased if emissions during the production of geopolymer mortar.

Sodium hydroxide (NaOH) and sodium silicate (Na_2SiO_3) solutions combined with various Na_2SiO_3 to NaOH mass ratios seem to be the most often used alkaline activators [3]. Similar to zeolites, geopolymers may be developed by a numerous different reaction step, from early pozzolanic polymerization to formation of final microstructure. Numerous variables depending on the chemical constituents of

B. Gopalakrishna · D. Pasla (✉)
School of Infrastructure, Indian Institute of Technology Bhubaneswar, Bhubaneswar, Odisha, India
e-mail: pdinakar@iitbbs.ac.in

B. Gopalakrishna
e-mail: bg13@iitbbs.ac.in

R. K. Pancharathi et al. (eds.), *Low Carbon Materials and Technologies for a Sustainable and Resilient Infrastructure*, Lecture Notes in Civil Engineering 440,
https://doi.org/10.1007/978-981-99-7464-1_3

the alkaline activator content (AAC), by-product materials, and curing conditions impact the development of compressive strength (CS) and the chemical response of geopolymer concrete. FA-based geopolymer mortar required to oven curing for polymerization of mortar. The molds were taken out of the oven and kept at room temperature while the samples conducted CS testing [4]. GGBS used in FA-based binder to develop ambient temperature curing geopolymer mortar, and increase CS compare to FA-based geopolymer mortar [5, 6].

MK was employed in the primary development of GP and is used as a raw material because of its pure alumina silicate content. Naturally, MK is a pozzolanic substance that is produced by calcining kaolinitic clay at temperatures between 500 and 800 °C. Comparing OPC with MK reveals that it is also more ecologically friendly since it needs a minor calcining temperature during manufacture and 80–90% of CO_2 lesser [7]. MK produces more geopolymerization and is a rich resource of Al_2O_3 and SiO_2, it is highly reactive combination with AAC. The CS of GPC mortar might be enhanced, according to Islam et al. [8] research, by increasing the amount of GGBS in the matrix that included FA. The mechanical characteristics of the MK-based GPC were increased by the addition of calcium (Ca) from different sources binder materials. In order to characterize alternate cementitious materials, Davidovits [9] developed a binder in 1978 called geopolymer. Concrete produced of geopolymers, such as fly ash, GGBS, and metakaolin, is an alternative to traditional concrete.

This study's primary objective was to observe into the production of GPC mortar's CS utilizing two locally accessible waste materials as binders: GGBS, FA, and MK. Investigation and reporting were done changing the proportions of these three binders' material how affected on the CS. By varying the binder doses while maintaining the same ratios for other elements including sand and AAC, the optimum CS of the cube specimen was found. FA, GGBS, and MK are environmentally friendly geopolymer cement pastes that don't release greenhouse gases during the polymerization process and are rich in silica and alumina. AAC may be used to provide the reaction between the materials a binding character. The geopolymer cement paste applicable to repair the heritage structures.

2 Investigation of Raw Materials and Mixing Procedure

2.1 Materials

Fly ash: In this study, the alumina silicate source material used to make the geopolymeric binder was FA. Low calcium FA from the NALCO plant in Odisha, India, was used in this investigation. The chemical content of FA as exanimated by X-ray fluorescence (XRF) analysis given in Table 1.

Ground Granulated Blast Furnace Slag (GGBS): GGBS is one of the sources of by-products needed to produce a cement-free binder. Because it has both pozzolanic

Table 1 XRF analysis of geopolymer cement binder

S. No	Oxide	Dry weight (%)		
		FA	GGBS	MK
1	SiO_2	50.13	35.32	53.3
2	Al_2O_3	27.28	15.63	40
3	Fe_2O_3	9.28	1.56	4.21
4	CaO	3.25	40.15	0.43
5	Na_2O	1.12	0.21	0.13
6	K_2O	2.82	0.52	0.51
7	MgO	1.52	6.54	0.09
8	Other oxides	1.76	1.02	0.64
9	LOI	2.9	0.41	0.69

and cementitious properties, GGBS differs from supplementary cementitious materials. The hydraulic action of GGBS is developed when it reacts with water [10]. The presence of calcium, alumina, and silicates in GGBS, which together constitute for about 90% of the substance, indicates that it satisfies the requirements for a pozzolanic material.

Table 2 lists its physical properties, while Table 1 illustrates the GGBS's chemical composition.

Metakaolin: The metakaolin (MK) utilized in this investigation manufacture from kaolin that had calcined for 3 h at 650 °C. It had a slight cream to it, and this is due to MK's mild acidity caused by dihydroxylation during calcination. It is an extremely reactive pozzolanic mineral additive with many possibilities in the development of geopolymer mortar composites.

Tables 2 and 1 give the physical properties and chemical composition of MK.

Fine aggregates: River natural sand is commonly used as fine aggregate grading standards and is usually clean. They're utilized in geopolymer cement paste to improve homogenous mix. Natural sand is that can only be obtained from local river and utilize in present experiment, and with a specific gravity of 2.63.

Table 2 Physical properties of geopolymer cement binder

Material properties	Units	FA	GGBS	MK
Color		Grayish white	Off white	Cream
Specific surface area	m^2/kg	390	430	15,000
Moisture	%	< 0.1	< 0.1	< 0.212
Specific gravity		2.1	2.83	2.59
Bulk density	kg/m^3	995	1200	357

Table 3 Chemical composition of NaOH and Na_2SiO_3

Properties	Sodium hydroxide solution	Sodium silicate solution
Molecular formula	NaOH	Na_2SiO_3
pH	13–14	11.42
Specific gravity	2.2	1.36
Molecular weight (g/mol)	40	122.06
Color	White	Clear colorless
Density (g/cm^3)	2.14	1.39
SiO_2	–	29.90
Na_2O (%)	45.71	10
H_2O (%)	54.29	60.8

Alkali activator solutions: Geopolymers are inorganic polymeric compounds produced from source materials rich in silicates and aluminates. The polymerization process has a complex aluminosilicate gel when these raw ingredients activate with an alkaline solution. The amount of alkaline solution employed influences the geopolymerization approach significantly. The effect ratio of Na_2SiO_3 to NaOH and the molarity of NaOH on the synthesis of a FA- and GGBS-based geopolymer mortar was studied. The samples with a Na_2SiO_3 to NaOH ratio of 1.5, and 16 M NaOH had the maximum CS. Soluble alkalis, the majority of which are sodium or potassium-based, help compensate for the alkaline activator solution. The most popular alkaline activator for GPC processing is sodium hydroxide (NaOH) in association with sodium silicate (Na_2SiO_3) [11]. Table 3 provides the characteristics of sodium-based alkali activator solutions utilized in this investigation, including chemical composition and specific gravity.

2.2 *Mixing Methodology*

There were four different mixes developed. The mix proportions for each geopolymer mortar are given in Table 4. The preparation of the mixtures involved various percentage of FA, MK, and GGBS content. The AAC and natural sand were maintained constant in order to study the impact of the binders. All of the mixes were maintained at a 1:3 binder to fine aggregate content ratio. The MK, FA, and GGBS ternary binders were mixed with the sand in a bowl mixer under low-speed conditions for about 5 min. In the following five minutes, the alkali activator solution was added gradually while maintaining the same speed. After that, the mortar was poured in three phases with the proper compaction using molds that were 70.6 mm in size. For every mix proportion, total of 12 specimens were cast. A rod and vibrating table were used to vibrate the samples with standard compaction in order to remove the entrained air and bubbles. Following casting, plastic film was used to cover the 100%

Table 4 Percentages of binder in mixtures

Mix ID	Binders (%)		
	Fly ash	GGBS	MK
M1	100	0	0
M2	70	30	0
M3	60	30	10
M4	55	30	15

FA specimens and the molds to prevent water loss during the 24-h, 60 °C oven curing process. After that, the remaining mix specimens were kept in ambient temperature conditions, until the day of the test.

3 Results and Discussions

3.1 Compressive Strength of Geopolymer Cement Paste Replacement with GGBS and MK

The CS of hardened GPC mortar is the primary indication of the performance of substitute source materials, since it provides a fundamental description of the quality of geopolymerization product. Figure 1 depicts the development of CS at 7, 28, 56, and 90 days. The methodology states that the specimens made completely of FA were oven cured at 60 °C temperature for 24 h. The time required for heat treatment to reach high strength may be shortened by curing at 60 °C temperature. Moreover, it can be observed in Fig. 1 that using 100% FA in GP mortar results in a reduced CS. When a proportion of FA is replaced with GGBS, the CS improves. The CS, however, was adversely affected by an increase in FA over 55%. Therefore, the quantity of Ca and Al_2O_3 had a more impact on the mortar's CS [12]. Figure 1 shows that mix 16M30G15MK, which comprised 15% MK, 30% GGBS, and 55% FA and was cured at room temperature, provided the maximum CS around 61 MPa could be reached, however, decreased CS when the MK was reduced from 15 to 10% and the FA was subsequently raised. As can be observed, this pattern maintained until the MK decreased to 0%. Furthermore, the inclusion of MK enhances the CS. This might be due to the fact that the Si/Al ratio changes throughout the geopolymerization process [13]. The majority of the specimens obtained between 40 and 49% of the 90-day CS, as can be shown by a comparison of the 7-day test results with that CS.

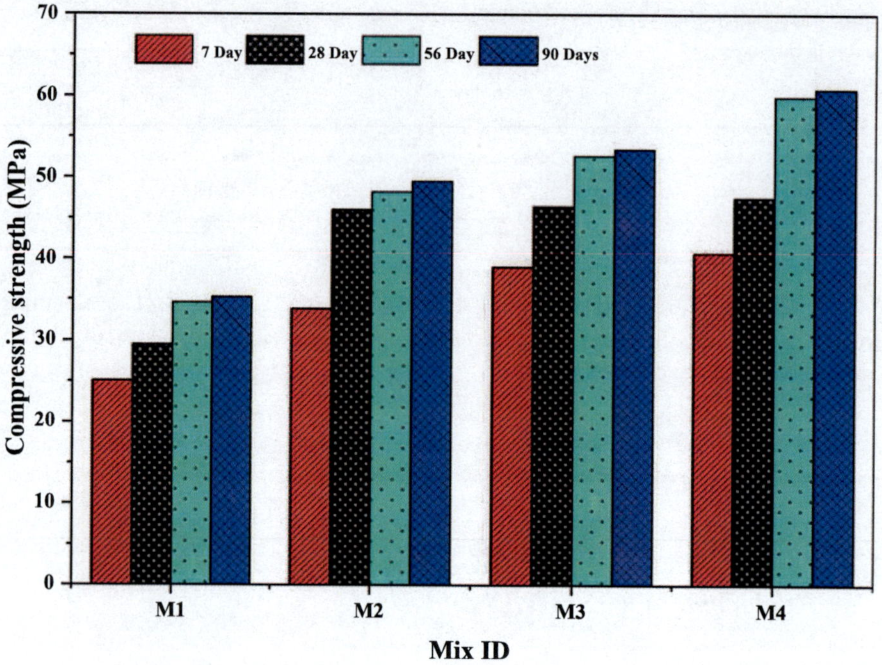

Fig. 1 Compressive strength of geopolymer mortar at various % of GGBS and MK

3.2 Evaporation Studies

The results of the gel and capillary porosity tests for various geopolymer binder mortar mixes are illustrated in Fig. 2. It is evident that the majority of the mortar has a comparable volume of gel porosity, ranging from 5.5 to 15.33%. The amount of MK used has a significant impact on the gel porosity, which is the primary factor affecting the porosity of the mortar. Moreover, a decrease in MK percentage leads to a reduction in capillary porosity, possibly due to a decrease in the amount of porous material in the mortar mix and an increase in the amount of binder material. This trend has already been observed in concrete [14, 15]. The Mix M4 has less capillary porosity than the Mix M1, as shown in Fig. 2. Furthermore, the addition of GGBS and MK significantly decreased the gel porosity of the system, which may be attributed to their ability to microfill and generate extra CASH and CSH gel and obtained the maximum CS. Although GGBS and MK are more effective at reducing gel porosity, their inclusion in the system had little impact on capillary pores. In the case of geopolymer mortar, a more thorough understanding of the evaporation process is necessary to determine porosity accurately.

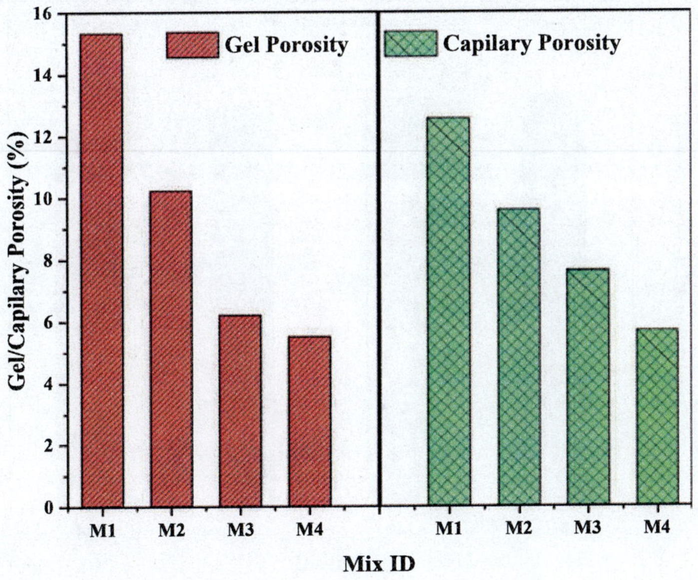

Fig. 2 Gel/capillary porosity of geopolymer mortar

The evaluation criteria established by the Concrete Society (CEB, 1989) were utilized to compare the obtained outcomes. Figure 3 depicts the results of the absorption investigation for all the geopolymer mortars. The figure illustrates the changes in absorption for geopolymer mortars with varying percentages of GGBS and MK replacements. The results also reveal that the M4 mix has an insignificant effect on the absorption of geopolymer mortar. The absorption characteristics of these mixes indicate that even though the strength and AAC binder ratio may differ, the absorption properties fall within a limited range.

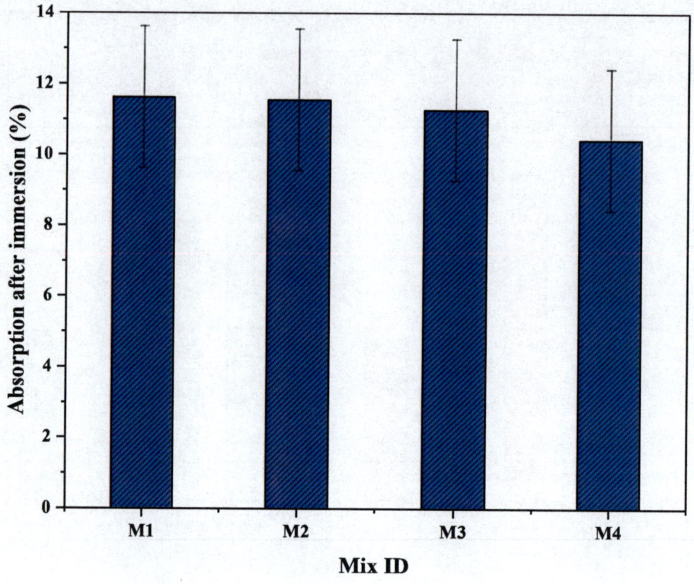

Fig. 3 Water absorption of geopolymer mortar

4 Conclusion

The CS of GPC mortar increased up to 30 MPa at the age of 28 days, whereas GPC mortar with 100% FA provides 60 °C oven curing at 24 h. In effort to accomplish the 48 MPa CS desired, the combination of FA and GGBS can be employed to develop GPC cement paste without the need of heat. The MK content up to 15%, GGBS up to 30%, and FA up to 55% exhibited the maximum CS and good durability among the mixes. Additionally, it can be inferred that the maximum combined proportions of MK, FA, and GGBS may be utilized to achieve a sustainable building material that can take the place of OPC in the manufacture of GPC mortars that are ecologically beneficial.

References

1. Bashar II, Alengaram UJ, Jumaat MZ, Islam A (2014) The effect of variation of molarity of alkali activator and fine aggregate content on the compressive strength of the fly ash: palm oil fuel ash based geopolymer mortar. Adv Mater Sci Eng 2014. https://doi.org/10.1155/2014/245473
2. Chana P (2011) Low carbon cements: the challenges and opportunities. In: First global future cement conference 2011
3. Hardjito D, Wallah SE, Sumajouw DMJ, BV Rangan (2004) On the development of fly ash-based geopolymer concrete. Mater J 101:467–472

4. Temuujin J, Van Riessen A, MacKenzie KJD (2010) Preparation and characterisation of fly ash based geopolymer mortars. Constr Build Mater 24:1906–1910. https://doi.org/10.1016/j.conbuildmat.2010.04.012
5. Rao MG, Jagadish V, Venu M et al (2020) Performance of fly ash and Ggbs based geopolymer concrete using single alkaline activator solution and its cost analysis. IOP Conf Ser Mater Sci Eng 998. https://doi.org/10.1088/1757-899X/998/1/012051
6. Gopalakrishna B, Dinakar P (2023) Mix design development of fly ash-GGBS based recycled aggregate geopolymer concrete. J Build Eng 63:105551. https://doi.org/10.1016/j.jobe.2022.105551
7. Davidovits J (2008) Geopolymer chemistry and applications. Geopolymer Institute
8. Islam A, Alengaram UJ, Jumaat MZ, Bashar II (2014) The development of compressive strength of ground granulated blast furnace slag-palm oil fuel ash-fly ash based geopolymer mortar. Mater Des 56:833–841. https://doi.org/10.1016/j.matdes.2013.11.080
9. Davidovits J (1994) Properties of geopolymer cements. In: First international conference on alkaline cements and concretes, pp 131–149
10. Yunsheng Z, Wei S, Zongjin L et al (2008) Impact properties of geopolymer based extrudates incorporated with fly ash and PVA short fiber. Constr Build Mater 22:370–383. https://doi.org/10.1016/j.conbuildmat.2006.08.006
11. Kong DLY, Sanjayan JG (2008) Damage behavior of geopolymer composites exposed to elevated temperatures. Cem Concr Compos 30:986–991. https://doi.org/10.1016/j.cemconcomp.2008.08.001
12. Khale D, Chaudhary R (2007) Mechanism of geopolymerization and factors influencing its development: a review. J Mater Sci 42:729–746. https://doi.org/10.1007/s10853-006-0401-4
13. Fernández-Jiménez A, Palomo A, Criado M (2005) Microstructure development of alkali-activated fly ash cement: a descriptive model. Cem Concr Res 35:1204–1209. https://doi.org/10.1016/J.CEMCONRES.2004.08.021
14. He YX (2011) Experimental research on pore structure of RCA and its impact on drying shrinkage. Adv Mater Res 335–336:1141–1144. https://doi.org/10.4028/www.scientific.net/AMR.335-336.1141
15. Nandanam K, Biswal US, Dinakar P (2021) Effect of fly ash, GGBS, and metakaolin on mechanical and durability properties of self-compacting concrete made with 100% coarse recycled aggregate. J Hazard Toxic Radioact Waste 25:04021002. https://doi.org/10.1061/(asce)hz.2153-5515.0000595

Microstructure and Mechanical Properties of Fly Ash and GGBS-Based Alkali-Activated Concrete

Koppoju Manasa and Andal Mudimby

1 Introduction

Concrete is considered a highly consuming material in the construction industry. Due to the rapid increase in urbanization, the demand to produce ordinary Portland cement has increased which contributes 5–7% of worldwide greenhouse gas emissions [1]. To reduce the environmental impact caused by the production of OPC, it is desirable to use the alternative binders extracted from the by-products of industrial wastes in place of conventional materials which can also reduce the consumption of natural resources as well as the land requirement for the disposal of waste [1]. In past decades, in the field of civil engineering, the inorganic polymers known as "geopolymers" are used as binders in the manufacturing of concrete. The major compositions of geopolymers are rich sources of silicon, aluminum, and calcium which are produced by geological origins and industrial by-products such as fly ash and GGBS [2]. Alkali-activated binders can significantly reduce the greenhouse gas emission caused by the production of OPC. When these alkali-activated binders react with the alkaline solution, polymerization takes place, resulting in the formation of a 3D-alumino-silicate network similar to the rock-forming minerals. alkali-activated concrete is an environmentally friendly construction material produced by the chemical reaction of inorganic molecules [1, 2]. The materials made of these inorganic binders are proven as excellent construction materials due to their high mechanical, chemical, and thermal-resisting properties. In the present study, the alkali-activated concrete was manufactured by replacing cement with 50% of fly ash and 50% of GGBS [3–5]. The alkaline solution used in the study is the combination of sodium silicate (Na_2SiO_3) and sodium hydroxide (NaOH) with a silicate-to-hydroxide (Na_2Sio_3/ NaOH) ratio of 2. Various types of research were focused on material-level tests

K. Manasa (✉) · A. Mudimby
Kakatiya University, Warangal, India
e-mail: koppojumanasa114@gmail.com

R. K. Pancharathi et al. (eds.), *Low Carbon Materials and Technologies for a Sustainable and Resilient Infrastructure*, Lecture Notes in Civil Engineering 440,
https://doi.org/10.1007/978-981-99-7464-1_4

on GPC, but fewer investigations are available on structural-level tests. Most of the advanced research on the microstructure analysis of GPC is limited compared to OPC concrete. The present study investigates the microstructure analysis of fly ash and GGBS-based alkali-activated concrete. The alkali-activated concrete beams are designed as per the universal mixture design criteria. The comparison of mechanical strength and microstructure analysis was made between cement-based concrete and geopolymer concrete.

2 Research Significance

Cement plays a crucial role in the production of concrete. As the demand for concrete increasing, the production rates of cement also increased by about 4.1 billion tons in 2020 [6]. The cement industry requires high energy and emits large amounts of CO_2, which is harmful to the environment.

To reduce these problems, alkali-activated concrete is an innovative construction material that is produced by the chemical reaction of inorganic molecules to reduce CO_2 emissions [7, 8]. Alkali-activated concrete is made of the geological origin or by-product material such as fly ash and ground granular blast furnace slag. In India, the thermal power station produces large quantities of fly ash of which 83% is utilized and the remaining is dumped as waste. The alkali-activated concrete undergoes a polymerization process when these-products react with the alkaline activators and improves the mechanical strength of the concrete.

3 Novelty of Work

In this study, to reduce the industrial waste volumes and environmental degradation, the cement is completely replaced with industrial by-products like fly ash and GGBS. The fly ash and GGBS-based alkali-activated concrete is manufactured by developing a material proportion using conventional methods of concrete. The alkaline solution which is a combination of sodium hydroxide and sodium silicate binds the different constituents of the concrete. The present study deals with evaluating the strength criteria of fly ash and GGBS-based alkali-activated concrete by destructive testing and microstructure analysis.

Table 1 Physical properties of the materials

Material	Properties	Results
Flyash	Specific gravity	2.26
	Fineness	9.8%
	Initial setting time	120 min
GGBS	Specific gravity	2.9
	Fineness	9.5%
	Initial setting time	25 min
Fine aggregate	Specific gravity	2.61
	Fineness modulus	2.89
Coarse aggregate	Specific gravity	2.68
	Fineness modulus	7.88
Flyash + GGBS	Fineness	9.8%
	Initial setting time	60 min

4 Experimental Work

4.1 Materials

Ground granulated blast furnace slag used in the study is obtained from "Venkateshwara traders pvt. Ltd". GGBS is partially replaced by 50% in the mix of AAC. Fly ash is obtained from NTPC Ramagundam thermal power plant. The fly ash used in the study is a low calcium, i.e., class F which is replaced by 50% in the mix of AAC. The fine aggregate used in the experimental work is locally available and free from clayey matter. The sand is tested for various properties like fineness modulus, specific gravity, bulk density, and water absorption. The locally available coarse aggregate is used in the study and tested for various properties like fineness modulus, specific gravity, bulk density, and water absorption. The various alkaline activators, such as NaOH and Na_2SiO_3, were obtained from local suppliers. NaOH used in the study is in the form of flakes with 98% purity. Table 1 represents the physical properties of the materials.

4.2 Alkaline Activator Solution

The alkaline-activated solution is the combination of sodium hydroxide of 12 M and sodium silicate the ratio of silicate-to-hydroxide ratio considered is 2. To prepare a 12 M solution of alkaline activator 480 g of NaOH flakes were dissolved in 1 L of distilled water. During the dissolution of NaOH flakes, a huge amount of heat is evolved after 24 h; the 2 L of Na_2SiO_3 solution is placed into 1 L of NaOH solution and mixed properly so that the desired proportions of alkaline activators were prepared it is desirable to prepare the alkaline solution prior to 24 h of the casting of a specimen.

4.3 Preparation of Alkali-Activated Concrete

In the process of preparation of alkali-activated concrete, the dry materials were mixed thoroughly for 3 min, the alkaline solution was added to the dry material and mix it for 5 min until a uniform mix was obtained. The fresh fly ash and GGB-based alkali-activated concrete was poured into the concrete mold immediately and compacted by a needle vibrator to remove the air voids. To determine the mechanical properties of alkali-activated concrete the cubes of 150 × 150 × 150 mm and cylinders of 150 × 300 mm molds were used. The specimens are left in the molds undisturbed at room temperature for about 24 h after casting. Figure 1 represents the preparation process of alkali-activated concrete.

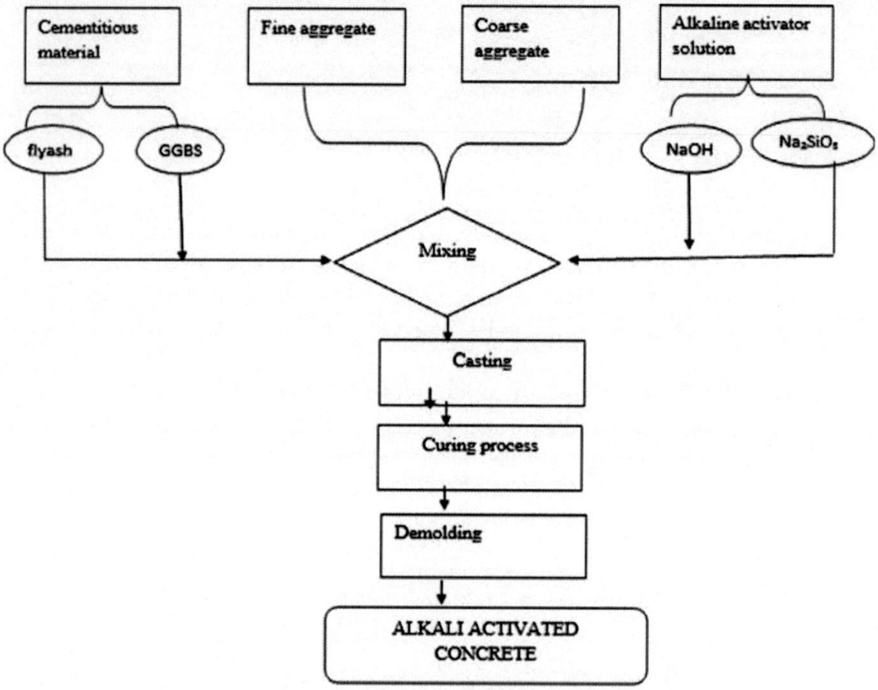

Fig. 1 Preparation process of alkali-activated concrete

Table 2 Mechanical properties of GPC and PCC

Grade of concrete	Compressive strength (28 days) N/mm^2	Flexural strength (28 days) N/mm^2	Split tensile strength (28 days) N/mm^2	Bond strength (28 days) N/mm^2	Modulus of elasticity N/mm^2
G40	56.63	7.651	4.87	8.11	6362.3 (stress–strain curve)
M40	48.23	5.23	3.45	6.19	As per IS 456:2000 $E = 5000 * \sqrt{f_{ck}}$ (N/mm^2) 34,723.9

5 Experimental Results

5.1 *Mechanical Properties*

The results include the mechanical characteristics of GPC after 28 days of curing. The samples were examined in order to measure their compressive strength, flexural strength, split tensile, and modulus of elasticity of GPC and PCC. The test outcomes of the experimental work are given below for the G40 and M40 grades of concrete. Table 2 represents the mechanical properties of GPC and PCC.

5.2 *EDX Results*

The presence of material composition and their percentage is validated by EDS analysis. Figures 2 and 3 represent the EDS plot for M40 and G40 concrete grades. O and Si are the major elements in M40-grade concrete and the percentage of O, Si, Al, and Ca are 48.74%, 27.15%, 8.2%, and 8.29% of total weight respectively. In GPC 40 O, Si, and Ca are the major elements, the percentage of O, Si, Ca, Na, and Al are 47.72%, 19.3%, 13.88%, 10.47%, and 6.89% of total weight respectively. The maximum percentage of calcium content results in the formation of CSH gel, this boosts the compressive strength, reduces porosity enabling the formation of dense gel structure. A long with O and Si small amounts of Ca, Na, Al, Mg are also observed by EDS analysis. Table 3 represents the element composition in G40 and M40.

5.3 *XRD Results*

The XRD analysis is performed at room temperature by applying powder X-ray diffraction with an X-ray wavelength of 0.154 mm Cu, K-beta radiation. Samples

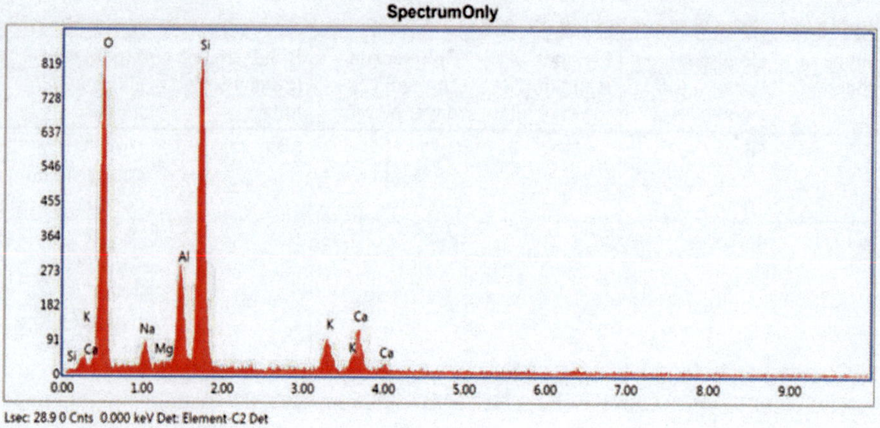

Fig. 2 EDS plot for M40

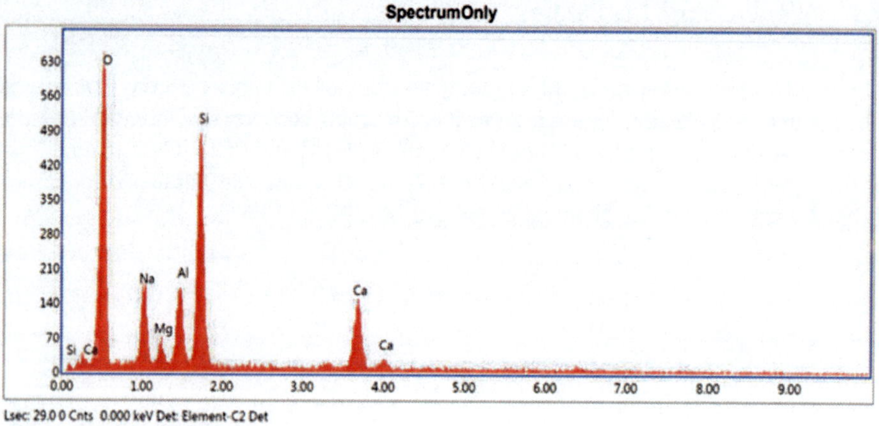

Fig. 3 EDS plot for G40

Table 3 Element composition in G40 and M40

Element (wt%)	G40	M40
O	47.72	48.74
Na	10.47	2.41
Mg	1.75	0.09
Al	6.89	8.20
Si	19.3	27.15
Ca	13.88	8.29

are scanned in a continuous mode with a scan speed of 5.0000 deg/min, Scan Axis: Theta/2-Theta, and can range: 10.0000–90.0000 deg. The X-ray powder diffraction plots the intensity of the scattered X-ray light against the angle difference of the deflected X-Rays. The X-ray powder diffraction was done on two samples of different grades PCC 40 and GPC 40 where G40 indicated geopolymer concrete for 28 days of ambient curing and PCC 40 indicated plain cement concrete for 28 days of ambient curing. The XRD diffraction plots for analyzing samples were displayed in Figs. 4 and 5, which were created using the origin software. Compounds contained in the concrete are identified by utilizing X-pert high score plus. Peaks of quartz (SiO_2) and silica were observed along with the alumina (Al), magnesium (Mg), calcite (Ca), and sodium (Na). The hardened geopolymer concrete identified the crystalline phases in the presence of quartz, silica, and calcite by their characteristic peaks. The strong peak was observed in the G40 grade of concrete at around 2θ. The peak intensity for PCC and geopolymer concrete is in a similar range from 2θ, 20° to 2θ, 30° and the peak intensities of C–S–H and N–A–S–H increase with an increase in the GGBS quantity in the mix. Similarly, the XRD test was conducted for a sample made with PCC concrete of M40 grade. The XRD graph shows the peaks of major crystalline constituents of quartz, silica, and calcite. The peaks of the graph are observed in between the range of 2θ, 20° and 2θ, 40°. The primary highest peak is observed at 2θ, 26°, at this peak the C–S–H phase of the concrete is observed, and the second highest peak is observed at 2θ, 21°. Tables 4 and 5 represents the phase composition of M40 and G40-grade concrete.

5.4 SEM Results

To examine the microstructure, the concrete samples were subjected to SEM examination after 28 days of curing. Figure 6 shows the SEM picture of an OPC concrete sample, which shows a significant number of white blocky crystals with tiny granular crystals on the surface, together with small bar-like particles and fine fissures. According to the XRD and EDS data, these white blocky crystals are quartz, gypsum, and mirabilite, and the fine granular crystals are silica, thenardite, and the bar-like particles are ettringite.

The scanning electron microscopy of the G40 mix at 28 days of curing mix shows that there are granular CSH gel formations accompanied by the pore system and finer cracks shown in Fig. 7. According to the XRD and EDS results, these granules represent CSH gel formations accompanied by the pore system. Even though the polymerization has reached the optimum stage there are still some unreacted particles of fly ash observed which indicates geopolymerization process was not up to cent percent [9]. This figure represents the denser structure formed as a result of the reaction of fly ash and GGBS with an alkaline activator along with some non-reacted fly ash sphere particles on the surface of the sample at the magnification of 10 μm. These unreacted fly ash particles form weaker links resulting in the propagation of microcracks were observed in Fig. 7.

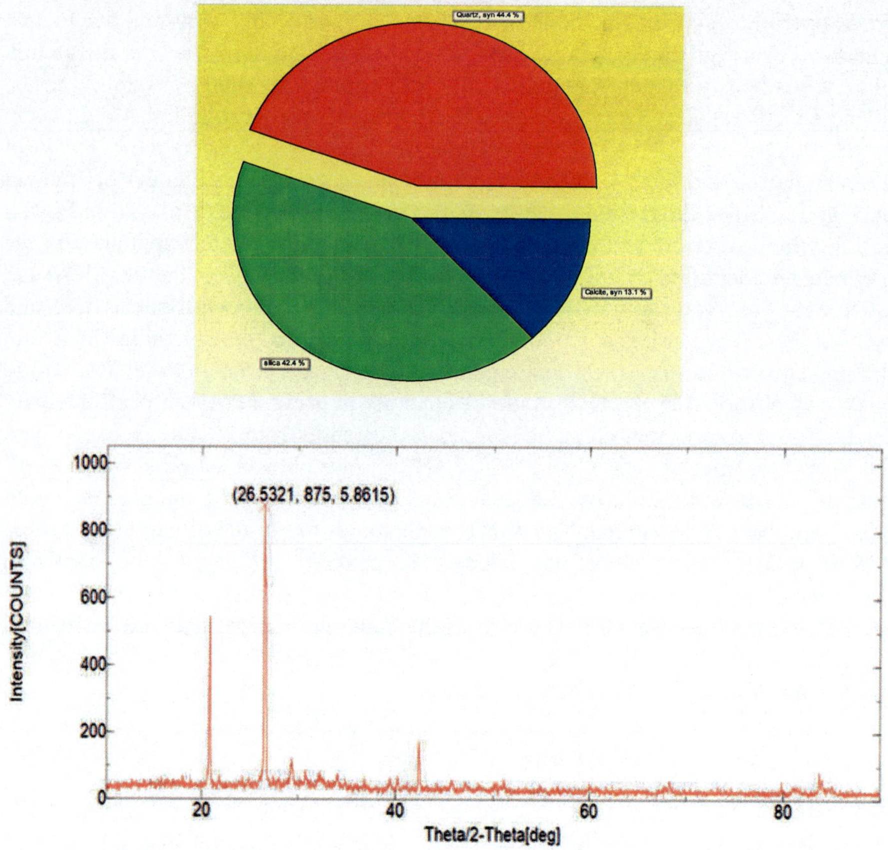

Fig. 4 XRD analysis of M40-grade concrete

6 Conclusions

In the present study, the microstructure and mechanical behavior of fly ash and GGBS-based GPC specimens are analyzed by conducting destructive testing, SEM, EDS, and XRD. Alkali-activated concrete has been prepared by completely replacing cement with 50% fly ash and 50% GGBS.

1. It is observed that mix becomes very harsh for the solution-to-binder ratio lower than 0.5 and requires superplasticizers to improve the workability of the mix.
2. The mechanical characteristics of AAC are comparably higher than the OPC concrete and show early strength.
3. The characteristic compressive strength of geopolymer concrete (f_{ck}) after 28 days under ambient curing is 58.63 N/mm^2

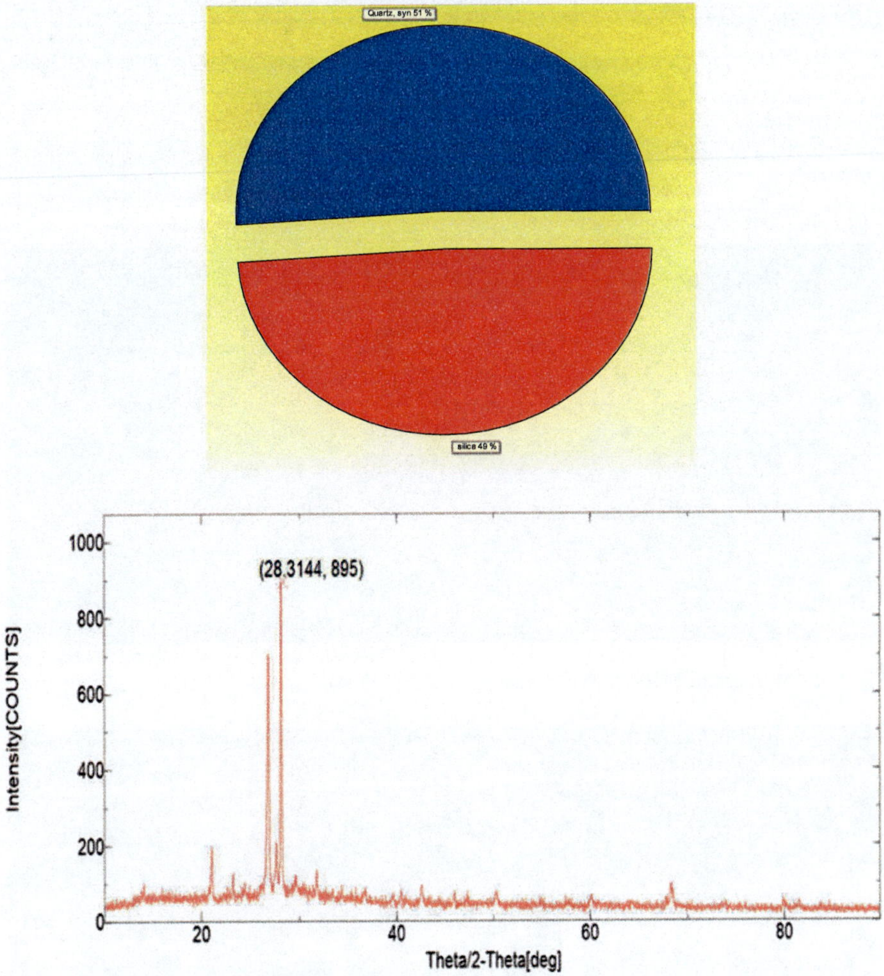

Fig. 5 XRD analysis of G40-grade concrete

Table 4 Phase composition of M40

S. No.	Visible	Ref. code	Compound name	Chemical formula	Score	Scale factor	SemiQuant (%)
1	True	00-046-1045	Quartz, syn	SiO_2	34	0.510	44
2	True	00-005-0586	Calcite, syn	$CaCO_3$	12	0.088	13
3	True	00-033-1161	Silica	SiO_2	34	0.513	42

Table 5 Phase composition of G40

S. No.	Visible	Ref. code	Compound name	Chemical formula	Score	Scale factor	SemiQuant (%)
1	True	00-033-1161	Silica	SiO_2	16	0.269	49
2	True	00-046-1045	Quartz, syn	SiO_2	14	0.269	51
1	True	00-033-1161	Silica	SiO_2	16	0.269	49

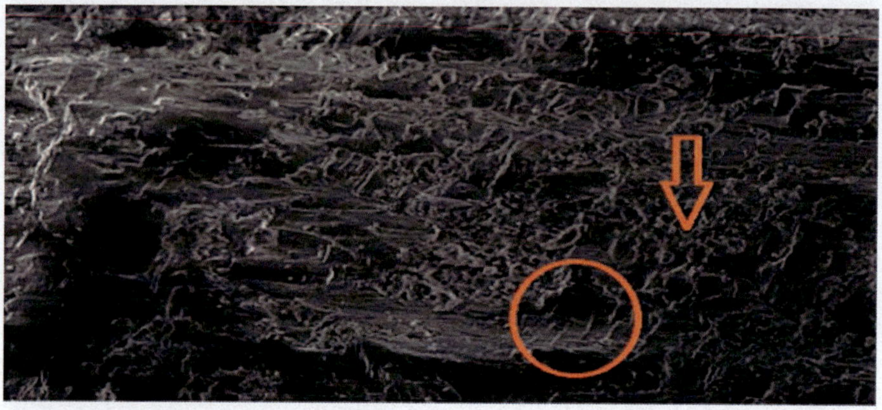

Fig. 6 SEM image of M40-grade concrete

Fig. 7 SEM image of G40-grade concrete

4. The results revealed that the mechanical properties of GPC are superior to OPC and bond strength increases with an increase in embedded length by a percentage of 45.85% and 67.75% for M40-grade concrete.
5. Geopolymer concrete has a split tensile strength of 4.87 N/mm^2 after 28 days under ambient curing.
6. Scanning electron microscopy of PCC and GPC has indicated improved densification was observed in alkali-activated concrete.

References

1. Manasa K, Andal M, Abhinay A (2022) Fracture parameters of flyash and GGBS based alkali activated concrete. Mater Today Proc 65(2):2053–2059. https://doi.org/10.1016/j.matpr.2022.06.246
2. Stephen A, Mahachi J (2021) Chloride ion penetration performance of recycled concrete with different geopolymers. Mater Today Proc 38(2):762–766
3. Mallikarjuna Rao G, Gunneswara Rao TD (2018) A quantitative method of approach in designing the mix proportions of fly ash and GGBS-based geopolymer concrete. Aust J Civ Eng 16:53–63
4. Mallikarjuna Rao G, Gunneswara Rao TD (2015) Final setting time and compressive strength of fly ash and GGBS-based geopolymer paste and mortar. Arab J Sci Eng 40(11):3067–3074
5. Gunneswara Rao TD, Seshu R (2015) Bond strength behaviour of geopolymer concrete. Malays J Civ Eng 27:371–381. https://doi.org/10.11113/mjce.v27.15931
6. Vikas G, Gunneswara Rao TD (2021) Setting time workability and strength properties of alkali activated fly ash and slag based geopolymer concrete activated with high silica modulus water glass. Iran J Sci Tech 45(3):1483–1492
7. Fu Q , Xu W, Zhao X, Bu MX, Yuan Q, Niu D (2021) The microstructure and durability of fly ash based geopolymer concrete: a review. Ceram Int 47:29550–29566
8. Venu M, Gunneswara Rao TD (2018) An experimental investigation of the stress-strain behaviour of geopolymer concrete. Slovak J Civ Eng 26(2):30–34
9. Kuunreddy SR, Balamurugan S (2020) Experimental and microstructural assessment of ternary blended geopolymer concrete with different Na_2SiO_3-to-NaOH volume ratios. Innovative Infrastruct Solutions 33. https://doi.org/10.1007/s41062-020-0279-z

Evaluation of Acid Resistance of Sustainable Binders Using Acid Consumption

Tom Damion and Piyush Chaunsali

1 Introduction

Portland cement has high carbon footprint associated with its manufacturing, resulting in 7–8% of global CO_2 emissions. Further, the acid resistance of Portland cement (PC) is poor [1, 2]. Acid attack is a critical problem associated with sewage carrying structures, and such a deterioration is called biogenic acid attack [3–5]. The acid attack can be considered as a surface deterioration, resulting in loss of cross section and strength [6]. The loads acting on a sewer structure include earth pressure, water pressure, and that due to traffic and building loads above it [7].

To improve the sustainability and durability of sewer structures, acid-resistant sustainable binders are required. Calcium aluminate cement (CAC) is known for its superior biogenic acid resistance [1, 8]. However, higher cost and loss in strength due to conversion reactions [9, 10] limit its widespread application [11]. High aluminium content is responsible for the superior acid resistance of high alumina binder [11]. Calcium sulfoaluminate (CSA) cement is a more sustainable binder with higher aluminium proportion as compared to PC [12–15].

Conventional acid immersion tests involve exposing the specimens in a fixed volume of dilute acid solution. The pH of exposure solution increases with time due to neutralisation [16, 17]. To obtain uninterrupted acid attack, the increment in pH should be controlled. Though the replacement of exposure solution reduces the pH, an efficient method would be titrating with a strong acid under continuous stirring. The dosing rate can be calculated from titration time and amount of acid dosed [18]. This procedure can be automated by an autotitrator as reported by Irico et al. [19]. Table 1 summarises the studies involving autotitrator. By using an autotitrator,

T. Damion (✉) · P. Chaunsali
Department of Civil Engineering, Indian Institute of Technology Madras, Chennai, India
e-mail: tomdamion@iisc.ac.in

R. K. Pancharathi et al. (eds.), *Low Carbon Materials and Technologies for a Sustainable and Resilient Infrastructure*, Lecture Notes in Civil Engineering 440,
https://doi.org/10.1007/978-981-99-7464-1_5

Table 1 Summary of literature review on titrator-aided acid attack study

Mellado et al. (2017)	Irico et al. (2020)	Gutberlet et al. (2015)	Huber et al. (2016)
10 × 10 × 60 mm prisms	150 × 150 × 40 mm concrete slices	3 mm thick, 30 mm dia. discs	3 mm thick, 30 mm dia. discs
Deionised water 200 ml	12 l and then sulfuric acid pumped	150 ml acid stirred with magnetic stirrer	150 ml acid stirred with magnetic stirrer
1 M HNO_3	2 M sulfuric acid	Sulfuric acid 0.5 M and HCl 1 M	0.5 M sulfuric acid
5 days exposure period	Weekly renewal till 35 days and monthly afterwards	28 days exposure period	28 days exposure period
pH 2	pH 2	pH 2, 3, 4	pH 2, 1

constant pH and continuous stirring can prevent the deposition of acid attack products on specimens, thus extending the study to less soluble salt forming acids [20].

A novel method based on acid neutralisation was reported for evaluating acid resistance [21]. Acid consumption was identified as a suitable criterion to rank the acid resistance of various binders. The statistical significance of acid consumption method, its limitations, and scope for future work are discussed in this article.

2 Materials and Methods

The binders used in the study included Portland cement (PC) of Grade 53 as per IS 269:2015, CSA cement-l (CSA (HY)), and calcium aluminate cement (CAC). The oxide compositions along with loss on ignition (LOI) of the binders considered are shown in Table 2.

The phase compositions of the raw binders determined by XRD analysis are shown in Table 3.

Figure 1 shows the XRD patterns of raw CAC. Two patterns are for 75 μm passed ground raw CAC (75 Mic.), and the bottom pattern corresponds to 53 μm passed ground raw CAC (53 Mic.). The three patterns correspond to scanning rates of 1.5 degree/min, 3 degree/min, and 6 degree/min from the bottom, respectively. In

Table 2 Oxide composition of binders

Binders	SiO_2	CaO	Al_2O_3	SO_3	Fe_2O_3	MgO	TiO_2	SrO	Na_2O	K_2O	LOI*
PC	21.2	58.1	6.9	2.8	5.1	1.0	0.5	0.0	0.1	0.8	3.1
CSA (HY)	14.0	39.8	20.9	14.5	3.6	2.8	1.1	0.1	0.2	0.5	1.8
CAC	8.3	32.5	43.6	4.3	2.4	0.7	3.8	0.1	0.3	0.0	3.4

LOI*: Loss on ignition

Table 3 Phase composition (%) of the raw binders

Phases/Binder	PC	CSA (HY)	CAC
Ye'elimite	0.0	35.5	5.9
Anhydrite	0.0	15.5	2.6
Gypsum/Bassanite	4.1	1.5	0.8
Dicalcium silicate	21.9	28.7	0.0
Dolomite	0.0	7.6	0.0
Brownmillerite	5.9	3.5	1.1
Mayenite	0.0	2.1	5.0
Tricalcium silicate	54.8	0.0	0.0
Tricalcium aluminate	4.7	0.0	0.0
Calcium monoaluminate (CA)	0.0	0.0	20.4
Grossite (CA_2)	0.0	0.0	13.9
Perovskite	0.0	0.0	7.7
Gibbsite	0.0	0.0	3.5
Aluminium oxide	0.0	0.0	0.6
Gehlenite	0.0	0.0	31.4
Lime	0.0	3.9	0.0
Quartz	1.8	1.7	0.0
Calcite	6.9	0.0	7.0

all these cases, a prominent peak was observed at 23.68° (2-theta degree), resulting due to the formation of ye'elimite. Hence, it can be assumed that the ye'elimite might have formed in the manufacture of this particular calcium aluminate cement. Again on referring to [22], it is evident that ye'elimite containing calcium aluminate cement is possible. The sulphate content in CAC was 4.3% which could explain the presence of 5.9% ye'elimite, 2.6% anhydrite, and 0.8% bassanite.

2.1 *Specimen Preparation*

Prismatic specimens (10 × 10 × 60 mm) were cast at water-to-cement (w/c) ratio of 0.4 (by wt.). The specimens were cured for 28 days in a fog room at 25 °C and 65% relative humidity environment. The cement was poured into measured volume of water and stirred for two minutes using a high shear mixer and then transferred to the moulds.

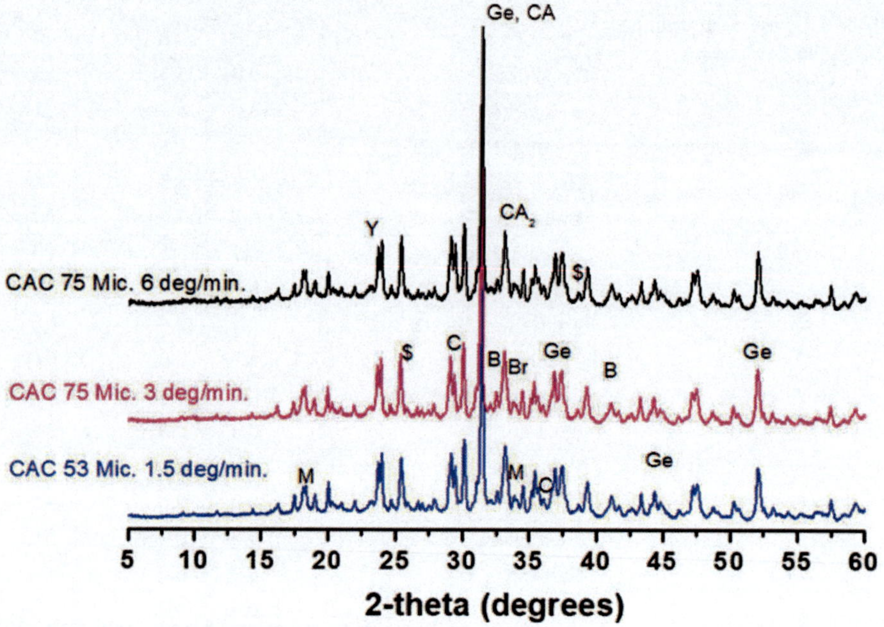

Fig. 1 XRD patterns of raw calcium aluminate cement (CAC) with different sample preparation procedure

2.2 *Acid Attack Tests*

The acid exposure was performed using an automatic titrator (Metrohm 916 Ti Touch). Titrator had a pH electrode, acid dispenser, temperature sensor, and a stirrer (Fig. 2). The cured specimens were subjected to constant pH experiment (STAT). Diluted sulfuric acid (5%) and citric acid (1 M) were used to maintain a constant pH of 1 and 3, respectively. Three paste specimens were used in a test. As described in [21], the solution was replaced after each day till five days.

2.3 *Mass and Dimensional Changes*

The change in mass and cross-sectional dimensions were measured. Normalised mass and cross-sectional area at a particular day were taken with respect to the control specimen's (28-day cured specimen) mass and area, respectively. Digital caliper having a sensitivity of 0.001 mm was used to measure the dimensions.

Fig. 2 Titrator running for experiment with citric acid

3 Results

3.1 Mineralogy and Mechanical Properties of Hydrated Binders

The 28-day flexural strengths of various cement mortar (cement-to-sand ratio 1:2.75) with w/c ratio of 0.5, determined as per ASTM C348, were: 5.41 MPa (SD: 0.13 MPa), 4.85 MPa (SD: 0.25 MPa), and 3.01 MPa (SD: 0.19 MPa) for PC, CSA (HY), and CAC cements, respectively. The results from static/constant pH test on monolithic specimens (STAT) are discussed below.

3.2 Sulfuric Acid

The mass loss and acid consumption during the test period are shown in Fig. 3. For STAT test with sulfuric acid at constant pH of 1, the order of mass loss of the binders was: CAC > CSA (HY) > PC (Fig. 3). The initial unaltered mass in PC was due to precipitation of gypsum on the surface, during early days of exposure [19, 23]. In

PC, gypsum deposition in pores caused pore blocking, creating a barrier for further attack [24–26].

The pH of acid solution increased by the associated neutralisation due to leached alkalis in the specimens. The increment was prevented by titrating the required amount of acid causing neutralisation. Acid consumption is the amount of dosed acid to maintain the pH at 1 (± 0.001). As seen from Fig. 3b, the sulfuric acid consumption to maintain the pH of 1 followed the order: CAC > CSA (HY) > PC. The barrier formation on the surface of PC specimens stopped continuous neutralisation of interior region. Hence, least acid consumption was in case of PC. It was noticed that the mass loss could be correlated to the amount of acid consumption.

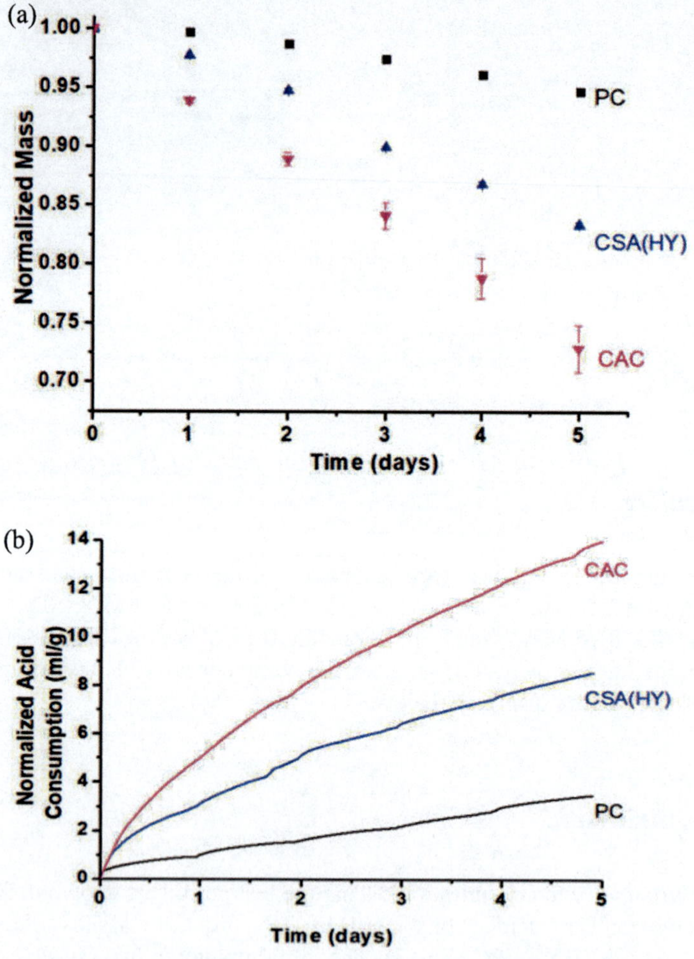

Fig. 3 **a** Normalised mass versus time, and **b** cumulative sulfuric acid consumption for STAT pH 1 sulfuric acid (5%) test (monolithic specimens)

The standard deviation associated with acid consumption is very small that it is not visible in Fig. 3b. The variability in mass loss was relatively higher than that for acid consumption.

3.3 Citric Acid

For the test with citric acid of constant pH of 3, mass loss in three binders followed the order: PC > CSA (HY) > CAC (Fig. 4). The higher damage in PC as compared to other binders could be attributed to the presence of large amount of portlandite in it. Portlandite transforms into expansive calcium citrate salt on reaction with citric acid. This calcium citrate salt was earlier identified as tri-calcium di-citrate hexahydrate [27], and not earlandite, as reported in the previous limited literature.

Citric acid consumption for maintaining the pH of 3 for the three binders followed the order: PC > CSA (HY) > CAC (Fig. 4). Aluminium hydroxide, one of the hydration products of CSA cement, has stability range up to pH 3–4. Because of the automated operation and precision of dosing ($\pm$ 0.002 ml), the acid consumption result was highly repeatable.

ANOVA was performed using JMP software for the normalised mass in case of STAT pH 1 sulfuric acid as shown in Fig. 5. From the confidence interval, the difference between binders was clear. The results were statistically significant.

4 Discussion

For the static pH tests, three specimens were tested in one run. The minimum number to satisfy statistical feasibility was three. For all the binders considered, volume of acid considered in monolithic specimen test was 250, and three specimens were of size 10 $\times$ 10 $\times$ 60 mm. There should be sufficient volume of solution for complete immersion of electrode. If more solution is taken, that will lead to quick overflow of vessel and automatic stoppage of the test. By trials, the optimum amount of solution was found to be 250 ml.

For sulfuric acid (pH 0.5) tests as suggested by ASTM 1898, small area-to-volume ratio (A/V) is preferred. Otherwise, more acid would be consumed leading to quick automatic stoppage of test. In that case, solution has to be removed in frequent time intervals, leading to some experimental disturbances.

The effect of stirring speed on the test was not included in this paper. However, the optimum stirring speed was found to be 900–1000 rpm based on trials. For comparative purposes, the stirring effect can be considered same for all the binders. The optimum stirring speed is the one without allowing any deep vortex or bubble formation. It should be ensured that the pH electrode and acid dosing point are kept apart.

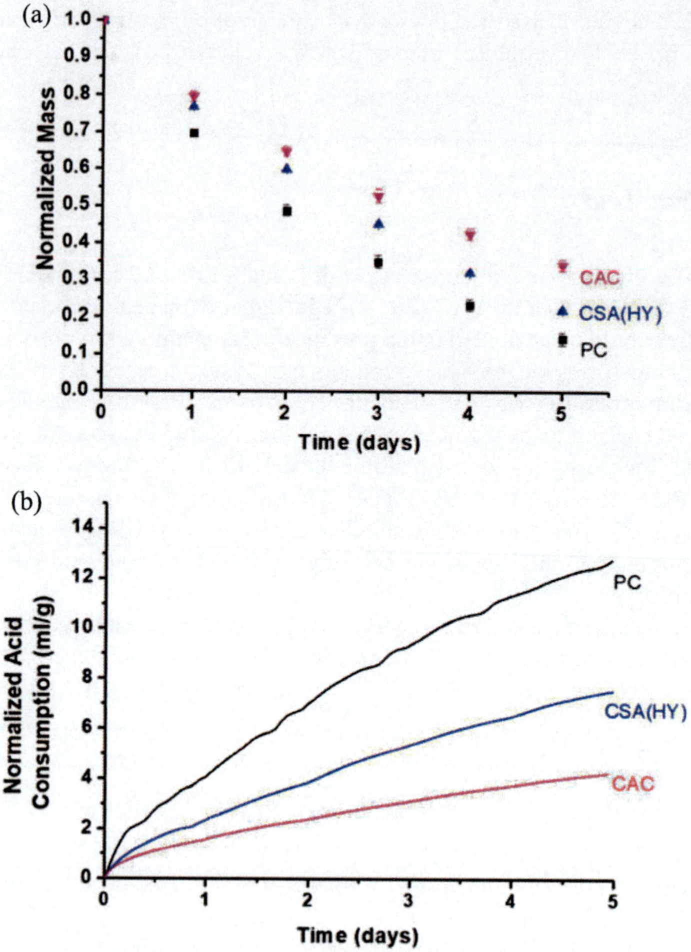

Fig. 4 **a** Residual mass across time, and **b** cumulative citric acid consumption of citric acid (1 M) in case of pH 3 test (monolithic specimens)

4.1 Prediction from Initial Test Period

It has been discussed that acid consumption was found to be directly proportional to the deterioration. The initial period of the test mainly contributes to the total acid consumption. Hence, the proportionate acid consumption can be calculated from the initial period of the test. Figure 6 is an enlarged version of monolithic test during the initial period. When the specimens were immersed into pH 1 or 3 acid solution, the pH increased, due to leaching and acid attack. Titrator then brings back the pH to initial value by dosing off the acid. The point at which original pH was reached after initial increment in pH was termed 'stabilising point (SP)'. The point of maximum

Oneway Analysis of Normalized Mass By Binder

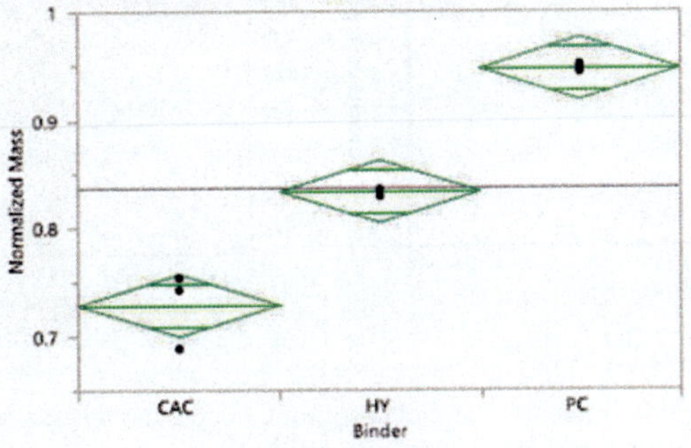

Oneway Anova
Summary of Fit

Rsquare	0.966086
Adj Rsquare	0.954781
Root Mean Square Error	0.020458
Mean of Response	0.8366
Observations (or Sum Wgts)	9

Analysis of Variance

Source	DF	Sum Squares	of Mean Square	F Ratio	Prob > F
Binder	2	0.07153744	0.035769	85.4591	<.0001*
Error	6	0.00251129	0.000419		
C. Total	8	0.07404873			

Fig. 5 ANOVA test results

pH is also the point after which the slope of acid consumption curve changes. The point corresponding to the maximum pH in the acid consumption curve was called 'divergence point (DP),' after this point acid consumption curves diverge and no longer collinear.

In Fig. 7, the acid consumption for the three binders is shown. The moles of H^+ ions in the case of sulfuric acid would be double the amount of acid dosed. However, for organic acids such as citric acid, that may not be the case because of their poor dissociation constants. Thus, 'acid' consumption (moles of acid molecule) could be an appropriate criterion to compare the results of acids having different dissociation constant. The acid consumption trend in the initial test period was similar to the entire test period. Hence, the acid resistance can be predicted based on the results from the initial test period.

ANOVA test was performed on the results of STAT pH 3 test using citric acid for initial period (Fig. 8). The difference among the binders was evident based on confidence interval. The narrow confidence interval could be attributed to a low standard

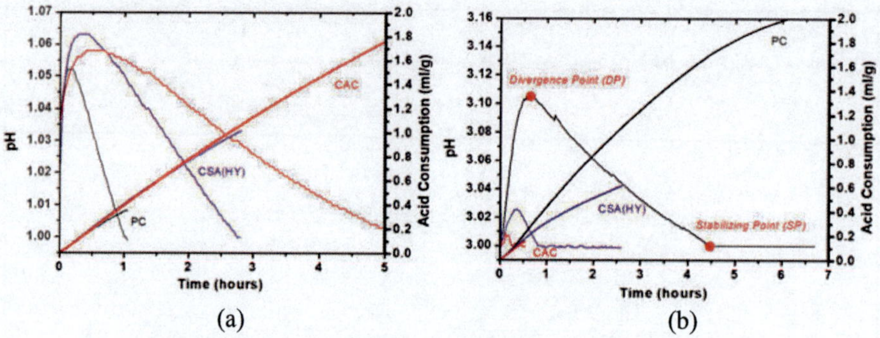

Fig. 6 Acid consumption (thick line) and pH evolution during the initial period of the constant pH test for **a** pH 1 sulfuric acid, and **b** pH 3 citric acid [21]

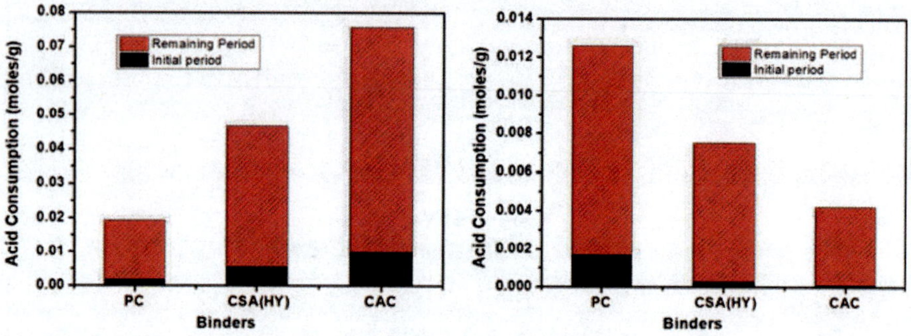

Fig. 7 Cumulative acid consumption during the entire period of test and during the initial period (up to SP) for **a** sulfuric acid (pH 1), and **b** citric acid (pH 3) [21]

deviation in acid consumption. This was due to the dosing precision (0.002 ml) of the autotitrator. It is thus statistically proven that the difference between binders can be predicted based on initial period data (within SP). Similarly, ANOVA was performed in case of sulfuric acid tests. In this case, the difference was obvious.

4.2 *Correlation Between ANC and Acid Resistance*

The criteria for evaluating acid resistance include retained cross-sectional area measurement or mass loss calculation. The correlation between acid resistance and acid consumption is shown in Fig. 9. It can be seen that the acid consumption was inversely proportional to acid resistance of the cement.

Statistical analysis was performed for various results presented. The acid consumption or ANC was determined from monolithic and powder, as in [21]. In

Oneway Analysis of Acid consumption till stabilizing point By Binders

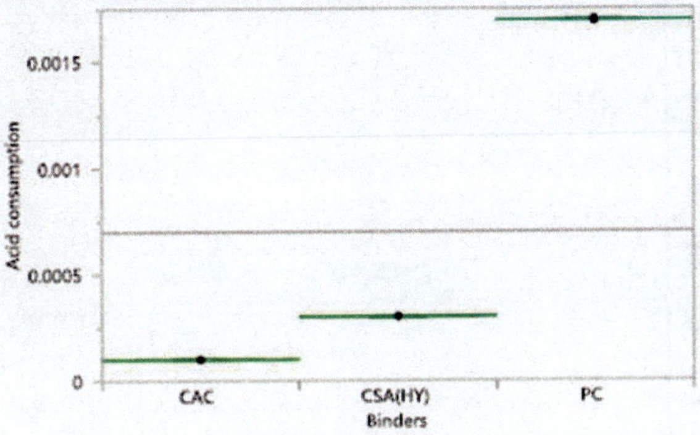

Oneway Anova

Summary of Fit

Rsquare	0.999985
Adj Rsquare	0.99998
Root Mean Square Error	3.367e-6
Mean of Response	0.000699
Observations (or Sum Wgts)	9

Analysis of Variance

Source	DF	Sum of Squares	Mean Square	F Ratio	Prob > F
Binders	2	4.54042e-6	2.2702e-6	200234.2	<.0001*
Error	6	6.8027e-11	1.134e-11		
C. Total	8	4.54049e-6			

Fig. 8 ANOVA test results in case of citric acid consumption till stabilising point

Fig. 10, the correlation of ANC between powder and monolithic test at (a) sulfuric acid (pH of 1), and (b) citric acid (pH of 3) is considered.

When acid consumption was used as a criterion for comparison of binders, it should be ensured that the concentration of diluted acid that is used as titrant is uniform in case of all binders. For comparing different binders, it would be better to use diluted acid from single dilution. The diluted acid may not be advisable to store for long period, particularly in case of volatile acids such as HCl. The method has been proved to be successful in comparative research. The reliability of absolute results should be validated with theoretical stoichiometric calculations.

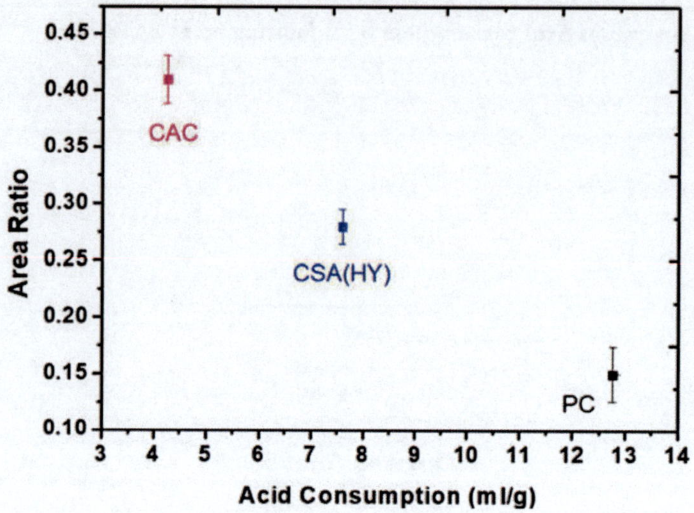

Fig. 9 Acid consumption versus area ratio (or acid resistance) after 5-day STAT pH 3 test with citric acid

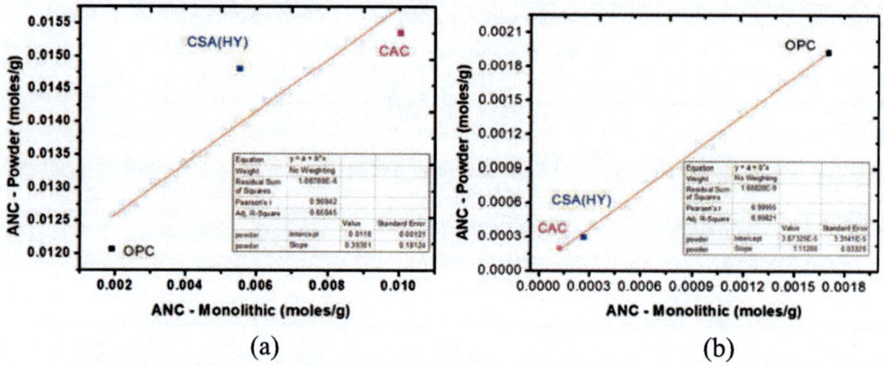

Fig. 10 Correlation coefficient for monolithic and powder tests **a** sulfuric acid **b** citric acid

5 Future Work

1. In this paper, insoluble salt forming acids were considered. The study should be extended to soluble salt forming acids such as acetic acid and hydrochloric acid.
2. The dosing rate particularly in powder titration experiment needs to be optimised (as it should consider the neutralisation reaction time). When the dosing rate is high, the pH decrement would be large.

6 Conclusions

The extent of deterioration of cementitious binders was proportional to the acid consumption. The criteria of acid consumption can be used to statistically differentiate the binders based on acid resistance. Below are the specific conclusions from this study:

- Alternative binders such as CSA and CAC outperformed PC in citric acid (pH of 3). The absence of portlandite in these binders, reduced formation of expansive calcium citrate salt.
- In sulfuric acid (pH of 1), the performance of PC was better than CSA (HY).
- The extent of deterioration can be determined based on the acid consumption corresponding to 'stabilising point' of pH vs time curve in initial period itself.

Acknowledgements The first author would like to acknowledge the doctoral scholarship received from the Ministry of Education, India, and post-doctoral equivalent fellowship from Indian Institute of Technology (IIT) Madras. All authors would like to acknowledge the resources provided by the Department of Civil Engineering at the IIT Madras towards the usage of experimental facilities in this study. The last author is also grateful for the financial support from the New Faculty Seed Grant (CE1920426NFSC008926) by Industrial Consultancy and Sponsored Research (ICSR) centre at IIT Madras. The Institute of Eminence Research Initiative project grant on Technologies for Low Carbon and Lean Construction from IIT Madras is also gratefully acknowledged.

References

1. Ehrich BS, Helard L, Letourneux R et al (1999) Biogenic and chemical sulfuric acid corrosion of mortars. J Mater Civ Eng 11:340–344
2. Bakharev T (2005) Resistance of geopolymer materials to acid attack. Cem Concr Res 35:658–670. https://doi.org/10.1016/j.cemconres.2004.06.005
3. Olmstead WH, Hamlin H (1900) Converting portions of the Los Angeles outfall sewer into a septic tank. Eng News Am Railw J XLIV:317–318
4. Parker C (1945) The corrosion of concrete: 1. the isolation of a species of bacterium associated with the corrosion of concrete exposed to atmospheres containing hydrogen sulphide. Aust J Exp Biol Med Sci 23:81–90. https://doi.org/10.1038/icb.1945.13
5. Parker C (1945) The corrosion of concrete: 2. the function of Thiobacillus concretivorus (Nov. Spec.) in the corrosion of concrete exposed to atmospheres containing hydrogen sulphide. Aust J Exp Biol Med Sci 23:91–98. https://doi.org/10.1038/icb.1945.14
6. Bensted J, Brough AR, Page MM (2007) Chemical degradation of concrete. Durab Concr Cem Compos 86–135. https://doi.org/10.1533/9781845693398.86
7. Vickridge IG (1997) Structural aspects of sewer rehabilitation. In: Sewer rehabilitation and new construction, pp 254–271
8. Sand W, Dumas T, Marchargent S (1994) Accelerated biogenic sulfuric-acid corrosion test for evaluating the performance of calcium-aluminate based concrete in sewage applications. ASTM Spec Tech Publ 234–249. https://doi.org/10.1520/stp12938s
9. Cuesta A, De La Torre AG, Santacruz I et al (2017) In situ hydration imaging study of a ye'elimite paste by ptychographic X-ray computed tomography. In: Proceedings of the 39 international conference on cement microscopy ICMA 2017

10. Majumdar AJ, Singh B, Edmonds RN (1990) Hydration of mixtures of "Ciment Fondu" aluminous cement and granulated blast furnace slag. Cem Concr Res 20:197–208. https://doi.org/10.1016/0008-8846(90)90072-6
11. Scrivener KL, Cabiron JL, Letourneux R (1999) High-performance concretes from calcium aluminate cements. Cem Concr Res 29:1215–1223. https://doi.org/10.1016/S0008-8846(99)00103-9
12. Beretka J, de Vito B, Santoro L et al (1993) Utilisation of industrial wastes and by-products for the synthesis of special cements. Resour Conserv Recycl 9:179–190. https://doi.org/10.1016/0921-3449(93)90002-W
13. Mather B (1970) Expansive cements
14. Ojovan MI, Lee WE, Kalmykov SN (2019) Immobilisation of radioactive waste in cement
15. Ribeiro MSS (1998) Expansive cement blend for use in shrinkage-compensating mortars. Mater Struct Constr 31:400–404. https://doi.org/10.1007/bf02480713
16. Onuaguluchi O, Banthia N (2022) The influence of $CaCl_2$-blended acrylic polymer on steel rebar corrosion and acid attack resistance of mortar. Corros Mater Degrad 3:160–177. https://doi.org/10.3390/cmd3010009
17. Ramaswamy KP, Santhanam M (2016) Degradation kinetics of cement-based materials in citric acid. In: Lecture notes in civil engineering, pp 891–905
18. Min H, Song Z (2018) Investigation on the sulfuric acid corrosion mechanism for concrete in soaking environment. Adv Mater Sci Eng 2018:1–10
19. Irico S, Meyst L De, Qvaeschning D et al (2020) Severe sulfuric acid attack on self-compacting concrete with granulometrically optimized blast-furnace slag-comparison of different test methods. Materials (Basel) 13. https://doi.org/10.3390/ma13061431
20. Stegemann JA, Shi C, Caldwell RJ (1997) Response of various solidification systems to acid addition. Elsevier B.V.
21. Damion T, Chaunsali P (2022) Evaluating acid resistance of Portland cement, calcium aluminate cement, and calcium sulfoaluminate based cement using acid neutralisation. Cem Concr Res 162:1–17. https://doi.org/10.2139/ssrn.4165828
22. Espinosa B, Fitzgerald, Winslow M, Walter, Charles A et al (2017) Long-workability calcium aluminate cement with hardening promoted by a temperature increase, and related use
23. Lloyd RR, Provis JL, Van Deventer JSJ (2012) Acid resistance of inorganic polymer binders. 1. Corrosion rate. Mater Struct Constr 45:1–14. https://doi.org/10.1617/s11527-011-9744-7
24. Qiu P, Zhang L, Li Y et al (2022) Permeability evolution model of coarse porous concrete under sulphuric acid corrosion. Constr Build Mater 326:126475. https://doi.org/10.1016/j.conbuildmat.2022.126475
25. Grandclerc A, Dangla P, Gueguen-Minerbe M, Chaussadent T (2018) Modelling of the sulfuric acid attack on different types of cementitious materials. Cem Concr Res 105:126–133. https://doi.org/10.1016/j.cemconres.2018.01.014
26. Valencia-Saavedra WG, Mejía de Gutiérrez R, Puertas F (2020) Performance of FA-based geopolymer concretes exposed to acetic and sulfuric acids. Constr Build Mater 257:119503. https://doi.org/10.1016/j.conbuildmat.2020.119503
27. Damion T, Cepuritis R, Chaunsali P (2022) Sulfuric acid and citric acid attack of calcium sulfoaluminate-based binders. Cem Concr Compos 130:104524. https://doi.org/10.1016/j.cemconcomp.2022.104524

CO_2 Curing for Enhanced Early Age Strength in Saw Dust Biochar Augmented Cement Mortars

Sunil Bhagat Tadi and Rathish Kumar Pancharathi

1 Introduction

The world is striving to meet its emission reduction, particularly the greenhouse gas emission (GHGE) and is clearly off course in meeting 1.5 °C goal [1], within reach. Cement industry is reported to have been contributing 6–8% of global CO_2 emissions. Cement finds its use in cement concrete (plain and reinforced), mortar, and paste/ grout in many forms, is the only material consumed by the mankind next to water [2]. The CO_2 emissions from the manufacturing of cement are reported to be 60% from calcination of raw ingredient—lime stone and remaining 40% from fossil fuel burning to facilitate kiln reactions and other processes such as crushing, grinding, and blending. To reduce the CO_2 emissions in cement manufacturing, alternative/ supplementary cementitious materials (SCM) such as fly ash, blast furnace slag (BFS), vegetable ashes (sugar cane bagasse ash, rice husk ash, etc.) were explored and are in wide use globally. In fact India has three times the requirement, 352 MTPA for 105 MTPA of SCMs are generated [3], but with uneven distribution and hence inefficient quantitative utilization. The uneven and diminishing availability of SCMs such as fly ash depending on the thermal power generation, BFS depending on the virgin steel production is further driving us to explore dependable source of SCMs such as calcined clay and other blended cements [4].

Use of SCMs besides reducing CO_2 emissions, amount to reduced cement/clinker use [5]. Use of cementitious products with SCMs has better performance to weathering agents [6]. To achieve carbon neutrality in the construction sector, it is not

S. B. Tadi (✉) · R. K. Pancharathi
Department of Civil Engineering, National Institute of Technology, Warangal, India
e-mail: bhagat@rguktn.ac.in

S. B. Tadi
Department of Civil Engineering, Rajiv Gandhi University of Knowledge Technologies, Nuzvid, India

R. K. Pancharathi et al. (eds.), *Low Carbon Materials and Technologies for a Sustainable and Resilient Infrastructure*, Lecture Notes in Civil Engineering 440,
https://doi.org/10.1007/978-981-99-7464-1_6

just clinker reduction, but utilization and/or sequestration of CO_2 must be considered as an option. Cementitious materials have the innate ability to absorb CO_2 from atmosphere, but at very slower rate. The CO_2 uptake is reported to be 20% of that calcination process emissions of making cement. Furthermore, 3% CO_2 uptake is reported considering reuse and recycling at the end of its design service period [7]. The CO_2, in liquid or gaseous or supersaturated state, can be effectively used either for treating recycled aggregate for reducing porosity, for mixing into the fresh concrete and as a curing method for enhanced early age performance [8].

Apart from processes presented above, "Biochar" is finding its prominence as a carbon negative technology [9]. Biochar is a highly porous and carbon-rich material that is obtained by the pyrolysis process, or heating of biomass in the absence or very limited presence of oxygen. The carbon negative effect per ton of biomass raw material can reach about 870 kg CO_2 equivalent (CO_2-e), of which 66% is related to the carbon retention effect of biomass charcoal. It was even found that biochar is highly advantageous in terms of cost-effectiveness and carbon sequestration capacity [10]. Research on biochar applications in construction materials is gaining a lot of momentum in the last decade. However, very limited research is available on the influence of CO_2 curing on biochar augmented concrete [11].

In this context, this study is focussed on determining the (i) Optimum replacement dosage of laboratory made biochar for cement (ii) Optimum duration of CO_2 curing on OPC based cement mortar and (iii) Influence of CO_2 cuing on biochar augmented concrete.

2 Materials

2.1 Basic Properties

Ordinary Portland cement (OPC) 53 grade as per IS 269:2015 is utilized for making cement mortar cubes in this experimental study. The properties and oxide composition based on X-ray fluorescence (XRF) of cement used is shown in Table 1. River sand is sieved to fit the requirement of standard sand as per IS 650:1991 with equal weights in three different size ranges 500 μm to 1 mm, 1 mm to 1.5 mm, and 1.5 mm to 2.0 mm.

2.2 Biochar Production Through Pyrolysis

Pyrolysis is the term used for the thermal degradation of organic compounds at temperatures between 250 and 900 °C in the absence of oxygen. This is an alternative method for turning waste biomass into products with added value like biochar, bio oil, and syngas. The compounds like cellulose, hemicellulose, and lignin which

Table 1 Properties of cement, biochar, and sand

For ordinary portland cement (OPC)		Chemical composition of OPC	
Specific gravity	3.1	Oxide	wt%
Bulk density	1442 kg/m^3	TiO_2	0.45
Normal consistency of cement	32%	NiO_2	< 0.01
Initial setting time	102 min	ZrO_2	0.01
Final setting time	345 min	ZnO	0.01
For saw dust biochar		SiO_2	19.23
Bulk density	542 kg/m^3	Al_2O_3	5.43
pH	8.86	BaO	0.01
%C	77.95	CaO	63.28
%H	2.021	Cr_2O_3	0.01
%N	2.44	Fe_2O_3	5.29
%S	0.074	K_2O	0.29
%O	17.515	MgO	2.21
H/C	0.026	MnO	0.18
C/N	31.94	SO_3	3.28
Fine aggregate	River sand		
Specific gravity	2.67		

are present in the biomass go through reaction processes like depolymerization, fragmentation, and cross-linking at specific temperatures during the process, producing a variety of products in the form of solid, liquid, and gas [12].

Saw dust collected from a local saw mill dump used as biomass feed stock for preparation of biochar. A pyrolysis reactor (see Fig. 1) with nitrogen (N_2) purging and vent for collection of bio oil and biochar is employed for manufacturing of SDBC. The feedstock is subjected to oven drying before undergoing slow pyrolysis process from room temperature to 550 °C at a ramp rate of 10 °C per minute in the N_2 environment. The heated feedstock is kept at the highest temperature for 1 h before allowing it to cool naturally. The temperature of 550 °C is chosen as the rate CO_2 adsorption on sawdust derived biochar reported to increase with increasing temperature from 450 to 650 °C [13].

The SDBC is found to have a yield of 29%. The ultimate analysis results (Table 1) reveals that the carbon content is high with lower H/C ratio inferring aromaticity caused by pyrolysis and therefore indicates higher stability of the material than the initial feedstock.

Powder XRD results of OPC (Fig. 2) indicate presence of typical chemical compounds, viz. Alite, Belite, and Allumino Ferrite phases of the cement. However, the presence of calcite at around $2\theta = 29°$ might be due to carbonation of cement sample before testing. The broad hump at $2\theta = 22°$ and 45° indicate the loss of crystalline structure of cellulose and progression of carbonization [14].

Fig. 1 Pyrolysis reactor for production of biochar

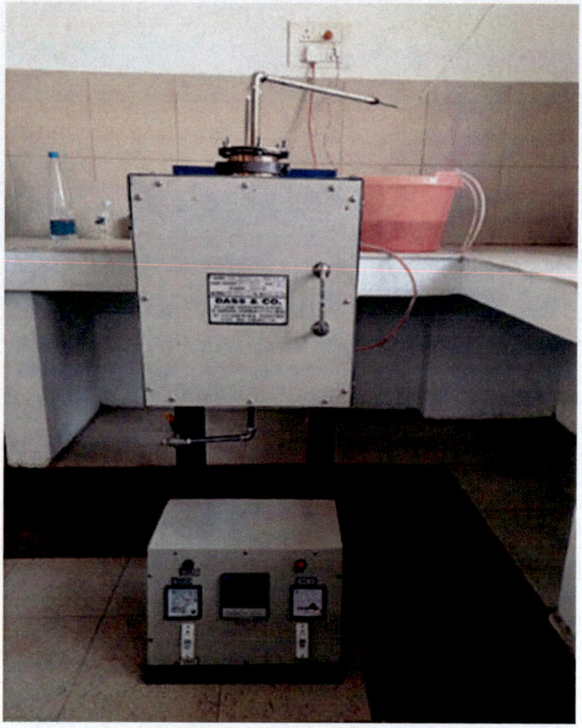

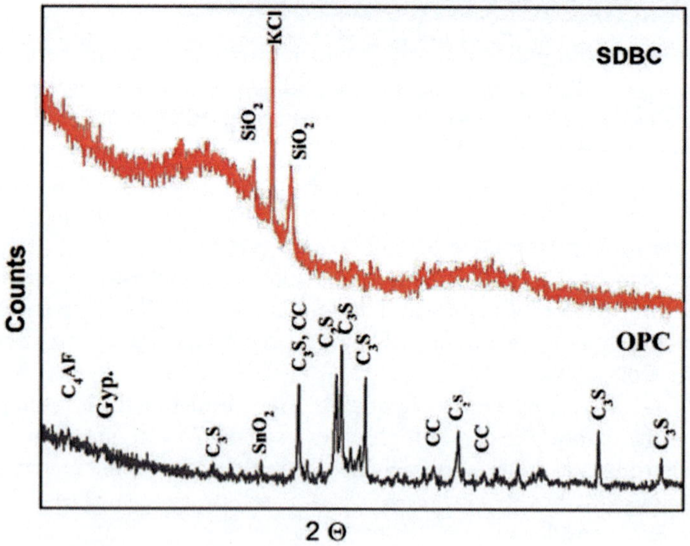

Fig. 2 XRD patter for OPC cement and saw dust biochar

3 Methodology

The experimental methodology is planned and executed with the objectives of finding the optimal CO_2 curing duration, fixing the dosage of SDBC in cement, and estimate the influence of CO_2 curing on the SDBC augmented cement mortar with optimized CO_2 curing duration and dosage of SDBC.

3.1 Preparation of Cement Mortar Specimens

For the entire study cement mortar of 1:3 (Cement: fine aggregate) with 0.5 water to cement ratio (w/c) is adopted, all the specimens were cast in cubes of size 70.6 mm conforming to IS:10086-1982, and each sample consists of six specimens. Mixing of the cement mortar is executed in a laboratory planetary mixer machine in which paddle rotates in 3 speeds: nudge (very slow), slow, and fast. Initially, the mixing bowl is cleaned with water and dried.

Based on the replacement level and the biochar, weights of all the components required per each cube are calculated. Nudging is used for dry mixing of cement, sand, and biochar for 2 min. Further mixed with adding water at slow speed for 2 more minutes and finally mixed at high speed for 1.5 min. This procedure is uniformly implemented for all the specimens. Vibratory tale is used for uniform compaction. The compacted specimens are further subjected to different curing regimes as explained in subsequent sections. The proportions of materials used for casting each cube of 70.6 mm size are presented in Table 2.

The mix proportions used for casting cement mortar cubes for optimizing the dosage of SDBC and tested at 28 days. Control mix proportions are used for evaluation of optimum duration of CO_2 curing and 2% SDBC is additionally used for understanding the effect of CO_2 curing on the biochar augmented cement mortar.

Table 2 Mix proportions implemented for making SDBC augmented cement mortars

Mix	Cement (g)	SDBC-biochar (g)	Fine aggregate (g)	Water (g)
Control	200	0	600	100
2% SDBC	196	4	600	100
4% SDBC	192	8	600	100
6% SDBC	188	12	600	100
8% SDBC	184	16	600	100

3.2 CO_2 Curing of Cement Mortar

Three days curing regimes are used to meet the desired objectives, and the same is shown in Table 3. CO_2 curing is executed in four stages based on literature review [15]. The set-up for carbonation curing is shown in Fig. 3. The chamber is capable of controlling relative humidity, temperature, and CO_2 concentration.

Stages of CO_2 Curing

(i) In-mold curing refers to the time in hours when the cast specimens are kept in the mold and 95% relative humidity is maintained.
(ii) Pre-conditioning stage refers to that stage of demolding the specimens, kept in ambient air to lose nearly 40% of the moisture mass to facilitate CO_2 diffusion into the samples.

Table 3 Curing regimes implemented in this study

Description	In-mold curing (h)	Pre-conditioning (h)	In CO_2 chamber (h)	Water curing (h)
Regime I	24	0	0	Till age of testing
Regime II	12	6	0	54
	12	6	2	52
	12	6	4	50
	12	6	6	48
Regime III	18	4	2	Till age of testing

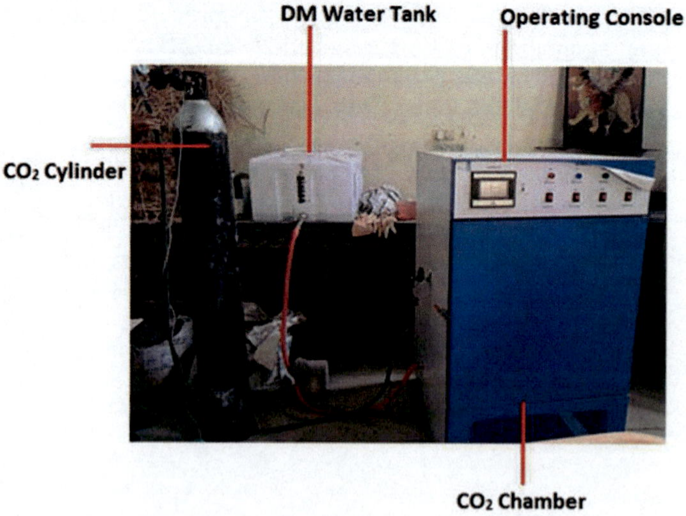

Fig. 3 Carbonation chamber for CO_2 curing

(iii) In this stage, the specimens are actually subjected to CO_2 gas in a closed chamber. A CO_2 concentration of 15%, relative humidity of 50%, and temperature of 25 °C are adopted in this study.
(iv) This stage involves subjecting the CO_2 cured specimens to normal water curing till the age of testing.

Water curing of specimens is implemented by complete immersion of the specimens in tap water till desired age of testing.

3.3 Testing and Characterization of Cement Mortar Cubes

After the desired period of curing, all the specimens are tested for their compressive strength in a universal testing machine (UTM) with a loading rate of 35 N/mm^2/min conforming to IS 4031 Part 6:1988, which is equivalent to a loading rate of 2.7 kN/s.

Further, powder X-ray diffraction (XRD) technique and scanning electron microscopy (SEM) are used for understanding the phase change and morphology of specimens subjected to CO_2 curing for 2, 4, and 6 h after pre-conditioning. Samples are collected from fractured surfaces of cubes immediately after testing. All the samples are subjected to solvent exchange in isopropyl alcohol for hydration stoppage [16]. Some of the collected samples are ground to pass through 75 μm size sieve for XRD while relatively flat broken pieces are used for SEM after sputter coating with gold palladium to avoid charge accumulation on the surface.

4 Results and Discussions

4.1 Determination of Optimum Dosage of Biochar in Cement Mortar

Cement mortar cubes with one control sample with 0% SDBC and three samples with 2%, 4%, and 8% SDBC dosages were cast and cured for 28 days under water following Regime I (Table 3). The results of the compressive strength test are presented in Table 4.

It can be observed that the dosage of SDBC resulted in a reduced mechanical performance. However, the strength reduction at 28 days for 2% SDBC is not significant with only 1.5%. This finding is not aligned with earlier reported results by Gupta et al. [17], where in the saw dust biochar dosage of 1 and 2% reported slightly higher strength than control mix at 28 days. It may be noted that the biochar used was ground and further super plasticizers were used in the earlier study.

However, the decrease in strength with higher dosage of SDBC is proved in this study owing to reduced density and increased porosity. As is expected the

Table 4 28 days compressive strength of cement mortars with various dosages

Mix	Compressive strength (MPa)	% Change
Control	51.68	–
2%SDBC	47.14	– 1.48
4%SDBC	38.00	– 18.20
6%SDBC	33.18	– 28.57
8%SDBC	30.84	– 33.61

porous structure and low density of SDBC without any mechanical treatment such as grinding post preparation might have caused lower strength. It may be stated that the SDBC used in this study resulted in lowest (1.5%) strength loss at 28-day water curing compared to other dosages which resulted in 18, 28, and 33% strength loss at 4, 6, and 8% dosages. Hence, the 2% SDBC dosage is considered for subsequent evaluation of effect of CO_2 curing on SDBC augmented cement mortar.

4.2 *Determination of Optimum CO_2 Curing Period for OPC Based Mortars*

Cement mortar cubes cast with OPC are allowed to cure in molds for 12 h till initial gain of strength to attain sufficient strength before demolding. After this in-mold curing period, the cubes are de-molded and exposed to ambient conditions for six hours to lose part of free water. This step helps for better diffusion of CO_2 into the structure of mortar cubes when exposed to CO_2 in chamber [18]. The cubes were exposed to CO_2 gas in the carbonation chamber for two hours. The relative humidity (RH) is maintained at 50%, and the temperature (T) is maintained at 25 °C in the chamber. The CO_2 concentration is put to 15% resembling typical flue gas CO_2 concentration [19]. After two hours of CO_2 exposure, the cubes are immersed in tap water filled tank till age of testing. This post carbonation hydration step is proved to be efficient for better performance of CO_2 cured cementitious products [20, 21]. The 3 day compressive strength of the cured specimens is reported in Table 5.

It can be observed that the two hour carbonation resulted in relatively high compressive strength compared to other mixes. This could be due to the carbonation reaction, where in, the hydration products of OPC, viz. Portlandite ($CaOH_2$)

Table 5 Three days compressive strength of cement mortar cubes under CO_2 curing

Carbonation duration (h)	Strength (MPa)	% Change
0	18.98	–
2	21.83	+ 15.01
4	19.57	+ 3.11
6	17.35	– 8.59

and unhydrated cement, (UC) undergo carbonation reaction in the presence of CO_2 and precipitate denser calcite ($CaCO_3$) [22]. It can also be seen that further exposure to CO_2 resulted in gradual decrease in compressive strength. This could be due to the decreased diffusivity of CO_2 into the microstructure as denser calcite is precipitating in the surface pore of the cement mortar and/or reduced duration of post carbonation hydration before testing after 3 days curing. The increase in early (3 day) compressive strength is significant with 15% increase with 2-h CO_2 curing, beyond which it showed reduction. The X-ray diffraction (XRD) technique and scanning electron microscopy (SEM) characterization techniques are used for validating the obtained results. Based on this study 2-h CO_2 curing is considered as optimum duration. Further study is executed with 2-h CO_2 curing duration and 2% SDBC dosage.

4.3 XRD and SEM Analysis for CO_2 Cured Specimen Samples

The hydration stopped specimens were ground for powder XRD phase identification from control, 2 h, 4 h, and 6 h CO_2 cured mortar specimens. The XRD patterns were obtained on a Bruker D8 Advance equipment with Cu-Kα radiation ($\lambda = 1.54$ A°). Samples were scanned from 10° to 70° 2θ values. Qualitative phase identification is done using high score software and is shown in Fig. 4.

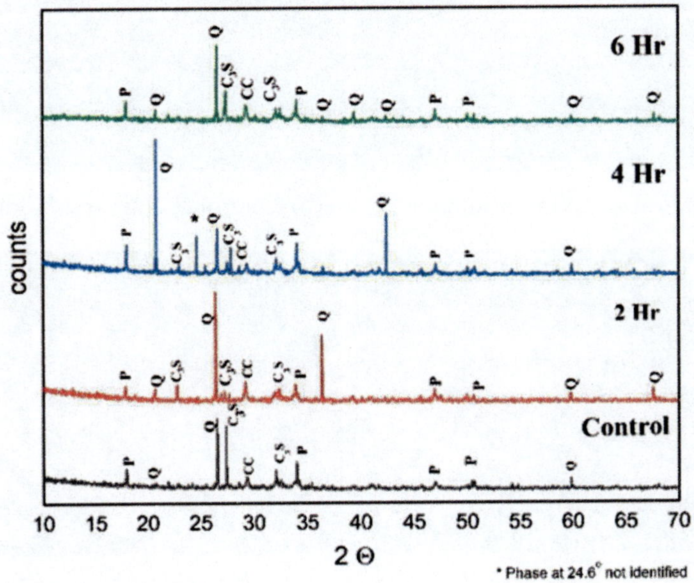

Fig. 4 XRD pattern for CO_2 cured cement mortars

The prominent peaks are identified to be of quartz (Q), Portlandite (P), calcium carbonate (CC), and calcium silicate (C_3S) in all the four cases. Quartz (Q), the mineral abundantly present in fine aggregate (river sand). River sand used as fine aggregate for making the cement mortar cubes. It is identified in the 2-h CO_2 cured cement mortar XRD pattern as the peak intensity of Portlindite is lower and of calcium carbonate (CC) is higher compared to the rest of the patterns. This indicates presence of more calcium carbonate which is a primary product of carbonation and the possible reason for higher strength in 2-h CO_2 cured cement mortar specimens.

The specimens (relatively flat solid cement mortar pieces from the fractured surfaces of broken cubes in compressive strength test) are sputter coated with gold palladium and tested in Zeiss Merlin Compact FE-SEM for visualizing the morphology of hydrates and/or carbonates of cement mortars that are subjected to two hour CO_2 curing and no CO_2 curing (i.e., control) under secondary electron mode of imaging is shown in Fig. 5.

Observation of the micrographs revealed a clear dense microstructure for the 2-h CO_2 cured surface specimens when compared to that of control. It is evident that more amount of Ettringite (E) with needle like morphology and Portlandite are dominant in control specimen micrographs while calcium carbonate or calcite is dominant in

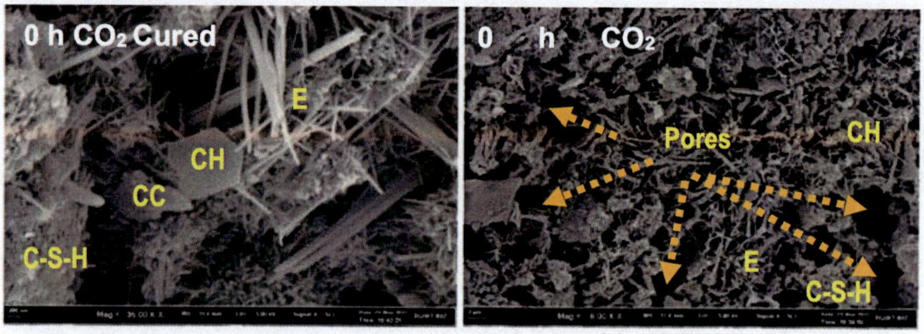

(a) Control at 0.2 μm resolution (b) Control at 2 μm resolution

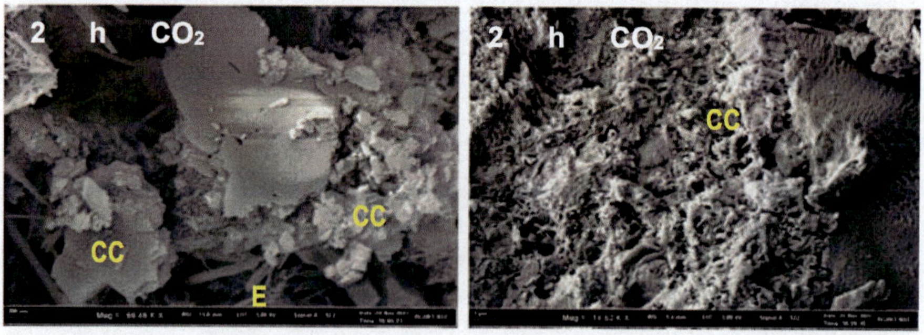

(c) 2h CO_2 cured sat 0.2 μm (d) 2h CO_2 cured at 2 μm resolution

Fig. 5 SEM micrographs of cement mortar specimen fractured surfaces

CO_2 cured specimen. These results are affirming the XRD results for CO_2 cured and control specimens.

4.4 *Study of the Effect of CO_2 Curing on SDBC Augmented Cement Mortar*

Considering the results obtained in the previous two studies, the 2-h CO_2 curing duration and 2% SDBC replacement for cement in 1:3 cement mortars are selected for understanding the influence of CO_2 curing on the SDBC augmented cement mortar. In this part of study, based on the observations made while demolding the specimens after 12 h, to avoid any minor damage to the specimens, the in-mold curing period is increased to 18 h from 12 h. Further, the pre-conditioning period is reduced to 4 h from 6 h. All the cubes are subjected to CO_2 curing for 2 h at 15% CO_2 concentration, 25 °C temperature and 50%RH. The testing for strength is carried out in 2 intervals. First, immediately after CO_2 curing (i.e. 1 Day) and second is after 2 more days of water curing. (3 Days). The results are presented in Table 6.

The one day strength attained an increase of 40% for CO_2 cured OPC mix (INCC) while the strength is decreased by 4.7% for SDBC augmented mortar. However, it is to be noted that the SDBC augmented mortars attained 12% strength improvement under water curing in one day. This finding reveals that the biochar improves early age strength. Overall, CO_2 curing is found to beneficiate more for OPC-based cement mortars than biochar augmented mortars. This trend is also observed to be consistent with the three day cured mortars strength with 18% improvement. Further, the relative increase in strength is found to be diminishing with post carbonation hydration for the mortars.

Table 6 Compressive strength of 2% SDBC cement mortar cubes under CO_2 curing

Designation	In-mold (h)	Pre-conditioning (h)	In CO_2 chamber (h)	In water (h)	Age at testing (Days)	Strength (MPa)
1NCW	18	0	0	6	1	8.42
1NCC	18	4	2	0	1	12.57
1BCW	18	0	0	6	1	9.44
1BCC	18	4	2	0	1	8.99
3NCW	18	0	0	54	3	19.55
3NCC	18	4	2	48	3	23.07
3BCW	18	0	0	54	3	16.57
3BCC	18	4	2	48	3	18.35

Legend 1-NC-W: 1 day strength—normal mix—water cured; 3-BC-C: 3 day strength—biochar mix—CO_2 cured

Post carbonation water curing is found to be beneficial for both water cured and CO_2 cured mortars. However, highest strength improvement is found in water cured specimens with 132% increase and CO_2 cured biochar augmented mortars with 104% increase. The efficacy of CO_2 curing is found to be prominent for OPC-based mortars than to SDBC augmented ones. The reason could be attributed to lack of calcium bearing species in the SDBC, which actually reacts and precipitates strength giving $CaCO_3$. The XRD for SDBC shown only coesite (SiO_2) and sylvine (KCl) peaks, but no calcium bearing phases. Concomitantly weathered source of feedstock for making SDBC could also be reason.

5 Conclusions from the Study

An experimental study is carried to evaluate the efficacy of CO_2 curing on enhancement of early age strength in biochar augmented concrete. SDBC is prepared in laboratory, and the optimum dosage of SDBC in cement mortars is first studied. Simultaneously the optimum CO_2 curing duration is also studied for OPC-based cement mortars. The third study involved use of the optimized SDBC dosage and CO_2 curing duration values to evaluate the influence of CO_2 curing on SDBC augmented cement mortars. The following conclusions are drawn in this study.

(1) Use of biochar and CO_2 curing has potential benefits in terms of early strength gain and sustainability.
(2) From the results of 28 days strength 2% of SDBC is found to attain nearly strength equal to control cement mortar sample.
(3) Subsequent to in-mold curing and pre-conditioning a 2 h duration of CO_2 curing at flue gas CO_2 concentration of 15%, relative humidity of 50% and at temperature 25 °C is found to be sufficient for high 1 day strength.
(4) XRD analysis and SEM morphology studies corroborated the reason for strength gain, revealed dense $CaCO_3$ formation in the microstructure.
(5) It is also observed that prolonged CO_2 curing duration is not likely to contribute to more strength gain beyond an optimum value.
(6) Although CO_2 curing is found to give higher early gain of strength in both OPC-based and SDBC augmented cement mortar, its efficacy is revealed to be more significant to OPC-based mortars than biochar augmented mortars.

References

1. IPCC (2022) Summary for policymakers. In: Global warming of 1.5 °C. Cambridge University Press, pp 1–24
2. Mehta PK, Monteiro PJM Concrete-microstructure, properties and materials, 3rd edn. Tata Mc Graw Hill

3. Gupta S, Chaudhary S (2022) State of the art review on supplementary cementitious materials in India—II: characteristics of SCMs, effect on concrete and environmental impact. J Clean Prod 357:131945. https://doi.org/10.1016/j.jclepro.2022.131945
4. Scrivener KL, John VM, Gartner EM (2018) Eco-efficient cements: potential economically viable solutions for a low-CO_2 cement-based materials industry. Cem Concr Res 114:2–26. https://doi.org/10.1016/j.cemconres.2018.03.015
5. Fennell PS, Davis SJ, Mohammed A (2021) Decarbonizing cement production. Joule 5:1305–1311. https://doi.org/10.1016/j.joule.2021.04.011
6. Black L (2016) Low clinker cement as a sustainable construction material, 2nd edn. Elsevier Ltd.
7. Sanjuán MÁ, Andrade C, Mora P, Zaragoza A (2020) Carbon dioxide uptake by cement-based materials: a Spanish case study. Appl Sci 10. https://doi.org/10.3390/app10010339
8. Li L, Wu M (2022) An overview of utilizing CO_2 for accelerated carbonation treatment in the concrete industry. J CO_2 Util 60:102000. https://doi.org/10.1016/j.jcou.2022.102000
9. Zhang Y, He M, Wang L et al (2022) Biochar as construction materials for achieving carbon neutrality. Biochar 4:1–25. https://doi.org/10.1007/s42773-022-00182-x
10. Maljaee H, Paiva H, Madadi R et al (2021) Effect of cement partial substitution by waste-based biochar in mortars properties. Constr Build Mater 301:124074. https://doi.org/10.1016/j.conbuildmat.2021.124074
11. Dixit A, Gupta S, Pang SD, Kua HW (2019) Waste valorisation using biochar for cement replacement and internal curing in ultra-high performance concrete. J Clean Prod 238:117876. https://doi.org/10.1016/j.jclepro.2019.117876
12. Yaashikaa PR, Kumar PS, Varjani S, Saravanan A (2020) A critical review on the biochar production techniques, characterization, stability and applications for circular bioeconomy. Biotechnol Rep 28:e00570. https://doi.org/10.1016/j.btre.2020.e00570
13. Ghani WAWAK, Mohd A, da Silva G et al (2013) Biochar production from waste rubber-wood-sawdust and its potential use in C sequestration: chemical and physical characterization. Ind Crops Prod 44:18–24. https://doi.org/10.1016/j.indcrop.2012.10.017
14. Gupta S, Kua HW (2019) Carbonaceous micro-filler for cement: effect of particle size and dosage of biochar on fresh and hardened properties of cement mortar. Sci Total Environ 662:952–962. https://doi.org/10.1016/j.scitotenv.2019.01.269
15. Chen T, Gao X (2019) Effect of carbonation curing regime on strength and microstructure of Portland cement paste. J CO_2 Util 34:74–86. https://doi.org/10.1016/j.jcou.2019.05.034
16. Scrivener K, Snellings R, Lothenbach B (2018) A practical guide to microstructural analysis of cementitious materials
17. Gupta S, Kua HW, Pang SD (2018) Biochar-mortar composite: manufacturing, evaluation of physical properties and economic viability. Constr Build Mater 167:874–889. https://doi.org/10.1016/j.conbuildmat.2018.02.104
18. Zhang D, Ghouleh Z, Shao Y (2017) Review on carbonation curing of cement-based materials. J CO_2 Util 21:119–131. https://doi.org/10.1016/j.jcou.2017.07.003
19. He Z, Wang S, Mahoutian M, Shao Y (2020) Flue gas carbonation of cement-based building products. J CO_2 Util 37:309–319. https://doi.org/10.1016/j.jcou.2020.01.001
20. He P, Shi C, Tu Z et al (2016) Effect of further water curing on compressive strength and microstructure of CO_2 -cured concrete. Cem Concr Compos 72:80–88. https://doi.org/10.1016/j.cemconcomp.2016.05.026
21. Sharma D, Goyal S (2018) Accelerated carbonation curing of cement mortars containing cement kiln dust: an effective way of CO_2 sequestration and carbon footprint reduction. J Clean Prod 192:844–854. https://doi.org/10.1016/j.jclepro.2018.05.027
22. Rostami V, Shao Y, Boyd AJ, He Z (2012) Microstructure of cement paste subject to early carbonation curing. Cem Concr Res 42:186–193. https://doi.org/10.1016/j.cemconres.2011.09.010

Thermodynamic Modeling of Cementitious Paste Containing Sugarcane Bagasse Ash and Rice Husk Ash

N. S. Ajeesh Kumar and K. L. Radhika

1 Introduction

Supplementary cementitious materials (SCMs) are widely used to promote sustainable construction methods by reducing CO_2 emissions and energy usage in cement production [1]. It consists of industrial and agricultural by-products, viz. fly ash, blast furnace slag, rice husk ash (RHA), sugar cane bagasse ash (SBA), etc. These materials have been found to improve durability against sulfate attack and chloride ingress over extended hydration durations [2] while maintaining similar mechanical properties to ordinary Portland cement (OPC). However, the mineralogical and chemical content of the cement can significantly affect the evolution of solid hydration phases and aqueous pore solutions [3–5]. Hydration studies of binders including SCMs at different ages have demonstrated their impact on elemental constituents, particle shape, speed of reaction, solid phase formation, and mechanical properties [3–11].

Thermodynamic and geochemical models are being developed to better understand the effect of various factors on the hydration of supplemental cementitious materials (SCMs) with cement. Geochemical speciation software, like TDM, can mimic the interactions between the solid and liquid components during the hydration of cement mixtures [10], giving important insights into the chemical impacts. TDM has been utilized in conjunction with empirical research to examine the effects of SCM type and amount, water phase change, and hydrate composition after lengthy reaction times [7, 12]. Moreover, thermodynamic models can predict the impacts of OPC's mineral and chemical content on hydration interactions with SCMs over

N. S. Ajeesh Kumar (✉) · K. L. Radhika
Department of Civil Engineering, University College of Engineering, Osmania University, Hyderabad, Telangana, India
e-mail: ajeeshk.phd@uceou.edu

R. K. Pancharathi et al. (eds.), *Low Carbon Materials and Technologies for a Sustainable and Resilient Infrastructure*, Lecture Notes in Civil Engineering 440,
https://doi.org/10.1007/978-981-99-7464-1_7

medium and long timescales, and their correctness is assessed using empirical results from hydration tests up to 90 days [3, 5, 6, 13].

2 Modeling Method

The soluble alkali sulfates in OPC readily dissolve in water, generating K (potassium), Na (sodium) as well as S (sulfur) into the water. Minerals that are less dissolvable like calcite, anhydrite, and gypsum partially dissolve until they are in equilibrium with the pore fluids [14]. While Si (silicon) as well as Ca (calcium) combine to produce the CSH state, Al with Fe reacts with the hydroxide, S, C, and Ca to form ettringite (AFt) phases, AFm phases, as well as other hydroxide forms [14]. The rate of hydration or dissolution of the clinker elements controls the quantity of Ca, Si, Al (aluminum), Fe (iron), and hydroxide [14] discharged into the solution as well as the rate at which AFt, AFm, C–A–H (calcium aluminate hydrate), and other compounds precipitate [2, 10, 12].

Cement hydration is presumed to happen through dissolution and precipitation mechanisms. The rates at which the phases in Portland cement hydrate are described by many models. The Parrot and Killoh [15] model employs a series of equations to describe the rate R of the hydration of the various clinker phases, in which the smallest value of R at the time t is regarded as the rate-controlling phase [14]. The subsequent significant formulas:

Nucleation and growth

$$R_t = \frac{K1}{N1}(1-\propto_t)(-\ln(1-\propto_t))^{(1-N1)} \tag{1}$$

Diffusion

$$R_t = \frac{K2 \times (1-\propto_t)^{2/3}}{1-(1-\propto_t)^{1/3}} \quad \text{or} \tag{2}$$

Formation of the hydration shell

$$R_t = K3 \times (1-\propto_t)^{N3} \tag{3}$$

The expression for the degree of hydration at time t in days is then given by $\propto_t = \propto_{t-1} + \Delta t \times R_{t-1}$. The coefficients from the Parrot and Killoh [15] model for OPC are $K1, K2, K3, N1$, and $N3$. The above three empirical expressions (Eqs. 1–3) provided by Parrot and Killoh [15] are employed in this paper.

The effect of w/c as calculated by $f(w/c) = \left(1 + 4.45 \times \left(\frac{w}{c}\right) - 3.33 \times \propto_t\right)^4$: for $\propto_t > 1.334 \times (w/c)$ [14, 15] using the information from Parrot and Killoh [15], the impact of surface area on early hydration is also taken into account.

The GEMS software in conjunction with the CEMDATA database is used in this study to perform thermodynamic modeling of cementitious systems. The molar amounts of the crystalline, aqueous, and gaseous components of processes as well as the reactivity of ions inside the pore solutions at an equilibrium state can be calculated using the GEMS-CEM platform.

The outputs of cementitious and pozzolanic processes were predicted using GEMS version 3.5 in conjunction with the CEMDATA version 18. Input data for the thermodynamic modeling (TDM) included the constituents and reactivity of the cementitious materials and pozzolanic materials, the water-to-cement ratio, as well as curing conditions. The hydrated pastes were simulated at 21 °C and 100% relative humidity, with a water-to-cement ratio of 0.5, and were comparable to those reported in previous studies [3–5]. Parrot's empirical model was utilized to simulate the reaction of cement, while the reactivity of rice husk ash (RHA) and sugarcane bagasse ash (SBA) were determined based on basic pozzolanic tests [3, 6]. The CSH Q model was utilized in this study due to its ability to represent the full spectrum of Ca/Si proportions.

Thermodynamic modeling is a valuable tool for predicting the behavior of cementitious systems, but it has certain assumptions and limitations. The model presumes that the system is homogeneous, however this may not be the case in reality. Another critical assumption is that the system is in thermodynamic equilibrium. However, some solids, such as amorphous CSH and AFm phases, are metastable. Additionally, hydrated cement is thermodynamically unstable when exposed to an open atmosphere since it reacts with CO_2 in the atmosphere and eventually degrades to calcite and similar solids.

In cement chemistry, precipitation, and dissolution processes usually occur rapidly enough to assume thermodynamic equilibrium. However, slow processes like clinker phase dissolution and hydrogarnet precipitation may not reach equilibrium within the specified timeline, requiring further consideration for accurate prediction of cementitious system behavior and properties.

Accurate thermodynamic modeling results depend on the database's quality and completeness. Although the updated Cemdata18 database is reliable for calculating the type, composition, amount, and volume of cement hydrates, there are still gaps, such as data on alkali, aluminum, and water uptake in CSH, that need to be filled.

Additionally, some phases, like quartz, dolomite, goethite, hematite, gibbsite, and talc, do not form at ambient conditions and in the timeframe considered, so they need to be deactivated in the model. Moreover, some phases like thaumasite only form at low temperatures, and their presence can significantly affect the behavior of cementitious systems.

The predictions of the thermodynamic model were contrasted with a variety of empirical data, including results from X-ray diffraction, differential thermal and thermogravimetric analyses of cementitious mixes, and mortar mechanical strengths [3–6]. Two OPCs with different chemical and mineral contents were modeled using thermodynamics. The SCMs under consideration were RHA and SBA.

The compositions of the OPCs and SCMs employed in the thermodynamic hydration models have been reported previously [3–6] and are reproduced in Table 1. Bogue equations were used to estimate the mineralogical composition after correction for $CaCO_3$ content (4% for OPC-1 and 3% for OPC-2). TGA was used to calculate the $CaCO_3$ and gypsum contents. The OPC with low levels of C_3A and alkali (0.4% Na_2O_{eq} and 4% C_3A) was designated as "OPC-1," whereas the PC with larger levels (0.7% Na_2O_{eq} and 10% C_3A) was designated as "OPC-2." The C_4AF content is also notably varied between the two OPCs (15% in OPC-1 and 7% in OPC-2). $CaCO_3$ levels were less than 1% in SBA and RHA. To mitigate the effect of particle shape on the interactions between two kinds of cement (OPC) and SCMs, the grain size distribution of the two OPCs was chosen to be similar. This was described in greater detail in previous studies [3–5]. In addition, reference mixtures consisting solely of OPC-1 or OPC-2 were used, as well as binary blends made up of OPC-1 combined with 20% RHA/SBA.

Table 1 Major phases composition

Percentages	OPC-1	OPC-2	RHA	SBA
Na_2O	0.2	0.1	–	–
K_2O	0.3	0.9	2.2	6.3
Na_2O_{eq}	0.4	0.7	–	–
CaO	60.3	63.3	0.6	3.1
SiO_2	17.4	19.6	93.1	80.8
Al_2O_3	4.7	5.5	0.1	5.1
Fe_2O_3	5.1	2.4	–	1.6
MgO	1.8	0.8	0.2	–
SO_3	3.2	3.2	–	1.5
LOI	4.2	2.5	3.2	0.4
Alite	50	53	–	–
Belite	12	16	–	–
Aluminate	4	10	–	–
Ferrite	15	7	–	–
Calcite	4	3	< 1	< 1
LOI-loss on ignition				

Major phases composition as well as the raw material's $CaCO_3$ content (in weight percentages) [3, 4, 13]

3 Results and Discussion

3.1 Hydration Modeling of Ordinary Portland Cement

The chemical reactions between the two OPCs and water were simulated as a function of hydration duration to examine the impact of various compositions on the type and quantity of hydrates generated. The predicted hydration for OPC-1 and OPC-2 is shown in Fig. 1. In either situation, the primary hydration phases—CSH gel, ettringite, Portlandite, and mono-carbonate (regarding OPC-2)—are generated within the first few hours of hydration. During this time, the anhydrous portions of the OPCs are gradually dissolved. Additionally, it is predicted that siliceous hydrogarnet will precipitate. For both OPCs, the hydrotalcite phase is also anticipated, but in much smaller amounts, i.e., assumed that up to 2 g develop for every 100 g of the binder. Calcite was empirically recognized by X-ray diffraction at various periods of hydration, which is shown in Fig. 2, which displays the XRD of two ordinary Portland cement at two and ninety days of hydration as estimated in [13]. It is calculated that calcite is present in both types of cement.

By modeling results given in Fig. 1, CSH gel was detected by thermogravimetric analysis [13] and Portlandite, ettringite, and $C_3(AF)S_{0.84}H$ (siliceous hydrogarnet) are easily evident by X-ray diffraction at two and ninety days of hydration. The key qualitative difference between the two ordinary Portland cement is the creation of mono-carbonate in OPC-2 as a result of the greater Al and lesser Fe quantity in OPC-2 than in OPC-1, which permits the reaction with the carbonates. The development of mono-carbonate is predicted by OPC-2 modeling on the first day of hydration. However, empirical confirmation of mono-carbonate appeared at later hydration periods, most likely due to the sluggish kinetics of mono-carbonate creation or due to its poor crystallization [12].

At 90 days of hydration, traces of mono-sulfate may have been found in OPC-1 by XRD, even though the signal at 12 2θ may also have been caused by the presence of ettringite. A minor weight loss at 180–195 °C is indicated by DTA data [8], which may also be attributed to mono-sulfate dehydration. The calculations do not suggest the creation of mono-sulfate since it is thermodynamically unstable when calcite is present. As seen in Fig. 3, quantitatively, the OPC-2 computations forecast greater CSH crystal growth (19%) and less $C_3(AF)S_{0.84}H$ (siliceous hydrogarnet, 26%) compared to the OPC-1 simulation.

Depending on the initial chemical components of an OPC, there is a distinct variation in the hydrates. The other phase content is remarkably comparable between the two PCs as seen in Table 2. The modeling's projected quantity of Portlandite and CSH gel is contrasted to the measured Portlandite and bound water quantity obtained from thermogravimetric analysis (data from [3, 5]), as well as with compressive strength (information from [3–6]). The predicted CSH gel and Portlandite compositions were estimated for 100 g of hydrated mixtures to be consistent with the TGA values.

The modeling calculations showed that the concentration of Portlandite was slightly overestimated for both types of ordinary Portland cement (OPCs). However,

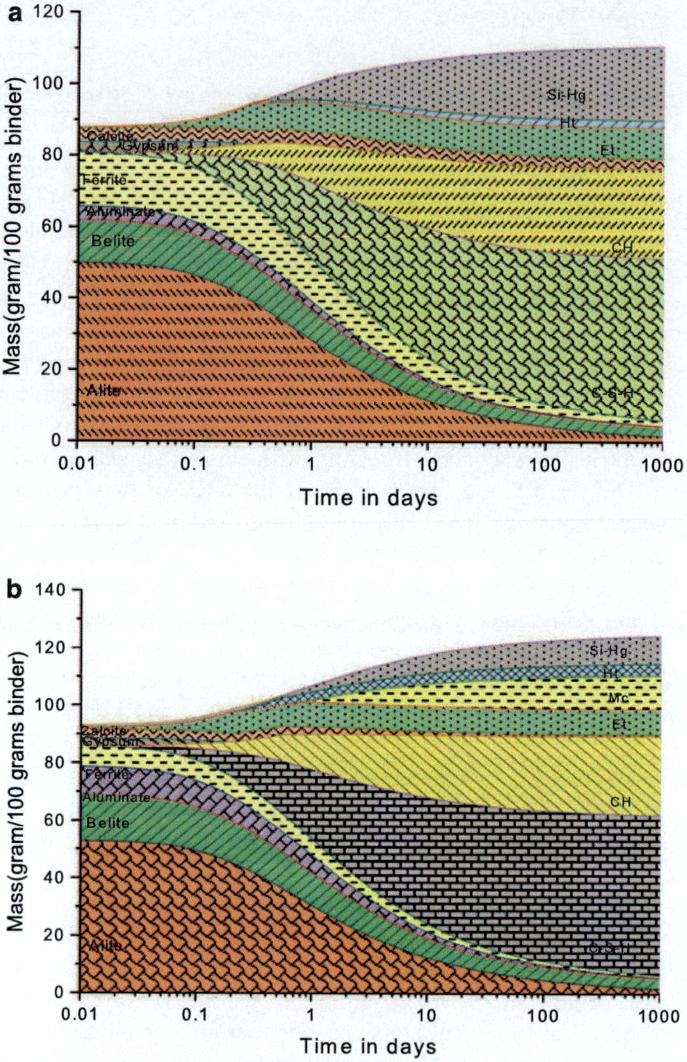

Fig. 1 **a** Hydration model of OPC-1 **b** hydration model of OPC-2

the simulations indicated that OPC-1 had a larger Portlandite content during all stages of hydration, compared to the experimental data. Even though the simulations predicted a smaller CSH gel or total volume and the experimental evaluation showed less bound water, the compressive strengths of OPC-1 were greater than those of OPC-2 at both the 28- and 90-day periods of hydration. This suggests that factors beyond space packing, such as the fine-tuning of the porous structure, may have an impact.

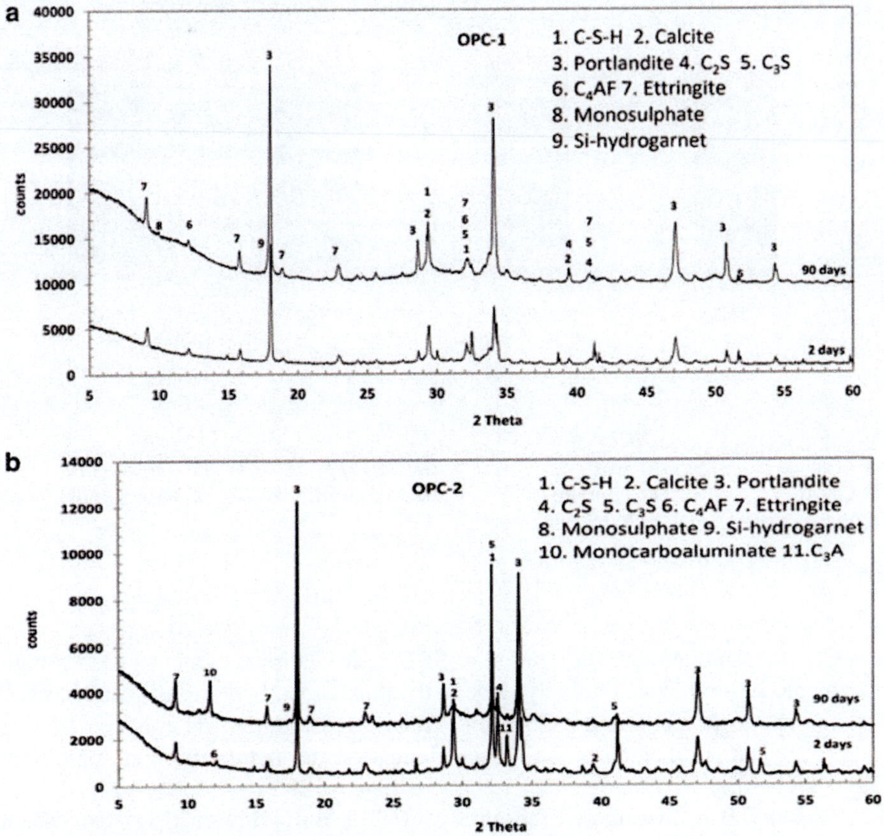

Fig. 2 **a** XRD pattern of OPC-1 [5] **b** XRD pattern of OPC-2 [5]

3.2 Hydration Modeling of Binary Blends with RHA/SBA

To assess the changes in solid phase composition based on the OPC contents, the hydration of blends containing RHA and SBA was modeled. The effects of RHA were investigated first (Fig. 4a). The thermodynamic modeling of OPC-1 and RHA blends predicted the development of similar principal hydration phases with minor numerical variations. High Ca/Si ratio phases were considered as "jennite-kind" CSH, while lower Ca/Si ratio phases were considered as "tobermorite-kind" CSH in TDM calculations [12]. The addition of RHA led to a decrease in Portlandite content and an increase in CSH gel formation. When more RHA was added, the additional silicon dioxide in the RHA reacted with the "jennite-kind" CSH, leading to progressively more "tobermorite-kind" CSH.

As shown in Fig. 3, the RHA-1 blend simulations predicted greater CSH crystal growth (48%) and less CH (Portlandite, 70%) compared to OPC-1 simulations. Similarly, the RHA-1 blend simulations predicted more CSH crystal growth (27%) and

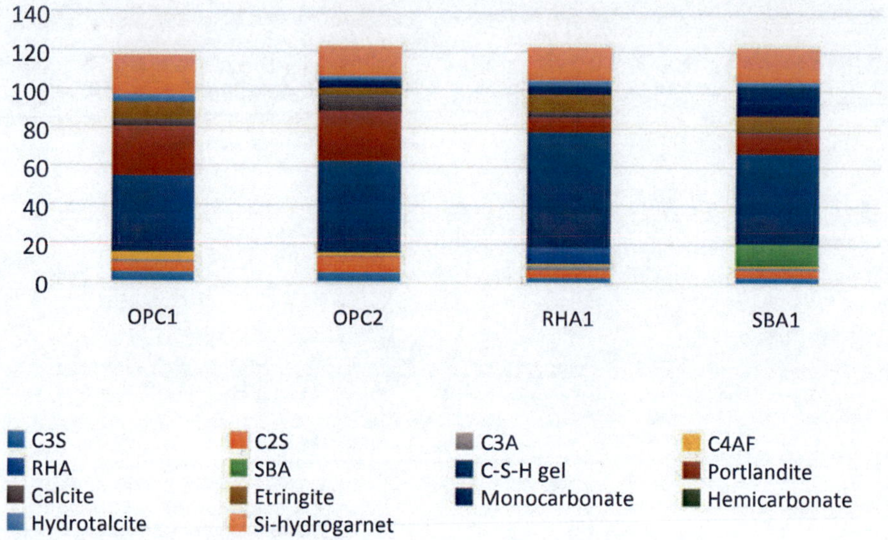

Fig. 3 Mass of the solid phases after ninety days

less CH (Portlandite, 28%) compared to SBA-1 simulations, owing to RHA's high pozzolanic activity. Quantitatively, especially at extended hydration ages, the RHA reaction results in a rise in the CSH gel and a decrease in the CH contents relative to OPC, which is reflected in the binary blend's greater compressive strength than OPC (Table 2).

The study also investigated the effects of SBA on hydration (Fig. 4b). When Portland cement and SBA are combined, the amount of Portlandite in the hydrated mixture is reduced, but the reduction is less than it is for RHA because SBA has relatively low reactivity. SBA contains 5% alumina, which results in the formation of modest levels of Al-rich phases when PC is added. The findings suggest that Portlandite is destabilized in the presence of modest levels of SBA, leading to the creation of more CSH with a lesser Ca/Si proportion [10]. Additionally, the mix forms mono-carbonates due to the incorporation of Al from SBA, which distinguishes it from OPC-1.

The RHA blend has a higher concentration of CSH gel, and a smaller bound water quantity has been determined (Table 2). This is because the additional CSH gel formed from the reaction with RHA has a lower bound water content than the CSH gel formed in OPC alone. The small size of RHA particles enables them to enter the pores of the CSH gel structure [3] resulting in a denser gel structure with lower porosity and higher strength. This denser structure can also reduce the amount of bound water in the gel.

Figure 5b shows that the total volume of hydrate in RHA-blended cement is lower compared to OPCs. Although the compressive strength is higher in RHA-blended mix, the increase in total hydrate volume is not proportional to the compressive

Table 2 CSH and Portlandite content forecasted with TDM, bound water and Portlandite content determined by TGA (%), compressive strength (MPa) [3–6]

Name	Duration in days	Model (percentage mass)		TG/DT (percentage mass)		Compressive strength (MPa)
		CSH	CH	Bound water	CH	
OPC-1	1	19.3	12.2	–	–	13.6
	2	23.2	14.4	13.8	14.1	–
	7	28.4	18.3	15.7	15.9	35.2
	28	32.7	21.1	18.8	17.4	51.4
	90	34.1	21.9	21.8	19.5	55.2
OPC-2	1	19.1	10.7	–	–	21
	2	25.7	13.9	13.8	12.8	–
	7	32.8	17.8	21.8	15.6	36.3
	28	37.3	20.1	23.4	17	43.1
	90	37.7	19.9	22.1	17.3	45.8
SBA-1	1	13.6	7.4	–	–	
	2	18.7	8.8	13.5	12.2	–
	7	30.2	9.2	15.3	14.3	31.28
	28	39.7	9.3	16.1	14.1	42.14
	90	40.5	9.7	17.2	14.1	55.1
RHA-1	1	14.5	6.2	–	–	
	2	20.8	7.5	11.7	10.9	–
	7	30.4	7.9	14.8	11	31.9
	28	45.3	7.1	15.1	10.8	54.7
	90	47.4	7.3	17.1	11.3	60.9

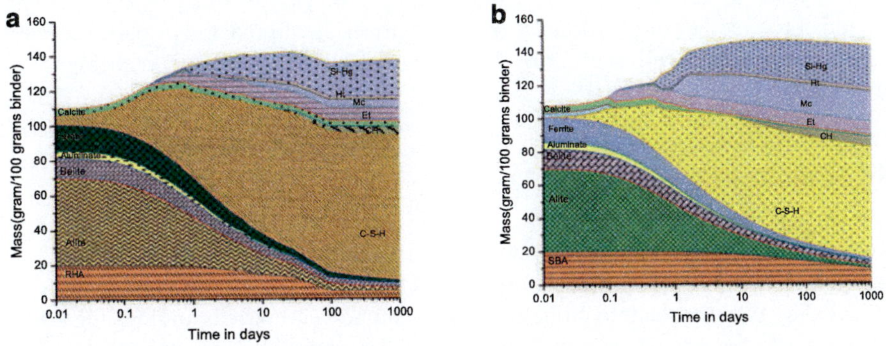

Fig. 4 **a** Hydration model of OPC-RHA **b** Hydration model of OPC-SBA. *Note* CH for $Ca(OH)_2$, Et for ettringite, Hc for hemi-carbonate, Ht for hydrotalcite, Mc for mono-carbonate, and Si Hg for siliceous hydrogarnet

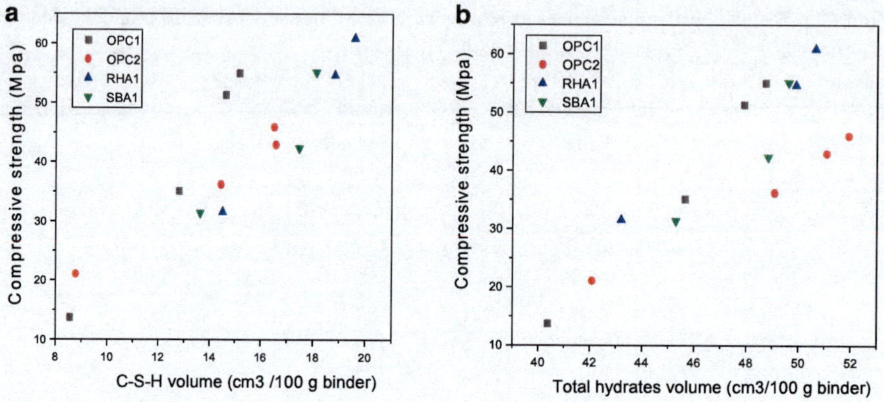

Fig. 5 **a** CSH gel volume versus comp. strength **b** total hydrate volume versus comp. strength

strength. This suggests that RHA-blended pastes contain more pores than pure Portland cement pastes, which may seem counterintuitive as PC-RHA blends exhibit higher strength than plain PC pastes. This could be attributed to the more complex pore structure of RHA-blended systems, which contains fewer large capillary pores and more fine pores than cement pastes. As a result, they exhibit higher compressive strength, despite having a higher total porosity.

The compressive strength recorded for the plain PCs and the mixes agrees with the estimated volume of CSH crystal in Fig. 5a, but a somewhat greater correlation is seen when the compressive strength is compared to the total volume of unreacted phases and crystals produced (Fig. 5b). This is because, the compressive strength of cement paste is influenced by various factors, including the amount and type of hydration products formed, the degree of hydration, and the paste microstructure. While the volume of CSH crystals is an essential factor in determining the strength of cement paste, the strength is also influenced by the total volume of all the phases and crystals produced during hydration. This comprises any unreacted or partly reacted cement particles as well as other hydration products including CH and ettringite.

4 Conclusions

The chemical and mineralogical nature of Portland cement can significantly affect the type and quantity of hydrates produced in OPC and blends with RHA and SBA, as revealed by thermodynamic modeling, which supports empirical findings regarding the produced hydrates. The modeling confirms primary differences between binary blends and plain cement observed experimentally, such as the larger formation of CSH gel and lower Portlandite composition in the mixes, and the generation of further mono-carbonate in sets depending on OPC composition.

There are a few slight differences between experimental and model results, which may be due to sample carbonation throughout hydration or aluminum absorption in CSH, which was not taken into account in the prediction. Additionally, the reduced CSH concentration in binary mixes anticipated by thermodynamic modeling correlates relatively poorly with the evaluated compressive strength, whereas the all-crystal volume, which includes both hydrates and unreacted matrixes, exhibits a higher association. This emphasizes that the overall volume of the solid matrix, rather than just the CSH content, is the major component impacting compressive strength. Additionally, it demonstrates that the binary mixes with RHA are more significantly impacted by the amount of PC substituted by SBA.

References

1. International Energy Agency. https://www.iea.org/reports/technology-roadmap-low-carbon-transition-in-the-cement-industry
2. Scrivener KL, John VM, Gartner EM (2018) Eco-efficient cements: potential economically viable solutions for a low-CO_2 cement-based materials industry. Cem Concr Res 114:2–26. https://doi.org/10.1016/j.cemconres.2018.03.015
3. Cordeiro GC, Kurtis KE (2017) Effect of mechanical processing on sugar cane bagasse ash pozzolanicity. Cem Concr Res 97:41–49. https://doi.org/10.1016/j.cemconres.2017.03.008
4. Fernandez A, Alonso MC, García-Calvo JL, Lothenbach B (2016) Influence of the synergy between mineral additions and Portland cement in the physical-mechanical properties of ternary binders. Materiales de Construccion 66. https://doi.org/10.3989/mc.2016.10815
5. Fernández, Calvo JLG, Alonso MC (2018) Ordinary Portland cement composition for the optimization of the synergies of supplementary cementitious materials of ternary binders in hydration processes. Cem Concr Compos 89:238–250. https://doi.org/10.1016/j.cemconcomp.2017.12.016
6. Cordeiro GC, Andreão PV, Tavares LM (2019) Pozzolanic properties of ultrafine sugar cane bagasse ash produced by controlled burning. Heliyon 5. https://doi.org/10.1016/j.heliyon.2019.e02566
7. Deschner F, Lothenbach B, Winnefeld F, Neubauer J (2013) Effect of temperature on the hydration of Portland cement blended with siliceous fly ash. Cem Concr Res 52:169–181. https://doi.org/10.1016/j.cemconres.2013.07.006
8. Haha MB, de Weerdt K, Lothenbach B (2010) Quantification of the degree of reaction of fly ash. Cem Concr Res 40:1620–1629. https://doi.org/10.1016/j.cemconres.2010.07.004
9. Lothenbach B (2010) Thermodynamic equilibrium calculations in cementitious systems. Mater Structures/Materiaux et Constructions 43:1413–1433. https://doi.org/10.1617/s11527-010-9592-x
10. Lothenbach B, Winnefeld F (2006) Thermodynamic modelling of the hydration of Portland cement. Cem Concr Res 36:209–226. https://doi.org/10.1016/j.cemconres.2005.03.001
11. Vollpracht A, Lothenbach B, Snellings R, Haufe J (2016) The pore solution of blended cements: a review. Mater Structures/Materiaux et Constructions 49:3341–3367. https://doi.org/10.1617/s11527-015-0724-1
12. Lothenbach B, le Saout G, Gallucci E, Scrivener K (2008) Influence of limestone on the hydration of Portland cements. Cem Concr Res 38:848–860. https://doi.org/10.1016/j.cemconres.2008.01.002
13. Alonso MC, Fernández Á, Calvo JLG (2018) Ordinary Portland cement composition for the optimization of the synergies of supplementary cementitious materials of ternary binders in hydration processes. Cem Concr Compos 89:238–250. https://doi.org/10.1016/j.cemconcomp.2017.12.016

14. Lothenbach B, Matschei T, Möschner G, Glasser FP (2008) Thermodynamic modelling of the effect of temperature on the hydration and porosity of Portland cement. Cem Concr Res 38:1–18. https://doi.org/10.1016/j.cemconres.2007.08.017
15. Parrot LJ, Killoh DC (1984) Prediction of cement hydration. Br Ceram Proc 35:41–53

Effect of RHA on High-Strength Geopolymer Mortar

C. S. Aishwarya, B. Dharshini, N. Shanmuga Priya, B. Swathi, and R. Vidjeapriya

1 Introduction

Geopolymer mortar or Alkali-activated binder (AAB) can be produced by using agricultural wastes, industrial by-products and alkaline activators. AAB mix has been considered as green material because of usage of waste materials in the preparation of that mix. As there is a need for reducing the usage of natural minerals, the demand for modern green construction techniques has been increased. Due to rapid industrialization and urbanization, the need for shelter has been increased which increases the demand for cement production. Manufacturing of cement has many environmental impacts like usage of natural minerals, emission of carbon dioxide, etc. Emission of CO_2 can occur in both direct and indirect ways. When calcium carbonate is heated, it will produce lime and CO_2 which is the direct way, and the indirect way is the emission of CO_2 during the manufacturing process. It has been estimated that about 10% of manmade CO_2 emission is due to the production of cement. Hence, there is a need for an alternative. One such alternative found is AAB material. The usage of AAB has given the solution for several problems like disposal of agricultural wastes, emission of CO_2 and overuse of natural resources [1]. The AAB has several properties like low shrinkage, excellent thermal stability, at relatively low production cost, mechanical strength, fire resistance, low thermal conductivity, acid resistance [2]. The main components of AAB are rich in aluminum (Al) and silicon (Si) [3]. In ABB mortar, GGBS and RHA are activated by using alkali activators to form a binder.

C. S. Aishwarya · B. Dharshini (✉) · N. Shanmuga Priya · B. Swathi · R. Vidjeapriya
College of Engineering, Guindy, Anna University, Chennai, India
e-mail: b.dharshini20@gmail.com

R. Vidjeapriya
e-mail: vidjeapriya@annauniv.edu

R. K. Pancharathi et al. (eds.), *Low Carbon Materials and Technologies for a Sustainable and Resilient Infrastructure*, Lecture Notes in Civil Engineering 440,
https://doi.org/10.1007/978-981-99-7464-1_8

Geopolymerization can be defined as a chemical reaction between aluminosilicate oxide of source material and alkali silicates of alkaline activator solution produces a 3D polymeric ring structure consisting of Si–O–Al bonds [4].

Study conducted by previous researches indicated that, compressive strength of concrete decreases with increase in the percentage of RHA. RHA and steel fibers reduces concrete workability, therefore superplasticizer is added to make concrete to be flowable [5]. The compressive strength of AAB increases when GGBS and calcium-based and sodium-based aluminosilicate complexes are high [6]. With the use of high concentration of sodium hydroxide and geopolymerization reaction becomes faster [7]. Fly ash replaced GGBS decreases the final setting time of AAB paste [8]. Comparing metakaolin and RHA, the RHA from the investigated offers a very promising ingredient for production of geopolymer mortar at ambient curing [9]. As molarity increases, microstructure becomes dense this results in high degree of geopolymerization [10]. In AAB concrete, most of the research work has focused on fly ash binders. However in this study, RHA and GGBS are source materials for AAB. RHA was replaced in different percentage in GGBS and its effects on the compressive strength is determined.

1.1 Need for Study

Every developing country is facing environmental pollution. Most extensively construction material is Ordinary Portland Cement, which causes major pollution during the manufacturing process. OPC emits Green House gases which accounts to approximately 7% of total Green House gas emission. This led to find an alternative for OPC is concrete construction. This caused evolution in the sustainability development of construction material from waste materials. Process by which the concrete is formed by the use of alkali activators and aluminosilicate materials like GGBS and RHA, where GGBS and RHA are a by-product of iron and steel manufacturing industries and agricultural waste, respectively [11]. This can be called as geopolymer, which can used as a replacement to OPC due to its large availability. Objective of this study is to obtain the optimum percentage of RHA required for achieving maximum strength in AAB and to develop high-strength GGBS–RHA-based AAB under ambient and oven curing.

2 Materials and Methods

2.1 Materials

Fine Aggregate. Manufacturing sand is used as fine aggregate. It is obtained by crushing of rocks into fine angular shaped particles. The properties of fine aggregates used for specimen are found as per IS 2386 (Part 3)-1963 [12] given in Table 1 and gradation curve is shown in Fig. 1.

Ground Granulated Blast Furnace Slag (GGBS). It is an aluminosilicate material from iron and steel manufacturing industries. GGBS is considered as environmental friendly material as it emits low levels of CO_2 during production. Therefore, GGBS can effectively replace cement in concrete production. It is rich in calcium oxide, silica and alumina content which enhances the strength of AAB by adding alkali activator. GGBS shows good resistance against chloride, sulfate or other chemical attacks. The specific gravity of GGBS that were used in this experiment was found to be 2.53 [13].

Rice Husk Ash (RHA). Agricultural by-products like sugarcane bagasse ash which is available abundantly in India is used as a precursor material [14–16]. Similarly, nowadays, rice husk ash (RHA) an agricultural waste material is used as a precursor

Table 1 Properties of fine aggregate

Properties	Value
Specific gravity	2.69
Fineness modulus	2.60
Loose bulk density	1398 kg/m^3
Compacted bulk density	1748 kg/m^3

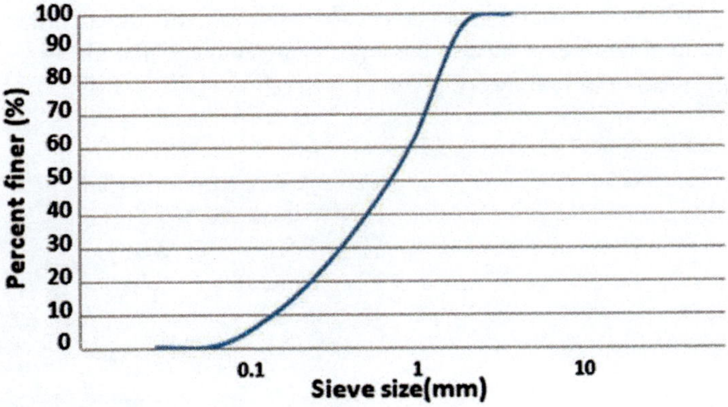

Fig. 1 Gradation curve

material. The characteristic of RHA depends on temperature at which it is burned and the duration of burning. The ash contains 90–95% of silica. The residual ash has amorphous silica which is highly porous and have pozzolanic properties. It has great surface area because of its micro-porous structure. This amorphous silica in RHA, when reacting with alkali activators produces binders for AAB. From material property testing, the average value of specific gravity of RHA used in this study was 2.25.

Alkali Activator. The alkali activators generally used in AAB are sodium hydroxide and sodium silicate solution. These alkali activators activate the aluminosilicate materials such RHA, GGBS and initiate geopolymerization process. 10 M alkali activator is used. Alkali activators with high molarity gives better result in strength, workability and less demolding time. Previous literatures indicated as RHA content increases, strength of the mortar reduces. In order to achieve high strength and dense geopolymerization gel, higher molarity was adopted. Normally, to achieve high-strength alkali-activated binder, a molarity ranging between 8 and 16 M is used. However, for the present study, only 10 M is used. Using 10 M, the cost becomes high but it can be compensated in the cost of binder as GGBS and RHA are used instead of cement and maintenance cost for alkali-activated binder.

2.2 *Methods*

Casting of Specimen. For preparing the AAB, first alkali activators should be prepared. 10 M of NaOH is prepared. After 24 h, the NaOH solution and Na_2SiO_3 are mixed in the ratio of 1:2 ($NaOH:Na_2SiO_3$). For control mix (G100) preparation, fine aggregates are mixed with GGBS in dry condition and then the alkaline solution is added to it as shown in Fig. 2. For mixes with RHA replacement of 5%, 10%, 15%, 20% and 25% in 100% of GGBS, fine aggregates are mixed with GGBS and RHA in dry condition followed by addition of alkaline solution. 12 cubes of sizes 70.6 mm × 70.6 mm × 70.6 mm are taken. After mixing, the prepared mortar is cast in the cubes as three layers by compacting manually as shown in Fig. 3. The mix designations used in this study are given in Table 2.

Fig. 2 Fresh geopolymer mortar

Fig. 3 Casting of geoploymer mortar in cubes

Table 2 Mix designation

Cube designation	Description
G100	100% GGBS (control mix)
G95R5	5% replaced RHA
G90R10	10% replaced RHA
G85R15	15% replaced RHA
G80R20	20% replaced RHA

Curing of Geopolymer Mortar. Curing of AAB is done in two ways, one by ambient curing and the other one by oven curing. Out of 12 cubes, 6 cubes are taken for ambient curing and remaining for oven curing for 7 and 28 days. After 24 h, specimens are demolded. For ambient curing, the specimens are kept in ambient condition as shown in Fig. 4. For oven curing, the specimens are kept in oven and heated at 60° C for 6 h as shown in Fig. 5. After 6 h, specimens were kept at room temperature until the day of testing.

Test on Fresh Mortar–Flow Table Test

Fig. 4 Specimens under ambient curing

Fig. 5 Specimens under oven curing

(a) A truncated cone was filled with mix and placed at the center of the flow table.
(b) After removal of truncated cone, the mix sample on table was dropped continuously for 25 times.
(c) Then samples diameter was measured using a scale as shown in Fig. 6. This test specifies the workability of the sample [17].

Testing of Hardened Mortar–Compression Strength Test

Compressive strength test was done on the cube specimens of 70.6 mm × 70.6 mm × 70.6 mm after 7 days and 28 days of curing in a compression testing machine (see Fig. 7). Load is noted at which crack occurs. The compressive strength is calculated by using the formula,

$$f_{ck} = P_c/A \left(\mathrm{N/mm^2}\right)$$

where

P_c load at the failure
A loaded area in mm^2

Fig. 6 Flow table test apparatus

Fig. 7 Compression testing machine

f_{ck} compressive strength in N/mm^2.

3 Results and Discussions

3.1 Flow Table Test

For AA/B = 0.45. Flow table test indicates the workability of the sample. Readings given in Table 3 and graph shown in Fig. 8 inferred that as addition of RHA increases, workability gets reduced when compared with control mix.

For AA/B = 0.5. Readings given in Table 4 and graph shown in Fig. 9 inferred that as addition of RHA increases, workability gets reduced. It was observed whatever may be the AA/B ratio, as RHA content increases, the flow gets reduced due to the fineness of RHA.

Table 3 Flow diameter for mortar with AA/B ratio as 0.45

Mix name	Flow value (in mm)	Percentage increase (%)
G100	227	127
G95R5	171	71
G90R10	142	42
G85R15	135	35
G80R20	120	20

Fig. 8 Flow diameter of GGBS-RHA-based AAB (AA/B = 0.45)

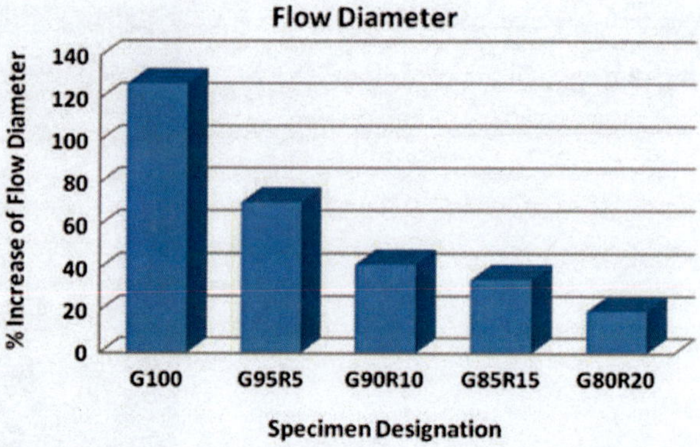

Table 4 Flow diameter for mortar with AA/B ratio as 0.50

Mix name	Flow value (in mm)	Percentage increase (%)
G95R5	232	132
G90R10	228	128
G85R15	216	116

Fig. 9 Flow diameter of GGBS-RHA-based AAB (AA/B = 0.5)

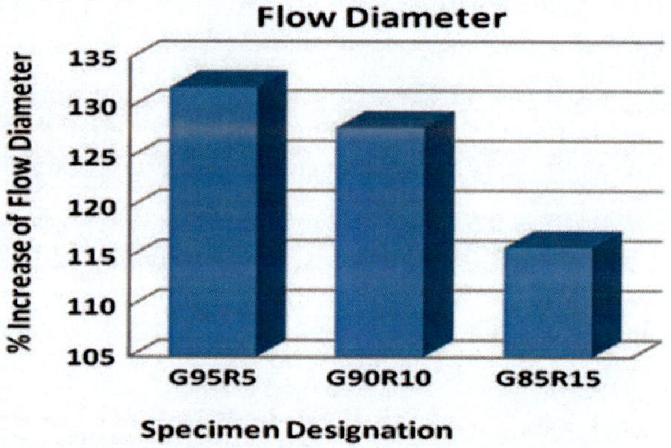

3.2 Compression Strength Test

For AA/B ratio = 0.45. From the readings given in Table 5 and graph shown in Figs. 10 and 11, it was observed from the results that replacement of 100% of GGBS by 10% RHA (G90R10) has produced the maximum strength of 86.5 N/mm^2 under 28th oven curing, which is higher than the other mixes. Initially 5% RHA, a reduction in the strength is observed. This was because the Si content in RHA was insufficient to form a dense C–A–S–H gel. Whereas for 10% of RHA, the maximum strength was observed as Si content was sufficient for the reaction between Si, Al and Ca ions to form a dense C–A–S–H gel. When the RHA content is increased to 15%, though the Si content was sufficient but a reduction in strength was observed because of relatively larger particle size and low specific gravity which cause the formation of a weaker zone in the geopolymeric paste.

Table 5 Compressive strength of AAB with AA/B = 0.45

Mix	Ambient curing		Oven curing	
	Compressive strength (N/mm^2)		Compressive strength (N/mm^2)	
	7th day	28th day	7th day	28th day
G100	68.2	71.6	72.6	72.9
G95R5	55.9	58.5	51.5	71.2
G90R10	60.7	69.9	74.2	86.5
G85R15	59.2	67.3	46.5	48.1
G80R20	44.9	59.9	36.8	39.7

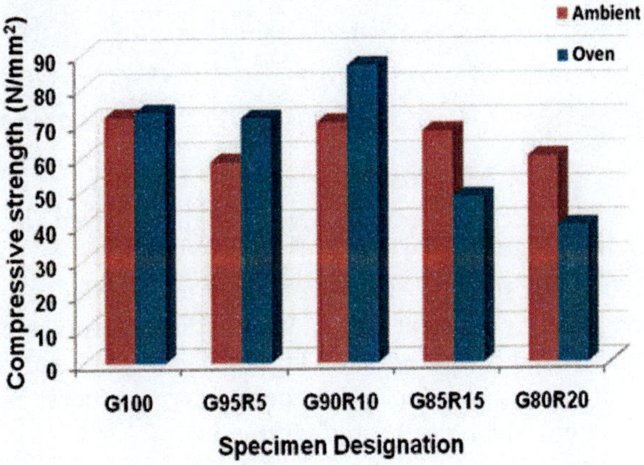

Fig. 10 28th day compressive strength test

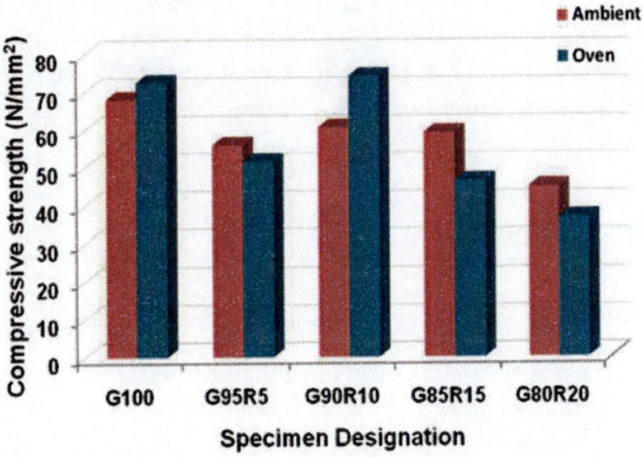

Fig. 11 7th day compressive strength test

Table 6 Compressive strength of AAB when AA/B = 0.5 Mix

Mix	Oven curing	
	Compressive strength (N/mm^2)	
	7th day	28th day
G95R5	59.3	59.7
G90R10	54.4	53.1
G85R15	47.2	57.7

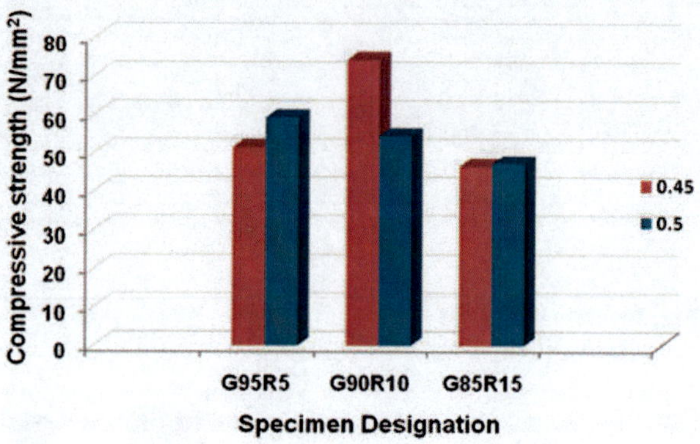

Fig. 12 Comparison of 7th day compressive strength test (for AA/B = 0.45 and 0.5)

The maximum compressive strength achieved for G90R10 mix under 28th day ambient curing is 69.9 N/mm^2, and for oven curing, the strength is 86.5 N/mm^2. It is found that strength of AAB obtained under oven curing is greater than the strength obtained by ambient curing. The 28th day average compressive strength of AAB under oven curing is 29.8% higher than the ambient curing for G90R10 mix. It can be inferred that inducing heat during curing process can enhance geopolymerization reaction, which results in high strength of AAB.

For AA/B ratio = 0.5. From the Table 6 and graph shown in Fig. 12, it can inferred that the maximum strength obtained at 28th day oven curing was 59.3 N/mm^2 when GGBS is replaced by 5% of RHA (G95R5). It is found that strength got decreased while increasing AA/B ratio from 0.45 to 0.5 (see Fig. 13) which is due to its effect in microstructure of alkali-activated binder in terms of geopolymerization.

4 Conclusions

Following conclusions are obtained from this study,

- That AAB under oven curing gives better compressive strength than AAB under ambient curing.

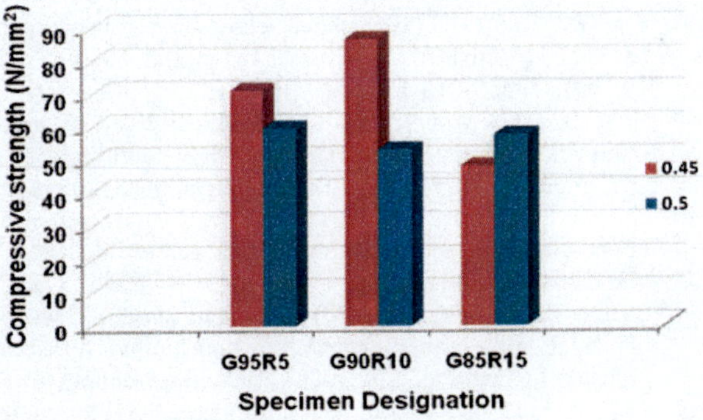

Fig. 13 Comparison of 28th day compressive strength test (for AA/B = 0.45 and 0.5)

- Inducing heat during curing process enhances the geopolymerization reaction, which results in high-strength AAB.
- As alkali activator to binder ratio increases from 0.45 to 0.5, compressive strength reduces due to the changes in the microstructure of the binder.
- The optimum percentage of RHA is 10% at 28th day oven curing with AA/B ratio of 0.45, for achieving higher strength in AAB.
- The alkali-activated RHA-based binder is a eco-friendly and a sustainable material that is a best alternative for Ordinary Portland Cement to achieve high strength.

Acknowledgements We would like to express our sincere gratitude to the Structural Engineering Laboratory, Division of Structural Engineering, Department of Civil Engineering, CEG, Anna University, Chennai-600025, for their support and thus making this study possible.

References

1. Nakshatra BS (2008) Fly ash-based alkali activated binder. A future construction material. Minerals 8. https://doi.org/10.3390/min8070299
2. He P, Jia D, Wang M, Zhou Y (2010) Improvement of high temperature mechanical properties of heat treated of geopolymer composites by sol-SiO_2 impregnation. J Eur Ceram Soc 30. https://doi.org/10.1016/j.jeurceramsoc.2010.07.031
3. Jindal BB (2018) Feasibility study of ambient cured geopolymer concrete—a review. Adv Concr Constr 6:387–405. https://doi.org/10.12989/acc.2018.6.4.387
4. Davidovits J (1999) Chemistry of geopolymeric systems terminology. Geopolymer'99 second international conference. Saint-Quentin, France, pp 9–39
5. Saloni P, Lim YY, Thong M (2021) Influence of Portland cement on performance of fine rice husk ash alkali activated binder concrete. Strength and permeability properties. Constr Build Mater 300. https://doi.org/10.1016/j.conbuildmat.2021.12432
6. Singham SS, Singh SP (2019) Fresh and hardened properties of flyash-slag based blended alkali activated binder paste and mortar. IJCSM 13. https://doi.org/10.1186/s40069-019-0360-1
7. Sharmin AU, Alengaram J, Jumaat MZ, Yusuf MO, Kabir SMA, Bashar II (2017) Influence of source materials and the role of oxide composition on the performance of ternary blended sustainable alkali activated binder mortar. Constr Build Mater 144:608–623. https://doi.org/10.1016/j.conbuildmat.2017.03.178

8. Rao GM, Rao TDG (2015) Final setting time and compressive strength of fly ash and GGBS-based alkali activated binder paste and mortar. Arab J Sci Eng 40:3067–3074. https://doi.org/10.1007/s13369-015-1757-z
9. EngyHany FN, Abdel-Wahab M, Sadek E (2020) Compressive strength of mortars incorporating alkali-activated materials as partial or full replacement of cement. Constr Build Mater 261. https://doi.org/10.1016/j.conbuildmat.2020.120518
10. Kaur K, Singh J, Kaur M (2018) Compressive strength of rice husk ash based geopolymer: the effect of alkaline activator. Constr Build Mater 169:188–192. https://doi.org/10.1016/j.conbuildmat.2018.02.200
11. Saeli M, David MT, Seabra MP, Labrincha JA (2019) Mix design and mechanical performance of geopolymeric binders and mortars using biomass fly ash and alkaline effluent from paper-pulp industry. J Clean Prod 208:1188–1197. https://doi.org/10.1016/j.jclepro.2018.10.213
12. IS 2386-3 (1963) Methods of test for aggregates for concrete, part 3: specific gravity, density, voids, absorption and bulking. https://archive.org/details/gov.in.is.2386.3.1963
13. IS 12089 (1987) Specification for granulated slag for the manufacture of Portland slag cement. https://archive.org/details/gov.in.is.12089.1987
14. Murugesan T, Vidjeapriya R, Bahurudeen A (2020) Development of sustainable alkali activated binder for construction using sugarcane bagasse ash and marble waste. Sugar Tech 22:885–895. https://doi.org/10.1007/s12355-020-00825-y
15. Akbar A, Farooq F, Shafique M, Aslam F, Alyousef R, Alabduljabbar H (2021) Sugarcane bagasse ash-based engineered geopolymer mortar incorporating propylene fibers. J Build Eng 33. https://doi.org/10.1016/j.jobe.2020.101492
16. Sousa LN, Figueiredo PF, França S, de Moura Solar Silva MV, Borges PH, Bezerra AC (2022) Effect of non-calcined sugarcane bagasse ash as an alternative precursor on the properties of alkali-activated pastes. Molecules 27. https://doi.org/10.3390/molecules27041185
17. Khalil MG, Elgabbas F, El-Feky MS, El-Shafie H (2020) Performance of alkali activated binder mortar cured under ambient temperature. Constr Build Mater 242. https://doi.org/10.1016/j.conbuildmat.2020.118090

Sustainable Concrete Making Materials and Special Concretes

Performance as the Criteria for the Durability in Concrete Mix Proportioning

B. Kameswara Rao and Rajasekhar Cheruvu

1 Introduction

Construction necessitates using a considerable amount of concrete and makes it the material with the second-highest consumption (only water consumes more). This chapter discusses the sustainability of concrete and assesses its carbon footprint from the life cycle and environmental impact perspectives, emphasizing cement, and cement substitute materials used in construction. The life cycle evaluation of concrete comprises data on the production of raw materials for concrete, concrete manufacture, the use of additional materials like fly ash and metakaolin, silica fume, ground-granulated blast-furnace slag, chemical admixtures, and concrete recycling. It describes reducing concrete's environmental effect by employing partial cement replacement, enhancing cement manufacturing, developing alternative binders, reusing waste by-products, and improving Durability. This chapter discusses the substitution of cement with fly ash, GGBS, and silica fume; tested the rapid chlorine penetration test to know the durability of concrete with different replacements binders and water-binder ratio. Corrosion of reinforcing steel caused by chloride intrusion is generally a prevailing environmental attack that causes concrete structures to deteriorate. Bridge deck overlays, parking garages, marine buildings, industrial structures, etc., are being damaged because of the poor concrete quality, which causes corrosion that incurs repair costs in the millions of dollars annually. Because of recurring expenditures on repairs, the durability issue has been focused on more in recent years.

Chlorides enter crack-free concrete via numerous methods, including capillary absorption, hydrostatic pressure, diffusion, and evaporative movement. Diffusion

B. Kameswara Rao (✉) · R. Cheruvu
Department of CE, Koneru Lakshmaiah Education Foundation, Vaddeswaram, Andhra Pradesh, India
e-mail: bkrao1@kluniversity.in

R. K. Pancharathi et al. (eds.), *Low Carbon Materials and Technologies for a Sustainable and Resilient Infrastructure*, Lecture Notes in Civil Engineering 440,
https://doi.org/10.1007/978-981-99-7464-1_9

is the most common of them. Diffusion happens when chloride absorption on the external of a concrete member is larger than the concentration on the internal. As a result, chloride ions pass through the concrete to the reinforcement steel level. Conditions of reinforcement corrosion are favorable in conjunction with alternate wetting and drying cycles in oxygen. The pore structure inside the concrete mainly determines the rate of chloride ion infiltration into the concrete. Factors influencing pore structure include mix design, degree of hydration, curing conditions, usage of extra cementitious ingredients, and building procedures. As a result, if chloride-induced corrosion is possible, the concrete must be tested for chloride permeability. We desire an effortless check to run for project conditions and quality-control purposes that can be completed quickly. The rapid chloride permeability test meets these objectives. Whiting invented the RCPT in 1981, and its results have been shown to match well with the traditional 90-day salt ponding test. AASHTO T 277 or ASTM C 1202 have standardized testing protocols.

2 Literature Survey

In the experimental study on the chloride penetration of concrete consisting of fly ash Class F, the authors stated that the chloride penetration is reduced with an enhancement in fly ash from the test [1]. The compressive strength, diffusion, resistivity, chloride permeability, and accelerated corrosion of the concrete mix of ordinary Portland cement and pozzolana Portland cement [2]. From the RCPT test, authors stated that 24% chloride penetration is less in PPC concrete than in OPC concrete. The compressive strength, diffusion, resistivity, chloride permeability, and accelerated corrosion of the concrete mix consisting of the fly ash stated that partial replacement of the fly ash enhances the durability properties of concrete [3]. The compressive strength and RCPT tests were accomplished on the high-volume fly ash (HVFA) concrete and compared with plain concrete. The RCPT test revealed that HVFA had greater resistance than plain concrete [4]. Conducting two tests on the concrete that had been partially replaced with GGBS: the conventional RCPT test and the chloride ion penetration test using silver nitrate. These tests were performed in both normal and marine environments. The authors also stated that concrete in the marine environment acts similarly to normal concrete when GGBS is replaced at 30–50% [5]. RCPT tests compared the concrete of nanomaterials to plain or normal cement concrete. They found that chloride ion penetration had been decreased by the substitution of cement with nanomaterials [6]. The compressive strength testing, water penetration depth, and chloride permeability on plain concrete fly ash-based concrete and silica fume-based concrete. The authors developed a high-quality concrete mix with crushed limestone aggregate with the above testing data [7]. The strength and durability tests for fly ash and silica fume specimens were compared with plain concrete. They also correlated concrete's compressive strength and durability properties with distinct replacements [8]. The durability assessment is chloride ion penetration in the concrete after substituting cement using Class F fly ash. The authors stated that the major point as the RCPT value has decreased by the higher amounts of class F fly ash compared by non-fly ash concrete [9]. RCPT tests on ordinary Portland cement-based carbonated concrete and pozzolana Portland cement-based carbonated concrete using three individual w/c ratios (0.40, 0.45, and 0.55). These samples were preserved in the accelerated carbonation chamber for 150 days at a constant 10% carbon dioxide (CO_2). It is seen that the low w/c of PPC-based concrete has great resistance in the carbonation and RCPT tests [10]. The mechanical and durability properties of LC^3 concrete were studied and compared with all OPC concrete values and fly ash-based modified cement. The resistivity of LC^3 concrete was greater than that of OPC and fly ash-based concrete [11]. The concrete's compressive strength and durability properties. The alternate cementitious materials used in this experiment are Alcofine and GGBS. Nine different mixes were cast in this experiment. They stated that using both cementitious materials enhanced the durability of the concrete [12]. The sulfate attack and RCPT, durability of concrete investigated the high volume of slag cement with a cement ratio: GGBFS is 50:50, having the w/b ratio from 0.55 to 0.27, and tested for different ages of curing. The results stated that the chloride penetration is very low for later concrete ages [13]. The strength and chloride penetration of the concrete with different percentages of cement with GGBFS. The specimens were cast for different curing ages. It was concluded that at 50% replacement of GGBS, it had got better strength and chloride penetration than the other mixes cast and tested [14]. The popular technique in the study of non-destructive testing techniques is RCPT, but it has its limitations in the field. So, they adopted the electrical resistivity technique. This compared the results between both techniques. From the results obtained, it was found that those with less chloride permeability have larger resistivity. Hence, electrical resistivity directly influences chloride penetration [15]. In the experimental study on concrete's strength and durability properties. They have done this test on high-strength concrete with a 50% concrete substitution. This 50% substitution outperformed the control mix regarding chloride penetration [16]. The chloride penetration per ASTM C 1202 with concrete was replaced with silica fume at different percentages. This test showed reduced chloride penetration when the cement was substituted with silica fume [17]. The investigation used concrete consisting of silica fume and rice husk ash as a substitution for slag cement. The durability of the concrete replaced with RHA follows the same trend as that of silica fume concrete. In rice-producing countries, RHA has an impact on substituting concrete as well as silica fume [18]. In the tests on high-performance concrete with the substitution of silica fume in cement. There were different percentages of substitution of cement, like 5, 7.5, 10, and 12.5%. The chloride penetration decreases as there is an enhancement in the substitution of cement with silica fume [19]. A few tests were performed on fly ash and silica fume-modified concrete for different curing conditions. The testing was done to evaluate RCPT, RCMT, and electrical resistance on the fly ash and silica fume-based concrete. The resistance to chloride ion penetration decreases as the percentage of both cementitious materials increases [20]. An article [21] titled "A review on performance-based specifications toward concrete durability" mentioned that an obvious shift is detected globally from "prescriptive" to "performance-oriented specifications, particularly in the large

Table 1 Physical properties of cement

Property	Value
Specific gravity	3.15
Bulk density (kg/m^3)	1440
Specific surface area (cm^2/gm)	2960
Standard consistency (%)	28
Initial setting time (min.)	135
Final setting time (min.)	240
3D strength (MPa)	37.8
7D strength (MPa)	48.0
28D strength (MPa)	61.0

infrastructure construction sector. The main objective of this research is to evaluate the "prescriptive approach" as well as the "performance-based approach". The boundaries of the "prescriptive approach" are emphasized and the successful application of the "performance-based approach". It was observed that "performance-based specifications" can assure the accomplishment of essential quality and longstanding durability.

3 Materials

In this experimental investigation, OPC 53 grade is used. Fly ash and GGBS are employed as a mineral. Fine and coarse aggregates are obtained from a local quarry. The fine aggregate is confined to zone II; the coarse aggregate used is 12 and 20 mm. Different tests, like specific gravity, sieve analysis, etc., were carried out on all the materials used in experimentation. The following tables discuss OPC's physical properties, chemical composition, cementitious materials, and superplasticizers. Aggregates were tested as per I.S.: 383-2016, fly ash values as per ASTM C618. Cement is tested per I.S.: 269-2015 and I.S.: 4031-1988.

Table 1 contains the physical properties, and Table 2 contains chemical composition of cement. Table 3 contains physical properties of the aggregates used in the investigation. Table 4 contains the physical properties of fly ash and GGBS. Tables 5 and 6 contain the chemical compositions of the fly ash and the GGBS, respectively.

4 Mix Proportions

This research considered three categories of binder contents (533.33, 400, and 320 kg/m^3), and water-binder ratios of 0.3, 0.4, and 0.5 were used to make concrete for testing. All fly ash and GGBS mixtures have a water content of 160 kg/m^3. The

Table 2 Chemical composition of cement

Chemicals	Value
Lime saturation factor [$CaO–(0.7SO_3)/(2.8SiO_2 + 1.2Al_2O_3 + 0.65Fe_2O_3)$]	0.85
Insoluble residue	3.26
Magnesium oxide (MgO)	0.99
Sulfur trioxide (SO_3)	2.26
Chloride	0.02
Loss of ignition (LOI)	3.38
Alumina ratio = (Al_2O_3/Fe_2O_3)	1.20

Table 3 Physical properties of aggregates

Material	Fine aggregate	Coarse aggregate
Specific gravity	2.65	2.64
Bulk density (kg/m^3)	1700	1550

Table 4 Physical properties of cementitious materials

Materials	Fly ash	GGBS
Specific gravity	2.25	2.91
Specific surface area (cm^2/gm)	3980	4000

Table 5 Chemical composition of fly ash

Oxides	Percentages
$SiO_2 + Al_2O_3 + Fe_2O_3$	89.98
Calcium oxide (CaO)	2.02
Magnesium oxide (MgO)	1.89
Sulfur trioxide (SO_3)	0.15
Potassium oxide (K_2O)	3.09
Titanium oxide (TiO_2)	1.11
Sodium trioxide (Na_2O_3)	0.28
Loss of ignition	0.295

Table 6 Chemical composition of GGBS

Oxides	Percentages
Silicon dioxide (SiO_2)	34.92
Aluminum oxide (Al_2O_3)	19.29
Ferric oxide (Fe_2O_3)	0.8
Calcium oxide (CaO)	34.22
Magnesium oxide (MgO)	8.97
Others	3.41

Table 7 Compressive strength

Fly ash	w/b	28D	56D	90D	GGBS	w/b	28D	56D	90D
0	0.3	55.6	59.1	63.1	0	0.3	55.6	59.1	63.1
20	0.3	54.7	60.4	66.7	20	0.3	56.9	66.7	71.1
30	0.3	52.9	63.6	69.3	30	0.3	54.7	67.6	72.9
40	0.3	51.1	60.4	65.8	40	0.3	53.3	64.4	69.8
50	0.3	45.8	54.2	60.9	50	0.3	48.0	57.8	62.7
60	0.3	40.0	51.6	58.7	60	0.3	38.2	53.3	58.7
70	0.3	36.9	49.8	56.0	70	0.3	32.9	49.8	55.6
0	0.4	46.2	50.7	55.1	0	0.4	46.2	50.7	55.1
20	0.4	44.0	54.2	58.2	20	0.4	47.1	54.7	60.4
30	0.4	43.1	55.1	59.6	30	0.4	44.9	57.3	63.1
40	0.4	40.9	52.4	56.9	40	0.4	43.6	53.3	60.0
50	0.4	38.2	48.4	53.3	50	0.4	38.7	50.7	56.9
60	0.4	33.8	44.9	50.2	60	0.4	34.7	44.9	51.6
70	0.4	30.7	41.3	48.4	70	0.4	32.0	41.3	48.4
0	0.5	38.7	41.8	44.4	0	0.5	38.7	41.8	44.4
20	0.5	36.0	42.7	46.7	20	0.5	40.0	44.9	55.6
30	0.5	33.3	44.9	47.6	30	0.5	36.0	46.7	58.2
40	0.5	30.7	41.8	45.8	40	0.5	34.2	44.0	54.2
50	0.5	28.9	36.0	40.4	50	0.5	32.0	40.4	50.7
60	0.5	25.3	32.0	37.3	60	0.5	28.9	38.2	47.1
70	0.5	23.1	29.33	34.7	70	0.5	28.0	35.6	43.6

mix was designed on a volumetric base, the binder content was fixed initially, later water-binder ratio set basing on it. The residual volume is balanced among fine and coarse aggregate in an inexact ratio of 34:66. Furthermore, the coarse aggregate was 12 mm aggregate and 20 mm aggregate in a 40:60 ratio. The binder content of the PCE-based superplasticizer (S.P.) addition was also tuned to produce an aim slump of 80–120 mm. The concretes be uniformly mixed in a pan mixer at 25 rpm and 27 °C in a pan mixer. The values of compressive strength are given in Table 7.

5 Experimental Program

D. Whiting of the Federal State Highway Authority created this method between 1981 and 1988. This was accepted in the early 1990s by the American Association of State and Highway Transportation Officials (AASHTO) and in the late 1990s by ASTM. Diffusion, migration, capillary suction, adsorption, and desorption are all handled by it. This test is simpler and faster to do. It is a widely used test procedure all over the world. This test is to establish the resistance of chloride ions in concrete. The test on concrete in the laboratory evaluates the resistance to rapid chloride ion penetration. This test is mainly for evaluating the materials and material proportions that were used in the actual design. The specimen's age is one of the critical factors governing the results. Initially, this method was used to evaluate alternative materials, but it was useful in quality control and acceptance testing over time. The apparatus is known as the rapid chloride permeability tester and tests as per the ASTM C 1202 pattern. Electrical indication of concrete's capacity to resist chloride ion penetration.

Though there are many advantages to using rapid chloride ion permeability test (RCPT), this method has some criticism and limitations. They are:

i. The specimen gets heated up due to high voltage, and the test may not represent the ambient condition.
ii. The other negatively charged ions also may get transported in the given electrical field.
iii. This may not be suitable for concretes with larger size aggregates.
iv. Chloride ions may get transported to weak interfaces and micro-cracks.

Despite the above limitations, the RCPT is the reasonably best method to evaluate the durability of concrete.

The instrument has four cells, connecting cables, temperature sensors, desiccators, vacuum pumps, and software. The rapid chloride permeability tester has four channels, and at a time, four concrete specimens can be tested by applying an electrical potential across the specimen. When the RCPT tester is started for testing, 60 V D.C. is involuntarily selected by the instrument. It shows actual voltage, current, and temperature throughout the test period. It also simultaneously keeps logging the data required as per standard. A user can change test parameters before the test is started. Finally, collected software data can be saved, representing total change or Coulombs passed across the specimens. There are some demerits on the merits of this test, given that the sample is heated because of the high voltage. The test may not accurately depict the environment. Other negatively charged ions may be carried in the given electrical field. This may not be suited for concrete using bigger aggregates. Ions of chloride may be transferred across weak surfaces and micro-cracks. A photograph showing experimental setup of RCPT and a schematic diagram of the RCPT cell is shown in Fig. 1.

6 Results and Discussion

The compressive strength results for the control concretes and fly ash concretes and the slag concretes at the ages of 28, 56, and 90 days are given in Table 7. A classification based on coulombs passed through specimens is reproduced in Table 8 (based on ASTM C 1202). Table 9 and Figs. 2, 3 and 4 given the charge values passed through fly ash-based concrete specimens with a water-binder ratio of 0.3. The different ages (28, 56, and 90 days) were adopted in experimentation which

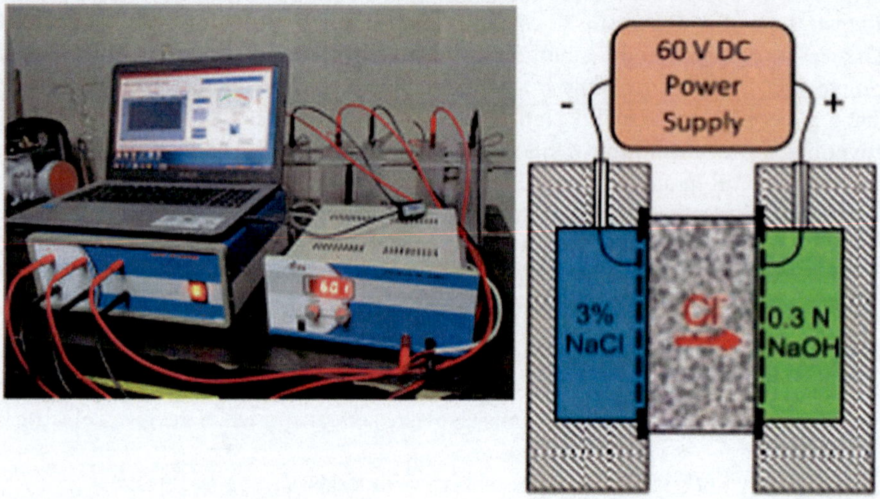

Fig. 1 Chloride ion penetration apparatus

is for of water curing. The values for 0%, 20%, 30%, 40%, 50%, 60%, and 70% have decreased by 47.68%, 62.57%, 65.81%, 69.62%, 73.88%, 72.03%, and 68.3%, respectively, from 28 to 90 days water curing. The least chloride ion permeability is obtained at 70% fly ash replacement with cement, and the charges passed through it are 635, 300, and 195 Coulombs when tested at the ages of concrete mentioned above.

Table 10 and Figs. 5, 6 and 7 given the charge values passed through fly ash-based concrete specimens with a water-binder ratio of 0.4. The different ages (28, 56, and 90 days) were adopted in experimentation which is for of water curing. The values for 0%, 20%, 30%, 40%, 50%, 60%, and 70% have decreased by 21.68%, 59.39%, 63.79%, 66.67%, 70.03%, 68.24%, and 70.34%, respectively, from 28 to 90 days water curing. The most negligible chloride ion permeability is obtained at 70% fly ash replacement with cement. The charge passed through it is 725, 375, and 215 coulombs when tested at the ages of concrete mentioned above.

Table 11 and Figs. 8, 9 and 10 given the charge values passed through specimens of fly ash-based concrete with a water-binder ratio of 0.5. The values for 0%, 20%, 30%, 40%, 50%, 60%, and 70% have decreased by 19.28%, 43.87%, 60.47%, 65.83%,

Table 8 Coulombs passed through specimens

Charge passed in Coulombs	Chloride ion penetrability
More than 4000	High
From 2000 to 4000	Moderate
From 1000 to 2000	Low
From 100 to 1000	Very low
Less than 100	Negligible

Table 9 Charge passed in Coulombs through 0.3 water/binder ratio specimens with fly ash

		RCPT (charge passed in Coulombs)		
w/b ratio	Fly ash-%	28 days	56 days	90 days
0.3	0	1617	1141	846
0.3	20	1162	655	435
0.3	30	1044	552	357
0.3	40	951	422	289
0.3	50	846	396	221
0.3	60	740	326	207
0.3	70	635	300	195

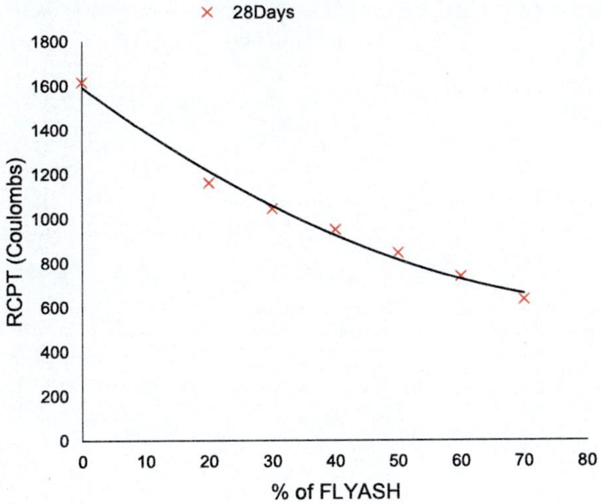

Fig. 2 w/b—0.3—fly ash—28 days

64.549%, 67.14%, and 64.24%, respectively, from 28 to 90 days water curing. The most negligible chloride ion permeability is obtained at 70% replacement of fly ash with cement, and the charges passed through it are 825, 395, and 295 coulombs when tested at the ages of concrete mentioned above, respectively.

Table 12 and Figs. 11, 12 and 13 given the charge values passed through GGBS-based concrete specimens with a water-binder ratio of 0.3. The different ages (28, 56, and 90 days) were adopted in experimentation which is for of water curing. The values for 0%, 20%, 30%, 40%, 50%, 60%, and 70% have decreased by 47.68%, 53.39%, 51.09%, 56.69%, 55.17%, 65.75%, and 61.93%, respectively, from 28 to 90 days water curing. The most negligible chloride ion permeability is obtained at 70% replacement of GGBS with cement. The charges passed through it are 465, 285, and 177 coulombs when tested after tested at the ages of concrete mentioned above, respectively.

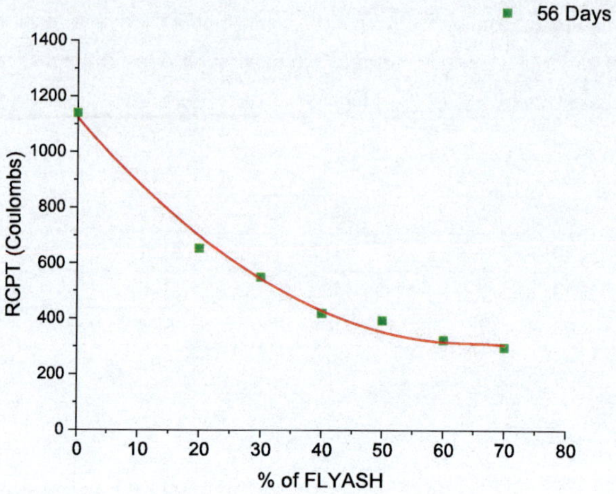

Fig. 3 w/b—0.3—fly ash—56 days

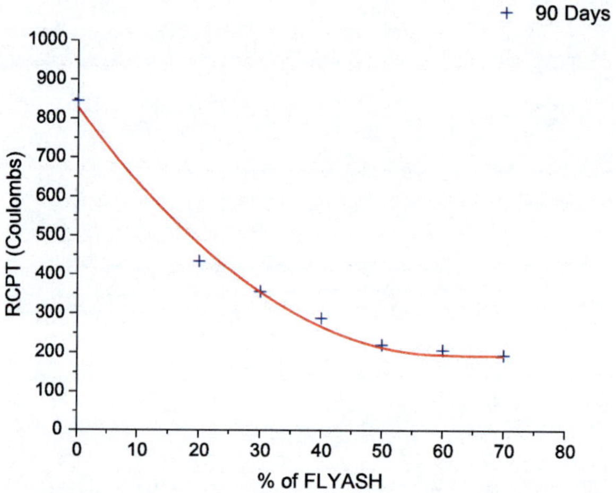

Fig. 4 w/b—0.3—fly ash—90 days

Table 13 and Figs. 14, 15 and 16 given the charge values passed through GGBS-based concrete specimens with a water-binder ratio of 0.4. The different ages (28, 56, and 90 days) were adopted in experimentation which is for of water curing. The values for 0%, 20%, 30%, 40%, 50%, 60%, and 70% have decreased by 21.68%, 62.14%, 64.37%, 69.31%, 72.88%, 77.11%, and 84.38%, respectively from 28 to 90 days water curing. The most negligible chloride ion permeability is obtained at 70% replacement of GGBS with cement. The charges passed through it are 1185, 350, and 185 coulombs when tested at the ages of concrete mentioned above, respectively.

Table 10 Charge passed in Coulombs through 0.4 water/binder ratio specimens with fly ash

		RCPT (charge passed in Coulombs)		
w/b ratio	Fly ash-%	28 days	56 days	90 days
0.4	0	1820	1526	1398
0.4	20	1214	839	493
0.4	30	1113	587	403
0.4	40	1038	528	346
0.4	50	941	456	282
0.4	60	866	400	275
0.4	70	725	375	215

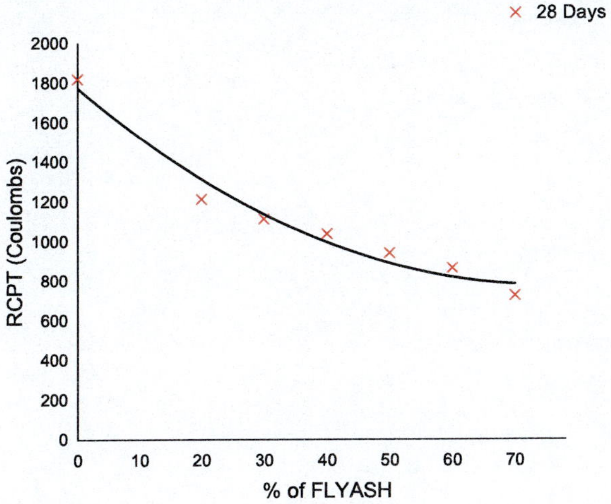

Fig. 5 w/b—0.4—fly ash—28 days

Table 14 and Figs. 17, 18 and 19 given the charge values passed through specimens of GGBS-based concrete with a water-binder ratio of 0.5. The different ages (28, 56, and 90 days) were adopted in experimentation which is for of water curing. The values for 0%, 20%, 30%, 40%, 50%, 60%, and 70% have decreased by 19.28%, 42.96%, 55.56%, 69.53%, 72.67%, 77.03%, and 78.26%, respectively, from 28 to 90 days water curing. The least chloride ion permeability is obtained at 70% replacement of GGBS with cement, and the charges passed through it are 1265, 485, and 275 coulombs when tested after 28, 56, and 90 days of curing, respectively.

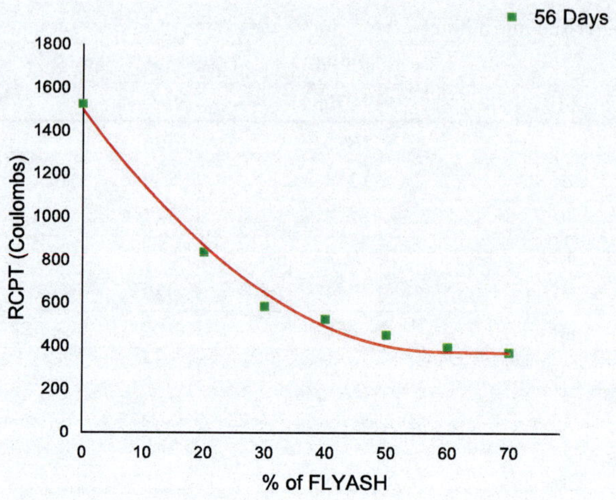

Fig. 6 w/b—0.4—fly ash—56 days

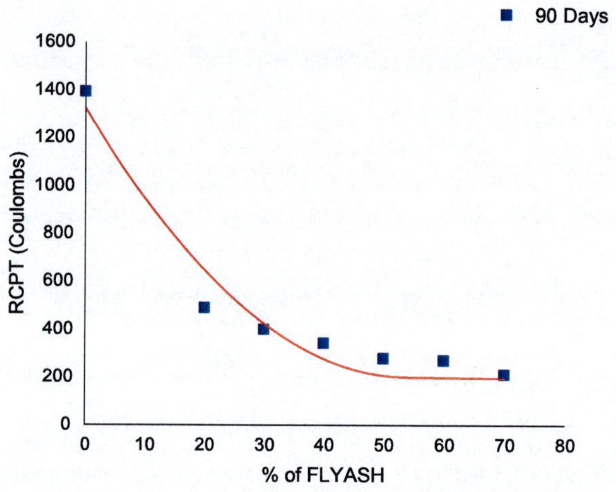

Fig. 7 w/b—0.4—fly ash—90 days

7 Conclusions

i. There is an increase in the permeability of chloride ions (number of coulombs) with the increase in the water-binder ratio.
ii. There is a decrease in permeability with the increase in the percentage of supplementary cementitious material (Fly ash and GGBS), which indicates the rise in terms of durability due to the reduction in the chloride ion permeability (number of coulombs).

Table 11 Charge passed in Coulombs through 0.5 water/binder ratio specimens with fly ash

		RCPT (charge passed in Coulombs)		
w/b ratio	Fly ash-%	28 days	56 days	90 days
0.5	0	2075	1741	1675
0.5	20	1395	946	783
0.5	30	1194	637	472
0.5	40	1150	552	393
0.5	50	1055	477	374
0.5	60	995	434	327
0.5	70	825	395	295

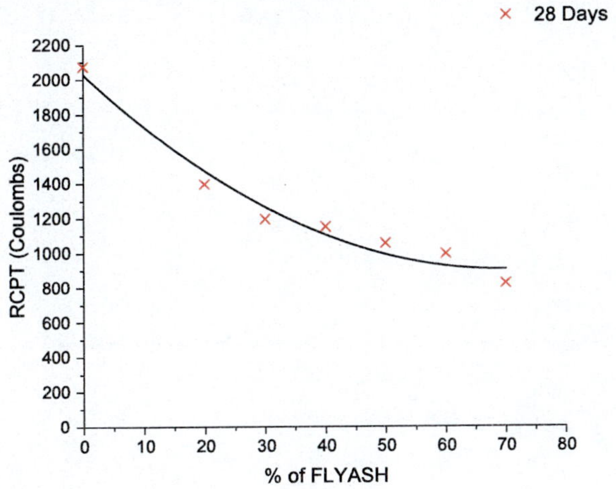

Fig. 8 w/b—0.5—fly ash—28 days

iii. The permeability is reduced in all the mixes with the increase in age. This reduction is due to the pore refinement in concrete due to the formation of secondary CSH gel due to the pozzolanic reaction.

iv. The above-published data will be helpful to the concrete mix designer to consider RCPT as the criterion for durability at the time of concrete mix proportioning.

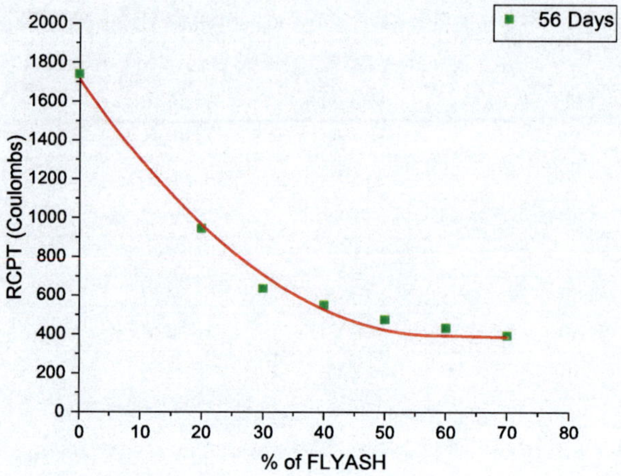

Fig. 9 w/b—0.5—fly ash—56 days

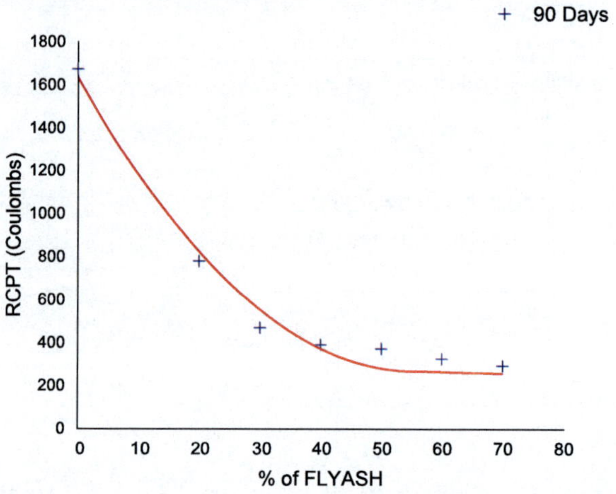

Fig. 10 w/b—0.5—fly ash—90 days

Table 12 Charge passed in Coulombs through 0.3 water/binder ratio specimens with GGBS

		RCPT (charge passed in Coulombs)		
w/b ratio	GGBS-%	28 days	56 days	90 days
0.3	0	1617	1141	846
0.3	20	1240	944	578
0.3	30	1055	876	516
0.3	40	900	734	427
0.3	50	774	559	347
0.3	60	587	385	201
0.3	70	465	285	177

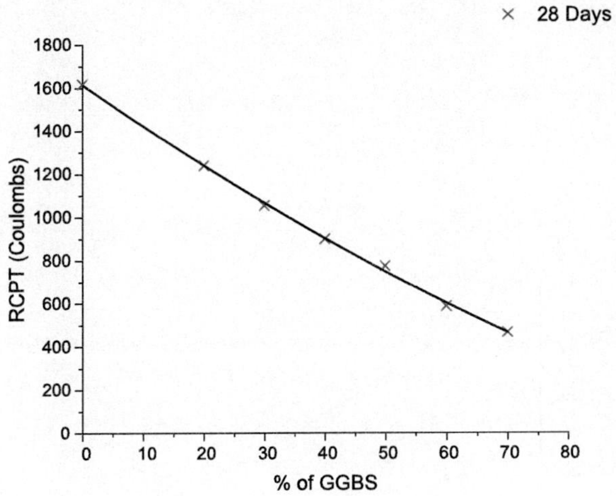

Fig. 11 w/b—0.3—GGBS—28 days

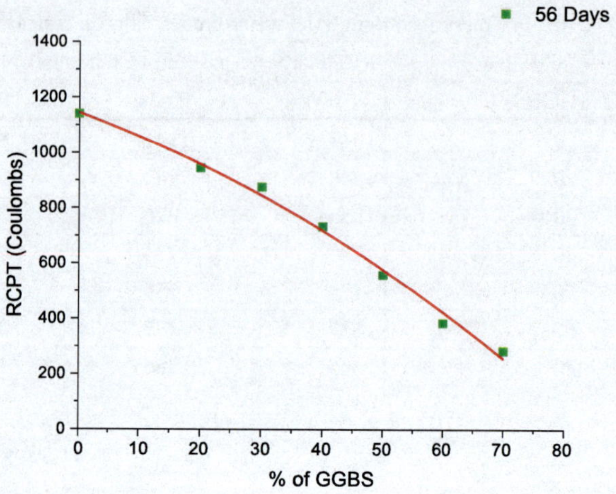

Fig. 12 w/b—0.3—GGBS—56 days

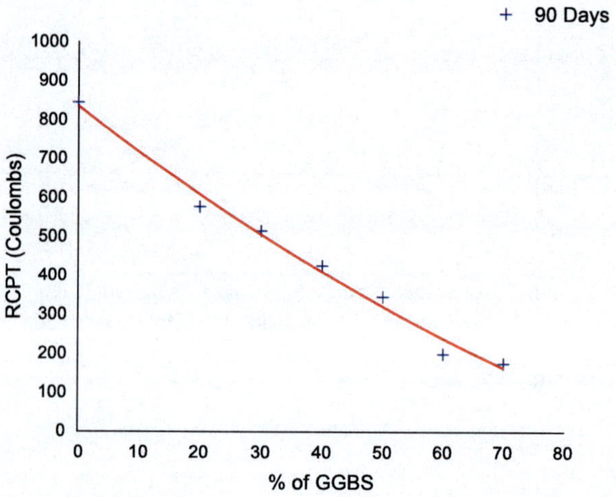

Fig. 13 w/b—0.3—GGBS—90 days

Table 13 Charge passed in coulombs through 0.4 w/b ratio specimens with GGBS

		RCPT (charge passed in Coulombs)		
w/b ratio	GGBS-%	28 days	56 days	90 days
0.4	0	1820	1526	1398
0.4	20	1785	969	689
0.4	30	1614	899	575
0.4	40	1486	774	456
0.4	50	1357	666	368
0.4	60	1228	413	281
0.4	70	1185	350	185

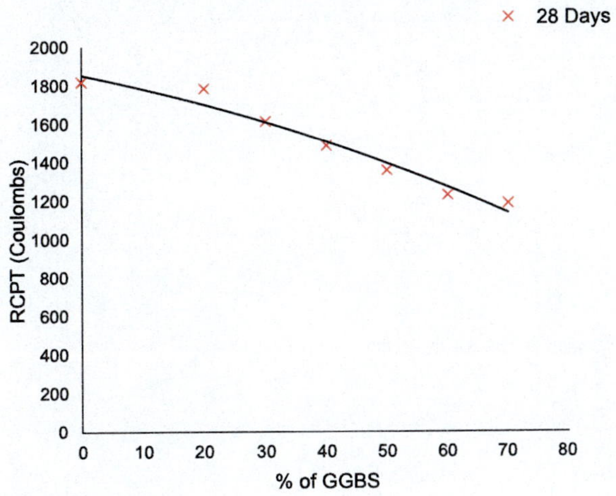

Fig. 14 w/b—0.4—GGBS—28 days

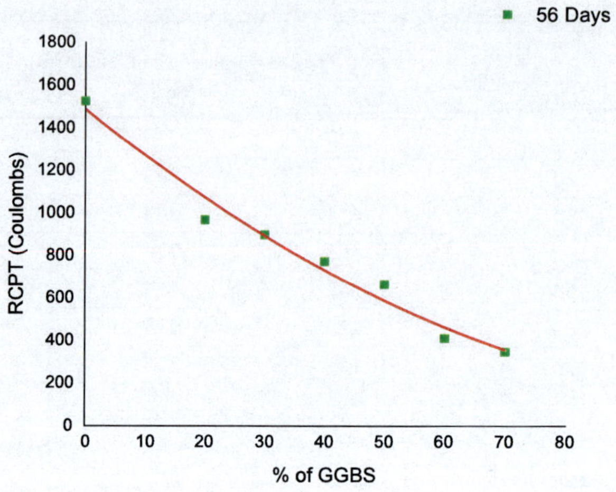

Fig. 15 w/b—0.4—GGBS—56 days

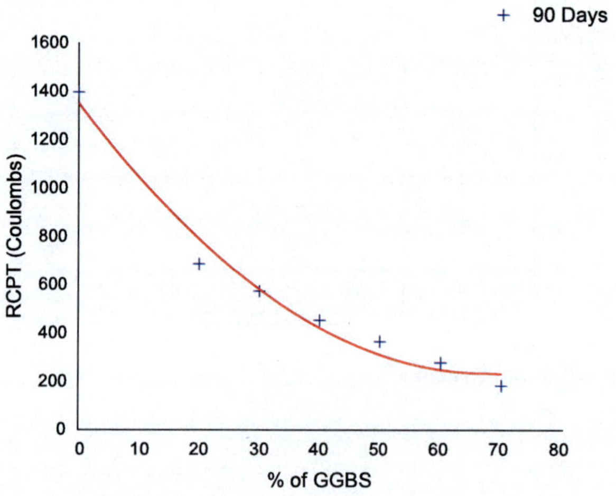

Fig. 16 w/b—0.4—GGBS—90 days

Table 14 Charge passed in Coulombs through 0.5 water/binder ratio specimens with GGBS

		RCPT (charge passed in Coulombs)		
w/b ratio	GGBS-%	28 days	56 days	90 days
0.5	0	2075	1741	1675
0.5	20	1855	1068	1058
0.5	30	1746	1018	776
0.5	40	1546	821	471
0.5	50	1445	734	395
0.5	60	1341	531	308
0.5	70	1265	485	275

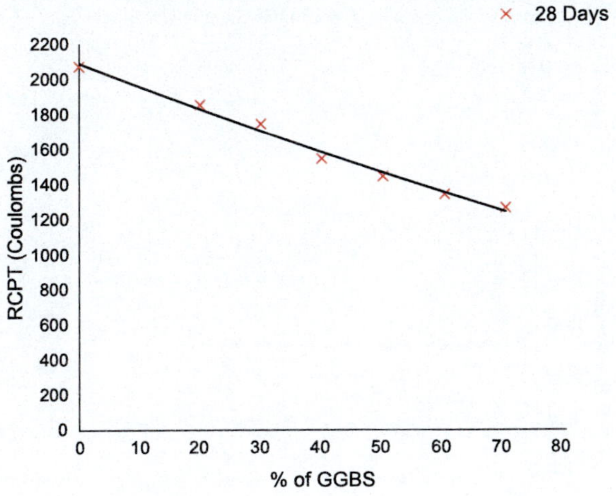

Fig. 17 w/b—0.5—GGBS—28 days

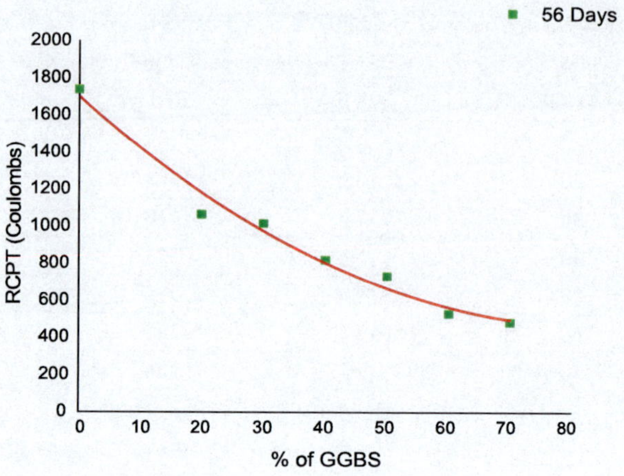

Fig. 18 w/b—0.5—GGBS—56 days

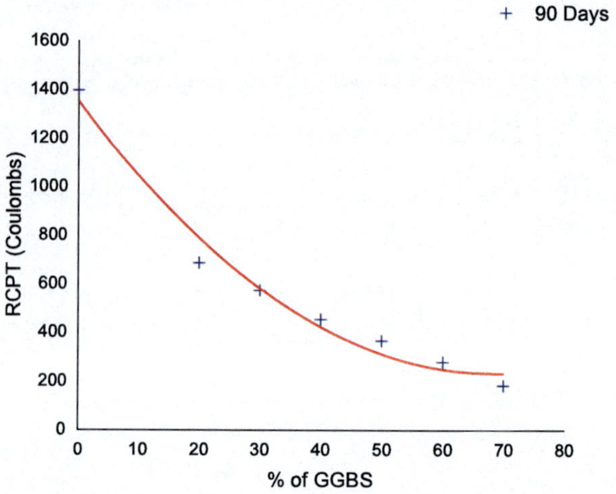

Fig. 19 w/b—0.5—GGBS—90 days

References

1. Madhavi T, Martin S, Luther SV, Kumar P, Raju S, Mathur D (2015) Chloride penetration in fly ash concrete. Int J Res Eng Technol 4(1):2321–7308
2. Prabakar J, Devadas Manoharan P, Neelamegam M (2011) Effect of fly ash on durability and performance of concrete. Indian Concr J 11:1–10
3. Nath P, Sarker P (2011) Effect of fly ash on the durability properties of high strength concrete. Procedia Eng 14:1149–1156
4. KrishnaTeja K, Kameswara Rao B (2018) Strength and durability of high-volume fly ash concrete. Int J Civ Eng Technol 9(6):109–116

5. Bhat AV, Tengli SK (2019) Behaviour of ground granulated blast furnace slag concrete in marine environment under chloride attack. Int J Innov Technol Exploring Eng 9(1):278–302
6. Carmichael J, Prince Arulraj M (2019) Rapid chloride permeability test on concrete with nanomaterials. Int J Eng Adv Technol 8(3):249–258
7. Ahmad S, Al-Kutti WA, Baghabra OS, Al-Amoudi MM (2008) Compliance criteria for quantity concrete. Constr Build Mater 22:1029–1036
8. Baghabra OS, Al-Amoudi WA, Al-Kutti SA, Maslehuddin M (2009) Correlation between compressive strength and certain durability indices of plain and blended cement concrete. Cem Concr Compos 31:672–676
9. Kanta Rao M, Naga Satish Kumar Ch (2019) Durability assessment of concrete with class-F fly ash by chloride ion permeability. Int J Recent Technol Eng 8(4):1–19
10. Das BB, Singh DN, Pandey SP (2012) Rapid chloride ion permeability of OPC and PPC based carbonated concrete. J Mater Civ Eng 21:228–239
11. Dhandapani Y, Sakthivel T, Santhanam M, Gettu R, Pillai RG (2018) Mechanical properties and durability performance of concretes with limestone calcined clay cement (LC3). Cem Concr Res 107:136–151
12. Balamuralikrishna R, Saravanam J (2019) Effect of alccofine and GGBS addition on the durability of concrete. Civ Eng J 5(6):1–14
13. Vijyagowri T, Sravana P, Srinivasa Rao P (2015) Rapid chloride permeability test on high volumes of slag concrete. J Struct Eng 4(1):1–17
14. Rama Mohan Rao P, Himath Kumar Y (2014) The effect of GGBS on the strength and chloride penetration of concrete. Int J Chem Sci 12(3):957–964
15. Lasker AA, Ghosh P (2019) Durability assessment of concrete by using electrical resistivity and rapid chloride permeability technique. Int J Eng Adv Technol 9(1):249–261
16. Raghavendra YB, Ramalinga Reddy Y, Hossiney N, Dinesh HT (2021) Properties of high strength concrete with reduced amount of portland cement—a case study. Cogent Eng 8(1):256–272
17. Lizarazo-Marriaga J, Lopez Yepez LG (2012) Effect of silica fumes addition on the chloride related transport properties of high-performance concrete. Medellin 21:105–110
18. Sahoo S, Parhi P, Chandra B (2012) Durability properties of concrete with silica fume and rice husk ash. Cleaner Eng Technol 11:1–14
19. Krishna Kumar S, Asif Abdul Vahab, Akhil Raj SR, Jayasree S, Job Thomas (2018) Effect of silica fumes on the resistance to chloride ion penetration in high-performance concrete. Am J Eng Res 2:01–05
20. Bagheri A, Zanganeh H, Alizadeh H, Shakerinia M, Marian MAS (2013) Comparing the performance of fine fly ash and silica fume in enhancing the properties of concretes containing fly ash. Constr Build Mater 47:1402–1408
21. Sangoju B, Gopal R, Bharatkumar Bhajantri H (2020) A review on performance-based specifications toward concrete durability. Struct Concr 22(12):1–22

Influence of Nanosilica and Microsilica on Mechanical and Microstructural Properties of Self-cured Fibre-Blended Concrete

J. Philips, V. Vandhana Devi, R. L. Lija, P. Leeba Grace, and S. Sathish

1 Introduction

Concrete is universal construction material [1] which used in a wide range of civil engineering projects, including bridges, buildings, concrete spires and pavement surfaces [2–5]. Cement is predominate binder in concrete. Due to urbanization, the requirement of concrete production increases which directly impacts the manufacturing of cement. Many studies have been conducted in an effort to lessen the impact of the cement industry on global warming, either by increasing the effectiveness of the process for making cement or by using supplemental cementitious materials, which replace regular cement to some extent [6, 7]. One of the most intriguing pozzolanic admixtures is the mineral silica inclusions in cement-based products due to their successful performance. A modest number of nanoparticles can be added to concrete to change the nanostructure of the cementitious material and increase durability. Due to its superior performance in concrete compared to conventional mineral addition, nanosilica has recently attracted particular interest. These silica materials also hasten the hydration process, because to their considerable specific surface area and nanoparticle size. Additionally, they trigger the pozzolanic reaction, which increases the morphological and mineralogical characteristics of the cementitious system by combining silica nanoparticles and calcium hydroxide to produce more C–S–H gel. The addition of silica nanoparticles to the cementitious system results in a microstructure with greater density and a better interfacial transition region, which enhances the mechanical performance of concrete.

Singh et al. [8] stated that while silica nanoparticles and silica fume help to increase strength in the early stages, at later ages, strength enhancement shows a

J. Philips (✉) · V. Vandhana Devi · R. L. Lija · P. Leeba Grace · S. Sathish
KCG College of Technology, Chennai, India
e-mail: Philips.civil@kcgcollege.com

R. K. Pancharathi et al. (eds.), *Low Carbon Materials and Technologies for a Sustainable and Resilient Infrastructure*, Lecture Notes in Civil Engineering 440,
https://doi.org/10.1007/978-981-99-7464-1_10

more or less comparable tendency. At the end of 28 days curing, the strength at compression of concrete made with fly ash and silica nanoparticles was 33% greater than that of control specimens. Silica nanoparticles produce higher early age strength as a result. This happens as a consequence of nanosilica changing the C–S–H gel structure from low-stiffed to high stiffed, which accelerates hydration rate at an early stage. Higher early age strength is due to the creation of highly rigid C–S–H, but because silica nanoparticles limit the microstructure's porosity, later age hydration may be hampered.

Khaloo et al. [9] researched that larger nanosilica particles enhance the mechanical qualities of concrete more effectively than smaller ones. Behfarnia et al. [10] demonstrated that the 28-day and 90-day compressive strengths are increased by 12% and 11%, respectively, by the addition of 3% nanosilica. Although the 28-day target strength of the cube by adding 5% nanosilica is even lower than that of the nominal sample, the strength was lowered by the larger percentage of nanosilica.

Lo et al. [11] represented that quantitative analysis was done on the impacts of the addition levels of nano-SiO_2 and replacement levels of used catalyst in a geopolymer. The sample with 0.5% more nanosilica was denser, more homogeneous, and had better particle-to-particle cementing than the other specimens, all of which enhanced the sample's strength attributes. Shaikh et al. [12] stated that compressive strength of nanosilica containing mortars increases with nanosilica content up to 2% and declines with higher nanosilica concentrations, such as 4% and 6%. High-nanosilica content addition had no effect on the progress of strength at either age.

Heikal et al. [13] stated that the heated concrete, including silica fume, performs poorly, especially at 10% replacement at temperatures higher than 350 °C. Moreover, silica fume raises internal steam pressure and matrix density at high temperatures. As a result, silica fume concretes quickly lose their strength, and at high temperatures, the structure becomes more broken and fragmented.

Hendi et al. [14, 15] observed that a high-strength self-consolidating concrete that can withstand the medium of sulfuric acid is designed using a combination of micro and nanosilica (colloidal silica). Within seven days of curing, colloidal silica had little effect on cement hydration, but their combination increased calcium hydroxide consumption. According to the findings, additional pozzolan replacement could result in less mass loss, while nanosilica has only a little impact on remaining compressive strength. Li et al. [16] observed that the workability would diminish with the addition of microsilica and nanosilica, but the loss in workability might be made up for by increasing the super plasticizer dosage. Yet, nanosilica has an even larger requirement for super plasticizer than microsilica does.

Aisheh et al. [17] observed that the mechanical properties of UHP-GPC are complexly affected by the addition of microsilica. When the microsilica dose was raised from 5 to 15%, mechanical properties were impaired [18]. These characteristics might be retained or even improved when the microsilica volume is greater than 15%. Due to a stronger bonding property between the matrix and steel fibre, when the microsilica volume of UHP-GPC was higher than 15%, mechanical strength and cracking characteristics improved.

Lou et al. [19] reviewed that the microstructure and interfacial transition zone between the aggregate and cement matrix are improved when silica fume is employed as an additional cementitious ingredient. This improves the performance of composites. The strength due to compressive load was improved to 21% by the use of silica fume up to 14% in place of cement. Compressive strength is increased physically as a result of silica fume and better granular material packing filling cement particle gaps. They reported that the ideal silica fume content for split tensile strength was between 15 and 25%, similar to the cube's strength. The splitting strength due to tensile was reduced as silica fume content was increased further. Said et al. [20] suggested that the incorporation of nanosilica, the strength typically rose up to 6% at all curing ages. The cementitious matrix's conductivity is significantly decreased and its pore structure is improved by the introduction of tiny doses of nanosilica.

When it relates to traditional concrete constructions, the process of strength-gaining is crucial to achieving the appropriate qualities. Concrete needs to cure for 28 days with the right amount of water in order to reach the desired strength. Internal curing is a contemporary method of curing concrete that, owing to the moisture content in it, cured itself. Water conservation and maintaining relative humidity levels are major challenges in hot, dry areas. The fundamental purpose of curing concrete is to keep the moisture level constant so that the cement can continue to hydrate.

According to numerous studies, [21–23] the inclusion of self-curing water-soluble compounds like polyethylene glycol significantly enhances the physical characteristics of concrete, such as moisture retention, absorption and permeability. Due to its deep microstructure, high-strength concrete is unable to absorb external curing water into its interior, which is necessary for the hydration process to be fully completed. Amin et al. [21] concluded that the compressive strength of concrete mixtures can be increased by self-curing with polyethylene glycol and porous ceramic waste aggregate up to 3% PEG dosages and 20% substitution ratios of coarse aggregates for porous ceramic waste aggregate.

Younis et al. [24] found that 1.5% PEG is the recommended dose and increased compression strength by 48.87% as compared to AC samples. Mokhtari and Madhkhan [25] stated that due to PEG 600 has a negative effect on concrete strength, the compressive and tensile strengths of concrete decreased when the mass fraction of SSPCM was increased.

Concrete has a few drawbacks, including its brittleness, which renders it susceptible to crack development and spread, as well as its poor tensile, strain, and energy absorbing capacities [26–28]. The non-reinforced concrete matrix may elastically deform under tension until it failed as a result of the growth of microcracks and the emergence of local macrocracks [29, 30]. Due to numerous influencing parameters, including fibre type, strength, modulus, aspect ratio, fibre content, aggregate size, and matrix strength, fibre reinforcing of concrete enhances the post-cracking behaviour of brittle concrete following the elastic stage [31–33]. In an effort, synthetic fibres like polypropylene and polyethylene were mixed, which resulted in increased concrete hardness, flexural strength and performance against impact loads [34, 35].

Ioannou et al. [36] briefed that when fly ash is added to fibre-reinforced microsilica cement mixtures, the setting periods are prolonged, Ettringite, C–S–H gel and calcium hydroxide contents fall, and the consistency peaks are lowered, notably in mixes with fly ash concentrations above 30%. Tayeh et al. [37, 38] found that the split tensile strength rose when 35% microsilica was combined with 0.5–3% polypropylene fibre. Comparing with control sample, the compressive strength was increased by roughly 20% by adding 35% microsilica with polypropylene fibre. Orouji and Najaf [39] concluded that by using polypropylene fibres, concrete's compressive strength was somewhat boosted; as a result, 1.5% fibre addition resulted in a 4.6% enhancement in strength. Yet, the behaviour of fibre-containing concrete was totally ductile, preventing specimen failure due to brittleness.

Despite numerous studies on the mechanical and freshly hardened characteristics of cementitious systems incorporating silica nanoparticles, there lack many studies in the published literature on how Silica Nanoparticles added to self-curing fibre-blended concrete affect the cementitious system's strength and microstructural features. So, the effect of nanosilica and microsilica addition on the strength qualities for self-cured fibre-blended concrete specimens has been investigated in the current work.

2 Materials

2.1 Ordinary Portland Cement (OPC)

For this experimental research, self-cured fibre-blended concrete was made using cementitious material of Grade 53 Ordinary Portland cement. The physical characteristics of OPC in Grade 53 were assessed in accordance with IS: 4031-1988, [40, 41] and the results were summarized in accordance with IS: 12269-1987 [42]. Table 1 contains a list physical attributes of the Grade 53 OPC utilized in this study.

Table 1 Physical attributes of OPC53 Grade

Attributes	Results
Standard consistency	35%
Fineness	4%
Specific gravity	3.14
Initial setting time	40 min
Final setting time	275 min

Table 2 Physical attributes of aggregates

Attributes	FA	CA
Density (kg/m^3)	1.62	1.20
Specific gravity	2.70	2.85
Fineness modulus	2.74	7.12

2.2 *Fine Aggregate (FA)*

Manufacturing sand with granules smaller than 4.75 mm was utilized as filler material in the fine aggregate category. The fine aggregate underwent a test in accordance with IS 383-1970 [43]. The physical attributes of manufacturing sand are summarized in Table 2.

2.3 *Coarse Aggregate (CA)*

In this investigation, a locally accessible crushed aggregate was used as filler in the coarse aggregate category. The aggregates are angular shaped with particle sizes ranging from 20 to 10 mm. Its characteristics in accordance with IS 383-1970 influence the concrete's strength, and it should be free of impurities of mineral deposits and contaminants. The results of the examination of the coarse aggregates in accordance with IS: 2386-1963 [44, 45] are summarized in Table 2.

2.4 *Chemical Admixture*

For increased concrete workability, Ceraplast300, a high-grade naphthalene-based super plasticizer (SP), is strongly advised. Ceraplast300 improves the workability of concrete without creating air intrusion and also coats the particles because to its excellent dispersion properties. The mix will be more homogeneous attributed to the consistent distribution, which lowers the possibility of segregation during placement. Table 3 provides a list of the Ceraplast300's as per IS 9103-1999 [46].

Table 3 Physical attributes of admixture

Attributes	Values
Type	Liquid
Colour	Dark brown
Solubility	Soluble in water
Specific gravity	1.22

Table 4 Physical attributes of polypropylene fibre

Attributes	Values
Density (g/cm^3)	0.91
Tensile strength (MPa)	300–750
Elongation (%)	15
Modulus of elasticity (GPa)	3.5–5
Aspect ratio	50

2.5 *Polypropylene Fibre (PPF)*

Fibre is a filament that is used to make vegetable tissue, mineral material, or textile.

Both polymeric and mineral bases can be found in synthetic fibres. The majority of the times, synthetic fibres are more resilient than other natural fibres. For this investigation polypropylene fibre was used with a dosage of 0.2% of volume of concrete. Table 4 lists the physical parameters of polypropylene fibre employed in this research.

2.6 *Mineral Admixture*

Mineral admixtures are typically added to concrete in higher proportions to increase its workability, resistance to heat cracking, alkali-aggregate elongation and sulphate attack, as well as facilitate a decrease in the amount of cement.

2.6.1 Nanosilica (NS)

Nanosilica, commonly known as quartz dust, is a substance that has a high SiO_2 content of above 99%. Besides lowering the quantity of cement required, the application of nanosilica completes the gradation curve of the aggregate mix in the region of the smallest sizes. Its goal is to provide a "filler effect," or to fill in spaces and so increase the concrete's compactness. Table 5 lists the physical parameters of nanosilica employed in this research.

Table 5 Physical attributes of mineral admixtures

Attributes	NS	MS
Appearance	Powder	Powder
Colour	White	White
Particle size	15 nm	150 nm
Specific gravity	1.16	2.2

Table 6 Physical attributes of PEG-200

Attributes	Values
Type	Liquid
Colour	Colourless
Solubility	Soluble in water
Specific gravity	1.13

2.6.2 Microsilica (MS)

A powdery residue of the production of silicon and ferro-silica from quartz, limestone, and iron is known as microsilica or silicafume. Electrostatic filters are used to catch the dust produced by electric furnaces. It is composed of mostly amorphous SiO_2 (82–96%) and very small, spherical particles (with a greater particles of less than 1 μm size). Table 5 lists the physical parameters of microsilica employed in this research.

2.7 *Polyethylene Glycol*

PEG, or polyethylene glycol, is a substance that aids in both water retention and reducing water evaporation from concrete surfaces. In order to lower the chemical reactivity of the molecules and form bonds of hydrogen with the water molecules, the polymers are incorporated to the mixture. This lowers the pressure of the vapour and slows the process of surface evaporation. For this investigation, PEG-200 was used with a dose of 1.0% of weight of binding materials. Table 6 lists the physical parameters of PEG-200 employed in this research.

3 Methodology

3.1 *Mix Design*

A self-cured fibre-blended concrete was prepared with various concentrations of mineral admixtures like nanosilica (NS) and microsilica (MS) for replacement of OPC. The nominal mix with a grade of M20 was prepared with cement content of 350 kg/m^3 and water cement ratio of 0.53 as per IS456-2000 [47]. The dosage of self-curing agent PEG-200 was a constant of 1% of weight of cementitious materials. The dosage of polypropylene (PF) fibre was a constant of 0.2% of concrete volume for all mix proportions other than nominal concrete mix. The nanosilica dosage was varying with 0.5, 1.0, 1.5 and 2.0% of weight of cement. Similarly, the concentration of microsilica was varying with 2, 4, 6 and 8% of weight of cement. There were 11

Table 7 Mix design proportions of fibre-blended concrete

Mix ID	Mix	OPC	NS	MS	FA	CA	Water	PF	PEG	SP
M0	NOMINAL	350	–	–	689	1258	186	0.00	0.00	0.00
M1	PEG	350	–	–	689	1258	186	0.00	3.50	0.00
M2	PEG + PF	350	–	–	689	1258	186	2.00	3.50	3.50
M3	0.5 NS + PEG + PF	348.2	1.8	–	689	1258	186	2.00	3.50	3.50
M4	1.0 NS + PEG + PF	346.5	3.5	–	689	1258	186	2.00	3.50	3.50
M5	1.5 NS + PEG + PF	344.7	5.3	–	689	1258	186	2.00	3.50	3.50
M6	2.0 NS + PEG + PF	343.0	7.0	–	689	1258	186	2.00	3.50	3.50
M7	2.0 MS + PEG + PF	343.0	–	7.0	689	1258	186	2.00	3.50	3.50
M8	4.0 MS + PEG + PF	336.0	–	14.0	689	1258	186	2.00	3.50	3.50
M9	6.0 MS + PEG + PF	329.0	–	21.0	689	1258	186	2.00	3.50	3.50
M10	8.0 MS + PEG + PF	322.0	–	28.0	689	1258	186	2.00	3.50	3.50

Mix ingredients (kg/m^3)

different mix proportions created in total, one of which was a nominal mix that cured using standard water curing and had no siliceous ingredients replaced. The various mix ratios according to IS 10262-2009 [48] are listed in Table 7.

3.2 *Casting and Testing Procedures*

For testing the characteristic compressive strength, the concrete specimens were cast in a cube mould of 150 mm size. The concrete mix proportions were cast and extracted from the casting mould after a period of twenty-four hours. Then the specimens were kept in a room temperature environment for 28 days of self-internal curing due to PEG-200. Similarly, for testing the splitting strength of concrete due to tensile load, the cylinder specimens of size 150 mm cross-sectional diameter and 300 mm height were cast and self-cured. For evaluating the strength due to flexure of plain concrete, the beam of lateral dimensions 100 mm × 100 mm and longitudinal size of 500 mm was cast and self-cured.

A testing machine which is used to check the strength due to compression load with a 2000 kN capacity was utilized to measure the cube's strength and the cylinder's strength. The cube's strength for compression load was measured using the guidelines outlined in IS: 516-1959 [49]. The standards described in IS: 5816-1999 [50] were used to evaluate the cylinder's tensile strength. The 400 kN-capable Universal Testing Machine (UTM) was utilized to evaluate the plain concrete beam's strength due to flexural load.

4 Results and Discussion

4.1 Mechanical Characteristics

4.1.1 Compressive Strength

The target compressive strength of nominal mix was achieved for M20 grade concrete. The compressive strength of mix (M3) with 0.5% replacement of nanosilica was 22.89 MPa and 33.56 MPa with an internal curing of 7 days and 28 days, respectively. The strength increment is 23.3% and 17.5% at 7 days and 28 days internal curing for 0.5% replacement of nanosilica, respectively. The compressive strength of mix (M10) with 8.0% replacement of microsilica was 22.56 MPa and 32.56 MPa with an internal curing of 7 days and 28 days, respectively. The strength increment is 21.6% and 14% at 7 days and 28 days internal curing for 8.0% replacement of microsilica, respectively. The effect of polypropylene fibre with internal curing was visible with strength variation of 7.2% and 10.5% at 7 days and 28 days, respectively, when compared to nominal mix. Table 8 gives the results of the compressive strength of various mixes (M0 to M11). Figure 1 shows that when the addition nanosilica decreases the compressive strength. Similarly, the addition of microsilica up to 8% dose increases the compressive strength. But past researchers mentioned that the optimum dose of microsilica for increment of strength was 8–10%. Further addition of microsilica decreases the strength of concrete mixtures.

4.1.2 Split Tensile Strength

The results of 28 days split tensile strength of various mixes were given in Table 9. The strength of the nominal mix (M0) was 3.06 MPa at 28 days of water curing. But the tensile strength of nominal mix with 28 days internal curing without any admixtures (M1) was 3.20 MPa. The strength variation was 4.6% higher than mix with normal water curing. The optimum result for split tensile strength from the Table 9 was 3.88 MPa for mix (M4) 0.5% nanosilica replacement with cement. The strength variation of mix (M4) was 26.8% higher than concrete the nominal mix (M0). The mixes with replacement of nanosilica particles were shown good mechanical characteristics due to the nanoparticles higher specific surface area and size of the nanoparticles. The nanoparticles occupied the gap in between the cement particles made the concrete denser. Polypropylene fibre has good tensile strength. When the

Table 8 Compressive strength of cube specimens

Mix IDs	7 days (MPa)	28 days (MPa)
M0	18.56	28.56
M1	19.56	30.37
M2	19.89	31.56
M3	22.89	33.56
M4	22.33	32.87
M5	22.00	32.67
M6	21.67	32.33
M7	19.56	30.89
M8	20.56	31.56
M9	21.89	31.89
M10	22.56	32.56

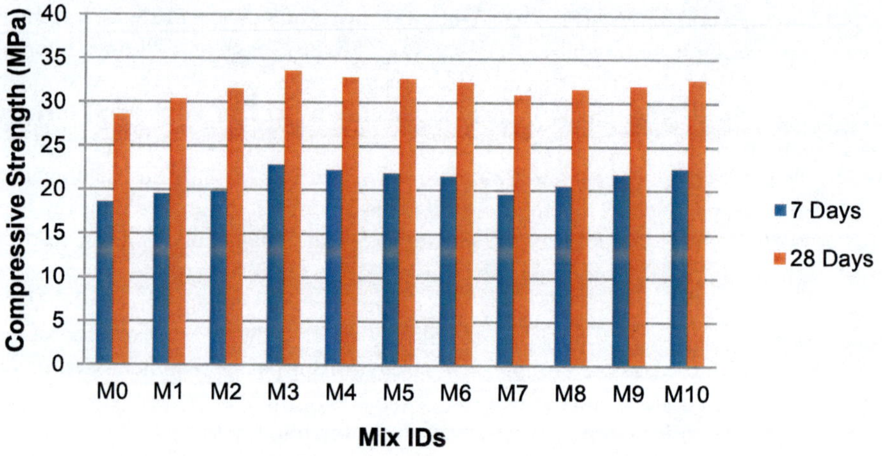

Fig. 1 Compressive strength variations

fibre is incorporated to concrete, it enhanced the tensile strength of concrete. Figure 2 represents the split tensile strength variations of various mixes.

4.1.3 Flexural Strength

The results of 28 days flexural strength of various mixes were given in Table 9. The flexural strength of mix (M2) with polypropylene incorporated concrete under internal curing without any admixtures was 4.90 MPa. The strength increment was phenomenal due to the incorporation of fibres. The small discrete fibres are acting as bridge to avoid the minor cracks. The cracks are the main reason for the

Table 9 Split tensile strength and flexural strength

Mix ID	Split tensile strength (MPa)	Flexural strength (MPa)
M0	3.06	3.56
M1	3.20	3.68
M2	3.65	4.90
M3	3.88	5.80
M4	3.85	5.62
M5	3.82	5.55
M6	3.80	5.36
M7	3.75	5.28
M8	3.78	5.30
M9	3.79	5.32
M10	3.82	5.40

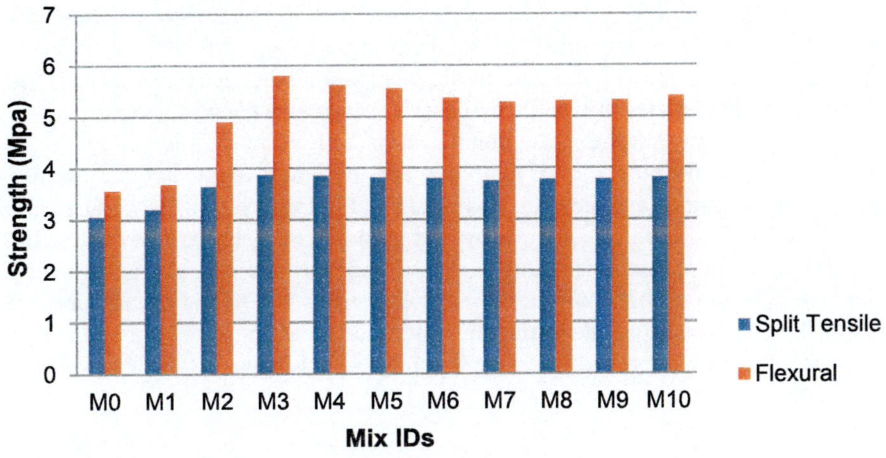

Fig. 2 Split tensile and flexural strength variations

minimisation of life span of any concrete structures. The optimum flexural strength obtained was 5.80 MPa for a 0.5% replacement of nanosilica (M3). The mixtures with microsilica replacement of cement showed good results in flexural strength behaviour. The optimum strength of 8% microsilica replacement with cement (M10) was 5.40 MPa. The optimum strength variation between nanosilica and microsilica was 7.4%. Figure 2 shows the variations in the modulus of rupture for various mixes.

4.2 Microstructural Characteristics

4.2.1 Scanning Electron Microscopy (SEM)

Concrete paste containing microsilica and nanosilica microstructure was compared using images from scanning electron microscopy (SEM). Figure 3a, b is SEM of the nominal mix and 0.5% nanosilica mix concrete, respectively. It was discovered that adding nanosilica improved the way the concrete matrix hydrated. By contrasting the microstructures of the control mixture and 0.5% nanosilica concrete, it was discovered that nanosilica not only packed the spaces between the concrete matrix's particles but also encouraged chemical reaction and produced C–S–H gel to seal the holes of the concrete slurry. Because of this, the tiny silica particles enhanced the concrete's mechanical and durability attributes. Figure 3b demonstrates that the C–S–H gel created by the chemical reaction of nanosilica particles has filled the pores of concrete paste in addition to the spaces between the particles in the concrete matrix. SEM images demonstrate that nanosilica particles have filled the pores of the concrete matrix similarly to microsilica particles, with the exception that they have also formed a laminate structure. This laminate structure appears to be the cause of the decreased resistance of concrete to the penetration of undesirable agents. This gel clings to the particles, fills in their spaces and connects them to one another.

Figure 4a demonstrates that the nominal mix (M0) sample's microstructure has substantial flaws, the outcomes of hydration are not tightly coupled with the aggregate, the ITZ is fragile, and there are clear fissures. Figure 4b shows that adding 0.5% nanosilica (M3) can enhance cement hydration to some level after less cement has been added to the concrete, resulting in denser and more uniform hydration results. Concrete now has a noticeably better microstructure. The hydration products are wrapped around the aggregate in the thick interface transition zone of the M3 specimen. Significantly improved interface bonding occurs between the slurry and the aggregate.

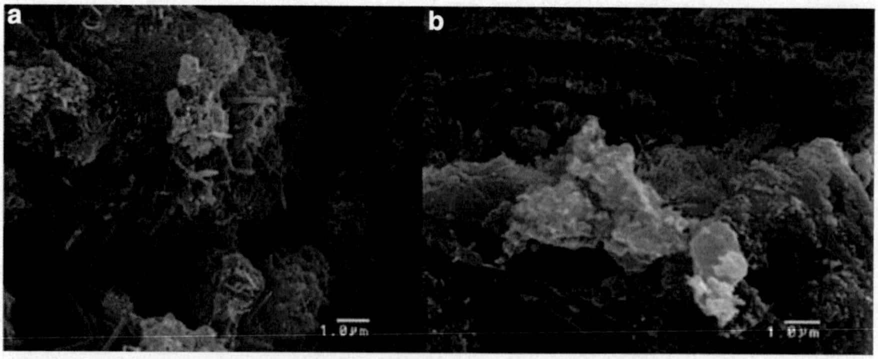

Fig. 3 **a** SEM image of nominal mix [51] **b** SEM image of nanosilica mix [51]

Fig. 4 **a, b** SEM image of ITZ of nominal and nanosilica replaced mix [52]

5 Conclusion

Examining the strength and microstructural behaviour of self-cured fibre-blended concrete that was made with various cementitious material substitutes was the aim of this investigation. The findings of various tests may be combined to provide a final conclusion:

- The cube's compressive strength and cylinder's splitting strength due to tensile load of (M3) mix with a nanosilica dosage of 0.5% are 33.56 MPa and 3.88 MPa, respectively, under the condition of reducing cement content in concrete, when the standard curing age is 28 days. These values are 17.5% and 26.8% higher than those of the nominal mix (M0) at 28 days. But for 7 days, the strength variation is 23.3% higher than 28 days strength. The outcomes demonstrate that adding the right quantity of nanosilica can greatly enhance concrete's mechanical qualities. The exceptionally high activity of nanosilica when used to replace some of the cement aids in promoting the process of hydration of concrete, which results in increase strength at an early stage.
- The compressive strength of mix (M10) with 8.0% substitution of microsilica was increased by 14% at 28 days compared to nominal mix. At 7 and 28 days, the strength variation of the nominal mix with its own curing agent PEG-200 without any mineral or chemical admixtures (M2) was 5.4% and 6.3% greater compared to the normal mix (M0).
- The most effective flexural strength was 5.80 MPa with a nanosilica substitution of 0.5% (M3). When compared to the nominal concrete mix (M0), the strength variation was 63% higher. The enhancement in strength performance was significantly impacted by PEG-200. The 8% microsilica replacement with cement's (M10) maximum strength was 5.40 MPa. In terms of strength, the optimal difference between nanosilica and microsilica was 7.4%. When compared to microsilica, the nanosilica proportions produced better results for the modulus of rupture.
- The mix (M4) with a 0.5% nanosilica substitution had the best split tensile strength value at 3.88 MPa. In comparison to the nominal mix (M0), the strength variation of the mix (M4) was 26.8% higher. Due of the nanoparticles' increased

specific surface area and smaller size, mixtures including nanosilica particles demonstrated good mechanical properties. The 8% microsilica replacement with cement's (M10) maximum spilt tensile strength was 3.82 MPa. The ideal strength disparity between nanosilica and microsilica was 1.5%.

- The pozzolan activity of nanosilica is greater. Because the pozzolanic reaction is more sufficient in the early stage, the initial strength enhancing impact caused by nanosilica concrete is greater. Smaller nanosilica particles and a weaker pozzolan reaction are caused by an increase in curing time. As a result, the later period's improving impact of nanosilica on the strength of concrete is diminished.
- From SEM analysis, the addition of nanosilica caused changes in the composition of the hydration products produced during the process of hydration, which improved the microstructure of the concrete and increased its compactness. The mechanical properties are ultimately improved as a result of this.

References

1. Abhilash PP, Dheeresh KN, Bhaskar S, Rajesh K, Veerendra K (2021) Effect of nano-silica in concrete; a review. Constr Build Mater. https://doi.org/10.1016/j.conbuildmat.2021.122347
2. Alonso D, Álvarez I, Reyes E, Moragues A (2017) New mortars fabricated by electrostatic dry deposition of nano and microsilica additions: enhanced properties. Constr Build Mater. https://doi.org/10.1016/j.conbuildmat.2017.01.011
3. Ina P, Jolanta P (2020) The synergistic effect of adding a blend of deflocculants and microsilica on the properties of high temperature resistant lightweight concrete with cenospheres. Constr Build Mater. https://doi.org/10.1016/j.conbuildmat.2019.116961
4. Kishor Kumar V, Rajan NK, Vandhana Devi V (2022) Relationship between Gaestel Index and rheology of blended bitumens. AIP Confer Proc 2452. https://doi.org/10.1063/5.0118092
5. Pravin Raaj E, Bhuvaneshwari K, Lakshmipathy R, Vandhana Devi V, Rico ILR (2022) Garlic peel surface modification and fixed-bed column investigations towards crystal violet dye. Adsorp Sci Tech. https://doi.org/10.1155/2022/6904842
6. Fan H, Li Y, Huang Y, Sang S, Li Y, Zhao L (2012) Microstructures and mechanical properties of Al_2O_3–ZrO_2–C refractories using silicon, microsilica or their combination as additive. Mater Sci Eng. https://doi.org/10.1016/j.msea.2012.03.013
7. Pandey A, Kumar B (2019) Evaluation of water absorption and chloride ion penetration of rice straw ash and microsilica admixed pavement quality concrete. Heliyon. https://doi.org/10.1016/j.heliyon.2019.e02256
8. Singh LP, Ali D, Tyagi I, Sharma U, Singh R, Hou P (2019) Durability studies of nano-engineered fly ash concrete. Constr Build Mater. https://doi.org/10.1016/j.conbuildmat.2018.11.0220
9. Khaloo A, Mobini MH, Hosseini P (2016) Influence of different types of nano-SiO_2 particles on properties of high-performance concrete. Constr Build Mater. https://doi.org/10.1016/j.conbuildmat.2016.03.041
10. Behfarnia K, Rostami M (2017) Effects of micro and nanoparticles of SiO_2 on the permeability of alkali activated slag concrete. Constr Build Mater. https://doi.org/10.1016/j.conbuildmat.2016.11.070
11. Lo WK, Long Lin K, Wui Cheng T, Min Chang Y, Ying Lan J (2017) Effect of nano-SiO_2 on the alkali-activated characteristics of spent catalyst metakaolin-based geopolymers. Constr Build Mater. https://doi.org/10.1016/j.conbuildmat.2017.03.152

12. Shaikh FUA, Supit SWM, Sarker PK (2014) A study on the effect of nano silica on compressive strength of high volume fly ash mortars and concretes. Mater Design. https://doi.org/10.1016/j.matdes.2014.04.025
13. Heikal M, Didamony HE, Sokkary TM, Ahmed IA (2013) Behavior of composite cement pastes containing microsilica and fly ash at elevated temperature. Constr Build Mater. https://doi.org/10.1016/j.conbuildmat.2012.09.069
14. Hendi A, Behravan A, Mostofinejad D, Kharazian HA, Sedaghatdoost A (2020) Performance of two types of concrete containing waste silica sources under $MgSO_4$ attack evaluated by durability index. Constr Build Mater. https://doi.org/10.1016/j.conbuildmat.2020.118140
15. Hendi A, Rahmani H, Mostofinejad D, Tavakolinia A, Khosravi M (2017) Simultaneous effects of microsilica and nanosilica on self-consolidating concrete in a sulfuric acid medium. Constr Build Mater. https://doi.org/10.1016/j.conbuildmat.2017.06.165
16. Li LG, Huang ZH, Zhu J, Kwan AKH, Chen HY (2017) Synergistic effects of micro-silica and nano-silica on strength and microstructure of mortar. Constr Build Mater. https://doi.org/10.1016/j.conbuildmat.2017.02.115
17. Aisheh YIA, Atrushi DS, Akeed MH, Qaidi S, Tayeh BA (2022) Influence of steel fibers and microsilica on the mechanical properties of ultra-high-performance geopolymer concrete. Case Studies in Constr Mater. https://doi.org/10.1016/j.cscm.2022.e01245
18. Ahmad S, Mohaisen KO, Adekunle SK, Dulaijan SUA, Maslehuddin M (2019) Influence of admixing natural pozzolan as partial replacement of cement and microsilica in UHPC mixtures. Constr Build Mater. https://doi.org/10.1016/j.conbuildmat.2018.11.260
19. Lou Y, Khan K, Amin MN, Ahmad W, Deifalla AF, Ahmad A (2023) Performance characteristics of cementitious composites modified with silica fume: a systematic review. Case Stud Constr Mater. https://doi.org/10.1016/j.cscm.2022.e01753
20. Said AM, Zeidan MS, Bassuoni MT, Tian Y (2012) Properties of concrete incorporating nano-silica. Constr Build Mater. https://doi.org/10.1016/j.conbuildmat.2012.06.044
21. Amin M, Zeyad AM, Tayeh BA, Agwa IS (2021) Engineering properties of self-cured normal and high strength concrete produced using polyethylene glycol and porous ceramic waste as coarse aggregate. Constr Build Mater. https://doi.org/10.1016/j.conbuildmat.2021.124243
22. Lija RL, Ananthi GBG, Roy K, Lim JBP (2020) Effect of super absorbent polymer on microstructural and mechanical properties of concrete blends using granite pulver. Struct Conc. https://doi.org/10.1002/suco.201900419
23. Lija RL, Ananthi GBG, Roy K, Lim JBP (2020) Influence of super absorbent polymer on mechanical, rheological, durability, and micro-structural properties of self-compacting concrete using non bio-degradable granite pulver. Struct. Conc. https://doi.org/10.1002/suco.201900470.
24. Younis MO, Amin M, Tahwia AM (2022) Durability and mechanical characteristics of sustainable self-curing concrete utilizing crushed ceramic and brick wastes. Case Stud Constr Mater. https://doi.org/10.1016/j.cscm.2022.e01251
25. Mokhtari S, Madhkhan M (2022) The performance effect of PEG-silica fume as shape-stabilized phase change materials on mechanical and thermal properties of lightweight concrete panels. Case Stud Constr Mater. https://doi.org/10.1016/j.cscm.2022.e01298
26. Yuan Z, Jia Y (2022) Mechanical properties and microstructure of glass fiber and polypropylene fiber reinforced concrete: an experimental study. Constr Build Mater. https://doi.org/10.1016/j.conbuildmat.2020.121048
27. Dong S, Zhou P, Guo R, Li C, Xian G (2022) Durability study of glass fiber reinforced polypropylene sheet under simulated seawater sea sand concrete environment. J Mater Res Tech. https://doi.org/10.1016/j.jmrt.2022.07.097
28. Hussain I, Ali B, Akhtar T, Jameel MS, Raza SS (2020) Comparison of mechanical properties of concrete and design thickness of pavement with different types of fiber-reinforcements (steel, glass, and polypropylene). Case Stud Constr Mater. https://doi.org/10.1016/j.cscm.2020.e00429
29. Aarthi K, Arunachalam K (2018) Reusing granite sawing waste in self-compacting concrete with polypropylene fiber at low-volume fractions. Struct Conc. https://doi.org/10.1002/suco.201800155

30. Yasouj SEM, Ghaderi A (2020) Experimental investigation of waste glass powder, basalt fibre, and carbon nanotube on the mechanical properties of concrete. Constr Build Mater. https://doi.org/10.1016/j.conbuildmat.2020.119115
31. Bakr MA, Mahmood HF, Mohammed AA (2022) Investigation of metakaolin and steel fiber addition on some mechanical and durability properties of roller compacted concrete. Case Stud Constr Mater. https://doi.org/10.1016/j.cscm.2022.e01136
32. Swathi V, Asadi SS (2022), Structural performance of hybrid fibres based concrete: mechanical, durability and microstructural properties. Sustain Futures. https://doi.org/10.1016/j.sftr.2022.100094
33. Islam MJ, Islam K, Shahjalal M, Khatun E, Islam S, Razzaque AB (2022) Influence of different types of fibers on the mechanical properties of recycled waste aggregate concrete. Constr Build Mater. https://doi.org/10.1016/j.conbuildmat.2022.127577
34. Li W, Shumuye ED, Shiying T, Wang Z, Zerfu K (2022) Eco-friendly fibre reinforced geopolymer concrete: a critical review on the microstructure and long-term durability properties. Case Stud Constr Mater. https://doi.org/10.1016/j.cscm.2022.e00894
35. Ahmad J, Zhou Z (2022) Mechanical properties of natural as well as synthetic fiber reinforced concrete: a review. Constr Build Mater. https://doi.org/10.1016/j.conbuildmat.2022.127353
36. Ioannou S, Chowdhury MS, Badr A (2018) Rheological, hydration and mechanical characteristics of microsilica fibre reinforced cement combinations with incremental fly ash contents. Constr Build Mater. https://doi.org/10.1016/j.conbuildmat.2018.10.039
37. Tayeh BA, Akeed MH, Qaidi S, Abu Bakar BH (2022) Influence of microsilica and polypropylene fibers on the fresh and mechanical properties of ultra-high performance geopolymer concrete. Case Stud Constr Mater. https://doi.org/10.1016/j.cscm.2022.e01367
38. Tayeh BA, Akeed MH, Qaidi S, Abu Bakar BH (2022) Influence of sand grain size distribution and supplementary cementitious materials on the compressive strength of ultrahigh-performance concrete. Case Stud Constr Mater. https://doi.org/10.1016/j.cscm.2022.e01495
39. Orouji M, Najaf E (2023) Effect of GFRP rebars and polypropylene fibers on flexural strength in high-performance concrete beams with glass powder and microsilica. Case Stud Constr Mater. https://doi.org/10.1016/j.cscm.2022.e01769
40. Methods of physical test for hydraulic cement, IS 4031: 1988(Part 3), BIS, New Delhi
41. Fineness test on cement, IS 4031: 1996 (Part 1), BIS, New Delhi
42. Specifications for ordinary Portland cement 53 Grade, IS 12269: 1987, BIS, New Delhi
43. Specifications for coarse and fine aggregates from natural sources for concrete, IS:383: 1970, BIS, New Delhi
44. Methods of test for aggregates on concrete, IS 2386: 1963 (Part 3), BIS, New Delhi
45. Fineness modulus test on aggregates, IS 2386: 1963 (Part 5), BIS, New Delhi
46. Specifications for concrete admixtures, IS 9103: 1999, BIS, New Delhi
47. Plain and reinforced concrete code of practice IS:456: 2000, BIS, New Delhi
48. Recommended guidelines for concrete mix design, IS 10262: 2009, BIS, New Delhi
49. Methods of test for strength of concrete, IS 516: 1959, BIS, New Delhi.
50. Methods of test for splitting tensile strength of concrete cylinders, IS 5816: 1999, BIS, New Delhi
51. Jo B, Kim C, Tae G, Park J (2007) Characteristics of cement mortar with nano-SiO_2 particles. Constr Build Mater. https://doi.org/10.1016/j.conbuildmat.2005.12.020
52. Liu C, Su X, Wu Y, Zheng Z, Yang B, Luo Y, Yang J (2021) Effect of nano-silica as cementitious materials reducing admixtures on the workability, mechanical properties and durability of concrete. Nanotech Rev 10:1395–1409. https://doi.org/10.1515/ntrev-2021-0097

Influence of Fly Ash on Mechanical Properties of Slag-Based Alkali-Activated Concrete with Low NaOH Concentrations

Mangalapuri Venkateswarlu and T. D. Gunneswara Rao

1 Introduction

The cement manufacturing process emits a large amount of CO_2, which has a negative impact on the environment. About 7% of carbon dioxide emissions are from cement industries. One ton of cement production emits about one ton of carbon dioxide into the atmosphere. To reduce these harmful effects (CO_2) on the environment, OPC needs new sustainable building materials. Use of locally available Supplementary Cement Materials (SCM) such as fly ash and GGBFS in concrete production reduces environmental pollution and landfill burden [1]. Activation of clinker-free supplementary cement materials with alkaline activator solution produces Alkali-Activated Concrete (AAC) and Geopolymer Concretes (GPC) [2–4]. GGBFS exhibits pozzolanic behavior and good binding properties in base media with low heat of hydration, and it gives better mechanical strength and high resistance to corrosive chemicals [5]. But there was setting time and workability problems with these GGBFS based alkali-activated concretes [6, 7]. The inclusion of low calcium fly ash as a binder result in less strength in room curing conditions [8, 9]. So, heat curing or oven curing over 600 °C is required to attain strength, this is difficulty to provide in in situ conditions. The NaOH solution molarity plays a prominent role in the dissolution of compounds present in source materials (fly ash and GGBFS) such as alumina and silica. Higher concentration of this chemical leads to higher alkaline medium which leads to faster dissolution and faster polymerization. Whereas using low concentrations of NaOH solution instead of high levels of sodium hydroxide

M. Venkateswarlu (✉) · T. D. Gunneswara Rao
Department of Civil Engineering, National Institute of Technology, Warangal, Telangana 506004, India
e-mail: mv718113@student.nitw.ac.in

T. D. Gunneswara Rao
e-mail: tdg@nitw.ac.in

R. K. Pancharathi et al. (eds.), *Low Carbon Materials and Technologies for a Sustainable and Resilient Infrastructure*, Lecture Notes in Civil Engineering 440,
https://doi.org/10.1007/978-981-99-7464-1_11

(NaOH) for concrete production can reduce the risk and cost associated with it [6]. Alkaline Ratio (AR) plays important role strength development high alkaline ratios lead to uneconomical mix and low alkaline ratio leads to poor mechanical strength [6–9]. Fly-ash-based GPC (FGPC) and Slag-based AAC (SAAC) are manufactured without Portland cement. Many studies have shown that GPC and AAC have mechanical properties comparable to OPC concrete.

The engineering aspects of alkali activated slag-fly ash concrete (AASFC) prepared using 4 M and 6 M NaOH solutions under ambient conditions were investigated by Lee et al. [10], and the results revealed that the compressive strengths of AASFC increased effectively with increasing slag (GGBFS) and sodium hydroxide molar concentrations, but modulus of elasticity values were slightly lower than those of conventional concretes (OPC based). Indirect tensile strengths also shown less values than OPCs. Under the same curing conditions, Fang et al. [11] observed similar types of behavior using 10 and 12 M NaOH solutions. Fernandez-Jimenez et al. [12] investigated the engineering features of heat-cured Fly ash-based Alkali-Activated Concrete (FAAC). According to the test results, FAAC has higher indirect tensile and flexural strengths than OPC. However, the modulus of elasticity of FAAC is lower than that of conventional concretes (OPCs). Bernal et al. [13] investigated the engineering aspects of SAAC in room curing. According to the test results, SAACs compressive strength was comparable to that of OPC; however, the indirect tensile strengths (split tensile and flexure) were slightly greater than conventional concretes (OPCs). Ryu et al. [14] investigated mechanical characteristics of FGPC under heat and air curing conditions with higher NaOH concentrations. According to the test results, mechanical strengths improved in heat-cured FGPCs over air-cured ones, and early strengths increased significantly due to higher NaOH concentrations. Increasing of GGBFS in fly-based geopolymer concretes increases the compressive strength under ambient curing conditions [15, 16]. The engineering characteristics of FGPC under heat-curing were investigated by Kamal et al. [17]. According to the test results, FGPC has better split tensile and flexural strengths than conventional concretes (OPCs), but it also has a similar MOE. Ivan et al. [18] developed equations for split tensile, flexural strengths in terms of compressive strength in FGPC and validated those equations with existed codes. And developed equations were satisfied the code provisions. Xie et al. [19] revealed that MOE in recycled aggregate GPC decreased with increasing fly ash content. However, increasing GGBFS reduces the ductility of the specimen while increasing its stiffness and resistance to deformation at peak stress. Furthermore, as GGBFS increased, ultimate strains decreased after peak stress. Peethamparan and Thomas [20] investigated the engineering aspects as well as stress–strain behaviors in fly ash, or GGBFS, as a solo binder in alkali-activated concrete (FAAC and SAAC). In the stress–strain curve, SAAC exhibits more brittle behavior than FAAC, and there is also more change observed before and after the peak stress. SAAC has higher modulus of elasticity readings than FACC. Venu and Gunneswara Rao [21] conducted experimental study on stress–strain behavior of fly ash-GGBFS-based GPC. From results, increasing GGBFS enhanced the elastic modulus while lowering maximum strains.

The majority of earlier investigations concentrated on the mechanical and fresh characteristics of AACs or GPCs made solely of fly ash, or GGBFS. Using both fly ash and GGBFS as binders, very little research has been done on GPC and AAC. Nearly all research has made use of NaOH solutions with higher concentrations. More research on AASFC is required because FAAC has excellent fresh properties but strength issues. Similarly, SAAC is stronger in terms of strength but has issues with workability. On the other hand, using low concentrations of NaOH solution instead of high levels of sodium hydroxide (NaOH) for concrete manufacture might lower the danger and cost connected with it. Understanding the engineering characteristics and stress–strain behaviors of AASFCs is critical in concrete structure design. Effectiveness of fly ash on engineering aspects and stress–strain behaviors of slag-based alkali-activated concrete with low NaOH concentrations at room temperature is investigated in this research.

2 Experimental Program

2.1 Materials

In this study, GGBFS and fly ash were used as binding material. GGBFS was taken from JSW cements Warangal and confirms to IS:12089-1987 [22], its specific gravity and specific surface are values are 2.9 and 355 m^2/kg, respectively. Fly ash was obtained from the Ramagundam thermal power plant in Telangana and confirms to IS:3812-1981 [23], its specific gravity and specific surface area of the fly ash used were 2.11 and 450 m^2/kg, respectively. Coarse aggregates of nominal size of 12.5 mm and this confirms IS:383 (1970) [24] were used, its specific gravity value is 2.79, the nominal size of fine aggregate is 4.75 mm and this confirms IS:383 (1970) [24], its specific gravity value is 2.62. In this study, Sulfonated naphthalene-based polymers were used as Superplasticizer (SP), that is, Conplast SP430 and this SP complying with IS:9103-999 [25] and BS: 5075 Part 1 [26] and which was obtained from Fosroc chemicals. And also, the dosage of SP 6% used in all the mixes.

2.2 Mixing of Alkali-Activated Solution and Mix Proportions of AASFC

Molarities of 1, 2, and 4 M of NaOH solution were used in mixes A, B, and C, respectively, but a fixed Alkaline Ratio (AR) of 1.5 was used in all mixes, which was decided by previous studies [6, 7, 27] as well as the relevant details are given in the Table 1. Both samples of NaOH and Na_2SiO_3 solutions were mixed before 24 h to obtain proper mixing.

Table 1 Mix proportioning of AASFC

Mix designation	Binder (B)		Alkaline solution (S) (kg)	Aggregates		S/B ratio
	GGBFS (kg)	Fly ash (kg)		Fine aggregate (kg)	Coarse aggregate (kg)	
MIX-A	300	–	165	581	1354	0.55
	240	60				
	180	120				
	120	180				
	60	240				
	–	300				
MIX-B	400	–	170	819	1001	0.45
	320	80				
	240	160				
	160	240				
	80	320				
	–	400				
MIX-C	400	–	200	810	990	0.50
	320	80				
	240	160				
	160	240				
	80	320				
	–	400				

Three main mixes were considered in the present study. They are MIX-A, B, and C. This whole mix proportioning took place in two stages. In the first stage, 100% GGBFS content was used in each mixture (MIX-A, B, and C). Similarly, these mixes are considered as reference mixes. MIX-A is designed for 20 MPa target compressive strength using 100% GGBFS as binder material. Similarly, MIX-B and MIX-C are designed for 40 MPa and 60 MPa target compressive strengths using 100% GGBFS as binders. As part of the second stage, the GGBFS content (MIX-A, B, and C) of each mix was replaced with 20%, 40%, 60%, 80%, and 100% fly ash content. Total 18 mixes were prepared. The density of alkali-activated concrete (AAC) for this mix proportioning was taken to be 2400 kg/m^3. And the mix proportioning was determined from that density and this density was taken from some previous studies [17, 27]. Finally, corresponding mix proportioning details are shown in Table 1.

2.3 AASFC Samples Preparation and Curing

Mixing process of AASFC is same as conventional concrete [11, 15]. First, mix binder materials GGBFS and fly ash well for 2–3 min. After that, add fine aggregate and coarse aggregate to the binder material and mix well for another 3–4 min. Alkaline solution and superplasticizers were then added. Mixing is allowed for another 5–6 min to get uniform and homogeneous mixture. Obtained mixture used for test the workability of the particular mix, and this concrete mixture was again thoroughly mixed and cast into the required molds. In the present study, used cube molds measuring 100 × 100 × 100 mm, cylinder molds of size100 mm diameter and 200 mm height. Prism molds of sizes 100 mm × 100 mm × 500 mm were used. Finally, after all samples were dry, they were removed from the molds and left to air cure at room temperature (ambient cure).

2.4 Tests Conducted

For the compressive strengths, cube specimens tested by using the Tinius-Olsen Testing Machine (TOTM), and its capacity is 2000 kN as per Indian standard: 516-1959 [28]. For the compressive strength, three cubes were tested. Testing of cylinder specimen under compression for stress–strain behavior shown in Fig. 1. The split tensile strength test was also performed on same testing machine as per Indian standard IS:5816 [29]. Three cylinders tested. Similarly, for flexural strength test, three-point loading test performed on prisms on a digital universal testing machine. Capacity of machine is 200 kN. This test is performed in accordance with ASTM C—293-02, 2002 [30]. Three samples were cast and tested in each mix proportioning. A total 108 cube specimens, 108-cylinder specimens and 56 prisms specimens were cast in this study.

3 Results and Discussion

3.1 Compressive Strength

The 7-day and 28-day compressive strength results of MIX-A, B, and C are shown in Figs. 2, 3 and 4. Compressive strength values decreased when the GGBFS content replaced by fly ash from 0 to 100% in all mixes. Previous studies using medium and high concentration NaOH solutions as activators revealed similar results [7, 8, 10, 11, 21]. The rate of compressive strength loss increased as the fly ash content in all the mixes was raised, and this was also the case when the fly ash replacement levels were above 40%. This might be because curing condition and low-concentration alkali activators were used in this investigation. The increase in compressive strength at

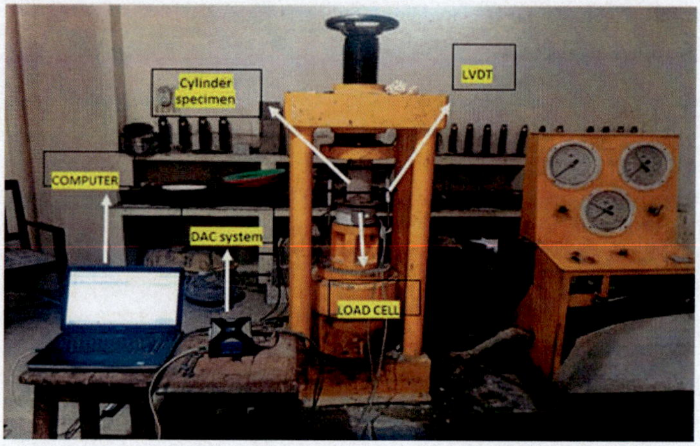

Fig. 1 Testing of cylinder specimen under compression for stress–strain behavior

higher levels may be due to the higher calcium content in GGBFS [7]. And inclusion of fly ash reduces Si/Al ratio in the mixture, indicating that determining the relative quantities of AlO4 and SiO4 generated in the geopolymer gel and determining the quantity of Si contained in the combination. Mixing with high fly ash content is associated with a low silicon/aluminum ratio. The compressive strength generally decreases with decreasing Si/Al ratio. Method of curing is also playing an important role in geopolymerization process [7]. However, for better rates of hydration to occur, higher fly ash levels demand higher activator concentrations and higher temperatures. However, substituting 40% fly ash content gave the expected target strengths in MIX-A and MIX-B, which were 22.44 and 40.01 MPa, respectively. The compressive strength values in MIX-A at 100/00, 80/20, 60/40, 40/60, 20/80, and 00/100 GGBFS and fly ash proportions is 27.03, 24.87, 22.44, 19.19, 9.46, and 4.87 MPa, respectively. similarly in MIX-B, at the same GGBFS and fly ash proportions, 49.51, 44, 40.01, 31.43, 17.57, and 6.08 MPa, respectively; and in MIX-C, 65.81, 58.22, 49.88, 41.04, 24.63, and 7.94 MPa, respectively.

3.2 *Split Tensile and Flexural Strengths*

The split tensile strength results of mixes A, B, and C are shown in Fig. 5. According to findings, as fly ash in all mixes increased from 0 to 40%, split tensile strengths decreased slightly. The decreased rate was very low. However, when more than 40% fly ash was added, the split tensile strength values dropped drastically (60–100%), and the decrement rate is higher. In comparison to 100% GGBFS mixtures, mixtures with 80/20 and 60/40 GGBFS/fly ash proportions little decrement in split tensile and flexural strengths. At remaining GGBFS-fly ash proportions, these values decreased highly. In Fig. 6, the flexural strength results of mixes A, B, and C are shown. As

Fig. 2 MIX-A compressive strength

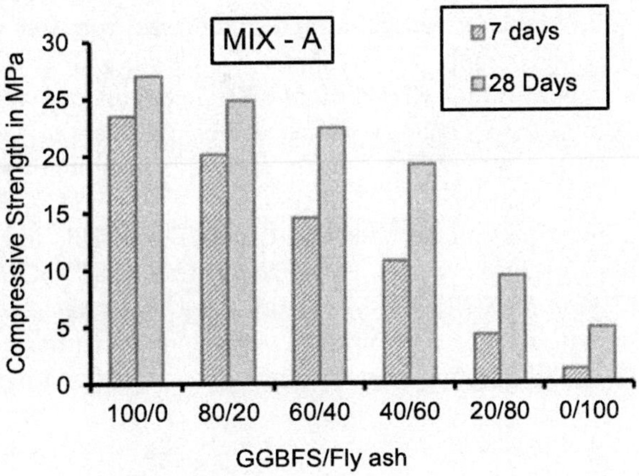

Fig. 3 MIX-B compressive strength

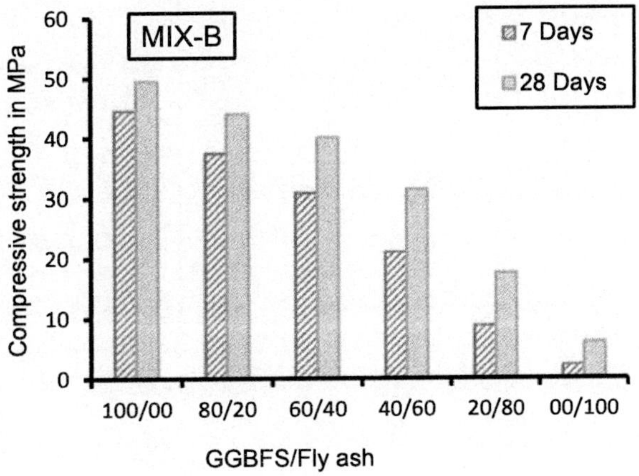

Fig. 4 MIX-C compressive strength

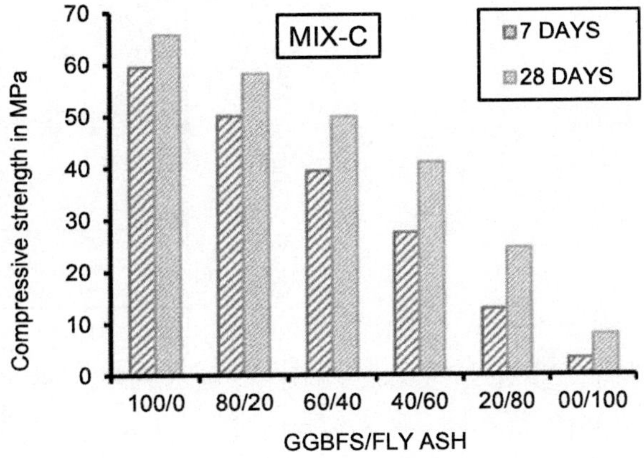

fly ash replacement in all mixes increased from 0 to 40% (100–60% GGBFS), the flexural strength values slightly decreased. Ivan et al. [18] also found that flexural strength values are higher for GGBFS content up to 60%. However, when more than 40% of the fly ash was replaced, flexural strength values decreased and the decrement rate was higher than for split tensile strength values. MIX-C had a lower % drop in split tensile and flexural strength than MIX-B and A, whereas MIX-B had a lower percentage reduction than MIX-A. This might be because MIX-C (4 M) uses a solution with a higher concentration of NaOH than the other combinations (2 M in MIX-B and 1 M in MIX-C). This proves that under ambient curing conditions, the NaOH concentration plays significant impact on mechanical characteristics of AASFC composites. Mechanical strengths and modulus of elasticity values are presented in Table 2.

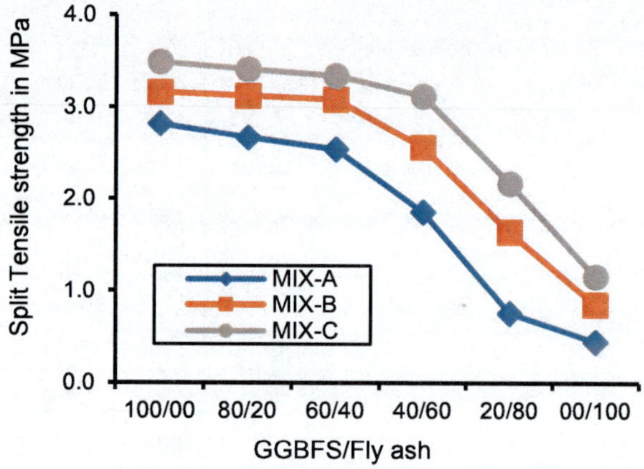

Fig. 5 Split tensile strength values

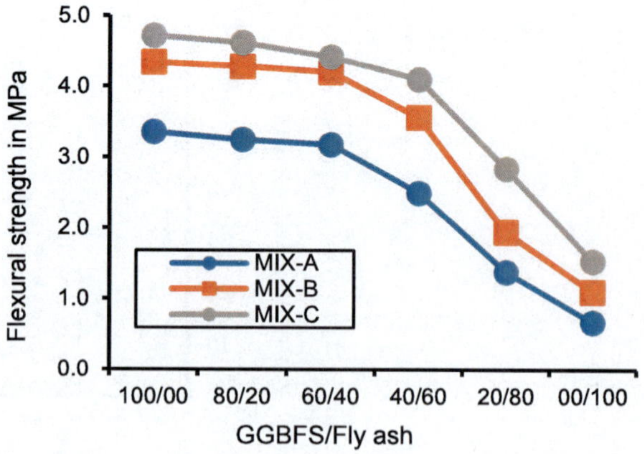

Fig. 6 Flexural strength values

Table 2 Mechanical strengths and modulus of elasticity values

Mix designation	GGBFS/fly ash Proportion	Compressive strength (MPa)	Split tensile strength (MPa)	Flexural strength (MPa)	Modulus of elasticity (MPa)
MIX-A	100/00	27.03	2.82	3.36	11.30
	80/20	24.87	2.67	3.25	9.91
	60/40	22.44	2.54	3.18	8.65
	40/60	19.19	1.86	2.50	7.21
	20/80	9.46	0.76	1.38	3.52
	00/100	4.87	0.45	0.67	1.42
MIX-B	100/00	49.51	3.16	4.34	24.29
	80/20	44.00	3.12	4.28	21.15
	60/40	40.01	3.08	4.20	18.75
	40/60	31.43	2.56	3.56	13.38
	20/80	17.57	1.64	1.94	7.27
	00/100	6.08	0.85	1.10	2.24
MIX-C	100/00	65.81	3.49	4.74	32.94
	80/20	58.22	3.41	4.62	30.86
	60/40	49.88	3.34	4.42	25.86
	40/60	41.04	3.12	4.10	20.51
	20/80	24.63	2.17	2.84	9.17
	00/100	7.94	1.16	1.53	3.26

3.3 *Correlation Studies on Mechanical Characteristics*

Correlation studies carried on mechanical strengths, and these correlations were compared with various codes and earlier research investigations.

According to Fig. 7a, the alkali-activated concrete specimens MIX-A, B, and C had split tensile strengths of 0.48, 0.46, and 0.45 times the square root of the compressive strengths, respectively. where flexural strength values are 0.62, 0.61, and 0.6 times the square root of compressive strength in mixes A, B, and C, respectively shown in Fig. 7b. In all of the mixes, the obtained split tensile and flexural strength values are linearly related to the square root of the compressive strength values; however, the proportionality constant value somewhat diminishes as the compressive strength or concrete grade rises. In Eqs. (1)–(6), the predicted empirical formulas are presented.

For split tensile strength

$$\text{Fts} = 0.48\sqrt{\text{fc}} \text{ and } R^2 = 0.95 \text{ for MIX-A } (4.87\text{ MPa} < \text{fc} < 27.03\text{ MPa}) \quad (1)$$

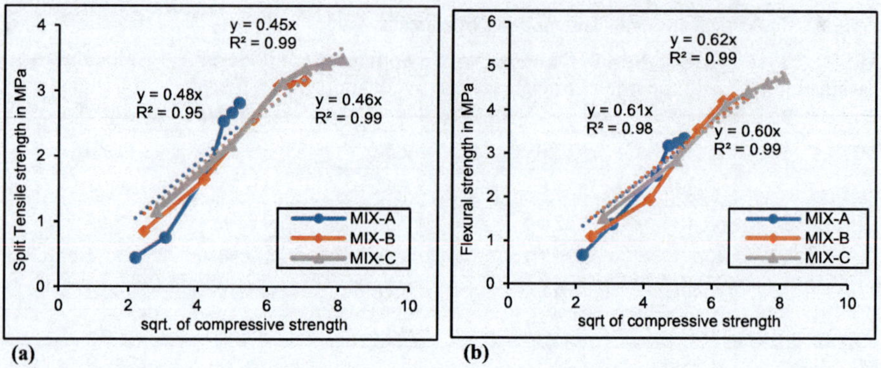

Fig. 7 **a** Split tensile strength versus sqrt. compressive strength. **b** Flexural strength versus sqrt. compressive strength

$$\text{Fts} = 0.46\sqrt{\text{fc}} \text{ and } R^2 = 0.99 \text{ for MIX-B } (6.08 \text{ MPa} < \text{fc} < 49.5 \text{ MPa}) \quad (2)$$

$$\text{Fts} = 0.45\sqrt{\text{fc}} \text{ and } \text{R}^2 = 0.99 \text{ for MIX-C } (7.94 \text{ MPa} < \text{fc} < 65.80 \text{ MPa}) \quad (3)$$

Similarly, for flexural strength

$$\text{Ffs} = 0.62\sqrt{\text{fc}} \text{ and } R^2 = 0.98 \text{ for MIX-A } (4.87 \text{ MPa} < \text{fc} < 27.03 \text{ MPa}) \quad (4)$$

$$\text{Ffs} = 0.61\sqrt{\text{fc}} \text{ and } R^2 = 0.99 \text{ for MIX-B } (6.08 \text{ MPa} < \text{fc} < 49.5 \text{ MPa}) \quad (5)$$

$$\text{Ffs} = 0.60\sqrt{\text{fc}} \text{ and } R^2 = 0.99 \text{ for MIX-C } (7.94 \text{ MPa} < \text{fc} < 65.80 \text{ MPa}) \quad (6)$$

where, fc = compressive strength of concrete in MPa, Fts = split tensile strength of concrete in MPa, Ffs = flexural strength of concrete in MPa

From Fig. 8a–c, split tensile strength values derived from experiments were in good accord with AS 3600 [31], Lee et al. [10], and Sofi et al. [32]; however, they were underestimated in comparison to the ACI-318 [33] building code and the Euro code [34] for ordinary Portland cement concrete (OPC). While the flexural strength values for alkali-activated concrete and OPC satisfy ACI-318 [33], AS 3600 [31], and Ivan et al. [18], they fall short of the values predicted by Nath and Sarkar [35] shown in Fig. 9a–c. The split tensile and flexural strength values obtained through experiment are consistent with most codes and previous studies cited in this study.

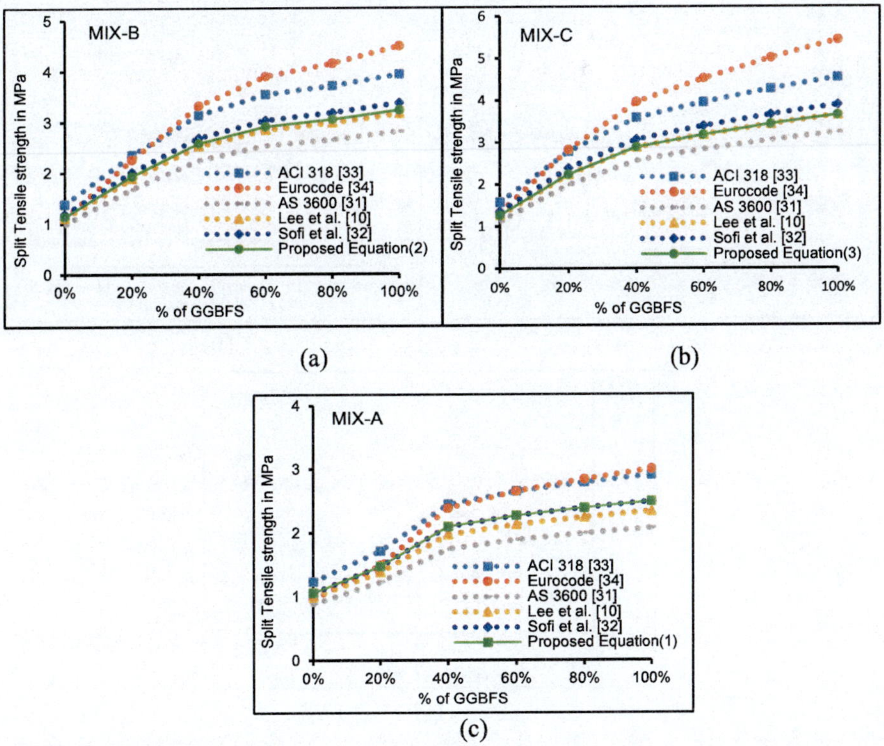

Fig. 8 **a–c** Comparison of obtained experimental split tensile strengths with various codes and previous studies

3.4 *Stress–Strain Behavior*

Figure 10a–c presents the experimental compressive stress-compressive strain behavior graphs for each mix. The data recorded in the DAC system are used to create these curves. Figure 2 illustrates the test setup for the stress strain behavior. Up to 65–70% of the ultimate load (i.e., 2/3rd of ultimate load), the load on the test specimens rose sharply during the test, then increased more slowly until it reached the ultimate load. The test was continued after determining the ultimate load of the specimens until the ultimate load dropped to between 50 and 60% in most of the mixes. At 65–70% of the ultimate load, a number of nonlinear vertical cracks were seen on the surface of the specimens. As the load increases further, these vertical cracks spread from top to bottom in the direction of loading, and these cracks lead to specimen failure. All of the mixes had linear elastic compressive stress–strain behavior up to a peak stress of 65–70% before changing to nonlinear behavior after the failure stress was reached. The load dropped after reaching its peak stress. This load dropped abruptly in some mixes while gradually dropping in others with increasing in compressive strain. When all the mixtures were examined, those that included fly ash

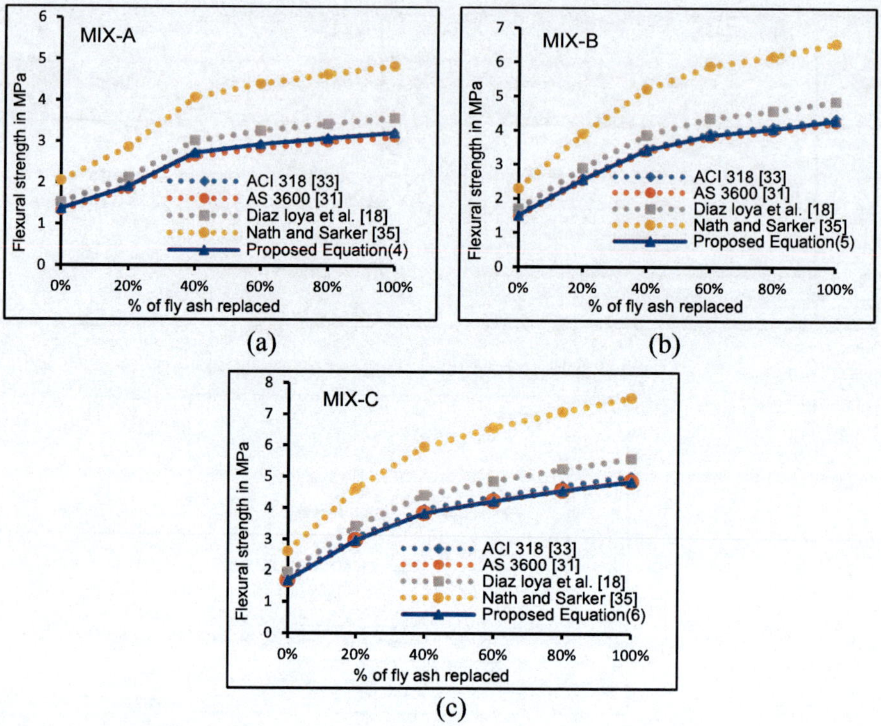

Fig. 9 a–c Comparison of obtained experimental flexural strengths with various codes and previous studies

replacements ranging from 0 to 40% showed greater strain-hardening behavior. Peak stress values in every mix dropped as fly ash content replacement increased. These peak stress values decreased greatly when more than 40% fly ash was replaced. Compared to Mix-B and Mix-C samples, Mix-A samples displayed lower strain values at peak stress with higher post-peak behavior. Similarly, Mix B showed less post-peak behavior than Mix A, and Mix C displayed less post-peak behavior than Mix B when all the mixes were compared. In comparison to Mix-A and mixes with different GGBFS to fly ash proportions, Mix-C and Mix-B specimens having 100% GGBFS displayed higher brittle behavior and quicker crack propagation. Increasing the GGBFS reduces the ductility of the specimen while increasing its stiffness and resistance to deformation at peak stress. Furthermore, as GGBFS increased, ultimate strains decreased after peak stress [19, 21].

This is because those mixes have more GGBFS volume than other mixes do. The percentage of decreasing peak stress and the corresponding strain up to 0–40% fly ash replacement is 14.50% and 14.60% in MIX-A, 19.96% and 9.31% in MIX-B, and 22.1% and 11.21% in MIX-C, respectively. Beyond 40% fly ash replacement levels, 83.98% and 24.77% are in MIX-A; 88.50% and 20.66% are in MIX-B; and 88.88%

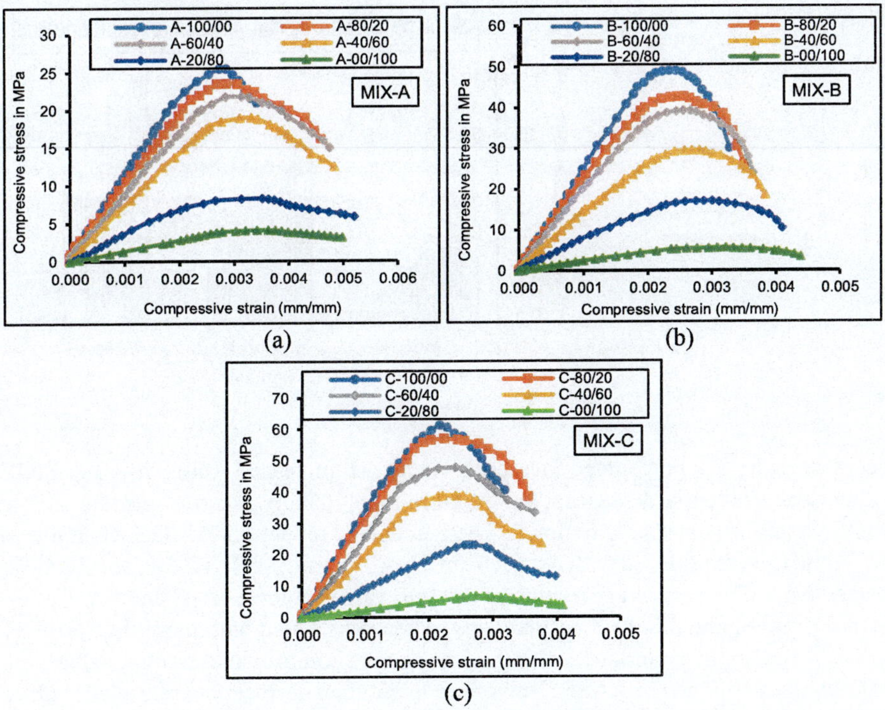

Fig. 10 **a–c** Compressive stress versus compressive strain in mixes A, B, and C

and 25.92% are in MIX-C. This shows that the AASFC's stress–strain behavior is greatly affected by greater replacement amounts of fly ash.

3.5 Modulus of Elasticity

Following ACI 318 [33], the Modulus of Elasticity (MOE) was calculated. By drawing a tangent from the origin to 45% stress at the peak stress and computing the modulus of elasticity value for each mix, the sample's modulus of elasticity is represented by the slope of the tangent. The modulus of elasticity values for mixes A, B, and C shown in Fig. 11.

When 0%, 20%, 40%, 60%, 80%, and 100% fly ash were substituted for GGBFS, the MOE values of MIX-A were 11.3, 9.91, 8.65, 7.21, 3.52, and 1.42 GPa, respectively. MIX-B displayed the following values when the above-mentioned quantities of fly ash replacement were present: 24.29, 21.15, 18.75, 13.38, 7.27, and 2.24 GPa. Similarly, 32.94, 30.86, 25.86, 20.51, 9.17, and 2.36 GPa were found in MIX-C.

As fly ash increased from 0 to 100% in each mix, the MOE values significantly dropped. One of the reasons for this could be that compressive strength values

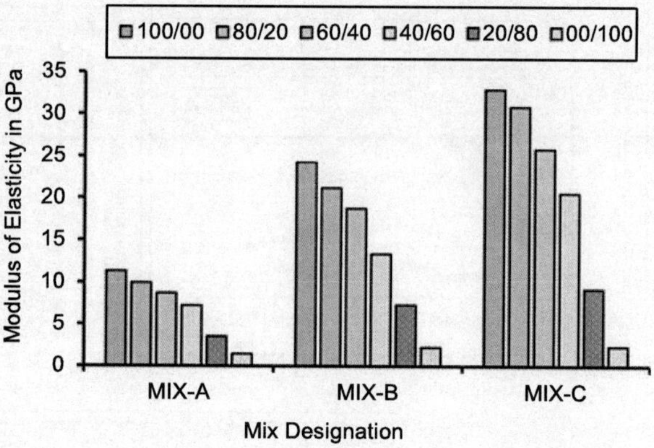

Fig. 11 Modulus of elasticity values in various mixes

decrease as fly ash percentage increases. As fly ash increased from 0 to 40%, MOE values also gradually decreased. When more than 40% fly ash was introduced, the MOE values fell extremely to down. The percentage reduction in MOE when mixes A, B, and C were replaced with fly ash from 0 to 40% was 23.5%, 22.8%, and 21.49%, respectively. The percentage reduction in MOE values of mixes A, B and C were 36–90%, 45–90%, and 38–90%, respectively, when substituted with more than 40% fly ash (60–100%). In accordance with their respective compressive strength values, the MOE values obtained in AASFC mixes are lower than normal concrete MOE values (OPC). Previous research on alkali-activated geopolymer concretes found similar results [10, 19, 21, 32].

4 Conclusions

In this study, the effect of fly ash on the mechanical properties and stress–strain behaviors of slag-based alkali-activated concrete was experimentally investigated using low NaOH concentrations at room curing temperature. These are the important things that have been identified from this study.

- Compressive strength, split tensile and Flexural strength values decreased when the GGBFS content replaced by fly ash from 0 to 100% in all mixes. Most of the composites attained 80–90% compressive strength of their maximum strength at early age (7 days) when 100% GGBFS was used as binder.
- The split tensile and flexural strengths got through experiment are consistent with most codes and previous studies cited in this study.
- Peak stress values in every mix dropped as fly ash content replacement increased. These peak stress values decreased greatly when more than 40% fly ash was replaced. Compared to MIX-A, B samples, MIX-C samples displayed lower strain values at peak stress with lower post-peak behavior. Similarly, MIX-B showed

less post-peak behavior than MIX-A and C displayed less post-peak behavior than MIX-B when all the mixes were compared.

- As fly ash increased from 0 to 40%, Modulus of Elasticity (MOE) values also gradually decreased. When more than 40% fly ash was introduced, the MOE values fell extremely to down. The percentage reduction in MOE when MIX-A, MIX-B, and C were replaced with fly ash 0–40% was 23.5%, 22.8%, and 21.49%, respectively. The percentage reduction in MOE values of MIX-A, B, and C were 36–90%, 45–90%, and 38–90%, respectively, when substituted with more than 40% fly ash (60–100%).

References

1. Duxson P, Provis JL, Lukey GC, Van Deventer JS (2007) The role of inorganic polymer technology in the development of green concrete. Cement Concr Res 37(12):1590–1597. https://doi.org/10.1016/j.cemconres.2007.08.018
2. Chindaprasirt Prinya T, Chareerat VS (2007) Workability and strength of coarse high calcium fly ash geopolymer. Cement Concr Compos 29(3):224–229. https://doi.org/10.1016/j.cemconcomp.2006.11.002
3. Wongpa J, Kiattikomol K, Jaturapitakkul C, Chindaprasirt P (2010) Compressive strength, modulus of elasticity, and water permeability of inorganic polymer concrete. Mater Des 31(10):4748–4754. https://doi.org/10.1016/j.matdes.2010.05.012
4. Bondar D, Lynsdale CJ, Milestone NB, Hassani N, Ramezanianpour AA (2011) Effect of type, form, and dosage of activators on strength of alkali-activated natural pozzolans. Cement Concr Compos 33(2):251–260. https://doi.org/10.1016/j.cemconcomp.2010.10.021
5. Puertas F, Martinez Ramirez S, Alonso S, Vazquez T (2000) Alkali-activated fly ash/slag cements: strength behaviour and hydration products. Cement Concr Res 30(10):1625–1632. https://doi.org/10.1016/S0008-8846(00)00298-2
6. Srikrishna TC, Rao TDG (2018) Appraisal on strength characteristics of alkali-activated GGBFS with low concentrations of sodium hydroxide. Iran J Sci Technol Trans Civil Eng 42(3):231–243. https://doi.org/10.1007/s40996-018-0113-4
7. Mallikarjuna Rao G, Gunneswara Rao TD (2015) Final setting time and compressive strength of fly ash and GGBS-based geopolymer paste and mortar. Arab J Sci Eng 40(11):3067–3074. https://doi.org/10.1007/s13369-015-1757-z
8. Mallikarjuna Rao G, Gunneswara Rao TD, Siva Nagi Reddy M, Rama Seshu D (2019) A study on the strength and performance of geopolymer concrete subjected to elevated temperatures. Recent Adv Struct Eng 1:869–889. https://doi.org/10.1007/978-981-13-0362-3_70
9. Chi M (2012) Effects of dosage of alkali-activated solution and curing conditions on the properties and durability of alkali-activated slag concrete. Constr Build Mater 35:240–245. https://doi.org/10.1016/j.conbuildmat.2012.04.005
10. Lee NK, Lee HK (2013) Setting and mechanical properties of alkali-activated fly ash/slag concrete manufactured at room temperature. Constr Build Mater 47:1201–1209. https://doi.org/10.1016/j.conbuildmat.2013.05.107
11. Fang G, Ho WK, Tu W, Zhang M (2018) Workability and mechanical properties of alkali-activated fly ash-slag concrete cured at ambient temperature. Constr Build Mater 172:476–487. https://doi.org/10.1016/j.conbuildmat.2018.04.008
12. Fernandez JA, Palomo A, Lopez HC (2006) Engineering properties of alkali activated fly ash concrete. ACI Mater J 103(2):106–112. https://doi.org/10.14359/15261

13. Bernal SA, de Gutiérrez RM, Pedraza AL, Provis JL, Rodriguez ED, Delvasto S (2011) Effect of binder content on the performance of alkali-activated slag concretes. Cem Concr Res 41(1):1–8. https://doi.org/10.1016/j.cemconres.2010.08.017
14. Ryu GS, Lee YB, Koh KT, Chung YS (2013) The mechanical properties of fly ash-based geopolymer concrete with alkaline activators. Constr Build Mater 47:409–418. https://doi.org/10.1016/j.conbuildmat.2013.05.069
15. Nath P, Sarker PK (2014) Effect of GGBFS on setting, workability and early strength properties of fly ash geopolymer concrete cured in ambient condition. Constr Build Mater 66:163–171. https://doi.org/10.1016/j.conbuildmat.2014.05.080
16. Goriparthi MR, Gunneswara Rao TD (2017) Effect of fly ash and GGBS combination on mechanical and durability properties of GPC. Adv Concr Constr 5(4):313. https://doi.org/10.12989/acc.2017.5.4.313
17. Kamal N (2016) Fly ash and GGBFS based powder-activated geopolymer binders: a viable sustainable alternative of portland cement in concrete industry. Mech Mater 103:110–122. https://doi.org/10.1016/j.mechmat.2016.09.012
18. Ivan D-L, Allouche EN, Vaidya S (2011) Mechanical properties of fly-ash-based geopolymer concrete. ACI Mater J 108(3):300. https://doi.org/10.14359/51682495
19. Xie J, Wang J, Rao R, Wang C, Fang C (2019) Effects of combined usage of GGBS and fly ash on workability and mechanical properties of alkali activated geopolymer concrete with recycled aggregate. Compos B Eng 164:179–190. https://doi.org/10.1016/j.compositesb.2018.11.067
20. Thomas RJ, Peethamparan S (2015) Alkali-activated concrete: Engineering properties and stress–strain behaviour. Constr Build Mater 93:49–56. https://doi.org/10.1016/j.conbuildmat.2015.04.039
21. Venu M, Gunneswara Rao TD (2018) An experimental investigation of the stress-strain behaviour of geopolymer concrete. Slovak J Civil Eng 26(2):30–34. https://doi.org/10.2478/sjce-2018-0011
22. IS:12089-1987 (1987) Specification for granulated slag for the manufacture of portland slag cement
23. IS 3812 (Part 1):2003 (2003) Indian standard pulverized fuel ash specification part 1 for use as Pozzolana in Cement, Cement Mortar and Concrete
24. IS 383 (2016) Indian standard coarse and fine aggregate for concrete. bureau of Indian standards
25. IS 9103:1999 (1999) Indian Standard concrete admixtures—specification
26. BS 5075-1: Concrete admixtures part 1: specification for accelerating and retarding water-reducing admixtures
27. Srikrishna TC, Rao TDG (2018) Mix design procedure for alkali-activated slag concrete using particle packing theory. J Mater Civ Eng 30(6):04018113. https://doi.org/10.1061/(ASCE)MT.1943-5533.0002296
28. IS:516-1959 (1959) Indian Standard methods of tests for strength of concrete
29. IS:5816 (1999) Indian standard splitting tensile strength of concrete method of test. Bureau of Indian Standards
30. ASTM C-293-02 (2002) Standard test method for flexural strength of concrete (using simple beam with center-point loading). International Standard Organization
31. Standards Australia (2001) Concrete structures, AS3600–2001
32. Sofi M, Van Deventer JSJ, Mendis PA, Lukey GC (2007) Engineering properties of inorganic polymer concretes (IPCs). Cem Concr Res 37(2):251–257. https://doi.org/10.1016/j.cemconres.2006.10.008
33. ACI M318-05 (2005) Building code requirements for structural concrete and commentary. American Concrete Institute
34. Code P (2005) Eurocode 2: design of concrete structures-part 1–1: general rules and rules for buildings. British Standard Institution, London
35. Nath P, Sarker PK (2017) Flexural strength and elastic modulus of ambient-cured blended low-calcium fly ash geopolymer concrete. Constr Build Mater 130:22–31. https://doi.org/10.1016/j.conbuildmat.2016.11.034

Study on Influence of Extra Water and Cement in the Development of Self-Compacting Geopolymer Concrete (SCGC)

T. Malleswari Devi and T. D. Gunneswara Rao

1 Introduction

Self-Compacting Geopolymer Concrete is one of the alternatives in place of Conventional concrete. The concept behind producing this concrete combines the benefits of geopolymerization and Self-Compacting principles. The materials used in SCGC are industrial-by-products like Fly ash, GGBFS, Rice Husk ash, Silica fume etc. by the utilization of such materials not only environmental pollution is reduced but also helps in reducing green house gases. Fly ash is rich in alumina and silica which is produced during coal burning in thermal power stations. In past, many research studies were conducted on utilization of low calcium fly ash as binder in geopolymer concrete and the outcomes are in favor of its utilization in a larger extent [1]. During the process of polymerization silica and alumina present in fly ash reacts with the alkaline solution to form alumino silicate hydrates [2]. As fly ash is a pozzolonic material, it requires more time to react and form hydrates. To obtain strength at an early stage, it is necessary to increase the temperature during curing period. The replacement of fly ash with GGBFS can reduce problem of curing at higher temperatures as GGBFS contains more amount of CaO-forming calcium silicate hydrate gel during polymerization [2]. By utilizing Fly ash and GGBFS, with alkaline solution geopolymer concrete can be produced to meet the requirements with required proportions [2]. Geopolymer concrete needs attention and workability requirements should be watched carefully. By introducing Self-Compacting properties in to geopolymerization, the concrete can be made flowable under its own weight [3]. By the addition of super plasticizer and extra water the workability properties of SCGC can

T. Malleswari Devi (✉) · T. D. Gunneswara Rao
National Institute of Technology, Warangal, Telangana, India
e-mail: malleswari@student.nitw.ac.in

T. Malleswari Devi
Chaitanya Bharathi Institute of Technology, Gandipet, Hyderabad, Telangana, India

R. K. Pancharathi et al. (eds.), *Low Carbon Materials and Technologies for a Sustainable and Resilient Infrastructure*, Lecture Notes in Civil Engineering 440,
https://doi.org/10.1007/978-981-99-7464-1_12

Table 1 Fly ash and GGBFS chemical compositions in % by mass

Chemical composition	Fly ash	GGBFS
Al_2O_3	21.35	17.92
SIO_2	63.17	34.81
SO_3	0.3	0.2
Fe_2O_3	4.2	0.7
MgO	1.20	
MgO	1.20	7.80
Na_2O	0.20	–
K_2O	1.60	–
LOI	1.00	1.41
pH	11.68	–

be improved [4]. The requirements of strength of SCGC in comparison with conventional concrete are investigated by addition of materials in addition to the normal procedures adopted for SCGC. This gives scope for the addition of small quantities of cement content to SCGC. In this study, an attempt was made to understand the fresh and hardened properties of SCGC by the inclusion of extra water and cement in the mixes.

2 Experimental Methodology

2.1 *Materials Used in SCGC Mix Proportions*

Fly ash and GGBFS together acting as binder content for preparing geopolymer concrete. Class F fly ash is procured from National thermal power plant, Ramagundam. GGBFS is procured from JSW cements, Warangal. The chemical composition of GGBFS and fly ash are shown in Table 1.

2.2 *Aggeragates*

Fine aggregate obtained from local river sand confirming Zone-II as per IS:383-2016. The sizes of fine aggregates are in the range of 4.75–0.15 mm. Coarse aggregates are obtained from crushing of granite rock. The size range of coarse aggregate is between 12.5 and 4.75 mm as in Self-Compacting concrete generally coarser materials of less size are preferred to reduce internal stresses. The physical properties of fine and coarse aggregates are shown in Table 2.

Table 2 Physical properties of fly ash and GGBFS

Type	Bulk density	Specific gravity	Water absorption (%)
Coarse aggregate	1542	2.574	0.33
Fine aggregate	1661	2.537	0.952

2.3 *Alkaline Solution*

Sodium silicate and sodium hydroxide are used as alkaline activators. Both the materials are procured from Aastra Chemicals, Chennai. Water, Sodium hydroxide, and Sodium silicate are used for preparing alkaline solution. The amounts of the above variables depend on molarity of sodium hydroxide and ratio of sodium silicate to sodium hydroxide. NaOH pellets are dissolved in distilled water and kept for 24 h so as to release the heat that is liberated in exothermic reaction. Sodium silicate is then added to sodium hydroxide to make alkaline solution and is mixed to concrete after 3–4 h.

2.4 *Super Plasticizer*

In order to obtain flowable concrete, high range Super plasticizer "Auramix 400" is used and it is procured from Fosrac, Secunderabad.

In order to develop SCGC, to satisfy the requirements of workability as per Indian Standard code and EFNARC guidelines and to achieve desired strength, Ex. water-to-cement ratio is taken. Cement used is of OPC 53 grade and is procured from local dealers. The Ex. water content is calculated as percentage of total binder content.

3 Mix Proportions

In order to develop SCGC, in the present study, four mix proportions, that is, SCGC1, SCGC2, SCGC3, SCGC4 are developed by using Fly ash and GGBFS in 50:50 proportions. The ratio of sodium silicate to sodium hydroxide is taken as 2. Alkaline solution/Binder ratio is taken as 0.55 and is maintained constant for all mixes. To suit the requirements of workability and strength properties, ratio called Ex. water/cement is added. Binder content of 400kg/m3 is taken as constant for all the mixes.

Fig. 1 L-box

3.1 SCGC Mix Batching, Casting and Testing

The mix proportions of SCGC for different mixes are measured using weigh batching. Alkaline solution is prepared using sodium hydroxide, sodium silicate, water. All the dry materials like coarse aggregate, Fine aggregate, fly ash, GGBFS, cement are added to the mixer and dry mixing of materials is done up to 2–3 mins. Alkaline solution, super plasticizer, extra water is then added to the mix and mixing is done continuously until uniform mixing is done for 2 min. The final mix thus obtained is tested for requirements of workability as per EFNARC guidelines. To characterize SCC, tests like Slump flow test, J_{ring} test, V-funnel test, L-box test are performed as shown in Figs. 1 and 2.

After performing workability tests, SCGC mix is then poured in to molds of cubical shape of size 150 × 150 × 150 mm. The cubes are removed from molds after 24 h and is kept under ambient curing conditions. The cubes are tested for Compressive strength at 3, 7, 28 days.

3.2 Mix Proportions of SCGC

Mix proportions of SCGC1, SCGC2, SCGC3, and SCGC4 are shown in Table 3.

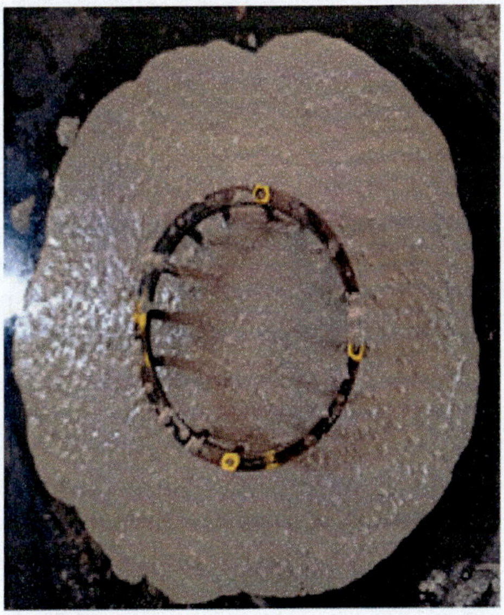

Fig. 2 Slump flow with J_{ring}

Table 3 SCGC mix proportions

Mix designation	Binder (kg/m^3)		Molarity of NaOH	Alkaline solution/binder	Coarse aggregate	Fine aggregate	Extra water (% of binder)	Cement	SP dosage (%)
	Fly ash	GGBFS							
SCGC1	200	200	6	0.55	964	788	25.5	80	9.1
SCGC2	200	200	8	0.55	964	788	29	81	8.9
SCGC3	200	200	10	0.55	964	788	29.45	81	8.2
SCGC4	200	200	12	0.55	964	788	29.60	81	8.8

4 Results and Discussions

4.1 Test Results of SCGC

In Fresh State. The SCGC mixes developed by varying Sodium hydroxide molarities with the addition of Extra water and cement are satisfying the workability requirements as per Indian standard code IS: 10262-2019. The mix proportions thus developed also satisfying EFNARC guidelines as per Table 4. The amount of extra water content varies as the molarity of Sodium hydroxide increases in alkaline solution to suit the requirements of workability criteria. In the fresh state of SCGC, it is observed that as NaOH concentration increases the diameter of slump flow increases. T500 sec values decreases as the molarity of Sodium hydroxide increases. This is because of decrease in viscosity of flow due to addition of extra water content in the mix although molarity is increasing. V-funnel flow is the parameter measuring viscosity of flow which is decreasing as the molarity of Sodium hydroxide increases. L-box value which is the ratio of H2/H1 increases with the increase in molarity of Sodium hydroxide. From the past research it is observed that as the molarity increases, the viscosity increases. To overcome this effect, Ex. water is added to the SCGC1 mix to SCGC2 mix depending on flowability of mix. From the above fresh properties of SCGC, it is clearly evident that the Viscosity of flow decreases with the addition of extra water even with increase of molarity of NaOH from 6M to 12M. The workability properties of SCGC mix proportions are shown from Figs. 3, 4, 5 and 6.

In Hardened State. From the past investigations, it is evident that increase in sodium hydroxide concentration increases Compression strength of SCGC. In the present study, cement content also added to the SCGC mixes along with increasing molarity of Sodium hydroxide. Compressive strength at 3, 7, 28 days is observed for SCGC1, SCGC2, SCGC3, SCGC4 mixes.

Testing of specimen and tested specimen are shown in Figs. 7 and 8. The compressive strength increased from SCGC1 mix to SCGC2 mix and then decreased to

Table 4 Acceptance criteria for SCC as per EFNARC guidelines

Acceptance limits	Slump flow in mm	T50cm slump flow time in sec	V-funnel flow time in sec	L-box (H2/H1) ratio	*J*-ring (mm)
Min	520	2	10	0.8	0
Max	900	5	27	1.0	10

Fig. 3 Effect of NaOH molarity on slump flow

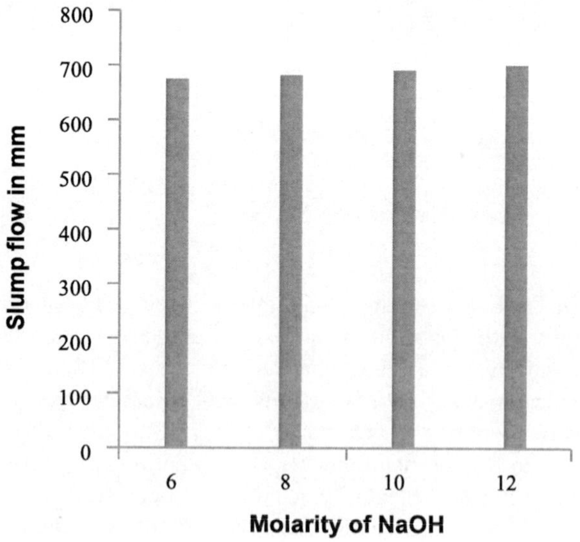

Fig. 4 Effect of NaOH molarity on T500 sec

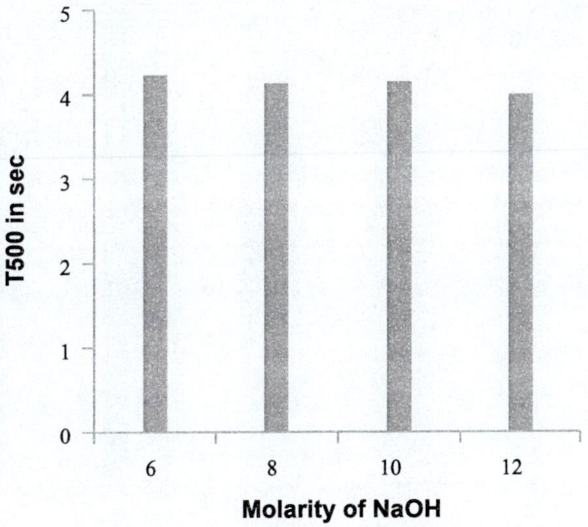

Fig. 5 Effect of NaOH on L-box ratio

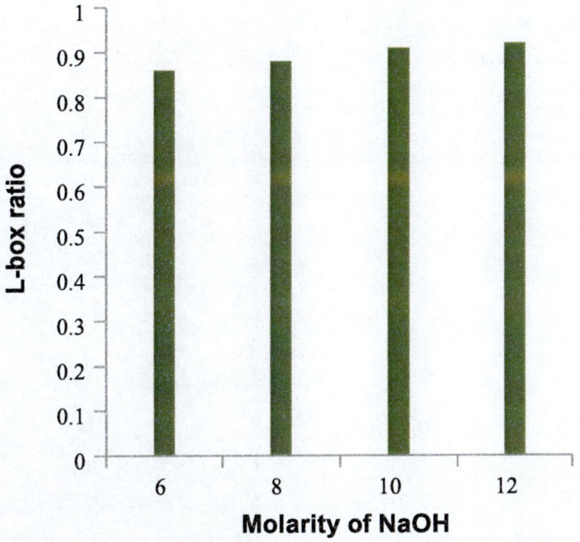

SCGC3 and further in SCGC4 compared to SCGC1 value. The variation in strength values are shown in Fig. 9.

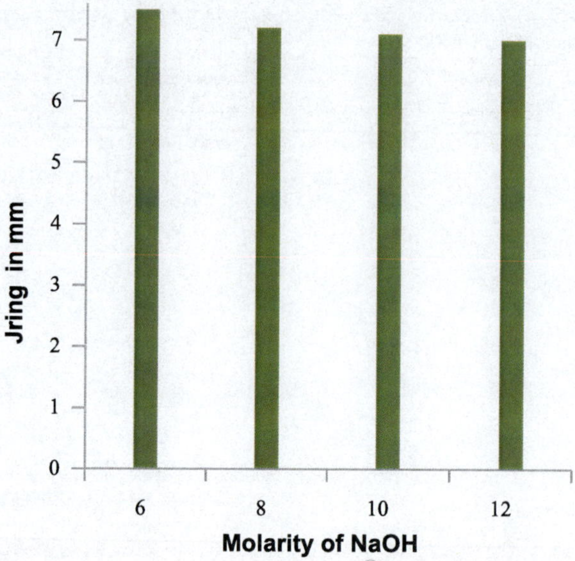

Fig. 6 Effect of NaOH molarity on J_{ring}

Fig. 7 Testing of specimens in compression

Fig. 8 Tested specimen in compression

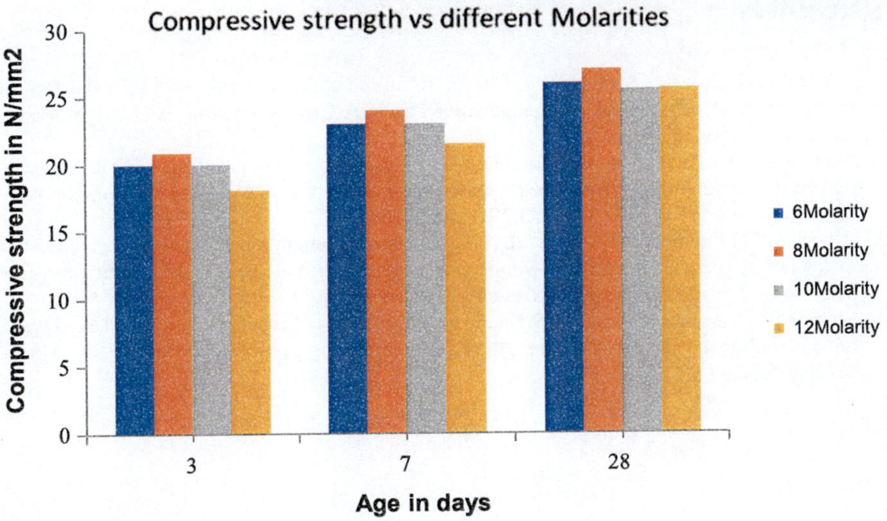

Fig. 9 Effect of NaOH molarity on of SCGC Compression strength

5 Conclusions

1. The inclusion of extra water and cement in developing SCGC helps to achieve required properties in fresh and hardened state.
2. The Slump flow is increased from SCGC1 to SCGC4 mixes as Sodium hydroxide concentration increases due to the addition of Extra water in the mixes.
3. Other workability properties like J_{ring}, T500 sec, V-funnel also matching the requirements as per Indian standard code IS:10262-2019 and EFNARC guidelines.
4. Addition of cement content in SCGC mixes increased Compression strength under ambient curing conditions.
5. The Compressive strength at 28 days is increasing from SCGC1 to SCGC2 by 2.7% and then strength decreases to 2.4% of that of SCGC1 in SCGC3. The strength of SCGC4 is decreased by 3.3% to that of SCGC1.
6. The SCGC2 mix is having higher strength compared to SCGC1, SCGC3, SCGC4. The SCGC2 mix with 8 M Sodium hydroxide with Extra water/cement ratio of 0.70. The strength of SCGC2 is 27 N/mm^2 in par with Conventional concrete by using less quantity of cement.
7. The results help in production of SCGC using industrial-by-products (80% of binder) with same strength requirements of conventional standard concrete using less amount of cement content (20% of binder) instead of using 100%OPC.

References

1. Nuruddin MF, Demie S, Shafiq N (2011) Effect of mix composition on workability and compressive strength of self-compacting geopolymer concrete. Can J Civil Eng 38(11):1196–1203. https://doi.org/10.1139/l11-07
2. Mallikarjuna Rao G, Gunneswara Rao TD (2018) A quantitative method of approach in designing the mix proportions of fly ash and GGBS-based geopolymer concrete. Aust J Civil Eng 16(1):53–63. https://doi.org/10.1080/14488353.2018.1450716
3. Ahmed MF, Nuruddin MF, Shafiq N (2011) Compressive strength and workability characteristics of low-calcium fly ash-based self-compacting geopolymer concrete. Int J Civil Environ Eng 5(2):64–70. https://doi.org/10.5281/zenodo.1330481
4. Memon FA, Nuruddin MF, Demie S, Shafiq N (2012) Effect of super plasticizer and extra water on workability and compressive strength of self-compacting geopolymer concrete. Res J Appl Sci Eng Technol 4(5):407–414

High-Temperature Behaviour of Concrete: A Review

S. Krishna Priya Rao and Tezeswi Tadepalli

1 Introduction

Concrete is chosen as the major building material, due to several benefits, including strength and durability properties. Concrete structures must satisfy the essential fire safety standards in line with building regulations [1, 2]. Fire remains the most hazardous environmental condition to which structures are exposed, implementing suitable fire protection precautions for the building's structural elements is an important element of building design [3]. In recent years, fire damages are frequently reported worldwide as a seriously threatening issue. When subjected to high temperatures, concrete undergoes structural damage through spalling and exposure of bar along with mechanical degradation of the concrete at the micro and meso levels. The density and porosity of the concrete, as well as its initial strength, moisture content, and heating rate, all impact the thermomechanical characteristics of a concrete structure when it is subjected to elevated temperatures. The range of temperatures related to building fires also has an impact on these characteristics. This chapter focuses on the thermally induced changes in concrete at elevated temperature.

2 Materials Behaviour of High-Temperature Concrete

Materials behaviour that will be studied in this chapter is supplementary cementitious materials, fibres, and aggregates. With the combination of cement gel structure, aggregate, and fibres, concrete acts as a heterogeneous material. The behaviour of the structure at various high temperatures is therefore difficult to characterize since

S. Krishna Priya Rao (✉) · T. Tadepalli
Department of Civil Engineering, NIT Warangal, Warangal 506004, India
e-mail: ks712021@student.nitw.ac.in

R. K. Pancharathi et al. (eds.), *Low Carbon Materials and Technologies for a Sustainable and Resilient Infrastructure*, Lecture Notes in Civil Engineering 440,
https://doi.org/10.1007/978-981-99-7464-1_13

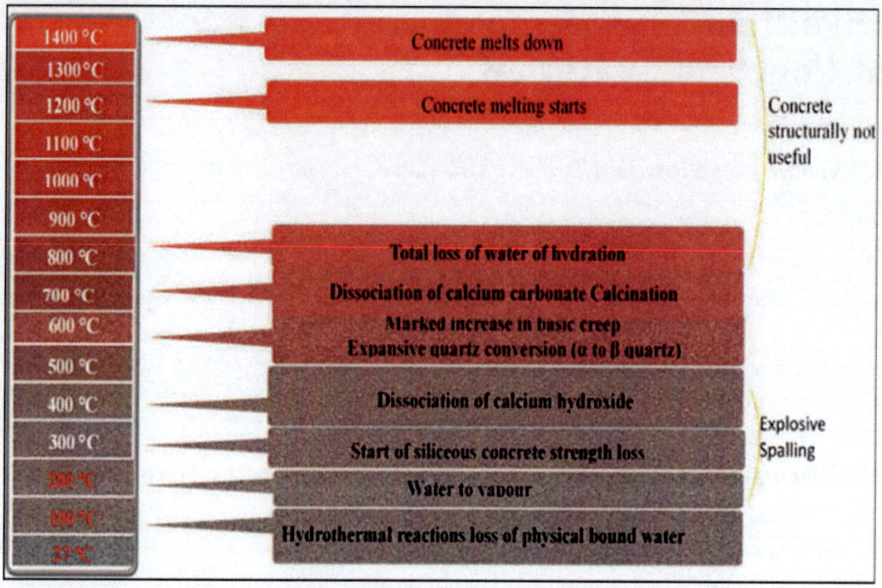

Fig. 1 Effects of increasing temperature in concrete at various stages [9]

each of these materials reacts independently when subjected to thermal conditions [4]. Several physicochemical changes take place as concrete is heated up, changing the material's thermo-mechanical characteristics [5]. In Fig. 1, effects of increasing temperature in concrete at various stages has been shown. From 20 to 80 °C, there is a gradual capillary water loss caused by the expansion of the water. Later on, from 80 to 100 °C dehydration of Ettringite and decomposition occurs where in physically bounded water in a cement matrix and aggregates evaporate, which leads to an increase in the microcracking due to capillary porosity. Concrete starts to dehydrate and decompose by losing its water and C–H–S gel at 100–200 °C [6]. Concrete strength is either maintained or increased at 300 °C as the cement gel layers become denser. However, concrete losses its strength and stiffness significantly over 300 °C [7]. From 400 to 600 °C, microcracking of cement paste takes place due to CH decomposition. In the range of 600–800 °C, it undergoes the second stage of C–H–S breakdown during which its compressive strength is substantially reduced. From 800 to 1200 °C temperature, an intense microcracking of the concrete takes place due to dehydrated phases [8].

2.1 Fibres

To avoid spalling at high temperatures, different types of fibre reinforcement based on steel, glass, carbon, basalt, jute, polyvinyl alcohol (PVA), polypropylene (PP), polyacrylonitrile (PAN), etc., are used. The highest preference, for the reduction of explosive spalling, is using hybrid-fibre reinforcement. The advantages of incorporating fibres in concrete structures improve mechanical and durability properties [10]. Studies on the influence of steel and polypropylene fibres on the mechanical characteristics of concrete after being exposed to high temperatures are shown in Table 1. In general, fibres made up of steel improve the mechanical properties of concrete when subjected to high temperatures, such as compression, flexural, and splitting tensile strength, as testing temperatures are lower than the steel fibres melting point. Hence, throughout the testing phase, ductility provides significant resistance to failure under strain [11]. Concrete reinforced with steel fibres resists thermal spalling better than concrete reinforced with polypropylene fibres [12].

Polypropylene fibre can be incorporated to enhance the material properties to resist cracking, and also improves the behaviour of concrete under tension. Polypropylene fibre has melting and ignition points around 150 °C and 400–500 °C. When polypropylene fibre-reinforced concrete is open to temperatures higher than 400 °C, the residual compression, flexural, and splitting tensile strengths decrease [13]. Accordingly, concrete reinforced with polypropylene fibres resists thermal spalling better over concrete without fibres [14]. At room temperature, ultra-high-performance concrete with the incorporation of steel and polypropylene fibre has slightly lower compressive strength than concrete with steel fibre additions, certainly as a result of the material's lower density [15]. To achieve high thermal resistance, polypropylene fibre should be used at an optimal dose of between 0.1% and 0.5%. As fibre length increases, reinforced concrete with polypropylene fibres ability to withstand high temperatures also increases [16].

2.2 *Supplementary Cementitious Materials*

Portland cement is partially replaced with low-cost, pollution-free, natural pozzolanic materials or industrial by-products, lowers the greenhouse gas emission during Portland cement production. To reduce cost, cement usage, industrial waste, and enormous CO_2 emissions, cement is largely substituted by supplemental cementitious materials [18, 19]. In order to enhance the characteristics of concrete, fly ash (FA), ground granulated blast-furnace slag (S), metakaolin (K), and silica fume (SF) are now often used as a partial substitution for cement in concrete [20]. Table 2 shows some of the supplementary cementitious materials that were partially replaced with cement when it is exposed to high temperatures.

2.3 *Behaviour of Aggregates at Elevated Temperatures*

Concrete is widely used construction material, and as a result, it has a significant effect on the environment and is essential for the construction industry [22]. At high

Table 1 Fibre proportions in concrete exposed to elevated temperatures [17]

Type of fibre	Test temperatures (°C)	Dimension of fibre	Replacement by % volume
Steel fibre	20, 15, 500	L: 35, 60 mm; D: 440, 750 μm	0, 0.5, 1
	20, 200, 400, 600, 800	L: 30 mm; D: 600 μm	0, 0.6
	20, 200, 400, 600, 800	L: 30 mm; D: 550 μm	0, 0.4
	20, 200, 400, 600, 800	L: 25 mm; D: 500 μm	0, 0.6
	20, 200, 400	L: 30 mm; D: 600 μm	0, 0.25, 0.5
	20, 400, 600, 800	L: 2 mm; D: 2000 μm	0, 1
	20, 300, 500, 800	L: 25 mm; D: 400 μm	0, 0.5, 1, 1.5, 2
	20, 600, 800	L: 25 mm; D: 42 μm	0, 1
	20, 100, 300, 500, 700	N/A	0, 0.5
	20, 600, 900	L: 12 mm; D: 50 μm	0, 1
	20, 200, 400, 600, 800	L: 32.6 mm; D: 950 μm	0, 1
	20, 350, 500, 600, 700	L: 30 mm; D: 500 μm	0, 2
Polypropylene fibre	20, 200, 400, 600, 800	L: 19 mm; D: 45 μm	0, 0.05, 0.1, 0.15
	20, 200, 400, 600, 800	L: 12 mm; D: 18 μm	0, 0.3
	20, 200, 300, 400, 800	N/A	0, 0.15, 0.2
	20, 200, 300, 400, 600, 800	L: 19 mm	0, 0.1
	20, 200, 400, 600, 800	L: 19 mm; D: 35 μm	0, 0.1
	20, 200, 400, 600, 800	L: 15 mm; D: 100 μm	0, 0.6
	20, 200, 400	L: 6, 30 mm; D: 60 μm	0, 0.25, 0.5
	20, 100, 200, 300, 600	L: 12 mm	0, 0.1, 0.2, 0.3
	20, 600, 800	L: 19 mm; D: 53 μm	0, 0.22
	20, 100, 300, 500, 700	L: 30 mm	0, 0.6
	20, 100, 450, 650	L: 12 mm; D: 18 μm	0, 0.5, 1, 1.5, 2
	20, 200, 400, 600	L: 15 mm; D: 100 μm	0, 0.5, 1
	20, 200, 400, 600, 800	L: 20 mm; D: 20 μm	0, 0.1, 0.3
	20, 600, 900	L: 12 mm; D: 50 μm	0, 0.1, 0.2, 0.3, 0.4
	20, 200, 400, 600	L: 6 mm; D: 18 μm	0, 0.1
	20, 100, 200, 300, 400, 500, 600, 700, 800, 900	L: 15 mm; D: 45 μm	0, 0.2
	20, 200, 300, 400, 500, 600, 700, 800, 900	L: 19 mm; D: 45 μm	0, 0.1, 0.2, 0.3
	ISO 834	L: 13 mm; D: 20 μm	0, 0.05, 0.1, 0.15, 0.2
	ISO 834	L: 3, 6, 12, 19, 30 mm; D: 40 μm	0, 0.05, 0.1, 0.15

Table 2 Some of the SCMs were partially replaced with cement exposed to high temperatures [21]

Supplementary cementitious materials	w/b ratio	Temperatures exposed (°C)	Replacement levels (%)
Fly ash	0.53, 0.56	20–550	0, 30
Fly ash	0.35	20–800	0, 30, 50, 70, 90
Fly ash	0.35	35–800	0, 40, 50, 60
Fly ash	0.3, 0.5	23–800	0, 25, 55
Fly ash	0.33	27–800	0, 20, 40, 60
Fly ash	0.3	20–800	0, 30, 40
Fly ash	0.77	20–800	0, 10, 20, 30
Fly ash	0.5	21–232	25
GGBS	0.45	27–350	0, 20, 40, 60
GGBS	0.41	150–700	0, 10, 30, 50
GGBS	0.35	20–800	0, 30, 50, 70, 90
GGBS	0.47	20–800	30, 50, 70
Silica fume	0.3, 0.4	20–600	0, 6, 10
Silica fume	0.3, 0.4	20–600	0, 6, 10
Silica fume	0.42, 0.45, 0.53, 0.55, 0.58	20–1000	0, 5, 10
Silica fume	0.3	20–800	0, 5, 10
Silica fume	0.77	20–800	0, 10, 20, 30
Silica fume	0.22, 0.33, 0.57	25–450	0, 10

temperatures, a thermal differentiation in cement paste and aggregate occurs, when aggregates expand by losing their thermal stability at high temperatures and cement paste shrinks. Microcracks begin to form in the ITZ and cement paste as a result, which lowers the concrete's ability to perform mechanically. As a result, the aggregate's thermal behaviour is dependent on their chemical composition, mineralogy, and petrographic origin [23, 24]. When subjected to high temperatures, concrete behaves in a way that has lower thermal strain coefficient and results in internal crystalline stresses and failure, which determines its mineralogical composition and differentiates it apart from other materials characteristics of thermal expansion [25]. When heated, the minerals go through several physical and chemical changes. Hence, there is neither thermal reaction nor weight loss, considering the absence of peaks in the DTA curves from Fig. 2 [26]. At elevated temperatures up to 600 °C, the limestone and dolomite are stable. At 700 °C, there is a decomposition of carbon aggregates into CaO and CO_2. The melting point of different types of aggregates are igneous rock is over 1000 °C, granites is over 1200–1250 °C, and basalt is over 1050 °C. This variation in melting temperature is because of the mineralogical composition [27].

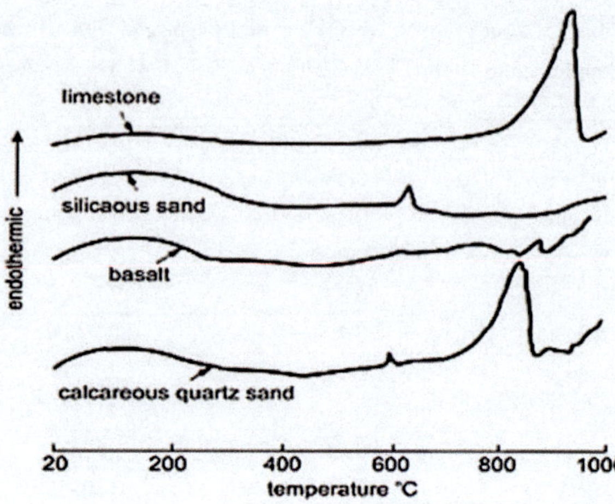

Fig. 2 Differential thermal analysis of aggregates, 10 °C/min heating rate [25]

3 Mechanical Characteristics of Concrete at High Temperature

At high temperatures, mechanical characteristics of constituent materials, like compression, flexure, tensile strength, elastic modulus, stress–strain response, and P-M interaction curve, determine whether reinforced concrete sections resist fire. At high temperatures, mechanical tests are typically executed on concrete samples which are generally cubical or cylindrical in shape of different sizes in comparison to specimen measurements at room temperature, in accordance with standards. A range of specimen sizes has been used in tests for mechanical properties since there are no established test standards for conducting high-temperature mechanical characterization [3].

3.1 Compressive Strength

At elevated temperatures, compressive strength is the main factor to consider in the design of fire resistance. The w/c ratio, ITZ phase, curing parameters, type, size of aggregates, and types of admixtures affect the compressive strength of concrete at ambient temperature [28]. Compressive strength of concrete specimens decreases at high temperatures. Removing some of the variables also affects the mechanical properties, such as mix proportions, test modalities, specimen size, stressed conditions, and hot/residual states, the data in Fig. 3 shows the residual compressive strength of unstressed cube specimens, at high temperatures [17]. The residual compressive strength can be used to distinguish between two phases despite inconsistent findings. The compressive strength decreases from ambient to 300 °C by up to 20% of its initial value. From 300 to 800 °C, the compressive strength also dramatically declines [29].

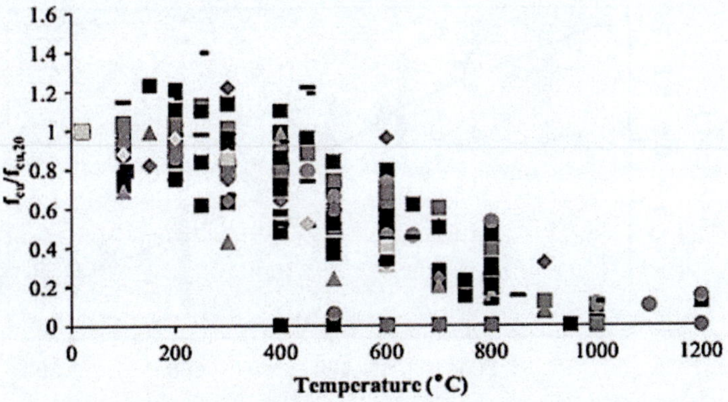

Fig. 3 At high temperatures, concrete's residual compressive strength [17]. $f_{t,20}$, Compressive strength at ambient temperature; $f_{t,\theta}$, Compressive strength at elevated temperature

3.2 *Flexural Strength*

In Fig. 4, it is clearly indicating that when concrete is at high temperatures, its flexural strength reduces. With increasing temperatures, concrete reduces its residual flexural strength over time. Depending on the replacement ratio, using cement-replacement materials also leads to a little improvement [30]. Concrete does not undergo any significant volumetric, physical, or chemical changes up to 100 °C before the free moisture evaporates. Evaporation of the free moisture occurs between 93 and 200 °C, during which there are minimal volumetric, chemical, and physical changes. The breakdown of the C–S–H and sulphoaluminate phases takes place between 200 °C and 300 °C, and surface cracking appears. Flexural strength decreases to 40–65% of its initial strength between 400 and 600 °C [31].

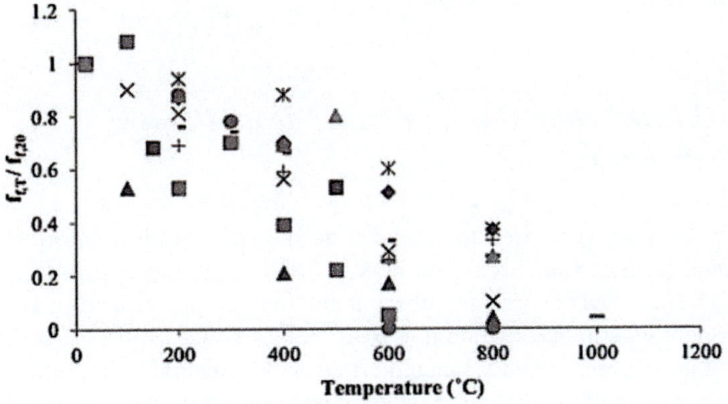

Fig. 4 At high temperatures, concrete's residual flexural strength [17]. $f_{t,20}$, Flexural strength at ambient temperature, $f_{t,\theta}$, Flexural strength at elevated temperature

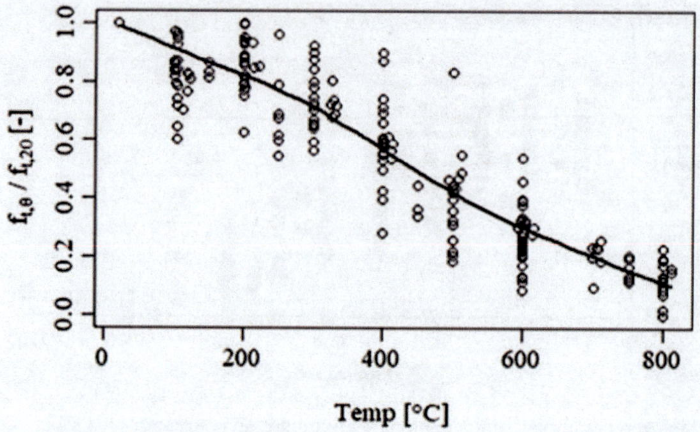

Fig. 5 Tensile strength experimental data show a slight multilinear declining tendency with increasing temperature [35]. $f_{t,20}$, Tensile strength at ambient temperature, $f_{t,\theta}$, Tensile strength at elevated temperature

3.3 *Tensile Strength*

The majority of the elements impact the tensile strength are homogeneous to those that affect its compressive strength, and also by incorporating steel fibres its tensile strength increases [3]. Tensile stresses are usually responsible for the crack's development in structure, and the formation of microcracks in structural members during a tension state damages the member [32]. Tensile strength can be particularly crucial in situations when spalling develops in concrete structures, when exposed to fire. Figure 5 shows that the tensile strength has a slight multilinear declining tendency with increasing temperature. Spalling can be decreased by having a higher tensile strength [33]. Permeability, type of fire exposure, and tensile strength are the variables that influence the spalling caused by fire [34].

3.4 *P–M Interaction Curve of Concrete at Elevated Temperature*

P–M interaction curve is used to characterize the structurally reinforced member that responds under axial forces and biaxial bending moments. To determine the strengths of reinforced concrete members at ambient temperature considers ultimate limit for concrete's compressive strain is 0.003 per ACI code [36] or Eurocode-2 [37]. Yet, several studies recommended methods that do not constrain the ultimate strain. A P-M curve is an important method of evaluating the structural performance of components made of reinforced concrete at high temperatures. Figure 6. shows the effect of fire on reinforced concrete columns. The thermo-mechanical properties and rebar at

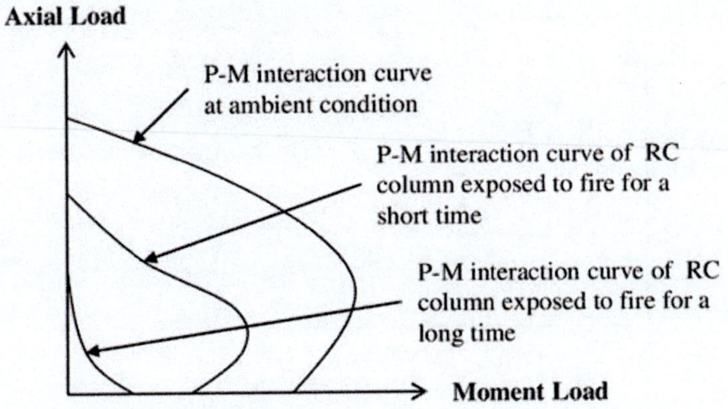

Fig. 6 Effect of fire on reinforced concrete columns [39]

high temperatures should be reflected in reinforced concrete columns fire resistance [38]. At high-temperature loads, its ultimate strain limit is less noticeable than it is when concrete is at ambient temperature. Due to reduction in strength, temperature, and creep strain, concrete that is exposed to both temperature and mechanical stresses exhibits dramatically higher strains. As a result, practical techniques like the 500 °C isothermal approach are used [2].

3.5 *Modulus of Elasticity*

Fire resistance influences the concrete's modulus of elasticity, which declines as temperature increases which as shown in Fig. 7. Exposure of concrete to high temperatures, hydrated cement products to disintegrate and bonds will break down at micro level, reducing the elastic modulus. The w/c ratio, age of concrete, conditioning technique, quantity of materials, kind of aggregates affects the modulus of elasticity [28]. The amount of moisture and microstructure of hydrated cement concrete are mainly responsible for the degradation in elastic modulus with temperature. Due to shrinkage and micro cracking caused by moisture loss, the modulus of concrete degrades up to 400 °C. Elastic modulus reduces from 400 to 750 °C as a result of the degradation of the aggregate-paste bond due to thermal variations and the early stages of $Ca(OH)_2$ and C–S–H disintegration [15]. Resistance to deformation of concrete is measured by the Modulus of elasticity, which helps to determine the development of stresses, deflections, and moments in the structural elements.

The physical characteristics of concrete deteriorate at high temperatures. Whereas, during the drying process, when the temperature rises, bond dissociation and a reduction in stiffness lead elastic modulus to decline [40]. By incorporating fibres, modulus of elasticity improves when compared to ordinary concrete. Figure 8 illustrates the residual modulus of elasticity varies for different fibre contents at high temperatures.

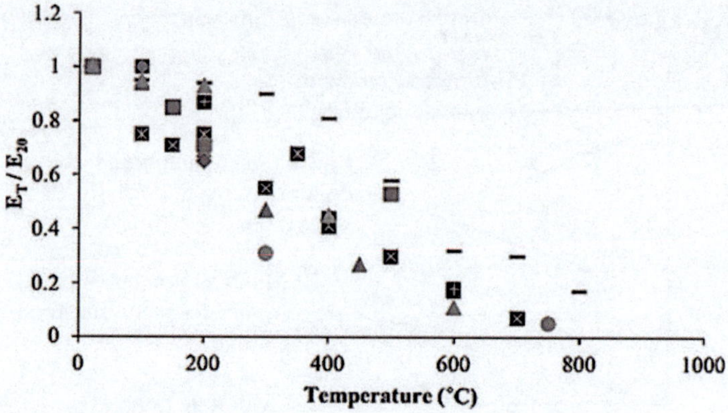

Fig. 7 At high temperatures, concrete's residual modulus of elasticity [17]

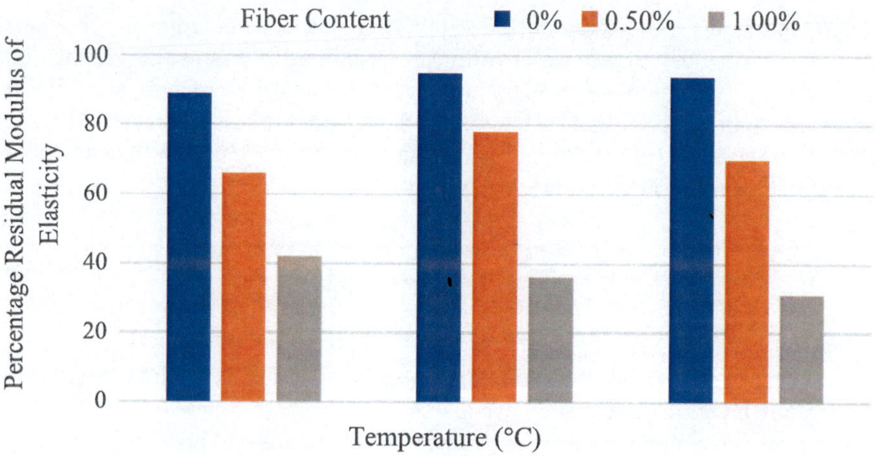

Fig. 8 At high temperatures from 0 to 600 °C, residual elasticity modulus for various fibre compositions [41]

Around 150 °C, it is noted that fibre-reinforced concrete has a minimal percentage decrease in elastic modulus. At 300–500 °C, loss in elastic modulus is increased by 40–70%. However, as the percentage of fibre is increased, the residual modulus of elasticity drastically decreases by 25%.

3.6 *Stress–Strain Response*

The reaction of mechanical characteristics is governed by the stress–strain relationship, and these properties are widely utilised as data in mathematical models [42].

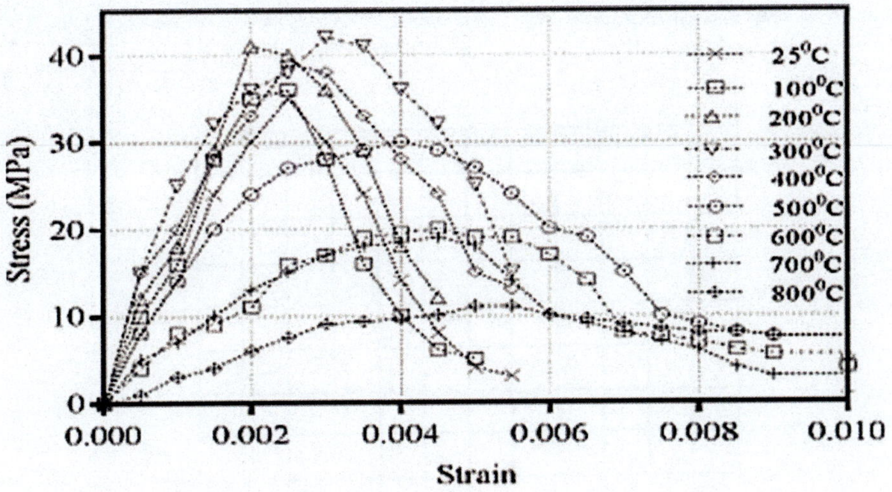

Fig. 9 At high temperatures, concrete's stress–strain response [28]

When temperature rises, material property of concrete degrades, which is a characteristic of concrete behaviour. Exposure of concrete to high temperatures, the stress–strain relationship peak, shifts downwards and rightwards and the curves become flatter, as shown in Fig. 9. These studies suggest that while concrete's peak stress and elasticity modulus decreases as temperature increases, peak strain increases.

4 Thermal Properties of Concrete at High Temperature

Thermal conductivity, specific heat, thermal diffusivity, mass loss, and spalling are among the thermal characteristics that affect concrete at high temperatures. The effectiveness of a substance that affects heat conduction is called thermal conductivity. The steady-state and transient circumstances are two distinct types of heat transfer across materials [43]. In a steady-state, the heat flow is continuous and time-independent. While in transient method, the temperature is time-dependent and fluctuates with time. To determine the thermal conductivity of concrete, transient approaches are often used over steady-state [44]. The thermal conductivity of typical concrete varies from 1.4 to 3.6 W/m °C at ambient temperature [45]. Due to concretes high specific heat, which is the quantity of heat per unit mass, a structure's temperature stability can be increased. Concrete density, type of aggregate, and amount of moisture content have the most impacts on specific heat. At elevated temperatures, differential thermal analyser (DTA) is often operated to assess specific heat at temperatures of 600 °C and above [46, 47]. The severity of spalling is decreased in the temperature range of 20–750 °C by adding polypropylene fibres to ultra-high-performance concrete without significantly altering the mechanical properties [15].

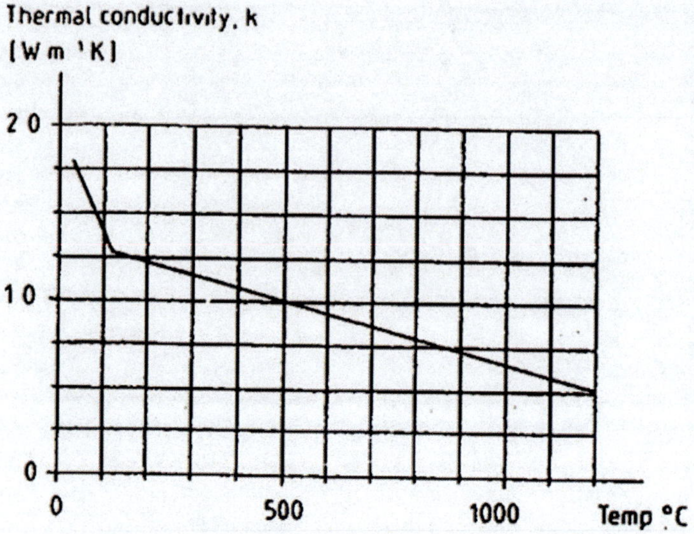

Fig. 10 Thermal conductivity versus temperature of normal weight concrete as per ASTM STP 882 [50]

4.1 Thermal Conductivity

When it comes to concrete, the thermal conductivity value varies. It generally differs with some of the factors like mix proportion, type of aggregates, moisture content, permeability, and density of concrete. The thermal conductivity increases along with concrete density, moisture content, and temperature [48]. Thermal conductivity for fibre-reinforced concretes has a comparable pattern to that of normal concrete when steel and polypropylene fibres are added. Therefore, the thermal conductivity is unaffected by fibres in the temperature ranging from 20 to 800 °C [33]. By incorporating various waste materials, such as recycled aggregates and building and demolition trash that may also contain ceramics, glass, and other materials, there are several elements that might affect the thermal conductivity of concrete [49]. Concrete's thermal conductivity drastically changes at 120 °C temperature, according to ASTM STP 882 standard curve given in below Fig. 10.

4.2 Specific Heat

Concrete undergoes various physical and chemical changes at high temperatures, and specific heat is considered in the form of thermal capacity. With an increasing w/c ratio, specific heat increases dramatically and is mostly influenced by moisture content [3]. According to Eurocode-2, Fig. 11 shows how specific heat capacity varies with temperature. At temperatures below 200 °C, moisture content has a significant

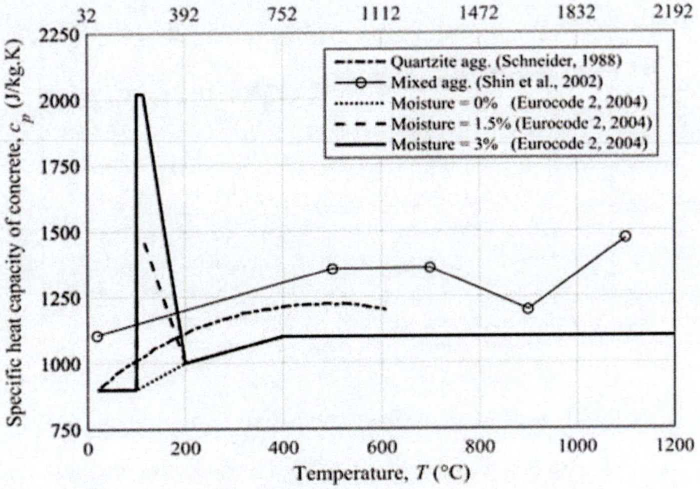

Fig. 11 Variations in concrete's specific heat capacity at high temperatures [51]

effect. When temperatures are lower than 800 °C, the aggregate type has negligible effect on heat capacity. Concrete specific heat capacity with moisture contents of 1.5% and 3%, respectively, increases by a factor of 2.2 and 1.6 at 100 °C, according to Eurocode 2.

4.3 Mass Loss

Concrete loses moisture when exposed to high temperatures, which causes a reduction in density. Concrete is often divided into normal and lightweight concrete based on its density. Moreover, the aggregate type has a major impact on the mass of concrete at elevated temperature [52]. The difference in concrete mass made of siliceous and carbonate aggregates up to 1000 °C temperatures is seen in Fig. 12. Up to approximately 600 °C, neither siliceous nor carbonate aggregate concretes experience significant mass loss [53].

4.4 Spalling of Concrete

The development of high pore pressure, thermal stresses, and a combination of both plays significant roles in the explosive spalling of concrete at high temperatures [54]. Due to the development of high pore pressure, the bound water in cement paste and the water in capillary pores in concrete both changed from liquid to vapour. A temperature gradient is created by heating the surface of concrete pushing moisture both out of the

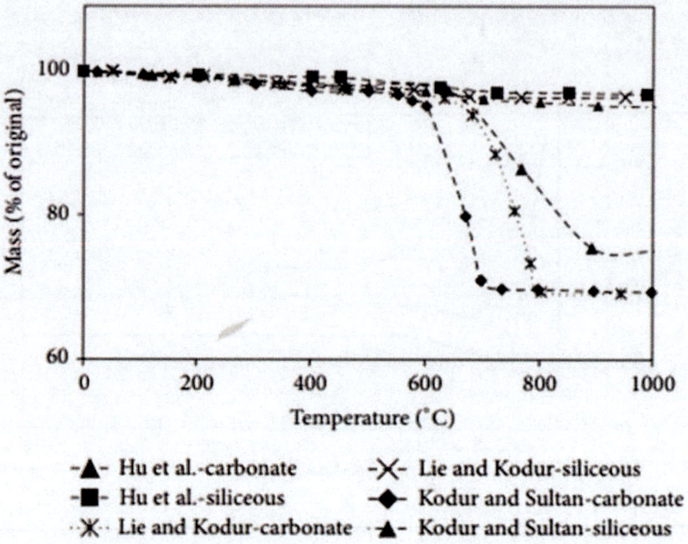

Fig. 12 Variations in concrete mass using various types of aggregates at high temperatures [3]

surface and into the inside. Depending on the permeability of the concrete, this causes pore pressures to increase to their highest level away from the surface and causes explosive spalling [55]. The inside of the structure will experience thermal strains as a result of temperature differences from the heated surface to the inner core areas of the concrete. As the heating rate picks up, these gradients will increase. Depending on the mechanical and thermal characteristics of the concrete, various strains caused by the temperature gradient will result in tensile and compressive stresses [54]. Figure 13 shows the explosive spalling of concrete caused by thermal stress as well as pore pressure.

5 Microstructural Properties

5.1 SEM Analysis

Concrete mix is exposed to different loads and high-temperature circumstances, the Scanning Electron Microscope (SEM) permits analysis of the mechanism of concrete mix's micro-cracks at micro-scale. Generally, crushed samples are often examined to determine the microstructural characteristics and the interface between the cement matrix and aggregates of the specimens. The concrete structure is internally denser as C–H–S gel is block shaped at ambient temperature. As concrete is exposed to 500–750 °C, there is an increase in the pore size as well as cracks in the matrix

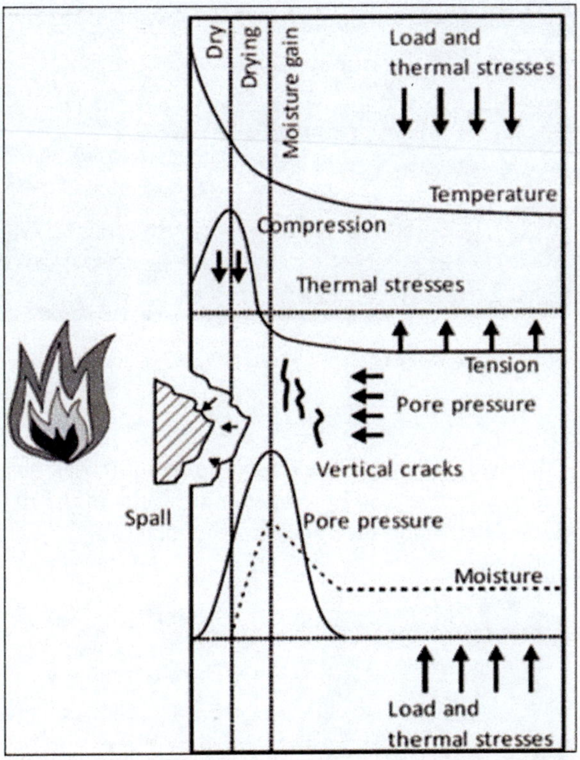

Fig. 13 Explosive spalling caused by thermal stress and pore pressure [54]

which leads to the deterioration of concrete due to larger pores and the formation of cracks at higher temperatures [56–58].

The microstructure characteristics of concrete are slightly enhanced by the inclusion of fibres. As shown in Fig. 14, steel fibre diameter changes with temperature as a result of massive amounts of water vapour building up in the porosity next to the steel fibre and matrix, which also causes rust and oxidation [59]. The polypropylene fibres loose their solid structure at 200 °C and create additional pores in the matrix [13].

5.2 *Porosity Analysis of Concrete*

When concrete is subjected to high temperatures, one method for determining the porosity and pore network at the meso and micro scale level is High-Resolution Computed Tomography (HRCT) and micro tomography. Figure 15 shows the micro tomography of concrete samples at different elevated temperatures i.e., at 27, 100, 400, 500, and 800 °C. With an increase in the temperature up to 800 °C, at the meso scale level, the porosity increases up to 800%, as observed by HRCT. At the micro

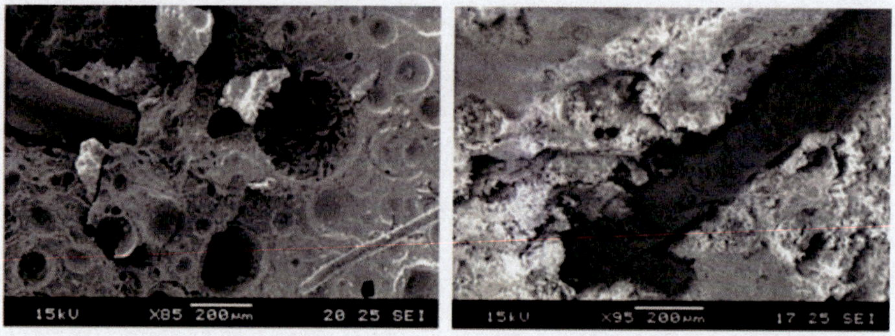

Fig.14 SEM image of steel-fibre-reinforced concrete that has been exposed to high temperatures [59]

scale level, porosity is increased up to 1400% as observed by Micro Tomography. Hence, a further rise in temperature results in the development of a pore network, which deteriorates the concrete specimen [60].

Fig. 15 Micro tomography samples at different elevated temperatures. **a** original sample specimen, **b** 27 °C, **c** 100 °C, **d** 400 °C, **e** 500 °C, **f** 800 °C [60]

6 Research Gaps

1. Most of the studies on mechanical and thermal properties of concrete consider temperatures up to 800 °C only, which is less than the temperature of actual fires.
2. There are limited studies on structure evolution at meso, micro, and nano scale levels when exposed to high temperatures beyond 1000 °C.
3. There is a lack of studies on non-destructive techniques like imaging-based analysis of concrete at elevated temperatures.
4. There are very few studies on thermal analysis like Thermal gravimetric analysis (TGA), Differential Thermal Analysis (DTA), and Dynamic Mechanical Analysis (DMA) of cementitious materials.

7 Conclusions/Summary

1. As the temperature increases, a general decrease is observed in the mechanical characteristics of concrete such as—compressive, flexural and tensile strength, modulus of elasticity and stress–strain response. A noticeable decrease in the strength of concrete is observed after 400 °C.
2. Spalling and microcracking in concrete are reduced with the inclusion of a small amount of steel and polypropylene fibres, which enhances mechanical and durability properties.
3. Mechanical and durability properties are enhanced by incorporating supplementary cementitious materials. Because the composite has smaller average particle size than cement, it enhances the microstructure of the concrete specimen by reducing porosity.
4. A variety of behaviour is observed in constituent materials at elevated temperatures, which depends upon the chemical composition, mineralogy, and crystallographic arrangement of aggregates.
5. The mix proportion, aggregate type, moisture content, permeability, and density of concrete are some of the parameters that impact thermal conductivity. Moisture content has a major effect on specific heat.
6. Thermal conductivity reduces with increasing temperature.
7. Due to loss of moisture content, concrete loses density (or mass) at elevated temperatures. Beyond 600 C temperature, a considerable mass loss occurs in concrete.

References

1. Joint ACI/TMS Committee 216 (2012) Masonry Society (U.S.) Code requirements for determining fire resistance of concrete and masonry construction assemblies: an ACI/TMS Standard
2. EN 1992-1-2 (1992) Eurocode 2: design of concrete structures—Part 1-2: general rules—structural fire design
3. Kodur V (2014) Properties of concrete at elevated temperatures. ISRN Civil Eng 2014:1–15
4. Khoury GA (2000) Effect of fire on concrete and concrete structures. Prog Struct Eng Mater 2(4):429–447. https://doi.org/10.1002/pse.51
5. Pimienta P, Alonso MC, McNamee RJ, Mindeguia JC (2017) Behaviour of high-performance concrete at high temperatures: some highlights. RILEM Tech Lett 2:45–52. https://doi.org/10.21809/rilemtechlett.2017.53
6. MC Alonso, U Schneider (2019) Degradation reactions in concretes exposed to high temperatures. In: Physical Properties and Behaviour of High-Performance Concrete at High Temperature, Springer, pp 5
7. Malik M, Bhattacharyya SK, Barai SV (2021) Thermal and mechanical properties of concrete and its constituents at elevated temperatures: a review. Constr Build Mater 270:121398
8. Naus DJ (2005) The effect of elevated temperature on concrete materials and structures—a literature review. Technical Report
9. Pavani HP, Tezeswi TP (2020) Micro-structure based multi-scale thermo-mechanical characterization of portland cements. Doctor of Philosophy in Civil Engineering
10. Fort J, Vejmelkova E, Pavlíkova M (2016) High-temperature testing of high-performance fiber reinforced concrete. In: AIP Conference Proceedings. American Institute of Physics Inc
11. Zheng W, Li H, Wang Y (2012) Compressive stress-strain relationship of steel fiber reinforced reactive powder concrete after exposure to elevated temperatures. Constr Build Mater 35:931–940. https://doi.org/10.1016/j.conbuildmat.2012.05.031
12. Chen B, Liu J (2004) Residual strength of hybrid-fiber-reinforced high-strength concrete after exposure to high temperatures. Cem Concr Res 34:1065–1069. https://doi.org/10.1016/j.cemconres.2003.11.010
13. Noumowe A (2005) Mechanical properties and microstructure of high strength concrete containing polypropylene fibres exposed to temperatures up to 200 °C. Cem Concr Res 35:2192–2198. https://doi.org/10.1016/j.cemconres.2005.03.007
14. Kim YS, Lee TG (2013) An experimental study on the residual mechanical properties of fiber reinforced concrete with high temperature and load. Mater Struct 46:607–620. https://doi.org/10.1617/s11527-012-9918-y
15. Banerji S, Kodur V (2022) Effect of temperature on mechanical properties of ultra-high-performance concrete. Fire Mater 46:287–301. https://doi.org/10.1002/fam.2979
16. Gao D, Yan D (2012) Li X (2012), Splitting strength of GGBFS concrete incorporating with steel fiber and polypropylene fiber after exposure to elevated temperatures. Fire Saf J 54:67–73
17. Ma Q, Guo R, Zhao Z, Lin Z, He K (2015) Mechanical properties of concrete at high temperature—a review. Constr Build Mater 93:371–383
18. Li X, Snellings R, Antoni M (2018) Reactivity tests for supplementary cementitious materials: RILEM TC 267-TRM phase 1. https://doi.org/10.1617/s11527-018-1269-x
19. Paul SC, Mbewe PBK, Kong SY, Savija B (2019) Agricultural solid waste as source of supplementary cementitious materials in developing countries. Materials 12:1112
20. Toutanji H, Delatte N, Aggoun S, Duval R, Danson A (2004) Effect of supplementary cementitious materials on the compressive strength and durability of short-term cured concrete. Cem Concr Res 34:311–319. https://doi.org/10.1016/j.cemconres.2003.08.017
21. Ramzi S, Hajiloo H (2022) The effects of supplementary cementitious materials (SCMs) on the residual mechanical properties of concrete after exposure to high temperatures—review. Buildings 13:103. https://doi.org/10.3390/buildings13010103

22. Habert G, Miller SA, John VM, Provis J, Favier A, Horvath A, Scrivener K (2020) Environmental impacts and decarbonization strategies in the cement and concrete industries. Nat Rev Earth Environ 1:559–573
23. Fib Bulletin 46 (2008) Fire design of concrete structures—structural behaviour and assessment. Technical Report, Lausanne, Swiss
24. Niry Razafinjato R, Beaucour AL, Hebert RL, Ledesert B, Bodet R, Noumowe A (2016) High temperature behaviour of a wide petrographic range of siliceous and calcareous aggregates for concretes. Constr Build Mater 123(2016):261–273. https://doi.org/10.1016/j.conbuildmat.2016.06.097
25. Hager I (2013) Behaviour of cement concrete at high temperature. Bull Pol Acad Sci Tech Sci 61:145–154. https://doi.org/10.2478/bpasts-2013-0013
26. Khoury GA (1992) Compressive strength of concrete at high temperatures: a reassessment. Mag Concr Res 44:291–309. https://doi.org/10.1680/macr.1992.44.161.291
27. Hager I, Tracz T, Sliwinski J, Krzemien K (2016) The influence of aggregate type on the physical and mechanical properties of high-performance concrete subjected to high temperature. Fire Mater 40:668–682. https://doi.org/10.1002/fam.2318
28. Anupama Krishna D, Priyadarsini RS, Narayanan S (2019) Effect of elevated temperatures on the mechanical properties of concrete. In: Procedia Structural Integrity. Elsevier B.V., pp 384–394
29. Fernandes B, Carré H, Mindeguia JC, Perlot C, La Borderie C (2021) Effect of elevated temperatures on concrete made with recycled concrete aggregates—an overview. J Build Eng 44:103235
30. Abed M, Nemes R, Lubloy E (2020) Performance of self-compacting high-performance concrete produced with waste materials after exposure to elevated temperature. J Mater Civ Eng. https://doi.org/10.1061/(asce)mt.1943-5533.0002989
31. Azree M, Mydin O, Roosli R (2012) Prediction of elevated temperature flexural strength of lightweight foamed concrete strengthened with polypropylene fibre and fly ash
32. Chan YN, Peng GF, Anson M (1999) Residual strength and pore structure of high strength concrete and normal strength concrete after exposure to high temperatures. Cement Concr Compos 21:23–27
33. Khaliq W, Kodur V (2012) High-temperature mechanical properties of high strength fly ash concrete with and without fibers. ACI Mater J 109(2012):665–674
34. Ali F, Nadjai A, Silcock G, Abu-Tair A (2004) Outcomes of a major research on fire resistance of concrete columns. Fire Saf J 39:433–445
35. Van der Merwe JE (2022) Evaluation of concrete tensile strength as a function of temperature. Constr Build Mater. https://doi.org/10.1016/j.conbuildmat.2022.127179
36. Building Code Requirements for Structural Concrete (ACI 318–14) (2014) Commentary on building code requirements for structural concrete (ACI 318R-14): an ACI Standard and Report from HIS
37. EN 1992-1-1 (2004) Eurocode 2: design of concrete structures—Part 1–1: general rules and rules for buildings
38. Kim HS (2021) Obtaining interaction diagram of concrete section exposed to fire by mathematical optimization. J Build Eng. https://doi.org/10.1016/j.jobe.2021.102993
39. Kang H, Cheon NR, Lee DH, Lee J, Kim KS, Kim HY (2017) P-M interaction curve for reinforced concrete columns exposed to elevated temperature. Comput Concr 19:537–544. https://doi.org/10.12989/cac.2017.19.5.537
40. Bernhart D (2004) The effect of support condition on the fire resistance of a reinforced concrete beam. DP. Sc. Thesis, Department of Civil Engineering, University of Canterbury, New Zealand. https://doi.org/10.26021/2919
41. Shallal MA, Sallal Rashid Al-Owaisy L (2007) Strength and elasticity of steel fiber reinforced concrete at high temperatures shear of concrete beams view project-technical report non-linear analysis of continuous composite beam subjected to fire view project. J Eng Dev 11:125–133
42. Abid M, Hou X, Zheng W, Hussain RR (2017) High temperature and residual properties of reactive powder concrete—a review. Constr Build Mater 147:339–351

43. ASTM E1530 (2006) Standard test method for evaluating the resistance to thermal transmission of materials by the guarded heat flow meter technique. ASTM International, West Conshohocken
44. Adl-Zarrabi B, Bostrom L, Wickstrom U (2006) Using the TPS method for determining the thermal properties of concrete and wood at elevated temperature. Fire Mater 30:359–369. https://doi.org/10.1002/fam.915
45. Bazant ZP, Zdenek P, Kaplan MF, Maurice F (1996) Concrete at high temperatures: material properties and mathematical models. Longman
46. Phan LT, Carino NJ (2000) Fire performance of high strength concrete: research needs
47. Howlader MK, Rashid MH, Mallick D, Haque T (2012) Effects of aggregate types on thermal properties of concrete assessment of heritage stone and brick masonry buildings under earthquake actions view project simulation of heritage structures view project
48. Kim KH, Jeon SE, Kim JK, Yang S (2003) An experimental study on thermal conductivity of concrete. Cem Concr Res 33:363–371
49. Misri Z, Ibrahim MHW, Awal ASMA, Desa MSM, Ghadzali NS (2018), Review on factors influencing thermal conductivity of concrete incorporating various type of waste materials. In: IOP Conference Series: Earth and Environmental Science. Institute of Physics Publishing
50. Ulf W (1985) Application of standard Fire curve for expressing natural fires for design purpose. Fire Saf Sci Eng ASTM STP 882
51. Deshpande AA (2019) Effects of elevated temperatures on concrete materials view project a multiscale study of concrete subjected to elevated temperatures. https://doi.org/10.13140/RG.2.2.24685.64489
52. Lie TT, Kodur VKR (2011) Mechanical properties of fibre-reinforced concrete at elevated temperatures
53. DiNenno PJ, Drysdale Dougal, Beyler CL (2002) SFPE handbook of fire protection engineering. National Fire Protection Association
54. So H-S (2016) Spalling prevention of high-performance concrete at high temperatures. In: High-Performance Concrete Technology and Applications. InTech
55. Stauder J (2017) Fire performance of ultra high-performance fibres reinforced concrete beams and influence of polypropylene fibres on explosive spalling. https://doi.org/10.13140/RG.2.1.3020.6321
56. Akca AH, Zihnioglu NO (2013) High-performance concrete under elevated temperatures. Constr Build Mater 44:317–328. https://doi.org/10.1016/j.conbuildmat.2013.03.005
57. Demirel B, Kelestemur O (2010) Effect of elevated temperature on the mechanical properties of concrete produced with finely ground pumice and silica fume. Fire Saf J 45:385–391. https://doi.org/10.1016/j.firesaf.2010.08.002
58. Arioz O (2007) Effects of elevated temperatures on properties of concrete. Fire Saf J 42:516–522. https://doi.org/10.1016/j.firesaf.2007.01.003
59. Zhang P, Kang L, Wang J, Guo J, Hu S, Ling Y (2020) Mechanical properties and explosive spalling behavior of steel-fiber-reinforced concrete exposed to high temperature—a review. Appl Sci (Switzerland) 10:2324
60. Praneeth PH, Tadepalli TP, Agrawal AK (2020) Characterisation of micro- and meso porosity in Portland cement at elevated temperatures. Mag Concr Res 72:304–313. https://doi.org/10.1680/jmacr.18.00321

Studies on Performance and Micro Structural Characteristics of Self-healing Concrete

D. J. Arpitha, A. R. Chandrashekar, and Kannam Praveen

1 Introduction

1.1 Self-healing Concrete

In recent years, a lot of advanced research is underway to heal cracks in concrete, extend concrete life, and avoid routine maintenance and concrete maintenance costs. One of the most important research projects is self- healing concrete [1].

Self-healing concrete is a type of concrete that, unlike ordinary concrete, automatically repairs or fills cracks that occur in the concrete. There are various methods of self-repair processes, mainly of two types known as intrinsic and extrinsic methods. In an inherent self-healing process, concrete or mortar repairs cracks without external additives. The presence of CaO in the material causes hydration and calcite precipitation to heal the cracks themselves [2].

Extrinsic healing process is something it requires an external enzyme to react with it and heal the cracks so a bacterium is induced into the mortar or concrete by encapsulation and direct methods. As shown in Fig. 1, the bacteria induced along with its food in the form of starch which when reacted with the water seeping from the cracks activates the dormant bacteria and chemical reaction takes place and formation of calcium carbonate takes place which fills up the cracks [3].

Selection of bacteria is the most important, not all the bacteria can be used as self-healing agents the selected bacteria must have the capabilities to survive in the worst environment and for long time as well. Mainly bacillus family bacteria are used such as *Bacillus megaterium*, *Bacillus subtilis*, *Bacillus Sphaericus*, etc.; these

D. J. Arpitha · A. R. Chandrashekar (✉)
Assistant Professor, BIET, Davanagere, India
e-mail: R21PCV04@reva.edu.in

A. R. Chandrashekar · K. Praveen
Associate Professor, REVA University, Bangalore, Karnataka, India

R. K. Pancharathi et al. (eds.), *Low Carbon Materials and Technologies for a Sustainable and Resilient Infrastructure*, Lecture Notes in Civil Engineering 440,
https://doi.org/10.1007/978-981-99-7464-1_14

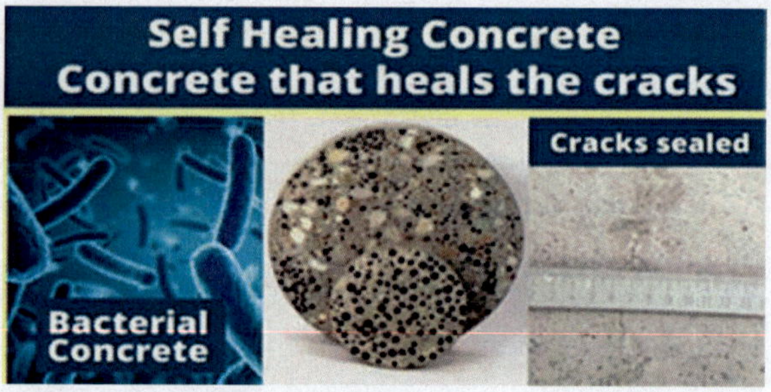

Fig. 1 Bacterial concrete

are the bacteria which can survive in extreme condition and good for self-healing process.

1.2 *Process of Self-healing*

Self-healing process, concrete is familiar for healing its own cracks through hydration of cement as cement contains calcium oxide that reacts with water in the presence of carbon dioxide leads to formation of calcium carbonate which will help in healing the cracks however not completely [4].

$$Ca\ (OH)_2 + CO_2 \rightarrow CaCO_3 + H_2O$$

There are two methods of inserting bacteria in concrete one is direct method and encapsulation.

- In this research in direct method the bacteria directly inserted during mixing itself, in encapsulation method.
- The bacteria induced in the form of clay pallets or capsules which is costly than direct method.

Bacillus subtilis bacteria of 109 population of 0%, 3%, 5% is induced in M30 and M70 mix of concrete using direct method, the mechanical properties such as compressive strength, tensile strength, flexural strength of conventional concrete and bacterial concrete are compared to see the effect of bacteria on the mechanical properties of concrete and the micro structural analysis is also carried out by scanning electron microscopy (SEM) analysis where we can find the surface's topography and composition of the sample and X-Ray Diffraction (XRD) test where we can find crystalline variants and to study particle size of nanomaterial's.

1.3 The Advantages of Self-healing Concrete

- The mechanical properties of the concrete such as compressive strength, tensile strength, flexural strength increases.
- The maintenance and cost of maintenance of the cement concrete decreases.
- It can last till centuries or decades so no need to replace during life time.

1.4 The Disadvantages of Self-healing Concrete

- The initial cost of construction using bacterial will be more.
- Required skill labourers for this work.
- Still there is no particular code to standardize the self-healing concrete.

2 Materials and Methodology

2.1 General

In this chapter the various materials and methodology used in the experimental programme are described in the below paragraphs.

Cement. Cement is a binder material used for construction that mixed with aggregates and water forms concrete. Concrete produced from Portland cement is one of the most versatile construction materials available in the world [5, 6].

For high strength concrete, OPC 53 grade is used. The min. compressive strength of OPC 53 Grade Cement should not be less than 53 N/mm^2. For the concrete of grade M-30 and above 8–10% can be saved by using OPC 53. Figure 2 represents the cement sample taken for the experimentation works.

Fig. 2 OPC 53 grade

Fig. 3 Fly-ash

Fly-ash. Fly-ash also known as 'Pulverized Fuel Ash' is waste product comes out after combustion of coal, generally we get fly ash from electrostatic precipitators. By using fly ash shown in Fig. 3 as the replacement of cement helps in reducing environmental pollution. Fly-ash can significantly improve the workability of concrete.

Silicafumes. Silica fume also known as condensed silica fume or micro silica is very fine, is produced by the by products of elemental silicon or silica-alloys. The specific of silica fumes is from 2.2 to 2.3.

Silica fume is one of best replacement for cement which increases the mechanical strength of concrete and also decreases the environmental pollution.

2.2 Methodology

- Calculation of mix design for M30 and M70.
- Concentration of bacteria (population) used = CFU (109).
- Percentage (dosage) of bacteria used = (0%, 3%, 5%).

Tests Concrete

Fresh properties of Concrete. When calculating the mix design, we first mix the ingredients according to the mix design and check the fresh physical properties such as workability, consistency, settlement, bleeding, etc. There are many possible checks such as set cone test, compaction factor, V-Bee test.

Slump cone test. After the concrete has been mixed, the wet concrete is poured into the hardening cone 25 times in three layers, after filling the cone is lifted slightly, the cone is placed upside down next to it and the hardening value is measured. According to IS 1199-1959, M30 concrete has zero slump and M70 concrete has true slump.

Tests on Hardened Concrete. According to IS:516-1959 (2004), the compressive strength test was performed on concrete cube specimens measuring 150 mm. After 7 and 28 days of curing, the cubes of each mix proportion (trial mix) were evaluated in a compression testing machine with a 2000 kN capacity. According to IS:5816-1999 (1999), the split tensile strength test was performed to determine the tensile strength of concrete cylinders with dimensions of 150 mm in diameter and 300 mm in length. The concrete beams used for the flexural strength test had dimensions of 500 mm in length, 150 mm in width, and height by applying a two-point loading to the concrete beam in accordance with the specifications of IS:516-1959 (2004) in a flexural strength testing equipment with a 500 kN capacity. On specimens that had undergone a 28-day water cure, split tensile and flexural strength tests were performed.

Durability Test. To access the performance of hardened concrete following test were carried out.

- Durability test (28 days).
- Acid attack test (by immersing in 5% of H_2SO_4 solutions for 28 days).
- Micro structural analysis.
- Scanning electron microscope (SEM).
- X-ray diffraction (XRD).

Mix Design of Concrete (As Per IS 10262:2019). Mix Design was carried out using IS 10262:2019. After number trails final mix proportions were determined [7]. The mix design for M30 and M70 are presented in the tables. Table 1 shows the mix proportion for M30 and Table 2 shows the mix proportion for M70.

Table 1 Mix design for M30

Cement (kg/m^3)	Fine aggregates (kg/m^3)	Coarse aggregates (kg/m^3)	Water (kg/m^3)	Water/cement ratio
413	706	1103	186	0.45

Table 2 Mix design for M70

Cement (kg/m^3)	Fine aggregates (kg/m^3)	Coarse aggregates (kg/m^3)	Water (kg/m^3)	Chemical admixture (kg/m^3)	Water–Binder Ratio	Fly ash (kg/m^3)	Silica fumes (kg/m^3)
428	589	1219	141	2.67	0.26	80.25	26.75

3 Result and Discussion

3.1 General

In this chapter, the experimental results were discussed. After casting and curing of concrete specimens. The hardened moulds were tested for mechanical, durability and micro structural properties.

3.2 Mechanical Properties

The performance of concrete is calculated by its mechanical properties it includes fresh properties and hardened properties.

Harden properties of the concrete viz., compressive strength, tensile strength, flexural strength can be calculated after curing the specimens for 28 in water and kept outside by that time specimens get harder and can go with tests of hardened properties [8].

Compressive strength: The compressive strength of concrete is calculated by casting M30 and M70 with and without bacteria concrete cubes of 150 mm × 150 mm × 150 mm and cured for 28 days [9], it is to calculate the maximum load which the specimen can bear, based on the literature-review, dosage of the bacteria to be added is calculated and the compressive strength results are compared. In this study the conventional concrete M30 and M70 are casted and M30 and M70 with 3% of *Bacillus subtilis* bacteria (which was calculated by weight of water) and M30 and M70 with 5% of *Bacillus subtilis* bacteria (which was calculated by weight of water), three each cube of M30, M70 with and without bacteria where casted for compressive strength and compared the conventional and bacterial concrete results. By comparing the results as shown in Figs. 4 and 5, the compressive strength is increased by 15%, 20% for M30 (3%) and M30 (5%), respectively, at 28 days, and compressive strength is increased by 15%, 20% for M70 (3%) and M70 (5%), respectively, at 28 days [10].

Tensile strength: Tensile strength of concrete refers to casting a cylindrical mould with a diameter of 150 mm and a length 300 mm, curing with Bacillus subtilis for 28 days (calculated by water weight), M30, M70 with 5% Bacillus subtilis (water weight), each three cylinders M30, compared the split strength of M70 with and without bacterial injection, and compared the results for conventional concrete and bacterial concrete. By comparing the results as shown in Figs. 6 and 7, after 28 days, M30 (3%) and M30 (5%) showed 14% and 18% increases in tensile strength, and M70 (3%) showed 14% and 18% increases in compressive strength increase [11].

Flexural strength: The concrete flexural strength is a mechanical property which can be measured by casting a beam of 100 mm × 100 mm × 500mm length are casted and cured for 28 days In this study the conventional concrete M30 and M70 are casted and

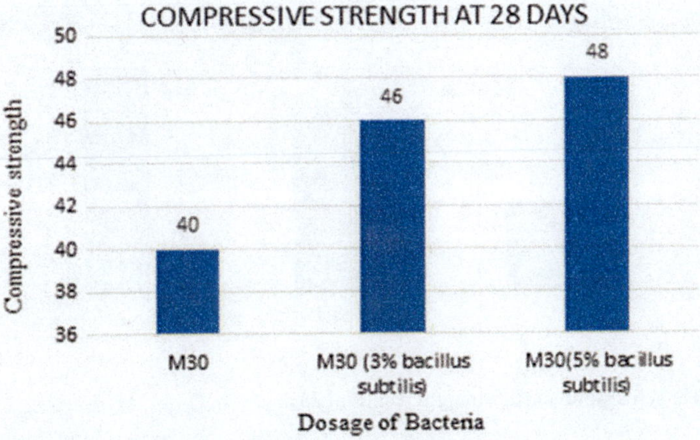

Fig. 4 M30 concrete compressive strength results at 28 days

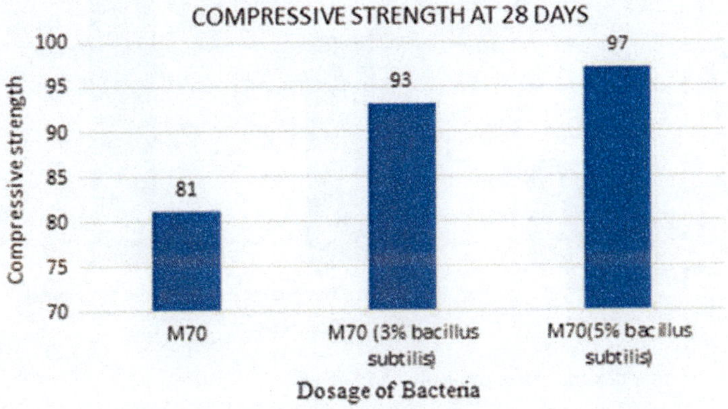

Fig. 5 M70 concrete compressive strength results at 28 days

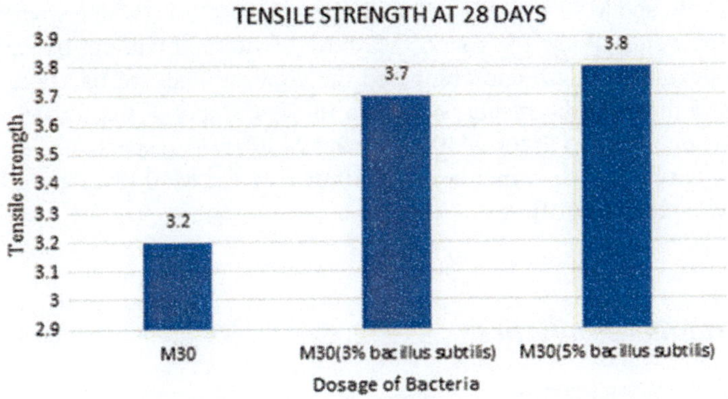

Fig. 6 M30 concrete split tensile strength results at 28 days

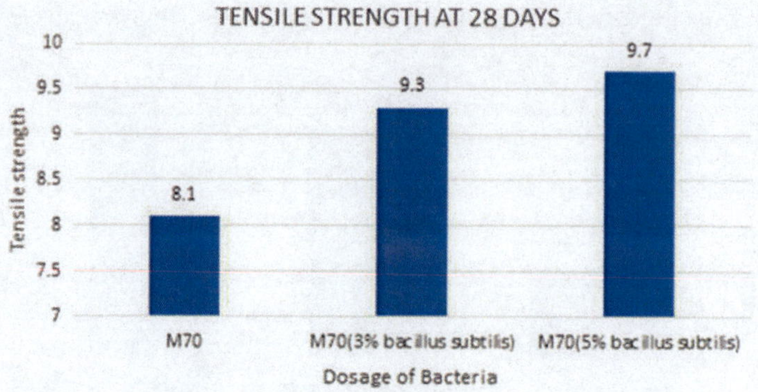

Fig. 7 M70 concrete split tensile strength results at 28 days

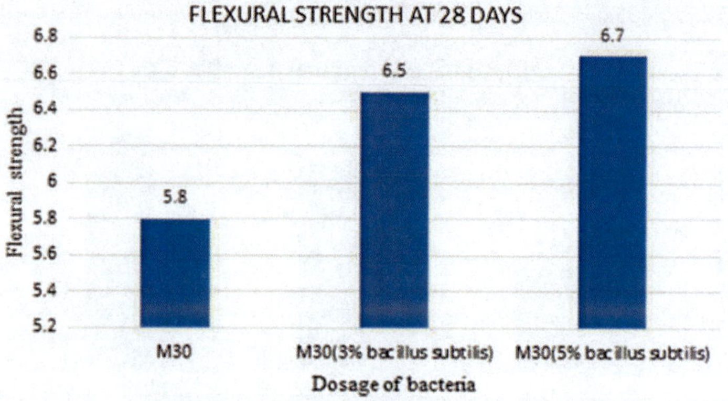

Fig. 8 M30 concrete flexural strength results at 28 days

M30 and M70 with 3% of bacillus subtilis bacteria (which was calculated by weight of water) and M30 and M70 with 5% of bacillus subtilis bacteria (which was calculated by weight of water), 3 each beams of M30, M70 with and without bacteria where casted for flexural strength and compared the conventional and bacterial concrete results. By comparing the results as shown in Figs. 8 and 9, the flexural strength is increased by 10%, 15% for M30 (3%) and M30 (5%) respectively at 28 days, and compressive strength is increased by 10%, 15% for M70 (3%) and M70 (5%) respectively at 28 days [12].

3.3 *Durability Properties*

Acid Attack Test: Chemical resistance of concrete cubes was tested by chemical attack by immersion in acid solution. Acid attack is caused by the reaction of the

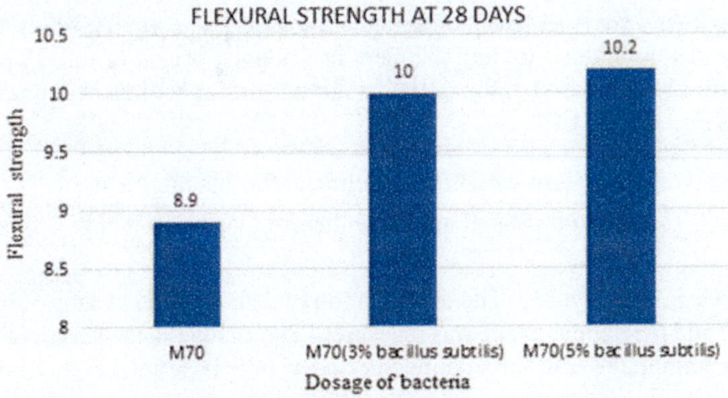

Fig. 9 M70 concrete flexural strength results at 28 days

acid with the calcium hydroxide portion of the cement paste, releasing a readily soluble by-product, the calcium salt.

These soluble calcium salts are easily removed from the cement paste, weakening the overall paste structure [13]. This fundamental reaction is

$$\text{Acid X} + \text{CH} \rightarrow \text{CX} + \text{H} \tag{1}$$

$$\text{Acid} + \text{Calcium Hydroxide} \rightarrow \text{Calcium Salt} + \text{Water}$$

The most aggressive and destructive cases of acid attack occur only when concrete is exposed to sulfuric acid (H_2SO_4). Calcium salts are formed by the reaction of sulfuric acid and calcium hydroxide to form calcium sulphate, which undergoes significant decomposition due to sulphate attack.

$$H_2SO_4 + Ca(OH)_2 \rightarrow Ca(SO)_4 + 2H_2O \tag{2}$$

At the end of the curing time, samples are taken out of the curing tank, and their surfaces are cleaned to remove any weak reaction products or loose material from the sample. Initial mass and diagonal dimension values are measured. For each batch, three samples of concrete cubes are immersed in 5% HCL and 5% H_2SO_4 solutions. At the end of 7, 14, 21, and 28 days, all through the curing time, steady acid concentration is maintained. After 28 days, each example is taken out of the tanks before testing, brushed with delicate nylon brushes and flushed in normal water. Changes in mass change, decrease of compressive strength and change in slanting aspects are observed. Acid attack results on SCC samples cured under various curing conditions are shown in the following three factors: namely, Acid Durability Loss Factor (ADLF), Acid Attack Factor (AATF) and Acid Strength Loss Factor (ASLF) are derived from ASTM C267 (2012) by reducing the durability aspects of specimens (strength, mass and geometry).

Acid Mass Loss Factor (AMLF): The Percentage Loss is estimated by placing the cubes in acid solution and finding the mass at a defined period of time. The change in mass with age compared to the initial weight of each specimen is defined as Acid Mass Loss Factor (AMLF).

$$\text{AMLF} = \frac{\text{Change in weight of specimen after placing in Acid}}{\text{Initialmass of specimen before placing in Acid}} \times 100 \quad (3)$$

Acid Attacking Factor (AAF): The continuation of deterioration at each corner of the struck face and the opposite face was measured. The change in the length of diagonal after immersion in the acid for a defined period of time is defined as Acid Attacking Factor (AAF).

$$\text{AAF} = \frac{\text{Change in dimension of diagonal after immersion in Acid}}{\text{Original diagonal dimension before immersion in Acid}} \times 100 \quad (4)$$

Acid Strength Loss Factor (ASLF): The relative strength present in concrete specimens after immersing in acid represents Acid Strength Loss Factor (ASLF). The relative strengths are always compared with respect to 28 days compressive strength values.

$$\text{ASL} = \frac{\text{Change in dimension of diagonal after immersion in Acid}}{\text{Original diagonal dimensions before immersion in Acid}} F \times 100 \quad (5)$$

To analyse the durability properties of self-healing concrete specimens, important durability properties like Acid attack are determined. The present chapter investigates the durability properties of specimens from both M30 and M70 grade concrete. Table 3 represents the loss in weight and loss of compressive strength because of acid attack test at 28 days.

From Table 4, Acid Mass Loss Factor, Acid Attacking Factor, Acid Strength Loss Factor, Acid Durability Loss Factor, the results shows that acid durability loss factor is more in M30 (5%) and M70 (5%) as the bacteria percentage increases the durability loss factor also increases.

Table 3 Weight loss due to acid attack test at 28 days

Grade	IW (kg)	FW (kg)	% of weight reduction	% of loss in CS
M30	7.53	6.92	8	58
M70	8.18	7.52	8	60
M30 (3% *Bacillus subtilis*)	8.2	7.38	10	51
M70 (3% *Bacillus subtilis*)	8.48	7.63	10	52
M30 (5% *Bacillus subtilis*)	8.76	7.70	12	54
M70 (5% *Bacillus subtilis*)	8.85	7.78	12	55

Table 4 Details of Acid Mass Loss Factor, Acid Attacking Factor, Acid Strength Loss Factor and Acid Durability Loss Factor

Grade	Acid Mass Loss Factor (AMLF)	Acid Attacking Factor (AAF)	Acid Strength Loss Factor (ASLF)	Acid Durability Loss Factor (ADLF)
M30	8.1	9.4	42	3198.247
M70	8.06	7.9	40	2549.633
M30 (3% *Bacillus subtilis*)	10.00	6.1	49	2989
M70 (3% *Bacillus subtilis*)	10.02	5.1	48	2453.774
M30 (5% *Bacillus subtilis*)	12.10	6.3	46	3506.712
M70 (5% *Bacillus subtilis*)	12.09	6.6	45	3590.847

Fig. 10 SEM analysis

3.4 *Micro-structural Properties*

Micro structural characterization is concerned with the study of microelements and chemical composition. From Fig. 10, whether the element present is crystalline or amorphous and what it consists of, as well as its chemical composition. In this project, it is mainly used to check the contained micro cracks and confirm the formation of $CaCO_3$ that fills the bacterial morphology and micro cracks. There are various methods for micro structural analysis, such as SEM (Scanning Electron Microscope) analysis, XRD (X-Ray diffraction), and EDX (energy dispersive X-ray analysis).

By performing EDX analysis shown in Figs. 11 and 12 at three different locations in the SEM image, different elements are obtained, including different weights and atomic percentages of SiK-OK-CaK. When this he reacts with H_2O (water), $CaCO_3$ is formed (crack healing).

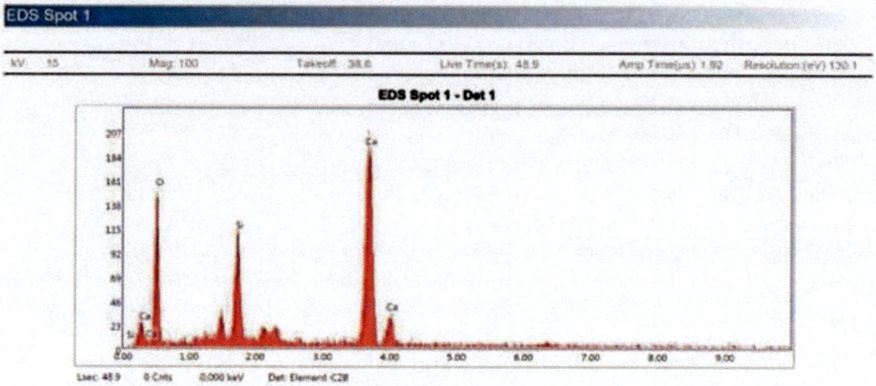

eZAF Smart Quant Results

Element	Weight %	Atomic %	Net Int.	Error %	Kratio	Z	A	F
O K	41.50	62.70	19.57	13.66	0.0777	1.0684	0.1720	1.0000
SiK	7.83	6.74	15.96	10.20	0.0642	0.9635	0.8281	1.0061
CaK	50.68	30.56	46.92	5.28	0.4670	0.9238	0.9951	1.0023

Fig. 11 EDX at Spot 1

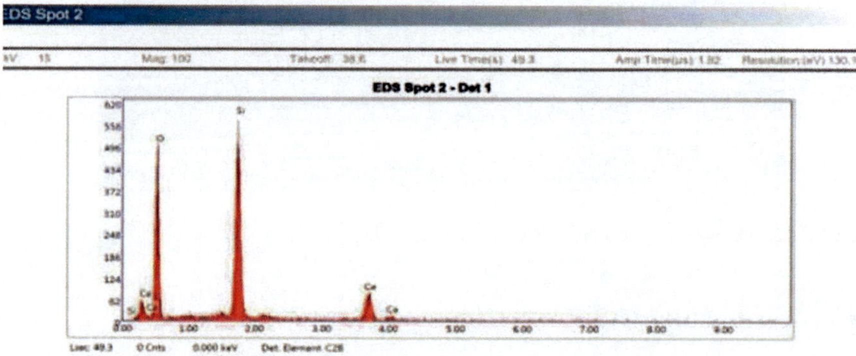

eZAF Smart Quant Results

Element	Weight %	Atomic %	Net Int.	Error %	Kratio	Z	A	F
O K	52.35	67.94	68.95	9.80	0.1889	1.0556	0.3418	1.0000
SiK	33.34	24.65	100.46	4.39	0.2782	0.9517	0.8749	1.0026
CaK	14.31	7.41	18.17	9.73	0.1247	0.8923	0.9696	1.0077

Fig. 12 EDX at Spot 2

4 Conclusion

This experiment is conducted for study of strength durability and micro structural behaviour of self-healing concrete, for M30 and M70 with and without Bacillus bacteria of 3% and 5%

1. The compressive strength for M30 (3%) and M30 (5%) at 28 days was increased by 15% and 20% respectively compared to M30.
2. The compressive strength for M70 (3%) and M70 (5%) at 28 days was increased by 15% and 20% respectively compared to M70.
3. The tensile strength for M30 (3%) and M30 (5%) at 28 days was increased by 14% and 18% respectively compared to M30.
4. The tensile strength for M70 (3%) and M30 (5%) at 28 days was increased by 14% and 18% respectively compared to M70.
5. The flexural strength for M30 (3%) and M30 (5%) at 28 days was increased by 15% and 20% respectively compared to M30.
6. The flexural strength for M30 (3%) and M30 (5%) at 28 days was increased by 15% and 20% respectively compared to M30.
7. The acid attack test shows weight reduced for M30 and M70 was 8%, for M30 (3%) and M70 (3%) was 10%, for M30 (5%) and M70 (5%) was 12%.
8. The SEM images shows us that the healing in M30 (5%) and M70 (5%) was faster compared to M30 (3%) and M70 (3%).

References

1. Ling H, Qian C (2017) Effects of self-healing cracks in bacterial concrete on the transmission of chloride during electromigration. Constr Build Mater 144:406–411
2. Mondal S, Ghosh AD (2019) Review on microbial induced calcite precipitation mechanisms leading to bacterial selection for microbial concrete. Constr Build Mater 225:67–75
3. Vijay K, Murmu M, Deo SV (2017) Bacteria based self healing concrete—a review. Constr Build Mater 152:1008–1014
4. Agarwal A, Bhusnur S, Chaudhary K, Priya TS (2020) Experimental investigation on bacterial concrete with micronized biomass silica. Mater Today Proc 22:2475–2481
5. IS:1199-1959 (1995) Methods of sampling and analysis of concrete. Indian standard. Bureau of Indian Standards, New Delhi
6. IS:12269-2003 (2003) Specification for 53 grade ordinary Portland cement. Indian Standard, Bureau of Indian Standards, New Delhi
7. IS:456-2000 (2000) Plain and reinforced concrete-code of practice. Indian standard. Bureau of Indian Standards, New Delhi
8. Akindahunsi AA, Adeyemo SM, Adeoye A (2021) The use of bacteria (Bacillus subtilis) in improving the mechanical properties of concrete. J Build Pathol Rehabil 6:1–8
9. IS:516-1959 (2004) Methods of test for strength of concrete. Indian standard. Bureau of Indian Standards, New Delhi
10. Abdulkareem M, Ayeronfe F, Abd Majid MZ, Sam ARM, Kim J-HJ (2019) Evaluation of effects of multi-varied atmospheric curing conditions on compressive strength of bacterial (*Bacillus subtilis*) cement mortar. Constr Build Mater 218:1–7

11. IS:5816-1999 (1999) Splitting tensile strength of concrete—method of test. Indian Standard, Bureau of Indian Standards, New Delhi
12. Reddy CMK, Ramesh B, Macrin D (2020) Influence of bacteria *Bacillus subtilis* and its effects on flexural strength of concrete. Mater Today Proc 33:4206–4211
13. Rao MVS, Reddy VS, Sasikala C (2017) Performance of microbial concrete developed using *Bacillus subtilus* JC3. J Inst Eng Ser A 98:501–510

Development of Self Compacting Geo Polymer Hybrid Fiber-Reinforced Concrete Using Highly Potential Sustainable Materials

B. Narendra Kumar, P. Pavan, and G. Vinod Kumar

1 Introduction

In the present world, the global warming is the major crisis for the sustainable development [1]. The utilization of waste products obtained from the power plant can be used as replacement to cement and decrease the environmental effect [2]. The research has led to the substitute of cement with the Fly Ash [3], which deals with the geo polymer concrete [4] incorporating SSC which is cement free [5]. The geo polymer concrete is difficult to place, but with ease of self-compacting concrete the SCGPC showed exceptionally good flow [6, 7]. In GPC, the silica and alumina content are more which undergoes chemical reaction and form C–S–H gel [8] called Geo polymerization. The other key factor for geo polymer concrete is alkali activated materials like sodium silicate and sodium hydroxide which acts as a binder [9–11]. The Cementitious materials like Fly Ash, GGBS and Metakaolin [12] reacts with alkali activated materials and improves the workability of the mix [13, 14]. The SCGPHFC incorporates combination of two fibers like steel fiber [15] and PVA [16] fiber. To design the SCGPHFC mix the serviceability of the concrete plays a key role for the workability, hardened properties, and durability properties which are interlinked with every property of the concrete. For, the workability of concrete the following test methods are determined which influences the strength and durability properties of concrete [17–19]. The strength properties for SCGPHFC are determined for the 7 days of ambient curing. Durability tests like abrasion resistance of concrete, acid resistance of concrete and rapid chloride permeability test (RCPT) of concrete were conducted [20–22].

B. N. Kumar (✉) · P. Pavan
Department of Civil Engineering, VNR VJIET, Hyderabad, India
e-mail: narendrakumar_b@vnrvjiet.in

G. V. Kumar
Siri Developers Pvt. Ltd., Hyderabad, India

R. K. Pancharathi et al. (eds.), *Low Carbon Materials and Technologies for a Sustainable and Resilient Infrastructure*, Lecture Notes in Civil Engineering 440,
https://doi.org/10.1007/978-981-99-7464-1_15

2 Significance of Research

This research focuses on incorporating cementitious materials such as GGBS, Metakaolin and Fly ash along with hybrid fibers like Steel fibers and PVA fibers to enhance the strength and performance properties of concrete. The aim is to produce green concrete having desirable strength and durability. The main intention of this work is to evaluate the performance of self-compacting geopolymer hybrid fiber-reinforced concrete by using steel fibers and PVA fibers in different proportions.

3 Research Gap

- The impact of supplementary cementitious materials with hybrid fibers on fresh and hardened qualities, as well as durability features, has not been fully determined in existing research surveys.
- Different doses of hybrid fibers in geopolymer concrete are being investigated.

4 Materials and Methods

Class F fly ash is used which is obtained from a by-products of industrial waste from a thermal power plant [23, 24] which consists mostly of alumina and silica which is classified according to the ASTM C 618 [3]. The GGBS was obtained from the steel industry supplied by local suppliers available in Hyderabad which have a specific gravity of 2.7 which is classified according to the ASTM C 9892018 [25–27]. The Metakaolin [MK] is supplied by local suppliers and was classified according to ASTM C 618 N as having specific gravity is 2.6 which is obtained from calcined kaolin [8] clay. The NaOH is taken in the form of pellets which is prepared for 8 M concentration with prior to one day of the mix. The sodium silicate was available in the form of liquid gel from the locally available with configuration of Na_2O, SiO_2 and water [28, 29]. The aggregates were used in the present work are locally available conforming to IS 383:2016 [30, 31]. The Super Plasticizer of Master Glenium ACE 30 which was available in BASF industries which is the admixture of poly carboxylic ether polymer with high early strength gain according to the IS 9103:1988 [32–34]. The steel fiber having the aspect ratio of 67 and density of 7800 kg/m^3. The PVA fiber is unique in their ability [16] to create the molecular bond with mortar with the aspect ratio [35, 36] of 315 and density of 1190 kg/m^3.

5 Mix Proportioning of SCGPHFC

The mix proportioning of SCGPHFC, the powder content and aggregates are constant for all the mixes. Overall, the six mixes were changing in the percentage of fibers with the total percentage of fibers is 1.5%. Table 1 shows the mix proportioning of SCGPHFC.

6 Mixing and Casting

For homogeneous mixing of concrete 80 L capacity electrically operated pan mixer was used mixing of concrete The total amount of aggregates and powder content are added simultaneously and alkali activator solution, water, and superplasticizer is added one after the other for the proper mix for uniform mixture as shown in Fig. 1. The mixed SCGPHFC was deposited into the molds. After 24 h, the specimens were detached and molds were kept for ambient temperature for 7, 28, and 90 days after performing all the workability.

7 Experimental Investigation on Workability

The workability of SCGPHFC was evaluated by performing the tests according to ASTM C 143–15 as shown below in the Figs. 2, 3 and 4.

The results of SCGPHFC mixes are depicted in Figs. 5, 6 and 7. It was observed that the mix of SCGPHFC 1, that is, mix without fiber showed the greater flow than other mixes with fiber. The SCGPHFC1 showed the values 650 mm for slump flow, 0.86 for L-Box test and 2.5, 7.5 for V-funnel tests whereas the optimum mix SCGPHFC 5 showed lowest values 540 mm for slump flow, 0.83 for L-Box test and 4.5,10 for V-funnel tests. Concrete flow is reduced due to non-homogeneity created by addition of fibers and with increase in the proportion of fiber content, the workability decreased as shown.

8 Compressive Strength

Compressive strength of concrete is conducted by cubes of size 150 mm as shown in Fig. 8, which was cast, cured, and tested at the end of 7, 28 and 90 days as shown in Table 2 as per IS 516 [37]. The SCGPHFC 5 mix showed highest value of compressive strength and for the next proportion a decline in strength was observed however the value was not less than SCGPC 1, that is, mix without fiber. Due to higher density of steel fiber, the gain in strength was more compared to PVA fiber.

Table 1 Mix designation

	FA (Kg/m^3)	GGBS (Kg/m^3)	MK (Kg/m^3)	NaOH (Kg/m^3)	Na_2SiO_3 (Kg/m^3)	Water	SP	Fiber	
								PVA	Steel
SCGPHFC 1	500	50	50	96	144	60	12	–	–
SCGPHFC 2	500	50	50	96	144	60	12	–	1.5%
SCGPHFC 3	500	50	50	96	144	60	12	0.25%	1.25%
SCGPHFC 4	500	50	50	96	144	60	12	0.5%	1%
SCGPHFC 5	500	50	50	96	144	60	12	0.75%	0.75%
SCGPHFC 6	500	50	50	96	144	60	12	1%	0.5%

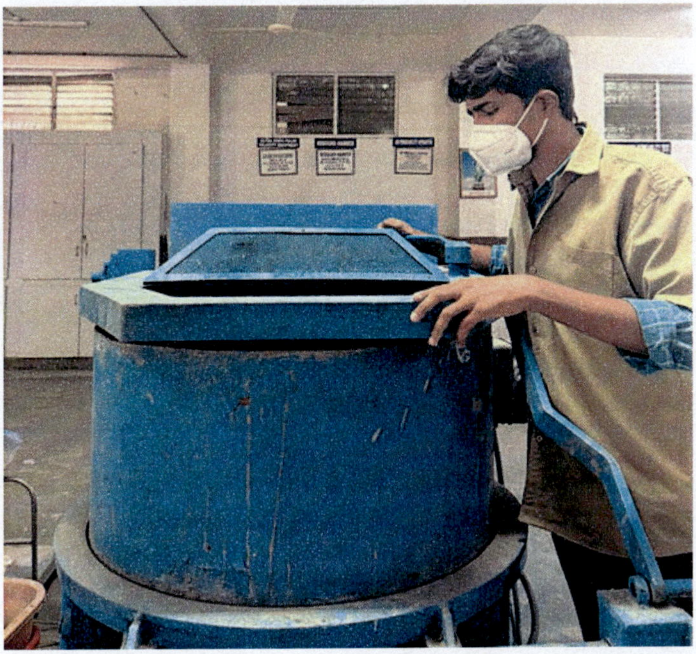

Fig. 1 Pan mixer

Fig. 2 Slump flow of concrete

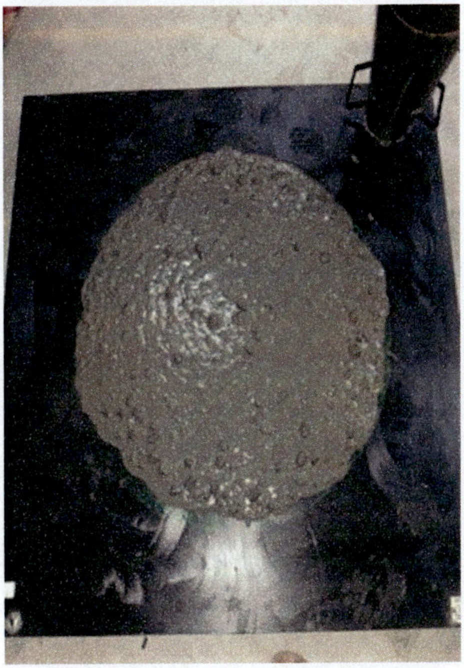

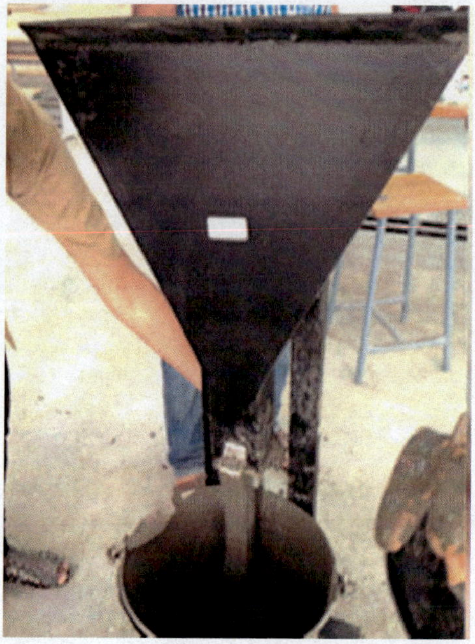

Fig. 3 V-Funnel test

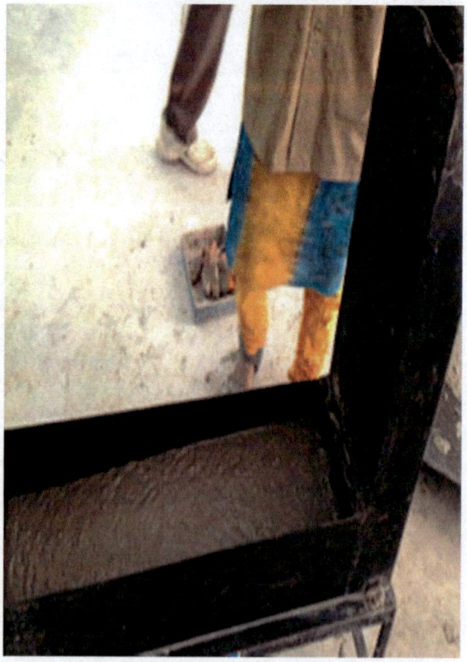

Fig. 4 L-Box test

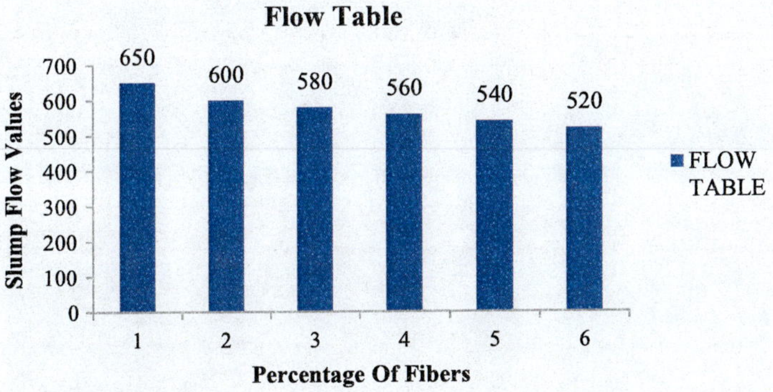

Fig. 5 Slump flow For SCGPHFC

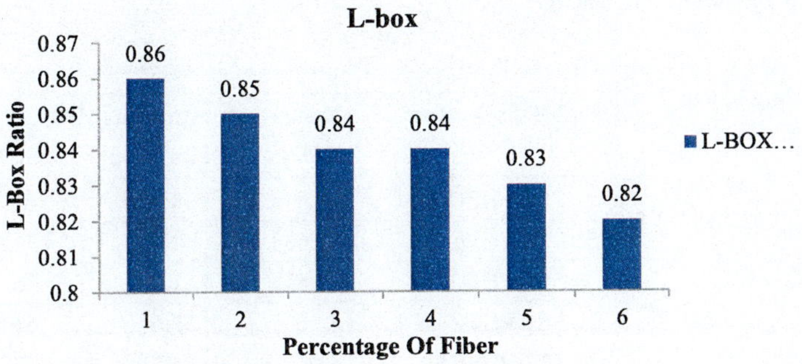

Fig. 6 L-Box values for SCGPHFC

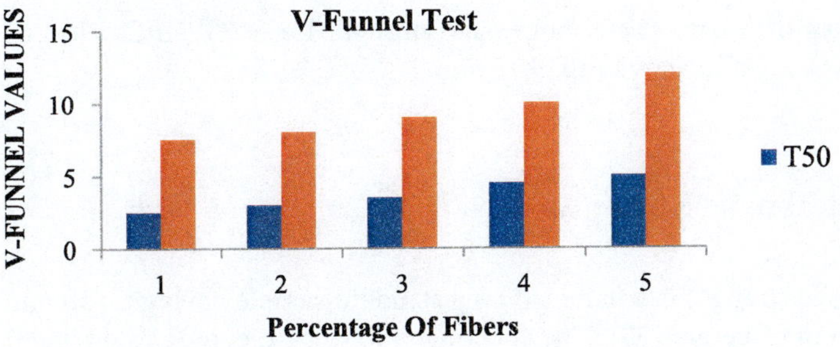

Fig. 7 V-Funnel test for SCGPHFC

Due to the constant powder content, a polymerization reaction takes place, which enhances the compressive strength by 26.3%, 29%, 30% at 7 days, 28 days, and 90 days when compared to the mix without fibers. Presence of equal proportions of fibers in the mix SCGPHFC 5, that is, 0.75% of PVA and steel fiber are responsible for having more strength properties compared to other mixes having different

Fig. 8 Compression testing machine

Table 2 Compressive strength

MIX selection	Compressive strength (MPa)		
	7 Days	28 Days	90 Days
SCGPHFC 1	29.8	38	41.2
SCGPHFC 2	37.63	49	53.59
SCGPHFC 3	39.85	51.8	56.19
SCGPHFC 4	42.63	52.6	61
SCGPHFC5	43.1	54.8	67
SCGPHFC6	40.8	52.3	55.6

percentages of fibers. The compressive strength for SCGPHFC 5 has 67 Mpa for 90 days as shown in above Fig. 9.

9 Split Tensile Strength

Split Tensile strength is conducted on a standard sample having a 150 mm diameter and 300 mm of height [38]. The specimens were cast, cured, tested and the experimental values are shown in the Table 3. The Split tensile strength test was performed according to IS 516 [37]. Higher value of tensile strength was attained for SCGPHFC 5 and for the later mix i.e. SCGPHFC 6 the strength decreased. Steel fiber has more tensile strength properties when compared to the PVA fiber. And thus the reduction in strength for SCGPHFC 6 is attributed to decrease in percentage of steel fiber.

The strength was increased by 58.2%, 57.5%, 58.1% at the end of 7 days, 28 days, 90 days due to the presence of fibers. The addition of fibers thus imparts good tensile strength to concrete. As the percentage of fiber increases the strength also increases

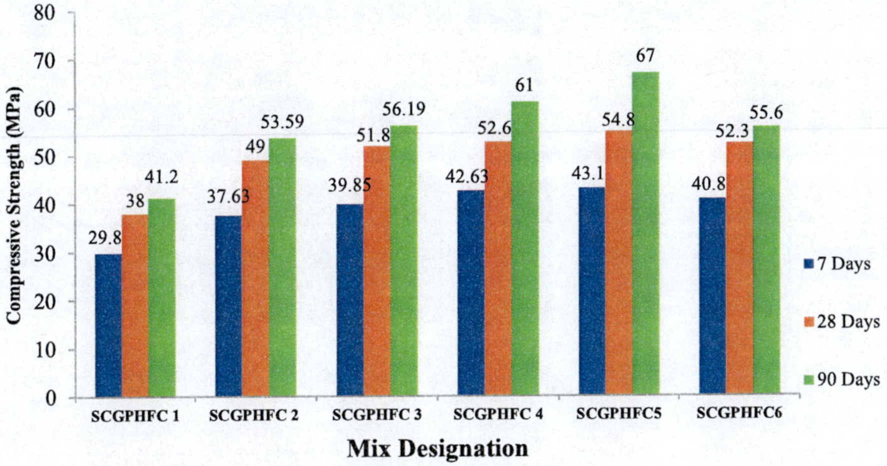

Fig. 9 Compressive strength for SCGPHFC

Table 3 Split tensile strength for SCGPHFC mixes

Mix designation	Split tensile strength (MPa)		
	7 Days	28 Days	90 Days
SCGPHFC 1	3.5	4.3	4.8
SCGPHFC 2	5.6	6.9	7.6
SCGPHFC 3	6	7.41	8.2
SCGPHFC 4	6.5	8	8.84
SCGPHFC 5	7	8.7	9.54
SCGPHFC 6	6	7.6	8.4

i.e. for SCGPHFC 5 due to the constant powder content, good geo-polymeization reaction occurs and addition of equal proportions of both fibers, that is, 0.75% of PVA and steel fiber, increased the strength properties compared to other mixes having different percentages of fibers. The tensile strength for SCGPHFC 5 was 9.54 Mpa as shown in Fig. 10.

10 Flexural Strength

The test is performed on samples of size 150 mm × 150 mm × 100 mm. These specimens are cast, cured, and tested at 7 days, 28 days and 90 days as shown in the Fig. 11 and Table 4. Based on the results, increase in flexural strength was observed to be highest for SCGPHFC 5 and for the later mix SCGPHFC 6 the strength was

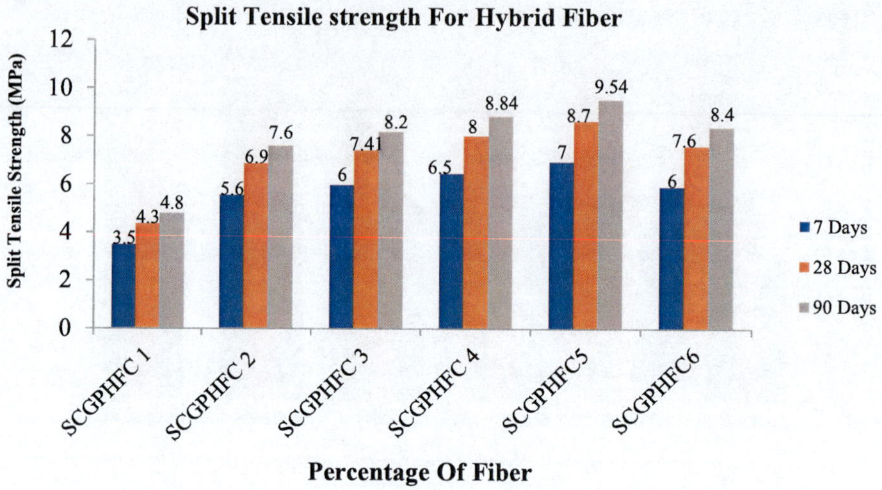

Fig. 10 Split tensile strength

decreased, however the value was not less than the mix without fiber. The reason for reduction in trength is due to lesser proportion of steel fiber in the latter mix.

The strength increases by 25.4%, 23.6%, 21.6% at the end of 7, 28, 90 days due to the presence of fibers. As the percentage of fiber increases the strength increases,

Fig. 11 Flexure testing machine

Table 4 Flexural strength

Mix designation			
	7 Days	28 Days	90 Days
SCGPHFC 1	4.01	5.7	5.92
SCGPHFC 2	6.34	9.01	9.37
SCGPHFC 3	6.84	9.73	10.11
SCGPHFC 4	7.38	10.5	10.91
SCGPHFC 5	7.97	11.34	12.78
SCGPHFC 6	7.01	9.97	10.36

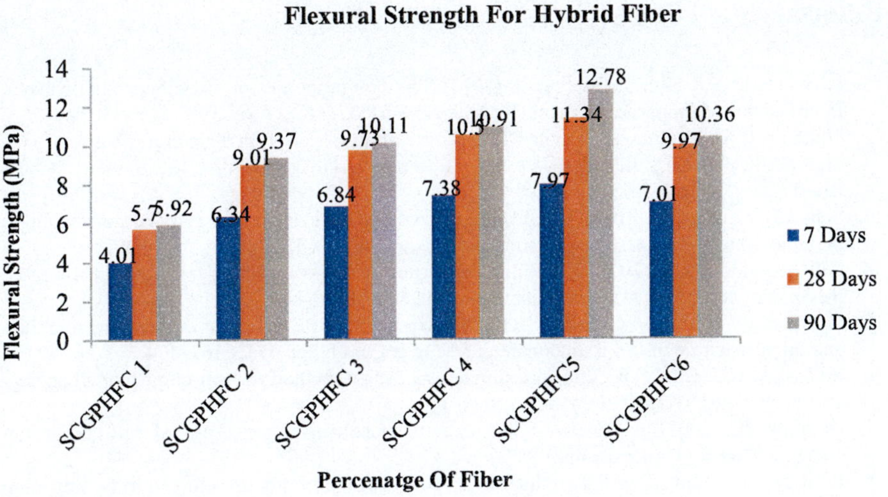

Fig. 12 Flexural strength

that is, for SCGPHFC 5 mix having 0.75% of PVA and steel fiber the concrete matrix properties are inhanced which increasing the flexural property compared to SCGPHFC 1. The strength for SCGPHFC 5 was 12.78 Mpaas shown in the Fig. 12.

11 Conclusions

Conclusions from the experimental investigation of Self Compacted Geo Polymer Hybrid Fiber Concrete are:

1. The mix proportions adopted in this experiment included by-products like fly ash, GGBS, metakaolin which satisfied the EFNARC guidelines. The workability of the concrete decreases as the dosage of fiber increases and fiber is said to induce non-homogeneity to concrete.

2. For the mix SCGPHFC 5 the Compressive strength, Split Tensile strength, and Flexural strength were maximum compared to the other mixes.
3. The compressive strength was increased by 16.7%, Split tensile strength was increased by 17.2% and flexural strength was increased by 12.67% for 7 days to 28 days, respectively.
4. The future scope of the work is the sustainable development of concrete with different mineral admixtures and different fibers in the practical applications of concrete.

References

1. Niewiadomski P, Hoła J, Ćwirzeń A (2018) Study on properties of self-compacting concrete modified with nanoparticles. Arch Civ Mech Eng 18:877–886
2. Toutanji H, Xu B, Gilbert J, Lavin T (2010) Properties of poly (vinyl alcohol) fiber reinforced high-performance organic aggregate cementitious material: converting brittle to plastic. Constr Build Mater 24(1):1–10
3. Saboo N, Shivhare S, Kori KK, Chandrappa AK (2019) Effect of fly ash and metakaolin on pervious concrete properties. Constr Build Mater 223:322–328
4. Gill AS, Siddique R (2018) Durability properties of self-compacting concrete incorporating metakaolin and rice husk ash. Constr Build Mater 176:323–332
5. Venkatesh C, Nerella R, Chand MSR (2020) Experimental investigation of strength, durability, and microstructure of red-mud concrete. J Korean Ceram Soc 57(2):167–174
6. Su N, Hsu KC, Chai HW (2001) A simple mix design method for self-compacting concrete. Cem Concr Res 31(12):1799–1807
7. Domone PL (2007) A review of the hardened mechanical properties of self-compacting concrete. Cement Concr Compos 29(1):1–12
8. Pająk M (2016) Investigation on flexural properties of hybrid fibre reinforced self-compacting concrete. Procedia Eng 161:121–126
9. Bakharev T, Sanjayan JG, Cheng YB (2003) Resistance of alkali-activated slag concrete to acid attack. Cem Concr Res 33(10):1607–1611
10. Gesoğlu M, Güneyisi E, Kocabağ ME, Bayram V, Mermerdaş K (2012) Fresh and hardened characteristics of self compacting concretes made with combined use of marble powder, limestone filler, and fly ash. Constr Build Mater 37:160–170
11. Mehta A, Siddique R, Ozbakkaloglu T, Shaikh FUA, Belarbi R (2020) Fly ash and ground granulated blast furnace slag-based alkali-activated concrete: mechanical, transport and microstructural properties. Constr Build Mater 257:119548
12. Nath P, Sarker PK (2017) Flexural strength and elastic modulus of ambient-cured blended low-calcium fly ash geopolymer concrete. Constr Build Mater 130:22–31
13. Uysal M, Yilmaz K, Ipek M (2012) The effect of mineral admixtures on mechanical properties, chloride ion permeability and impermeability of self-compacting concrete. Constr Build Mater 27(1):263–270
14. Felekoğlu B, Türkel S, Baradan B (2007) Effect of water/cement ratio on the fresh and hardened properties of self-compacting concrete. Build Environ 42(4):1795–1802
15. Carro-López D, González-Fonteboa B, Martínez-Abella F, González-Taboada I, de Brito J, Varela-Puga F (2017) Proportioning, microstructure and fresh properties of self-compacting concrete with recycled sand. Procedia Eng 171:645–657
16. Parra C, Valcuende M, Gómez F (2011) Splitting tensile strength and modulus of elasticity of self-compacting concrete. Constr Build Mater 25(1):201–207

17. Madandoust R, Mousavi SY (2012) Fresh and hardened properties of self-compacting concrete containing metakaolin. Constr Build Mater 35:752–760
18. Chindaprasirt P, Chareerat T, Sirivivatnanon V (2007) Workability and strength of coarse high calcium fly ash geopolymer. Cement Concr Compos 29(3):224–229
19. Nath P, Sarker PK (2014) Effect of GGBFS on setting, workability and early strength properties of fly ash geopolymer concrete cured in ambient condition. Constr Build Mater 66:163–171
20. Atiş CD, Karahan O (2009) Properties of steel fiber reinforced fly ash concrete. Constr Build Mater 23(1):392–399
21. Olivito RS, Zuccarello FA (2010) An experimental study on the tensile strength of steel fiber reinforced concrete. Compos B Eng 41(3):246–255
22. Noushini A, Samali B, Vessalas K (2013) Effect of polyvinyl alcohol (PVA) fibre on dynamic and material properties of fibre reinforced concrete. Constr Build Mater 49:374–383
23. Boukendakdji O, Kadri EH, Kenai S (2012) Effects of granulated blast furnace slag and superplasticizer type on the fresh properties and compressive strength of self-compacting concrete. Cement Concr Compos 34(4):583–590
24. Meena A, Singh N, Singh SP (2023) High-volume fly ash self consolidating concrete with coal bottom ash and recycled concrete aggregates: fresh, mechanical and microstructural properties. J Build Eng 63:105447
25. Singh RB, Singh B (2018) Rheological behaviour of different grades of self-compacting concrete containing recycled aggregates. Constr Build Mater 161:354–364
26. IS 2386-1963 Methods of test for aggregates for concrete
27. IS: 9103-1999 Specification for concrete
28. IS: 516-1959 Method of test for strength for concrete
29. EFNARC (2005) Specifications and guidelines for self-compacting concrete
30. Ramujee K, Raju MP (2017) Mechanical properties of geoploymer concrete compositure materials today proceedings 2937–2945
31. Dinakar P, Sethy KP, Sahoo UC (2013) Design of self-compacting concrete with ground granulated blast furnace slag. Mater Des 43:161–169
32. Gharzouni A, Vidal L, Essaidi N, Joussein E, Rossignol S (2016) Recycling of geopolymer waste: influence on geopolymer formation and mechanical properties. Mater Des 94:221–229
33. Güneyisi E, Gesoglu M, Algın Z, Yazıcı H (2016) Rheological and fresh properties of self-compacting concretes containing coarse and fine recycled concrete aggregates. Constr Build Mater 113:622–630
34. Al Bakria AM, Kamarudin H, BinHussain M, Nizar IK, Zarina Y, Rafiza AR (2011) The effect of curing temperature on physical and chemical properties of geopolymers. Phys Procedia 22:286–291
35. Karthik A, Sudalaimani K, Kumar CV (2017) Investigation on mechanical properties of fly ash-ground granulated blast furnace slag based self curing bio-geopolymer concrete. Constr Build Mater 149:338–349. SION
36. Kim JY (2015) A study on the preparation method of geopolymeric concrete using specifically modified silicate and inorganic binding materials and its compressive strength characteristics. J Korean Ceram Soc 52(2):150–153
37. Barluenga G, Palomar I, Puentes J (2015) Hardened properties and microstructure of SCC with mineral additions. Constr Build Mater 94:728–736
38. Girish S, Ranganath RV, Vengala J (2010) Influence of powder and paste on flow properties of SCC. Constr Build Mater 24(12):2481–2488
39. Manjunath R, Narasimhan MC (2018) An experimental investigation on self-compacting alkali activated slag concrete mixes. J Build Eng 17:1–12
40. Karthik A, Sudalaimani K, Kumar CV (2017) Investigation on mechanical properties of fly ash-ground granulated blast furnace slag based self curing bio-geopolymer concrete. Constr Build Mater 149:338–349
41. Ma K, Feng J, Long G, Xie Y, Chen X (2017) Improved mix design method of self-compacting concrete based on coarse aggregate average diameter and slump flow. Constr Build Mater 143:566–573

42. Bellum RR, Muniraj K, Madduru SRC (2020) Influence of slag on mechanical and durability properties of fly ash-based geopolymer concrete. J Korean Ceram Soc 57(5):530–545
43. Lee SH, Kim GS, Lee SH, Kim GS (2017) Self-cementitious hydration of circulating fluidized bed combustion fly ash. J Korean Ceram Soc 54(2):128–136

Innovative Construction Techniques and Management

Selection of Sustainable Repair Mortars for Heritage Structures: A Hybrid MCDM Approach

Nikhil Kumar Degloorkar and Rathish Kumar Pancharathi

1 Introduction

Multi-attribute decision-making (MADM) and multi-objective decision-making (MODM) methods are broadly classified into multi-criteria decision-making methods. In MADM methods, a set of finite predefined alternatives is considered from which the best alternative will be selected, considering certain defined criteria. A sorting and selection of a decision-making method is involved in this process. Few examples of MADM methods include pairwise comparison methods like analytical hierarchy process (AHP), analytical network process (ANP), distance-based method like technique for order preference by similarity to ideal solution (TOPSIS), compromising methods like VIseKriterijumska Optimizacija I Kompromisno Resenje (VIKOR), outranking methods like elimination and choice translating reality (ELECTRE) and preference ranking organization method for enrichment of evaluations (PROMETHEE). In MODM methods, optimization is performed for multiple objectives considering different constraints. This method is generally used in arriving at a product design. Few examples of MODM methods include linear programming (LP), non-integer linear programming (NILP), mixed integer linear programming (MILP), goal programming (GP), etc. [1–3].

N. K. Degloorkar (✉) · R. K. Pancharathi
Department of Civil Engineering, National Institute of Technology Warangal, Hanamkonda, Telangana 506004, India
e-mail: nikhildegloorkar@gmail.com; nikhilkumar.civil@anurag.edu.in

R. K. Pancharathi
e-mail: rateeshp@nitw.ac.in

N. K. Degloorkar
Department of Civil Engineering, Anurag University, Hyderabad, Telangana 500088, India

R. K. Pancharathi et al. (eds.), *Low Carbon Materials and Technologies for a Sustainable and Resilient Infrastructure*, Lecture Notes in Civil Engineering 440,
https://doi.org/10.1007/978-981-99-7464-1_16

The selection of material for a particular product is a major decision-making problem in any construction or manufacturing industry. Decision-making has become easy due to the availability of many databases regarding the requirements of a material to cater to a product, along with advancements in information technology. The consistency of data with a set of conflicting criteria is a huge concern, and various MADM techniques prove to be handy in this process. Additionally, a fuzzy approach was incorporated into MADM methods to avoid vagueness and uncertainty in data [4].

The importance of sustainable material selection that has less impact on the environment was discussed by Holloway [5]. The use of MCDM techniques for material selection in engineering applications was indicated to be gradually growing by Jahan et al. [6]. Fuzzy AHP analysis was used by Stanojevic et al. [7] to arrive at the key indicators for increasing energy efficiency in heritage structures. It was established by Nadkarni et al. [8] that MCDM application in heritage structures was predominantly focused on the alternative and adaptive use of materials. AHP was also established as the most widely used MCDM technique for the determination of the weightages of criteria. The best concrete repair strategy was decided by employing AHP analysis and considering physical and chemical requirements of repair material by Jeong-Yun Do et al. [9]. Anojkumar et al. [10] applied a hybrid MCDM approach in selecting the best pipe material for the sugar industry. In selecting sustainable materials in UAE, Govindan et al. [11] used a hybrid MCDM model. Khorshidi et al. [12] used TOPSIS and preference selection index (PSI) methods and applied a comparative analysis approach in selecting the desirable combination of aluminium silicate composites based on strength and workability characteristics. Preference ranking methods like PROMETHEE-I and PROMETHEE-II were used by Prasenjit et al. [13] in material selection. ELECTRE-I, a partial ranking, and ELECTRE-II, a complete ranking method, were employed by Chatterjee et al. [14] for material selection decision-making problems. Preference selection index (PSI) method was suggested by Maniya and Bhatt [15] for material selection problems where the relative importance of criteria was avoided. The selection of the desired natural fibre composite material for passenger vehicle design was done using the AHP method. Kumar et al. [16] proposed TOPSIS and PSI methods for arriving at the best suitable self-curing self-compacting mortars considering criteria such as slump flow, V-funnel time, compressive strength, sorptivity coefficient and acid durability loss factors.

It is evident from the literature mentioned above that the material selection predominantly used a hybrid MCDM approach, and the weightages of criteria were determined using the AHP method. In the present study, the AHP + TOPSIS and AHP + PROMETHEE-II hybrid MCDM approaches were utilized to select a suitable lime–GGBS-based mortar for repairing heritage structures. Criteria for selection included physical factors (porosity, water absorption, water absorption through capillarity), mechanical factors (compressive and transverse strengths), durable factors (drying shrinkage, alkali resistance, salt crystallization cycles) and sustainable factors (energy demand, total emissions, environmental load) based on the authors' previous investigations [17, 18].

2 Research Significance

Most heritage structures all over the world were constructed using locally available materials, and investigations have been carried out on ancient mortars worldwide. However, there are limited scientific studies on ancient mortars of heritage structures in the southern part of India [19, 20]. Lime was the predominant binder used in these heritage structures, along with additions of natural pozzolanic materials such as brick earth and volcanic tuff.

3 Materials

Binder components used in the study include hydrated lime (L) with specific gravity 2.24 and ground granulated blast-furnace slag (GGBS-G) with specific gravity 2.73. Zone II sand, which conforms to IS:386-2016 [21] and has a specific gravity of 2.68 and density of 1.46 g/cm^3, is used. Potable water, according to IS:456-2005 [22] and IS:10500-2012 [23], is used. The study considered mortars with binder:sand proportions of 1:3 (A) and 1:1 (B) by weight, with lime being replaced with GGBS from 0 to 75% by weight. Table 1 shows the legend and mortar mix proportions.

Table 1 Legend and mortars mix proportions

Legend	Hydrated lime (L) (in Kg)	Replacement of lime with GGBS (in %)	GGBS (G) (in Kg)	Binder:sand	Sand (in Kg)	Potable water (in Kg)
LA	405	0	0	1:3	1215	305
LG30A	304	30	130		1302	313
LG50A	227	50	227		1362	310
LG66A	156	66	310		1398	307
LG75A	119	75	355		1422	293
LB	692	0	0	1:1	692	416
LG30B	520	30	223		743	414
LG50B	398	50	398		796	390
LG66B	279	66	557		836	381
LG75B	214	75	641		855	376

4 Methods and Discussions

A hybrid MCDM model, which consists of AHP + TOPSIS and AHP + PROMETHEE methods, is proposed and depicted in Fig. 1. Different attributes related to the problem are identified based on expert experience and literature survey. The alternatives are lime-based mortars shown in Table 1, from which the best alternative mortar is to be obtained for repairing heritage structures using this proposed model. In this study, various attributes were identified as per authors' previous investigations [17, 18], such as compressive strength (N/mm^2) (A1), transverse strength (N/mm^2) (A2), salt crystallization cycles (Nos) (A3), shrinkage (mm/m) after 70 days (A4), alkali resistance (%) (A5), water absorption (%) (A6), porosity (%) (A7), capillary absorption ($Kg/m^2.min^{0.5}$) (A8), energy consumption (MJ) (A9), total emissions ($Kg.CO_2$ equivalent) (A10), environmental load (points) (A11) and bulk density (Kg/m^3) (A12).

The attributes are compared pairwise using analytical hierarchy process (AHP), and their consistency is measured with the consistency ratio (CR). Consistency of the data is considered achieved if CR ≤ 0.1. Then, the weightages of the attributes are computed. Favourable and unfavourable attributes are identified and listed in Table 2. The decision matrix showing the data values of all the alternate mortars in relation to the attributes is presented in Table 3.

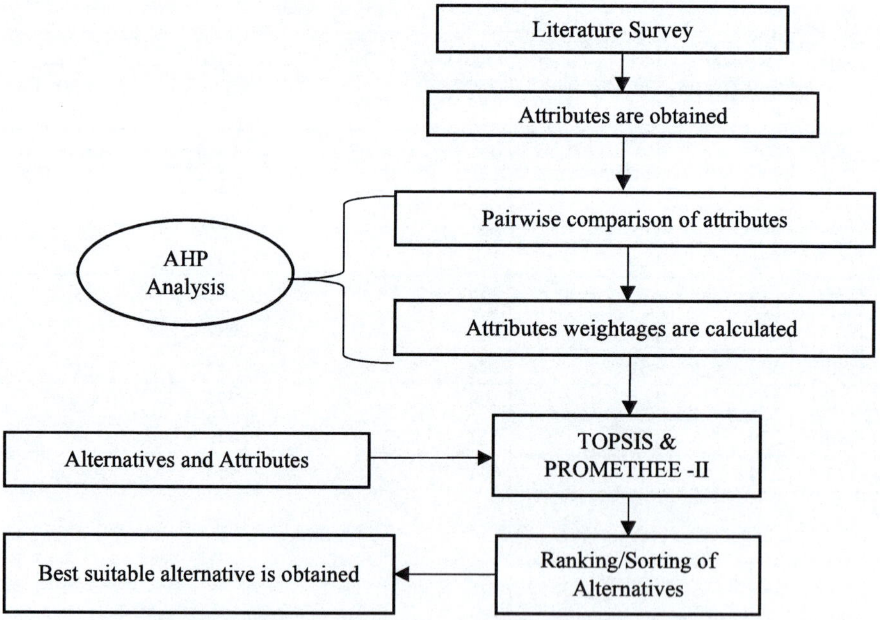

Fig. 1 Hybrid MCDM approach for the selection of best suitable lime-based mortar

Table 2 Attributes and their weightages using AHP analysis

Attribute	Favourable/Unfavourable	Weights	Consistency ratio (CR)
Compressive strength (N/mm^2) (A1)	Favourable	0.217	$CR = 0.07 \leq 0.1$ (Hence, the data are reliable and consistent)
Transverse strength (N/mm^2) (A2)	Favourable	0.164	
Salt crystallization cycles (Nos) (A3)	Favourable	0.113	
Shrinkage (mm/m) after 70 days (A4)	Unfavourable	0.097	
Alkali resistance (%) (A5)	Unfavourable	0.085	
Water absorption (%) (A6)	Unfavourable	0.074	
Porosity (%) (A7)	Unfavourable	0.065	
Capillary absorption ($Kg/m^2.min^{0.5}$) (A8)	Unfavourable	0.056	
Energy consumption (MJ) (A9)	Unfavourable	0.045	
Total emissions ($Kg.CO_2$ equivalent) (A10)	Unfavourable	0.038	
Environmental load (points) (A11)	Unfavourable	0.032	
Bulk density (Kg/m^3) (A12)	Favourable	0.014	

4.1 TOPSIS Method

In the TOPSIS method depicted in Fig. 2, vector normalization of data values for all alternatives with respect to attributes is performed, and the results are presented in Table 4. The weighted normalized matrix is obtained by multiplying the weightages of attributes with these normalized values. The positive ideal solution (PIS) and negative ideal solution (NIS) are obtained and presented in Table 5. The suitable alternate mortar is chosen as the one closest to the positive ideal solution. The square-root mean distance of alternatives from PIS and NIS, denoted as (S_i^+) and (S_i^-), respectively, is calculated and is plotted as shown in Fig. 3. The relative closeness coefficient (C_i) is then computed using these distances as illustrated in Table 6. The alternative with the highest (C_i) is selected as the best alternate mortar, representing its distance from NIS. Table 6 presents the sorting and ranking of all the alternatives based on their relative closeness coefficients. It is observed that lime-based mortars (LG66B and LG66A) showed high relative closeness coefficient values of 0.908 and 0.542, respectively, amongst 1:1 and 1:3 (binder:sand) mortars.

Table 3 Decision matrix of alternatives

S. No	Alternatives	A1	A2	A3	A4	A5	A6	A7	A8	A9	A10	A11	A12
1	LA	1.39	0.53	9	3.35	25.5	28.45	45.07	1.62	3847	503	123,314	1924
2	LG30A	2.84	1.08	15	2.92	12.95	22.05	35.06	1.021	3797	436	107,029	2046
3	LG50A	3.95	1.50	16	2.38	12.71	21.81	34.54	0.928	3744	385	94,377	2124
4	LG66A	4.49	1.71	17	2.18	11.72	20.82	33.88	0.781	3651	335	81,996	2166
5	LG75A	3.68	1.40	17	2.08	12.94	22.04	35.47	0.58	3614	309	75,736	2184
6	LB	1.78	0.68	8	4.45	24.27	27.22	42.98	1.30	4633	727	178,475	1799
7	LG30B	7.89	1.93	13	3.94	10.85	19.95	33.08	0.74	4417	605	148,545	1899
8	LG50B	8.38	2.19	14	3.11	10.61	19.71	32.56	0.65	4333	524	128,400	1981
9	LG66B	9.02	2.46	15	3.01	9.42	18.52	31.23	0.51	4203	441	107,996	2051
10	LG75B	8.45	2.32	14	2.93	10.64	19.74	32.82	0.39	4120	394	96,667	2084

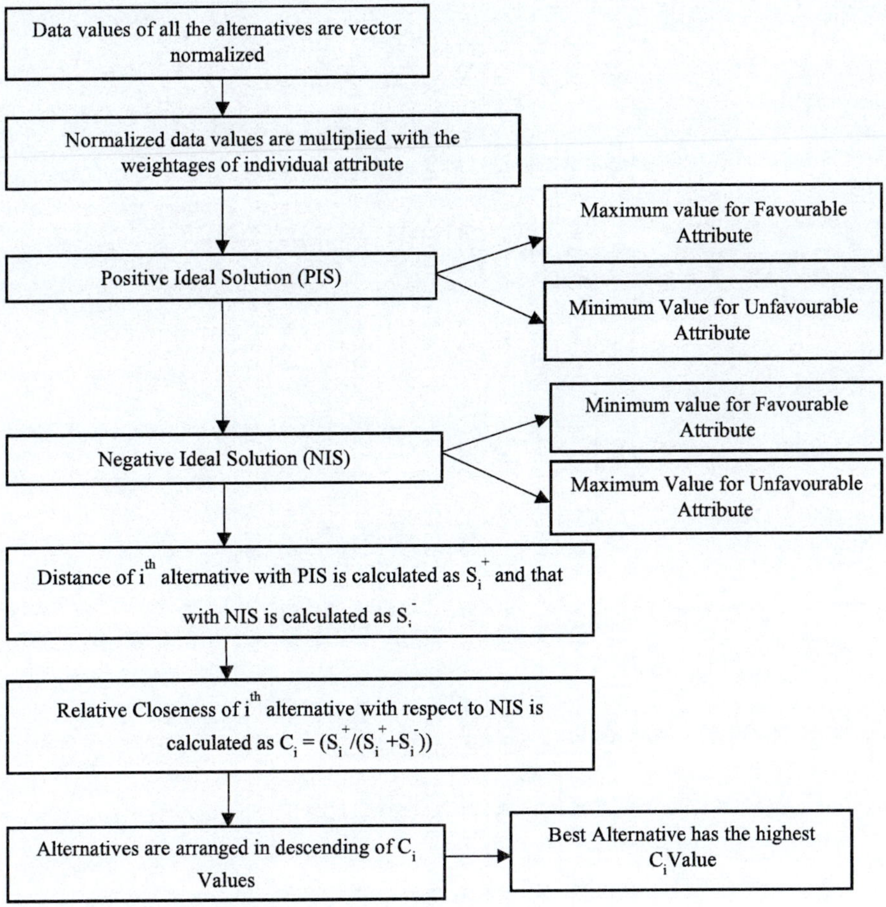

Fig. 2 TOPSIS methodology

4.2 PROMETHEE-II Method

The PROMETHEE-II method is a complete ranking method in which the data values obtained for all the alternatives with respect to all the attributes are normalized using minimum and maximum data values from Table 7, as shown in Fig. 4. The normalized data values are presented in Table 8. The establishment of preference of one alternative with another is done, and then weightages of attributes are multiplied with these preference values to obtain aggregate preference values. Positive ($\emptyset_a^+$) and negative flows ($\emptyset_a^-$) indicating the preference and non-preference of a particular alternative with all the other alternatives are calculated and are plotted as shown in Fig. 5. Net outranking flow ($\emptyset_a$), which is the difference of positive and negative flow, is calculated for all the alternatives. Alternatives are then arranged in the

Table 4 Normalized matrix in TOPSIS method

S. No	Alternatives	A1	A2	A3	A4	A5	A6	A7	A8	A9	A10	A11	A12
1	LA	0.074	0.098	0.202	0.340	0.531	0.404	0.397	0.554	0.300	0.330	0.330	0.300
2	LG30A	0.152	0.200	0.336	0.296	0.270	0.313	0.309	0.349	0.296	0.287	0.287	0.319
3	LG50A	0.212	0.279	0.359	0.241	0.265	0.310	0.304	0.317	0.292	0.253	0.253	0.331
4	LG66A	0.241	0.317	0.381	0.221	0.244	0.296	0.298	0.267	0.285	0.220	0.220	0.338
5	LG75A	0.197	0.260	0.381	0.211	0.269	0.313	0.312	0.198	0.282	0.203	0.203	0.340
6	LB	0.096	0.126	0.179	0.452	0.505	0.387	0.378	0.445	0.362	0.478	0.478	0.280
7	LG30B	0.423	0.358	0.291	0.400	0.226	0.284	0.291	0.253	0.345	0.398	0.398	0.296
8	LG50B	0.449	0.406	0.314	0.316	0.221	0.280	0.287	0.222	0.338	0.344	0.344	0.309
9	LG66B	0.484	0.457	0.336	0.305	0.196	0.263	0.275	0.174	0.328	0.289	0.289	0.320
10	LG75B	0.453	0.432	0.314	0.297	0.222	0.281	0.289	0.135	0.322	0.259	0.259	0.325

Table 5 Positive and negative ideal solutions (PIS and NIS)

Attribute	Positive ideal solution (PIS)	Negative ideal solution (NIS)
Compressive strength (N/mm^2) (A1)	0.105	0.016
Transverse strength (N/mm^2) (A2)	0.075	0.016
Salt crystallization cycles (Nos) (A3)	0.043	0.020
Shrinkage (mm/m) after 70 days (A4)	0.020	0.044
Alkali resistance (%) (A5)	0.017	0.045
Water absorption (%) (A6)	0.019	0.030
Porosity (%) (A7)	0.018	0.026
Capillary absorption ($Kg/m^2.min^{0.5}$) (A8)	0.008	0.031
Energy consumption (MJ) (A9)	0.013	0.016
Total emissions ($Kg.CO_2$ equivalent) (A10)	0.008	0.018
Environmental load (points) (A11)	0.007	0.015
Bulk density (Kg/m^3) (A12)	0.005	0.004

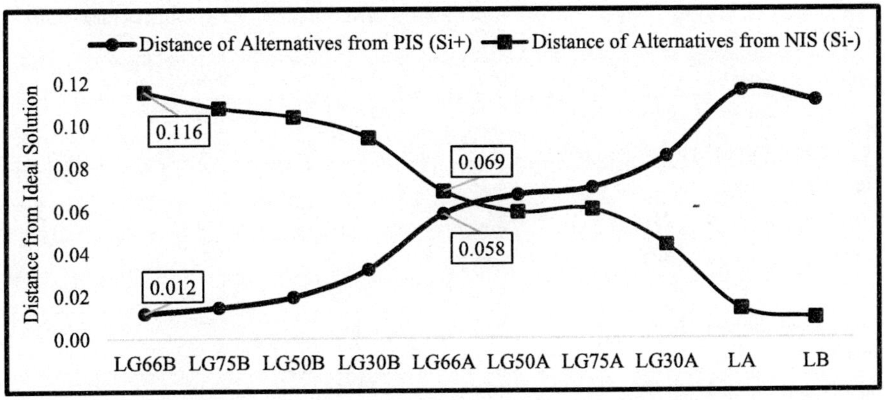

Fig. 3 Distances of alternatives mortars from positive and negative ideal solutions

descending order of these net outranking flows as shown in Table 9, and the alternative having the highest net outranking flow is considered as the best suitable alternate mortar. It can be observed that lime-based mortars (LG66B and LG66A) showed high net outranking flow values of 0.317 and 0.160, amongst 1:1 and 1:3 (binder:sand) mortars, respectively.

Table 6 Relative closeness coefficient and ranking of alternatives from TOPSIS method

Alternatives	Distance from PIS (S_i^+)	Distance from NIS (S_i^-)	Relative closeness coefficient $C_i = S_i^-/(S_i^+ + S_i^-)$	Rank
LG66B	0.012	0.116	0.908	1
LG75B	0.014	0.108	0.883	2
LG50B	0.019	0.104	0.843	3
LG30B	0.032	0.094	0.745	4
LG66A	0.058	0.069	0.542	5
LG50A	0.067	0.059	0.468	6
LG75A	0.071	0.060	0.461	7
LG30A	0.085	0.044	0.339	8
LA	0.116	0.014	0.105	9
LB	0.111	0.009	0.078	10

Table 7 Minimum and maximum data values of attributes amongst all the alternative mortars

Attribute	Maximum value (max)	Minimum value (min)	Difference (max–min)
Compressive strength (N/mm^2) (A1)	9.02	1.39	7.64
Transverse strength (N/mm^2) (A2)	2.46	0.53	1.94
Salt crystallization cycles (Nos) (A3)	17	8	9
Shrinkage (mm/m) after 70 days (A4)	4.45	2.08	2.37
Alkali resistance (%) (A5)	25.50	9.42	16.08
Water absorption (%) (A6)	28.45	18.52	9.93
Porosity (%) (A7)	45.07	31.23	13.84
Capillary absorption ($Kg/m^2.min^{0.5}$) (A8)	1.62	0.39	1.23
Energy consumption (MJ) (A9)	4633	3614	1019
Total emissions ($Kg.CO_2$ equivalent) (A10)	727	309	418
Environmental load (points) (A11)	178,475	75,736	102,739
Bulk density (Kg/m^3) (A12)	2184	1799	385

5 Conclusions

The study utilized hybrid MCDM approaches AHP + TOPSIS and AHP + PROMETHEE-II to determine the best suitable lime-based mortars for the repair of heritage structures. From the study, the following conclusions were derived:

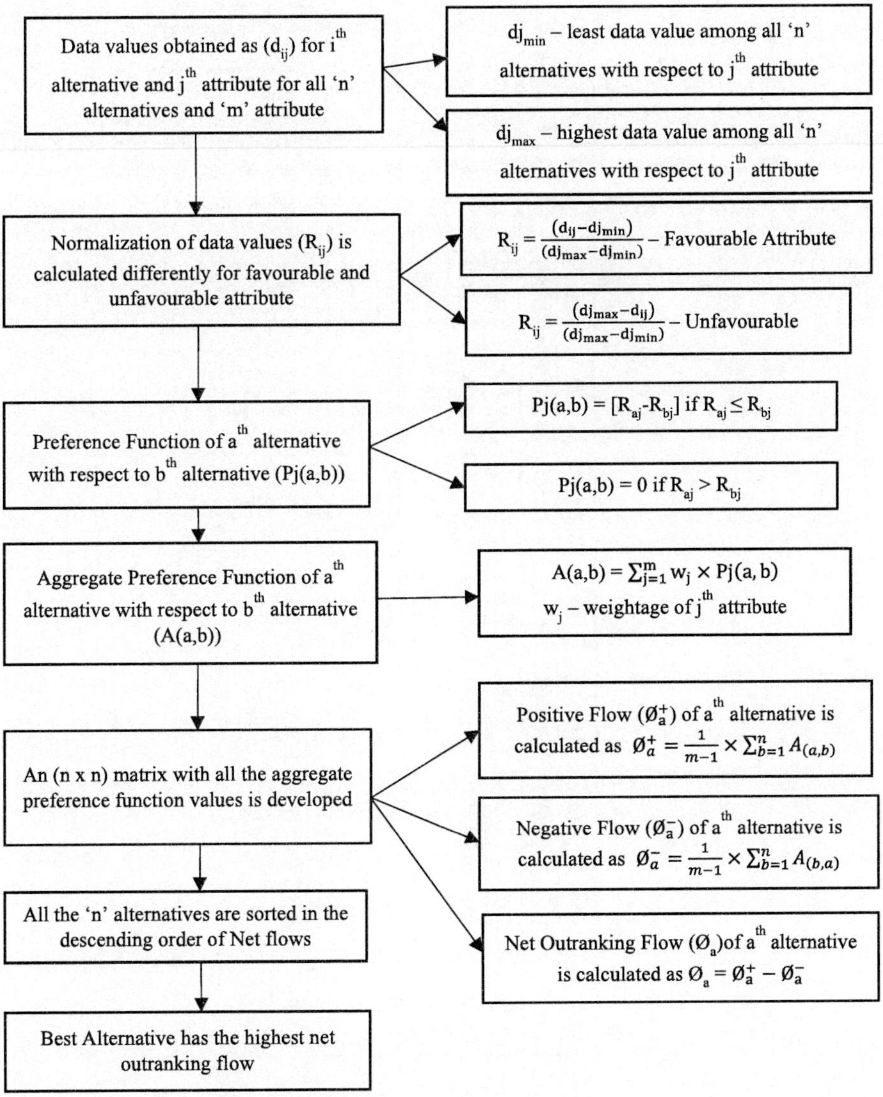

Fig. 4 PROMETHEE-II methodology

- Twelve different attributes namely compressive strength (A1), transverse Strength (A2), salt crystallization cycles (A3), shrinkage (A4), alkali resistance (A5), water absorption (A6), porosity (A7), capillary absorption (A8), energy consumption (A9), total emissions (A10), environmental load (A11) and bulk density (A12) were considered in the study.

Table 8 Normalized matrix in PROMETHEE-II method

Alternatives	A1	A2	A3	A4	A5	A6	A7	A8	A9	A10	A11	A12
LA	0.000	0.000	0.111	0.464	0.000	0.000	0.000	0.000	0.771	0.537	0.537	0.325
LG30A	0.190	0.286	0.778	0.646	0.780	0.645	0.723	0.489	0.820	0.696	0.695	0.642
LG50A	0.335	0.503	0.889	0.873	0.795	0.669	0.761	0.564	0.872	0.819	0.819	0.844
LG66A	0.406	0.609	1.000	0.958	0.857	0.768	0.809	0.684	0.964	0.939	0.939	0.953
LG75A	0.301	0.451	1.000	1.000	0.781	0.646	0.694	0.850	1.000	1.000	1.000	1.000
LB	0.052	0.079	0.000	0.000	0.076	0.124	0.151	0.259	0.000	0.000	0.000	0.000
LG30B	0.852	0.723	0.556	0.215	0.911	0.856	0.866	0.717	0.212	0.291	0.291	0.260
LG50B	0.916	0.857	0.667	0.565	0.926	0.880	0.904	0.793	0.294	0.487	0.487	0.473
LG66B	1.000	1.000	0.778	0.608	1.000	1.000	1.000	0.907	0.422	0.686	0.686	0.655
LG75B	0.925	0.929	0.667	0.641	0.924	0.877	0.885	1.000	0.503	0.796	0.796	0.740

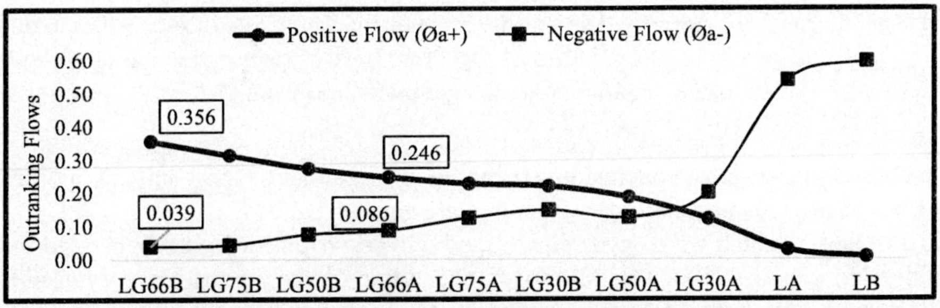

Fig. 5 Positive and negative outranking flows of alternate mortars

Table 9 Net outranking flow and ranking of alternatives from PROMETHEE-II method

Alternatives	Sum of $\boldsymbol{\emptyset_a}^-$	Sum of $\boldsymbol{\emptyset_a}^-$	Sum of $\boldsymbol{\emptyset_a} = \boldsymbol{\emptyset_a}^+ - \boldsymbol{\emptyset_a}^+$	Rank
LG66B	0.356	0.039	0.317	1
LG75B	0.313	0.043	0.270	2
LG50B	0.272	0.075	0.196	3
LG66A	0.246	0.086	0.160	4
LG75A	0.226	0.124	0.103	5
LG30B	0.219	0.147	0.071	6
LG50A	0.185	0.126	0.060	7
LG30A	0.121	0.200	−0.079	8
LA	0.028	0.538	−0.509	9
LB	0.007	0.595	−0.588	10

- A good consistency in pairwise comparison between attributes was achieved through analytical hierarchy process (AHP), with a consistency ratio of 0.07, which is less than 0.10.
- The weightages of attributes A1, A2, A3, A4, A5, A6, A7, A8, A9, A10, A11 and A12 were obtained as 0.217, 0.164, 0.113, 0.097, 0.085, 0.074, 0.065, 0.056, 0.045, 0.038, 0.032 and 0.014, respectively, using AHP analysis.
- The highest relative closeness coefficient values (C_i) of 0.908 and 0.542 amongst 1:1 and 1:3 (binder:sand) proportion mortars were attained by lime–GGBS-based mortars LG66B and LG66A using the hybrid AHP + TOPSIS method.
- The distances of LG66B from positive ideal solution (PIS) and negative ideal solution (NIS) were $S_i^+ = 0.012$ and $S_i^- = 0.116$, respectively. Similarly, the distances of alternative LG66A from positive ideal solution (PIS) and negative ideal solution (NIS) were $S_i^+ = 0.058$ and $S_i^- = 0.069$, respectively.
- Amongst 1:1 and 1:3 (binder:sand) proportion mortars, the highest net outranking flow values ($\emptyset_a$) of 0.317 and 0.160 were attained by lime–GGBS-based mortars LG66B and LG66A, respectively, using the hybrid AHP + PROMETHEE-II methods.

- The alternative LG66B had a positive or entering flow value of $Ø_a^+ = 0.356$ and a negative or leaving flow value of $Ø_a^- = 0.039$. Likewise, for the alternative LG66A, positive and negative outranking flow values were $Ø_a^+ = 0.246$ and $Ø_a^- = 0.086$, respectively.

Based on the above conclusions, it can be inferred that the best suitable mortars for repairing masonry heritage structures are lime–GGBS-based mortars (LG66A and LG66B), which were determined based on various physical, mechanical, durability and sustainable factors. Moreover, depending on the type of masonry used in the heritage structure, LG66A is recommended for brick masonry, and LG66BO is recommended for stone masonry.

References

1. Saaty TL (1980) The analytic hierarchy process. McGraw Hill, New York
2. Vincke P (1992) Multicriteria decision aid. John wiley and sons Inc., New York
3. Vulević T, Dragović N (2017) Multi-criteria decision analysis for sub-watersheds ranking via the PROMETHEE method. Int Soil Water Conserv Res 5:50–55. https://doi.org/10.1016/j.iswcr.2017.01.003
4. Huang WF (2014) An improved TOPSIS method for material selection model. Adv Mater Res 951:120–123. https://doi.org/10.4028/www.scientific.net/AMR.951.120
5. Holloway L (1998) Materials selection for optimal environmental impact in mechanical design. Mater Des 19:133–143. https://doi.org/10.1016/S0261-3069(98)00031-4
6. Jahan A, Ismail MY, Sapuan SM, Mustapha F (2010) Material screening and choosing methods—a review. Mater Des 31:696–705. https://doi.org/10.1016/j.matdes.2009.08.013
7. Stanojević AD, Milošević MR, Milošević DM et al (2021) Developing multi-criteria model for the protection of built heritage from the aspect of energy retrofitting. Energy Build 250:111285. https://doi.org/10.1016/j.enbuild.2021.111285
8. Nadkarni RR, Puthuvayi B (2020) A comprehensive literature review of multi-criteria decision making methods in heritage buildings. J Build Eng 32:101814. https://doi.org/10.1016/j.jobe.2020.101814
9. Do J-Y, Kim D-K (2012) AHP-based evaluation model for optimal selection process of patching materials for concrete repair: focused on quantitative requirements. Int J Concr Struct Mater 6:87–100. https://doi.org/10.1007/s40069-012-0009-9
10. Anojkumar L, Ilangkumaran M, Sasirekha V (2014) Comparative analysis of MCDM methods for pipe material selection in sugar industry. Expert Syst Appl 41:2964–2980. https://doi.org/10.1016/j.eswa.2013.10.028
11. Govindan K, Madan Shankar K, Kannan D (2016) Sustainable material selection for construction industry—A hybrid multi criteria decision making approach. Renew Sustain Energy Rev 55:1274–1288. https://doi.org/10.1016/j.rser.2015.07.100
12. Khorshidi R, Hassani A (2013) Comparative analysis between TOPSIS and PSI methods of materials selection to achieve a desirable combination of strength and workability in Al/SiC composite. Mater Des 52:999–1010. https://doi.org/10.1016/j.matdes.2013.06.011
13. Chatterjee P, Chakraborty S (2012) Material selection using preferential ranking methods. Mater Des 35:384–393. https://doi.org/10.1016/j.matdes.2011.09.027
14. Chatterjee P, Athawale VM, Chakraborty S (2009) Selection of materials using compromise ranking and outranking methods. Mater Des 30:4043–4053. https://doi.org/10.1016/j.matdes.2009.05.016

15. Maniya K, Bhatt MG (2010) A selection of material using a novel type decision-making method: preference selection index method. Mater Des 31:1785–1789. https://doi.org/10.1016/j.matdes.2009.11.020
16. Kumar DN, Kumar PR, Reddi SA, Madduru SRC (2022) Prioritisation of self-curing self-compacting mortars with hydrophilic chemicals using MCDM techniques. Int J Comput Mater Sci Surf Eng 11:1. https://doi.org/10.1504/IJCMSSE.2022.125762
17. Kumar DN, Pancharathi RK (2022) Life cycle assessment to identify sustainable lime-pozzolana binders for repair of heritage structures. In: Advances in sustainable materials and resilient infrastructure. pp 41–59
18. Nikhil Kumar D, Rathish Kumar P (2022) Investigations on alternate lime-pozzolana based mortars for repair of heritage structures. Constr Build Mater 341:127776. https://doi.org/10.1016/j.conbuildmat.2022.127776
19. Degloorkar NK, Pancharathi RK (2020) Investigation of microstructure characterization of mortars from 800 years old heritage structures in Southern part of India. J Archaeol Sci Reports 34:102634. https://doi.org/10.1016/j.jasrep.2020.102634
20. Degloorkar NK, Pancharathi RK (2020) Characterization of ancient mortar for sustainability of an 800 year old heritage site in India. Mater Today Proc 32:734–739. https://doi.org/10.1016/j.matpr.2020.03.472
21. IS:383 (2016) Coarse and fine aggregate for concrete-specification. Bureau of Indian Standards, New Delhi
22. IS:456 (2005) Plain and reinforced concrete—Code of practice. Bureau of Indian Standards, New Delhi
23. IS:10500 (2012) Drinking Water—Specification. Bureau of Indian Standards, New Delhi

Effect of Window Glazing Materials on Life Cycle Energy Performance of Building: A Case Study

Akshitha Kasula and Suchith Reddy Arukala

1 Introduction

The availability of energy resources is essential to everyday activities. Reducing energy consumption is one of the major goals of energy-efficient consumption. There are various types of buildings, including residential, commercial, and industrial. It is observed that buildings are essential to our country's socio-economic development but consume a lot of energy. Buildings are responsible for consuming nearly 47% of all main energy and large amounts of greenhouse gases are released into the atmosphere, which is why there is an increase in global warming. Hence, energy analysis is essential [1]. Structures require energy from the time they are built until they are demolished. As the energy consumption of a building can be minimized through alternate designs, it is more significant to examine it during the design phase rather than during construction. Although the number of commercial structures built with a glass facade has expanded rapidly, numerous examinations have revealed that the use of heating and cooling in these structures is excessive, indicating low energy efficiency. Construction companies need to develop sustainable buildings which conserve and utilize energy efficiently to meet this challenge.

A. Kasula · S. R. Arukala (✉)
Department of Civil Engineering, Kakatiya Institute of Technology and Science, Warangal, Telangana 506015, India
e-mail: Asr.ce@kitsw.ac.in

A. Kasula
e-mail: M20sc022@kitsw.ac.in

R. K. Pancharathi et al. (eds.), *Low Carbon Materials and Technologies for a Sustainable and Resilient Infrastructure*, Lecture Notes in Civil Engineering 440,
https://doi.org/10.1007/978-981-99-7464-1_17

1.1 Definition of Life Cycle Energy and Its Components

The term "life cycle energy" refers to a method of accounting for all energy inputs into a structure throughout the course of its existence. An analysis of this system includes phases such as construction, use, and demolition. Building materials are produced, transported, and technical equipment is installed during the production process to establish and restore structures. Operational activities are those related to the building's use throughout its life cycle. These activities include actions such as maintaining pleasant conditions within buildings, using water, and operating appliances. The demolition step includes demolishing the structure and moving the disassembled debris to landfills.

1.2 Significance of BIM

A building's information model includes geometric information, building components, material properties, cost of the materials used, HVAC equipment, and electrical systems. The major reasons for implementing BIM tools are to automate the modelling process, make construction documents more accurate, improve communication between the parties involved in the design stage and construction stage, and automatically reflect updates in all aspects after one view has been modified in order to reduce field coordination issues [2]. There are several tools referred to BIM tools such as Allplan, ArchiCAD, Microstation, Revit, and Vectorworks. GBS is integrated with Revit through Autodesk, requires only free activation of Autodesk ID to facilitate energy modelling in this CAD tool. According to research, ArchiCAD and Revit are the most common BIM authoring tools, with Revit being more user friendly. According to literature reviews, there is a research gap in current studies. [3–5], such that when data are transferred from BIM to BEM, users discover that certain data (HVAC data) are missing and must be manually added. Energy simulation is performed in energy plus by extracting the IFC file from Revit [5]. The present work covers energy modelling using Green Building Studio by importing data from the Revit BIM tool in gbXML format.

1.3 Green Building Studio

The Green Building Studio is known as an online-based energy evaluation tool that gives designers access to whole-building energy assessment, enhancing energy effectiveness, and ensuring carbon neutrality before construction begins. In GBS, energy usage is calculated hourly using the DOE-2 simulation engine. The DOE-2 program simulates an hourly building's layout, construction, operating schedules, conditioning systems (lighting, HVAC, etc.), and utility rates, along with weather data, to estimate utility bills for the building on an hourly basis.

1.4 Characteristics of Glass

Solar Factor: The solar factor (or g-value) is a measurement of the amount of heat that passes through the glass.

Shade Coefficient: The shading coefficient (SC) is a measurement of the heat uptake from solar radiation through glass.

U-value: An insulating capacity of glass can be determined by its U value, which measures how well it can resist heat during high temperatures (excluding direct sun rays).

Light Reflectance and Transmission: Glass allows light to pass through it. Reflection and absorption can alter the amount of light transmitted. The exterior wall is considered as a single glazing with the parameters mentioned in Table 1.

2 Research Methodology

Climate conditions were identified as a key factor that affects the energy performance of buildings, particularly those that use external facade glazing [7]. As shown in the flowchart of research methodology in Fig. 1, a case study research design was used to investigate the impact of window glazing materials on the life cycle energy performance of the Prestige Polygon building located in Chennai, India. Literature reviews and observations were conducted to collect data on sustainable building

Table 1 Parameters of glass used in the building

Parameter	Glass (ET125 clear cosmos)
Solar factor	0.27
SHGC	1.00
U-value	1.18 BTU/h.ft^2.°F
Light transmission	34%
External reflection	26%
Internal reflection	9%
Height	2.35 m for first floor to 13th floor, 4.60 m for ground floor
Type	External single glazing with 12 mm thickness

design and materials, as well as the existing building's energy usage and the glazing materials used. Based on the data collected, the two best alternatives were identified.

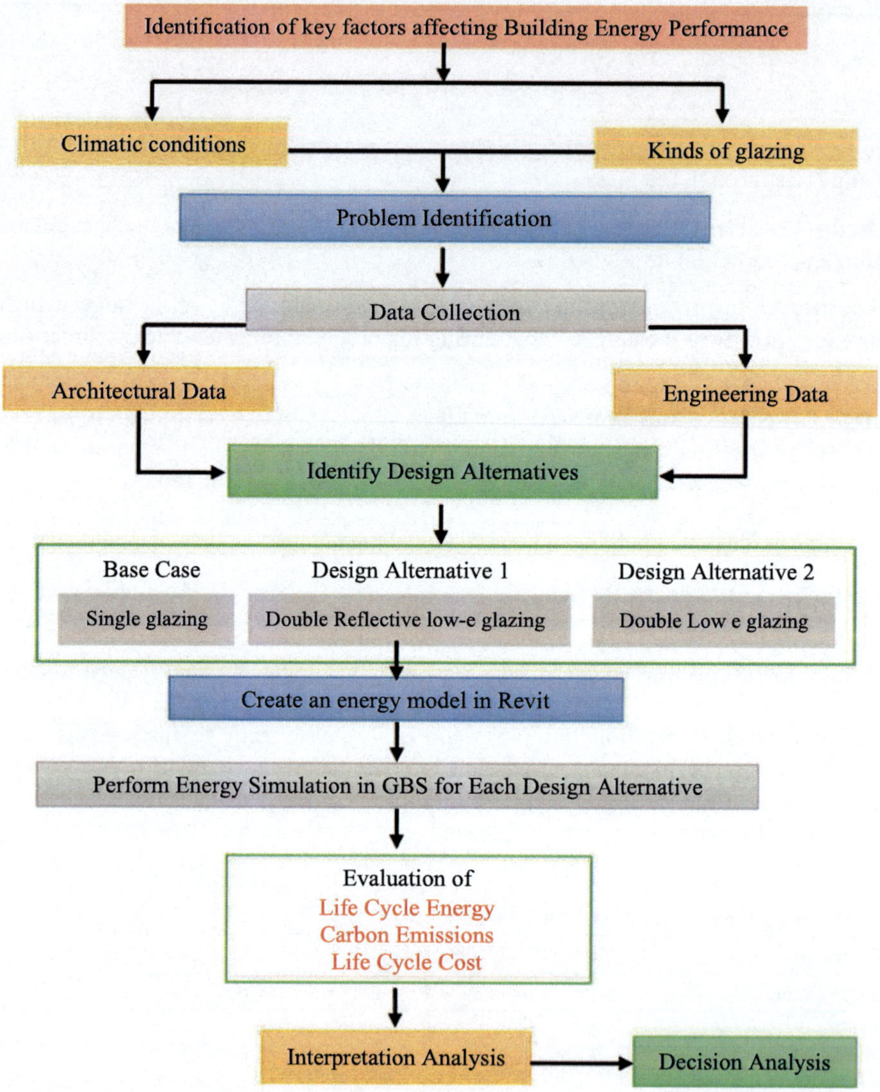

Fig. 1 Flowchart of life cycle energy analysis

Table 2 The optical properties of the selected glass types

	Double reflective with low-e glazing	Double low-e glazing
Solar factor	0.15	0.17
SHGC	0.39	0.42
U-value (BTU/h. ft^2.°F)	0.22	0.35
Light transmission %	13	24
External reflection %	28	20
Internal reflection %	31	22

3 Case Study

According to Ramachandran et al. [6], Chennai produces greater greenhouse gas emissions and carbon dioxide equivalent per person. Chennai has a tropical wet and dry climate, also known as a tropical savanna climate. The city experiences high temperatures and relatively high humidity throughout the year, with an average annual temperature of around 28 °C (82°F). The study found that buildings with external facade glazing have a greater impact on their energy performance than buildings without it. It was discovered that a certain number of buildings consumed a significant amount of electricity, from which the Prestige Polygon building was selected. The building is located at latitude 13.042N and longitude 80.131E and has 13 storey. The external façade of the building is made of Saint Gobain glass. Energy simulation is performed and compared by comparing the present glazing with the other two types of glazing and their parameters. Furthermore, a cost analysis is carried out on the building to determine which type of glass is most cost-effective to utilize.

3.1 Design Alternatives

Different cases are chosen for energy analysis in order to determine the maximum energy savings. By replacing the typical glass form of single glazing with other types of glass and by considering climatic conditions, two glazing are found to be more suitable [8], as shown in Table 2. The types of glazing considered are represented in Fig. 2.

3.2 Embodied Energy of the Building

The components of a building consist of a variety of materials, only the building's principal envelope components, such as facade glazing, walls, floors, and roof, are examined. The total embodied energy values of the multi-storey building are shown

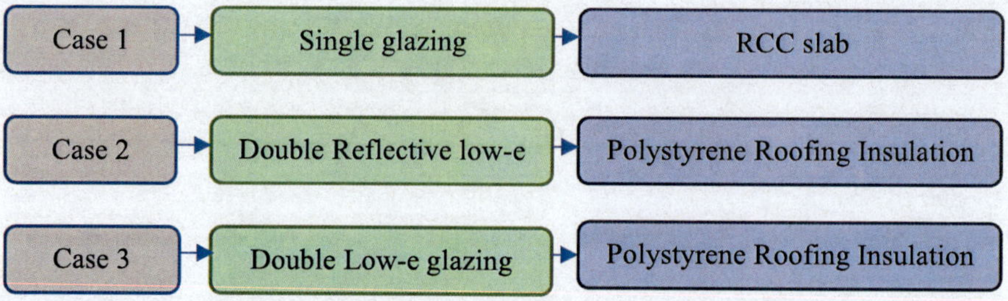

Fig. 2 Overview of three case studies

Table 3 The total embodied energy of the office building

Building component	Embodied energy (MJ)
Wall	2,166,504.663
Roof slab	3,221,267.35
Floor slab	38,404,432.35
Insulation	487,947.89
Single glass	14,606,253.12
Reflective low-e glass	21,964,150.24
Low-e glass	22,316,343.47

in Table 3. The data for energy content were taken from embodied and operational energy of urban residential buildings in India [8]. Embodied energy is the product of quantity of the building component and its embodied energy coefficients.

3.3 Modelling Process in Revit

The building data were obtained from the website Prestige Group, and it consists of the building's plan and elevation features. The study has created a 3D model in Revit using this data. The construction and energy setting inputs are represented in Table 4. The total area of the building is 607,859 ft^2, of which the building area is 478,462 ft^2.

Case 1 is performed using the exterior wall as a single glazing with the parameters as mentioned in Table 1. As it is an office building, the operating hours are 24/7. The HVAC system used is Central VAV, Electric Resistance Heat, Chiller 5.96 COP, which refers to a water-cooled centrifugal chiller having a coefficient of 5.96, and a high-efficiency motor, water pump is all included. Heating and cooling temperatures are supplied based on the city's average temperatures.

Table 4 Properties of the Revit-modelled building

	Building component	Properties
Construction	Square footage	47,462
	Exterior wall	Single glazing
	Interior wall	Gypsum board/metal stud/gypsum board (R-21)
	Roof	RCC slab
	Ceiling	Acoustic ceiling tile (R-1.6)
	Space	Office—70 ft^2/head Lobby—120 ft^2/head
Energy settings	Location	Chennai, IN
	Building type	Office
	Building operating schedule	24/7
	HVAC system	Central VAV, electric resistance heat, chiller 5.96 COP
	Thermostat set points	Heating—31 °C, Cooling—22 °C
	Outdoor air per person	8.00 L/s
	Export category	Spaces
	Project phase	New construction
	Building service	VAV-single duct
	Building infiltration class	Medium
	Export default values	Yes

3.4 *Energy Model in Revit*

After creating a 3D model, it is analysed for energy optimization. Energy settings are critical for defining the building's data. Figure 3 shows how to use the generate options to create an energy model as shown in Fig. 4. When the energy simulation is performed in energy plus software, other options such as optimizing and system analysis are used to export the model to Revit as a gbXML file.

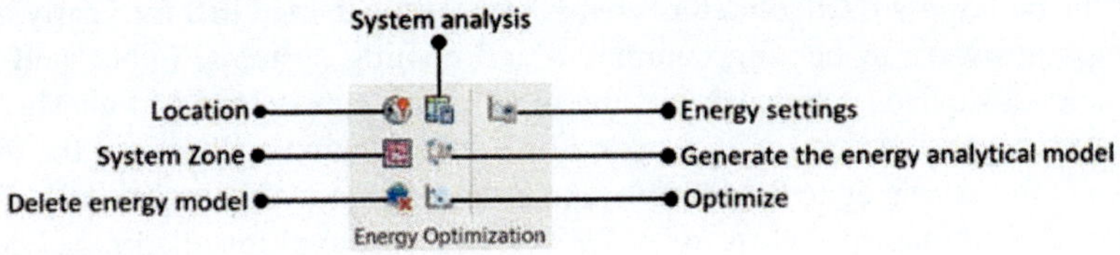

Fig.3 Energy settings in Revit

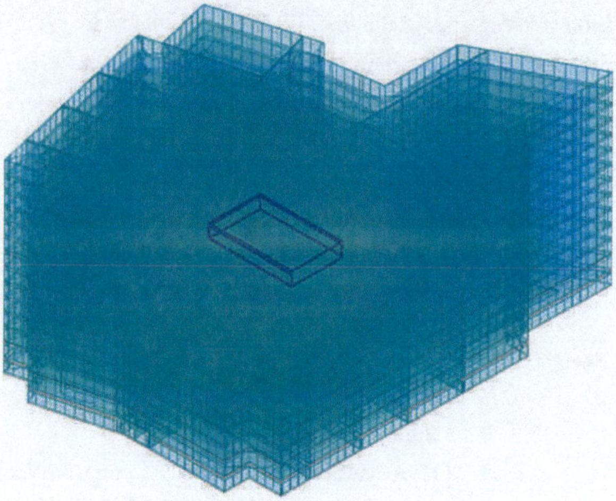

Fig. 4 Energy model of prestige polygon building

3.5 Energy Simulation in Green Building Studio (GBS)

GBS is a free, user-friendly software. The creation of a project involves four steps: (1) Enter project details such as the project name, building type, and project type. (2) Indicate the building's location using maps; (3) Accept the terms and conditions; and (4) Select the data access preference. Building properties, spaces, zones, surfaces, openings, and HVAC are all default values in the GBS.

4 Results and Discussion

4.1 Base Case—Single Clear Glazing

The total floor area is 44,068 ft^2. The number of people assumed on each floor is 1650 people which is lower than typical value. The annual energy and life cycle cost of the building are calculated in energy simulation using GBS for yearly power consumption owing to heating equipment and cooling systems, lights, and other miscellaneous devices. The peak cooling month and hour is found to be May 2:00 PM. It was suggested that using a period of 9-h ventilation will reduce the annual electric energy saving up to 8,844 kWh, and usage of photovoltaic potential will lead to a high annual energy savings of 967,835 kWh with total installed panel cost of Rs. 4,522,240. The results obtained for base case are represented in Fig. 5.

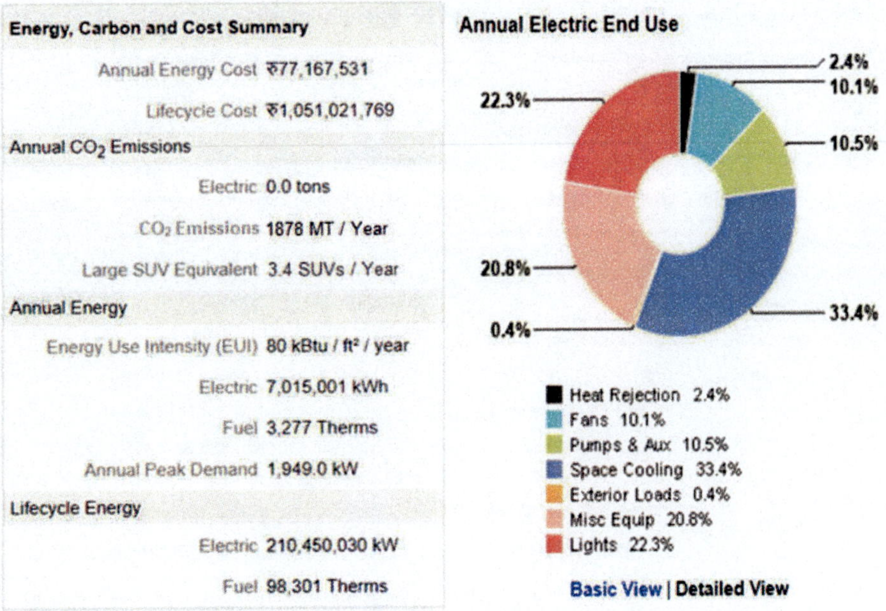

Fig. 5 Energy and carbon results for single glazed building

4.2 *Design Alternative 1—Double Reflective with Low-e Glazing*

The first design alternative for base case is by replacing single clear glazing with double reflective with low-e glazing. The energy use intensity for double reflective low-e glazing is 65 kBtu/ft^2/year, with electric consumption of 5,678,265 kWh and fuel consumption of 3,277 therms. It was suggested that using a period of 9-h ventilation will reduce the annual electric energy saving up to 5453 kWh. The results obtained are represented in Fig. 6. The space cooling consumes more electric usage of nearly 28.4%, and exterior loads consume very less amount of electricity of approximately 0.6%.

The annual energy simulated of double glazed is 5,678,265 *kWh* which is 19.05% less than single glazed building, ultimately the life cycle energy is also 19.05% less than the base case. Hence, the cost of the electricity is 19.05% less than base case.

4.3 *Design Alternative 2—Double Low-e Glazing*

The second design alternative for base case is by replacing single clear glazing with double low-e glazing. The cost analysis shows that double low-e glazing has an

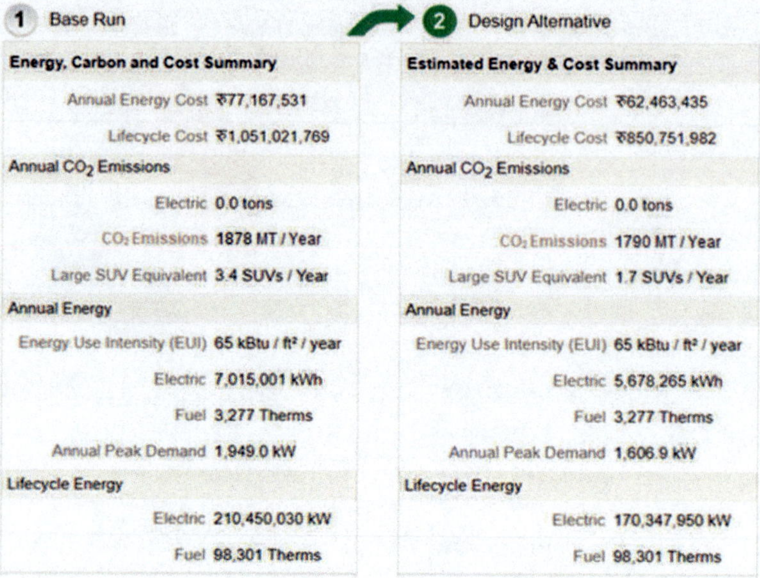

Fig. 6 Energy and carbon results for double reflective with low-e glazed building

annual energy cost of $62,503,640 and a lifecycle cost of $851,299,574, which is higher than single clear glazing but provides better energy efficiency, resulting in lower energy consumption and CO_2 emissions. The results are represented in Fig. 7.

The peak cooling month and hour is found to be May 2:00 PM. It was suggested that using a period of 9-h ventilation will reduce the annual electric energy saving up to 5488 kWh. The annual electric usage of HVAC and other miscellaneous equipment are represented in Fig. 8.

4.4 Carbon Emission Analysis

Due to the magnitude of the possible risk posed by global environmental change, majority of construction projects under consideration are focused on this effect. Even though standards and laws are established to reduce carbon emissions from new constructions, the proportional importance of the consequences of the carbon footprint grows. The carbon emissions per year is represented in the below bar chart such that single glazing has more carbon emission, followed by double reflective with low-e glazing and with least emission by low-e because low-e films are more reflective by reflecting waves of infrared waves (Fig. 9).

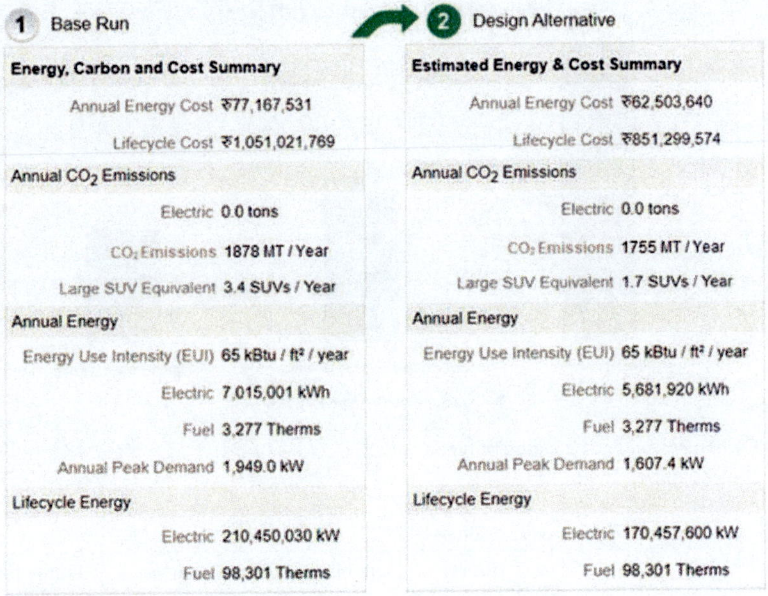

Fig.7 Energy and carbon results for double low-e glazed building

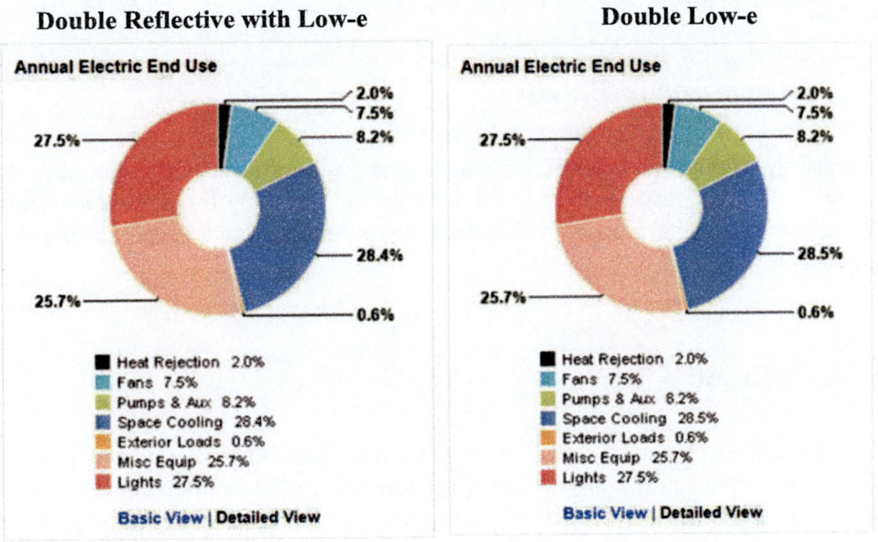

Fig. 8 Energy end-use charts

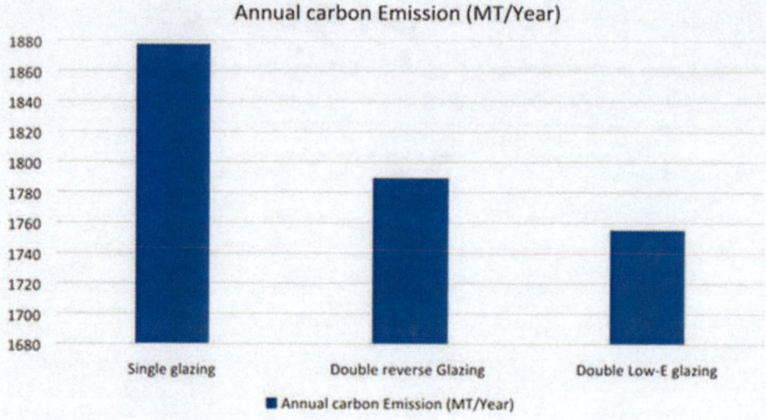

Fig. 9 Carbon emission of prestige polygon

Table 5 Comparison of three case studies

	Single	Double reflective with low-e	Double low-e
Life cycle energy (kW) × 10^6	210 0.45	170.34	170.45
Carbon emission (MT) × 10^3	93.900	89.500	87.750
Life cycle cost (crores)	105.10	85.07	85.13

4.5 Comparison of Results

The energy results, carbon emission results, and cost of the materials are compared for three different cases as below. The findings are primarily focused on choosing cost-effective glazing, glazing with a lower environmental impact, and glazing that saves energy (Table 5).

5 Conclusion

India has diverse climatic conditions, ranging from hot and dry to humid and tropical, which necessitates designing buildings to withstand various environmental conditions. Buildings must withstand temperature, humidity, monsoons, wind, and solar radiation to ensure occupant comfort, health, and safety. Ignoring local climatic conditions during building construction can result in negative impacts, such as increased energy consumption, reduced indoor air quality, higher maintenance costs, and safety hazards.

- In India, single glazing is commonly used for tropical monsoon climates, while double-pane glazing is suitable for tropical wet climates. Triple-pane glazing is ideal for Alpine climates, while low-e coatings work well for tropical semi-arid climates. Tinted or reflective glazing is suitable for arid climates, which have low humidity and high temperatures. These types of windows and coatings are designed to withstand different environmental conditions, including high temperatures, heavy rainfall, snowfall, and the sun's rays, to ensure building occupant comfort and energy efficiency.
- The study conducted in Green Building Studio indicates that double reflective low-e and double low-e glazing systems have 19.05% and 19.00% lower life cycle energy consumption, respectively, than single glazing. This is due to the insulation provided by low-e coatings, which reduces energy needed for heating and cooling. The special coating in double reflective low-e and double low-e glazing reflects heat back into the building, resulting in lower energy use and improved thermal performance. The study emphasizes the importance of considering the life cycle energy consumption of building components in designing and constructing green buildings.
- The results of the carbon emission shows that the carbon emissions of double reflective low-e glazing are 4.69% lower than those of single glazing, while the carbon emissions of double low-e glazing are 6.55% lower than those of single glazing. This indicates that double reflective with low-e glazing has a lower impact on the environment compared to other two glazing. In terms of life cycle cost, the life cycle cost of double reflective low-e glazing is 19.05% lower than that of single glazing, while the life cycle cost of double low-e glazing is 19.00% lower than that of single glazing. The lower life cycle cost is likely due to reduced energy use and associated energy cost savings.
- Finally, the results of this study indicate that the use of double reflective low-e glazing can significantly improve the energy efficiency of buildings by reducing life cycle energy consumption, carbon emissions, and life cycle cost. The lower carbon emission and life cycle energy consumption make it a suitable choice for building construction and energy performance improvement.

References

1. Cao X, Xilei D, Liu J (2016) Building energy-consumption status worldwide and the state-of-the-art technologies for zero-energy buildings during the past decade. Energy Build 128:198–213
2. Young NW Jr, Jones SA, Bernstein HM, Gudgel JE (2009) The business value of BIM. McGraw Hill, New York
3. Epstein E (2012) Implementing successful building information modelling. Artech House. ISBN-10: 9781608071395
4. Karen K, Douglas N (2014) Automated energy performance visualization for BIM, building information modeling: BIM in current and future practice. Wiley, pp 119–128
5. O'Neill R, Window A, Kenway S, Dargusch P (2020), Integrated operational and lifecycle modelling of energy, carbon and cost for building façades. J Cleaner Prod 125370

6. Ramachandra TV, Aithal BH, Sreejith K (2015) GHG footprint of major cities in India. Renew Sustain Energy Rev 44:473–495
7. Jincy MR (2022) Optical and solar properties of glazing. WFM Media
8. Praseeda KI, Venkatarama Reddy BV (2019) Embodied and operational energy of urban residential buildings in India. Energy and Build 110:211–219

Development of Framework for Achieving Optimum Thermal Insulation for Building Infrastructures

Alekhya Chetty and Suchith Reddy Arukala

1 Introduction

1.1 General

Building sectors utilize large amount of energy from its construction to end of its life. The International Energy Agency (IEA) estimates that there was a 92% increase in global energy use between 1971 and 2014. According to the United Nations Environment Program, buildings are responsible for 40% of all global energy use and more than 30% of global greenhouse gas emissions [1]. This high consumption of energy utilizes fossil fuels which results in the depletion of virgin materials. Material production results in harmful emissions and knocking down of building accounts in production of construction waste. Creating comfortable indoor environment by minimizing the consumption of energy has become challenging. The heat energy released from solar radiation is reached to the exterior walls by means of convection and radiation, and then this heat is transferred to the interior by means of conduction, finally this energy is circulated in the room by convection mechanism.

There are two forms of heat related to the storage of thermal energy; they are sensible heat and latent heat. Sensible heat is the heat energy that a substance absorbs to change its temperature without altering its phase. The heat energy absorbed/released during the transition of phase (solid to liquid (or) liquid to solid (or) liquid to vapour) of material by maintaining a constant temperature is known as latent heat.

A. Chetty · S. R. Arukala (✉)
Department of Civil Engineering, Kakatiya Institute of Technology and Science, Warangal, Telangana 506015, India
e-mail: asr.ce@kitsw.ac.in

A. Chetty
e-mail: m21sc001@kitsw.ac.in

R. K. Pancharathi et al. (eds.), *Low Carbon Materials and Technologies for a Sustainable and Resilient Infrastructure*, Lecture Notes in Civil Engineering 440,
https://doi.org/10.1007/978-981-99-7464-1_18

Buildings with adequate thermal insulation use less energy and have significantly smaller systems. Thermal insulation can provide long-lasting thermal comfort without the need of mechanical air conditioning systems, especially in the off-season [2]. Buildings in cold cities may turn down the heat losses by using the right insulation thickness and other energy-saving techniques, which reduces carbon dioxide emissions by 50% [3]. The quality of material possessing insulation property relies on its versatility to different climatic conditions. If properly chosen, the insulation material can also reduce the noise and can be fire-resistant [4].

The building envelopes account for 60–80% of the structure's overall heat transmission loss. In order to enhance interior air quality and lower building energy consumption, it is critical to increase the thermal efficiency of the building envelope, particularly the wall body. Numerous climatic conditions have led to the development of wall energy conservation optimization, and in-depth assessment studies have to be published. Prior review studies on insulating materials for buildings, however, tended to place more emphasis on thermo-physical characteristics and less emphasis on other important features like energy embodied, embedded carbon, acoustic, hygroscopic, and flame resistant. Additionally, none of the preceding assessments looked at the effectiveness of these insulating materials in diverse climate zones. The various thermal, energy, environmental, and performance requirements of the available insulating materials should therefore be studied. All types of materials may not be available in all climatic conditions, so available materials at that location must be made use of.

Adopting MCDM techniques will assist to keep expenses to a minimum and prevent the usage of improper materials. MCDM provides a framework for choosing, classifying, and prioritizing materials as well as aiding in entire evaluation. It addresses the requirement for a numerate structure in the materials selection process [5].

The objective of this investigation is to create a theoretical framework for improving building energy efficiency while considering the ideal building insulation materials for India's various climate zones by utilizing heat transfer mechanisms and select the optimum material that possess less effect to the environment, by applying MCDM techniques.

1.2 Thermal Characteristics of Materials

Thermal insulation creates obstruction to the flow of heat into the structure such that comfortable indoor environment can be created. Thermal performance of a building not only depends on the thermal conductivity and U-value but also on many properties such as material thickness, thermal mass, specific heat, thermal lag, position of the insulating material, material density, etc.

Thermal conductivity (*k*): Thermal conductivity is the term used to describe a material's capacity to conduct heat.

The rate of heat transmission through conduction, perpendicular to a material's unit cross-sectional area, is another way to quantify thermal conductivity. Using Fourier's law, this may be compound.

$$Q = -KA\frac{dt}{dx} \tag{1}$$

where Q is the rate of heat flow (W), k is the material's thermal conductivity (W/m K), A is the cross-sectional area (m^2), dt is the sample's temperature differential, and dx is the sample's variation in thickness (m). The unit of measurement of thermal conductivity is W/m K.

Thermal transmittance (*U*-value): The capacity of a material to conduct heat. It is the opposite of R-value. The measurement is in W/m^2 K.

Thermal resistance (*R*-value): It is the characteristic of a material to oppose the flow of heat. It is reliant on temperature as well as thickness of material. The unit of measurement is m^2 K/W.

Thermal mass: It is the capacity of a material to hold heat energy. The materials with high thermal mass delay the transfer of heat from one side of the material to other, thereby creating comfortable environment.

Specific heat: The amount of heat energy necessary to increase the temperature of the mass by 1 °C. The unit of measurement is J/kg. °C.

Thermal lag: The rate at which a material releases trapped heat is referred as its thermal lag. A material with high thermal mass and low thermal conductivity is said to possess large thermal lag.

1.3 Application of Insulation Material in Building Envelope

Large volumes of solar radiation reach the building envelope, to achieve desirable thermal comfort conditions; walls' ability to store and transfer heat is crucial. An essential method of energy conservation in buildings is wall insulation. The building's heating and cooling requirements might determine the choice of the wall thickness, materials used, and surface finishes. The main goal in a hot area is to decrease heat transfer into buildings; hence, proper thermal insulation and air spaces in walls are essential.

Thermal insulation blocks: In these, available cement blocks or light weight blocks like AAC are insulated with polystyrene boards. These type of insulation saves the labour cost and are easy for inspection and handling. For existing beams and columns, separate insulation has to be provided.

Insulation filled in cavity walls: In double-wall construction, insulation material of desired thickness like polystyrene, rock wool, etc., is inserted in the cavity between walls. The cavity walls should be water-proof.

Walls with external thermal insulation: Insulating material like expanded polystyrene is attached to the surface of wall by means of a special mortar. This material is covered by a fibre-reinforced mesh and further protected by a thin layer of weather-resistant mortar/plaster. It provides protection without any joints, and hence, thermal bridging is eliminated.

Insulation for interior walls: Insulating material such as expanded polystyrene or extruded polystyrene is attached to the inner walls and is covered by gypsum boards. It provides protection without any joints, and hence, thermal bridging is eliminated.

2 Literature Review

2.1 Parameters to Be Considered for Selection of Insulation Materials

Selection of an insulation material does not depend upon a single criterion, various parameters such as location of building (climatic conditions), available insulation materials and their properties, type, and function of building, optimum thickness of the material are to be considered.

Selection of appropriate insulation material.

The building envelope, also referred to as the exterior facade, is made up of fenestration and opaque systems. For instance, walls, roofs, slabs on grade, basement walls, and opaque doors are opaque components. Fenestration systems are made up of windows, skylights, ventilators, and doors with more than half of their surface area covered in glazing. The building's envelope guards the residents and interiors from the elements and other external factors like noise and pollution. The inhabitants' aesthetic and thermal comfort, as well as the building's energy usage, is all significantly influenced by the envelope design. The building's climatic zone determines the requirements for the building envelope. The five climatic zones that the ECBC specifies have very different weather patterns as mentioned in Table 1. For each temperature zone, there are varied needs for thermal comfort in buildings, and the manner in which those requirements reflect physically in architectural design.

Table 1 Classification of different climatic zones in India

Climate region	Typical temperature (°C)				Mean relative humidness
	Summer noon (high)	Summer night (low)	Winter noon (high)	Winter night (low)	
Hot and dry	40–45	20–30	5–25	0–10	Very low 25–40%
Warm and humid	30–35	25–30	25–30	20–25	High 70–90%
Temperate	30–34	17–24	27–33	16–18	High 60–85%
Cold (sunny/cloudy)	17–30	4–21	−7 to 8	−14 to 4	Low:10–50%, High:70–80%
Composite	32–43	27–32	10–25	4–10	Varying dry periods = 20–50%, Wet periods = 50–95%

(*Source* ECBC user guide, 2009 [6])

2.2 *Location of Building*

Building type and Function: The pattern of building differs from one another based on their preferred usage. Every building type is categorized based on their level of occupancy, usage of operation energy, and thermal zoning. A building's thermal performance is significantly influenced by the building envelope's components. Based on the function of buildings (residential, commercial, industrial purposes), the energy consumption varies, and the comfort levels in a building have to be maintained without any variation in the indoor temperature regardless of the climate.

Available insulation materials: The major insulation materials utilized nowadays for building walls are inorganic insulation materials and organic insulation materials. Inorganic thermal insulation materials have a high thermal conductivity and are non-combustible, in contrast to organic thermal insulation materials that are flammable but have a low thermal conductivity. The selection of insulation material for any building type depends not only on thermal transmittance but also several factors like ease of construction, cost of material, safety, environmental impact, and availability of materials.

The thermal conductivity of some of the insulating materials is taken from the literature and ECBC; their density, specific heat, vapour diffusion resistance, sound absorption coefficient, and response to fire are mentioned in Table.2. Regular insulating materials with low thermal conductivity can be installed to increase indoor comfort, but this also creates a heat island. A different method to improve the thermal mass of construction materials is to use latent heat storage materials. They possess a special capacity for storing latent heat through phase transitions and releasing heat during the transition at a practically constant temperature.

Thickness of insulation layer: Thermal insulation is crucial because it reduces the space conditioning demands when a structure needs mechanical air conditioning. The easiest way to link building energy use with external temperature is through the degree-day's technique. In order to ensure thermal comfort and regulate interior air quality for inhabitants and industrial processes, the base temperature is a balancing point temperature at which mechanical systems should be turned on/off in accordance with outdoor temperature. The formula used to estimate cooling degree-days and heating degree-days is given by Amber [35]. The optimum thickness of the insulation layer is determined by the heat and cooling load on the building envelope and is given by Kaynakli [36] and Mahlia [37].

Energy Performance Index (Bureau of Energy Efficiency [38]**)**: Energy Performance Index (EPI) is the most important statistic for comparing the energy use of buildings. Consumption of energy annually (kWh) per building square meter is expressed as kWh/m^2/year or kWh/person/year.

$$\text{EPI} = \frac{\text{Annual energy consumption in kWh}}{\text{Total built - up area (excluding basements unconditioned)}} \tag{2}$$

The rating for Dwelling units in different climatic zones of India is mentioned in BEE [38].

$$\text{EPI ratio} = \frac{\text{EPI of building proposed}}{\text{EPI of a standard building}}. \tag{3}$$

3 Methodology

Based on a thorough review of the body of published research, this section discusses the thermal, economical, environmental, hygroscopic, acoustic, and flame-resistant qualities and functionality of a variety of insulating materials. Environmental elements are critical for reducing CO_2 emissions in the built environment, while thermal qualities are necessary for attaining financial and operational energy savings. The schematic flow chart of the present work is given in Fig. 1.

3.1 Performance of Different Insulating Materials

Hygroscopic characteristics are essential for controlling the relative humidity inside in a humid environment. If it is necessary to keep a certain level of noise in a given

Table 2 Thermal properties of insulating materials

Insulation type	Thermal conductivity (W/m K)		Literature	Specific heat (kJ/kg. °C)	Vapour diffusion resistance	Sound absorption coefficient	Reaction to fire	References
	Density (kg/m^3)	ECBC						
Glass wool (A_1)	10–100	0.0351	0.030–0.050	0.8–1	1–1.3	0.45–0.8	A1	[7–9]
Rock wool (A_2)	40–200	0.04	0.035–0.050	0.8–1	1–1.3	0.29–0.9	A1–A2	[4, 7, 8, 20]
Expanded polystyrene (A_3)	18–50	–	0.032–0.053	1.25	20–100	0.22–0.36	E	[4, 7, 8, 10]
Extruded polystyrene (A_4)	32–40	0.0321	0.025–0.035	1.45–1.7	80–170	0.2–0.65	E	[4, 7, 10]
Polyurethane (A_5)	30–160	0.0269	0.022–0.035	1.3–1.45	50–100	0.67	D-F	[4, 10–12]
Polyisocyanurate (A_6)	30–45	0.0364	0.03–0.045	1.4–1.5	55–150	–	B	[10, 13]
Foamed glass (A_7)	100–200	–	0.038–0.055	0.21	–	–	A1	[14, 15]
Phenolic foam (A_8)	40–160	–	0.018–0.024	1.3–1.4	35	0.3–0.5	B-C	[16–18]
Cork (A_9)	100–120	–	0.037–0.043	1.5–1.7	May-30	0.39–0.85	E	[19–21]
Cellulose (A_{10})	30–80	–	0.040–0.050	1.3–1.6	1.7–3.0	0.53–0.9	E	[8]
Aerogel (A_{11})	70–150	–	0.013–0.021	1	2.0–5.5	0.54–0.78	A1/C	[22–35]
Vacuum insulation panels (A_{12})	160–230	–	0.003–0.008	0.8	2.66–7.03	0.1–0.3	A1	[13, 26]
Gas-filled panels (A_{13})	38	-	0.011–0.020	–	–	–	–	[27, 28]
Sheep wool (A_{14})	Oct-20	–	0.038–0.054	1.3–1.7	04-May	0.056–1	E	[29]
Rice husk (A_{15})	130–170	–	0.048–0.080	1.2–2.7	2	0.15–0.66	A	[30, 31]
Wheat husk (A_{16})	480	–	0.0001	0.718	02-Dec	0.29–0.78	–	[32–34]

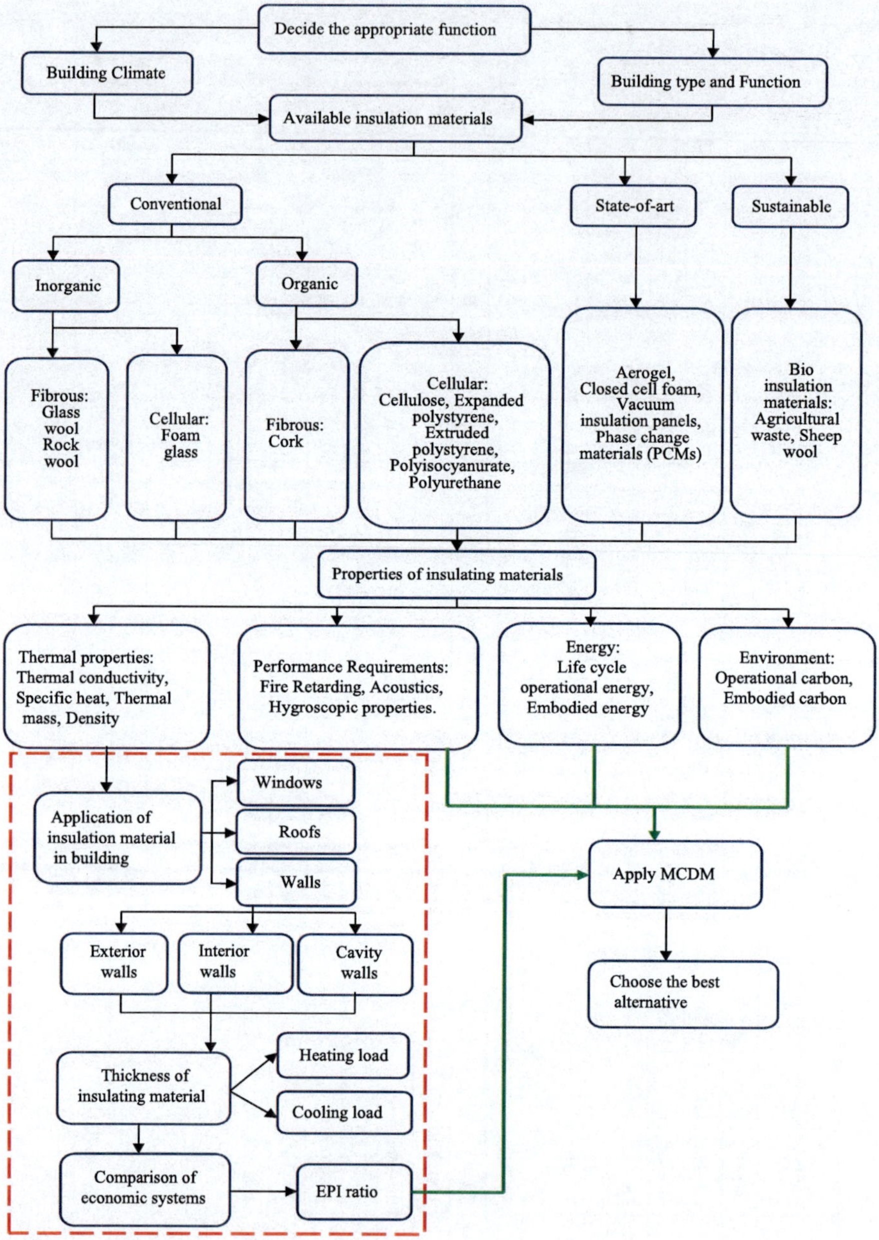

Fig. 1 Schematic flow chart of research methodology

area, acoustic qualities are very crucial. For the building's occupants to be safe in the case of a fire, the fire-retardant property is essential.

Thermal properties: Two essential thermal properties of insulating materials are thermal conductivity and volumetric heat capacity. Ideal insulation materials should transmit heat through the envelope less than common construction materials by having a lower thermal conductivity. Thermal conductivity of a material can be calculated using steady-state methods and transient methods. The method to be chosen is based on the type of material used.

The ability of a substance to store thermal energy is measured by its volumetric heat capacity; commonly referred to as thermal mass. A calorimeter is typically used to assess specific heat capacity. The American standard ASTM C1784-13 describes the Standard Test Method for Using a Heat Flow Meter Apparatus for Measuring Thermal Storage Properties. The density of material can be determined by dividing the mass of sample with its volume.

Thermal comfort assessment: It is important to consider both energy consumption and overheating when designing buildings, especially in areas where heating is a major source of energy and where high levels of insulation are used to keep the winter heat within the structures. One method is to use overall operating energy consumption and the peak summer cooling energy demand as optimization criteria. When used properly, the thermal mass of buildings can also lower the need for temporary cooling. This theory claims that because sustainable insulation can hold more heat than other types of insulation, it might lower peak cooling needs and lower the chance of overheating.

Incorporation of PCM in buildings is not feasible in warm and humid climatic zone. Better outcomes for cold and temperate regions come from PCM incorporation at the midst or the inner portion of the building envelopes [39].

Life cycle cost (LCC) performance: Total costs can be calculated by aggregating costs from the manufacturing (C_{M}), construction (C_{C}), operation (C_{O}), and end-of-life phases (C_{EOL}), then subtracting benefits beyond the system boundary (B_{R}).

$$C = C_{\mathrm{M}} + C_{\mathrm{c}} + C_{\mathrm{o}} + C_{\mathrm{EOL}} - B_{\mathrm{R}} \tag{4}$$

The whole cost of building materials, including the cost of insulation, and equal operating energy cost savings were taken into account for computing the LCCs. Therefore, it is challenging to independently compare the LCC of various insulating materials.

Performance at operational levels and energy emissions: A functional unit (FU) which considers unit area (A) of an insulation panel with a thickness that provides unit thermal resistance (R) and a design life duration of 50 years as recommended by the European Commission is chosen in order to compare the performance of various insulation materials. This option enables a fairly homogeneous operational performance when comparing insulating materials. By dividing the operating energy consumption by the emission factor for the region in which the buildings are located, emissions were calculated. The many fuel sources that are used to produce energy have a big influence on emissions. As a result, in different temperature zones, it is impossible to draw a connection between wall thermal resistance and carbon emissions. Some of the software like TRNSYS®, EnergyPlusTM, DesignBuilder®, and Esp-r® may be used to evaluate PCM/insulation performance in buildings.

Embodied carbon and embodied energy: Embodied energy is a unit of measurement for a substance's lifetime energy consumption (extraction, manufacture, transportation, installation, and disposal), while embodied carbon refers to the emissions that come from those processes. Compared to embodied energy, conventional buildings use greater operational energy. Today energy-effective and net-zero-energy structures require less operating energy than embedded energy, causing embodied energy to become a substantial component, as a result of the use of relatively energy-intensive environmentally friendly materials and equipment. Of all types of insulation, sustainable insulation offers the lowest embodied energy values.

Life cycle environment performance: An organization's environmental performance is measured, examined, reported on, and communicated in accordance with standards established by management through a formal process known as environmental performance evaluation (EPE). EPI, environmental management accounting (EMA), environmental management systems (EMS), life cycle analysis (LCA), and Eco-labelling are some of the tools and techniques utilized in EPE.

Hygroscopic properties: The thermal efficiency of a material decreases by 7.5% for every 1% increase in moisture. Therefore, materials' hygroscopic properties might affect indoor humidity, and it is crucial to understand them. They are beneficial in both damp environments with poor ventilation and hot, humid conditions. Water vapour diffusion resistance is an indicator of the material's resistance to allow water vapour to pass through while taking the thickness of the material into account. It is a temperature-dependent, dimensionless material attribute. The ability of the insulating material to control water vapour ingress over time increases with the diffusion resistance value. The gravimetric (dish) method is used to determine the water vapour transmission rate for sheet materials in accordance with ISO 2528.

Acoustical qualities: Although the operational energy usage is unrelated to the acoustic properties of insulators, it is a crucial component of the quality of the interior environment. Acoustical insulators can improve indoor comfort by reducing the propagation of noise through the building envelope. The sequential means of the sound absorbing coefficients at four frequencies 250, 500, 1000, and 2000 Hz are known as noise reduction coefficients (NRC). NRC is used to compare the acoustic properties of insulating materials. The effectiveness of a material's ability to absorb sound is measured using the sound absorptivity (α). It is the ratio of absorbing to incident energy. If the acoustic energy is completely absorbed, then $\alpha = 1$.The standing wave tube approach, the transfer function strategy, and the reverberation room technique are different methods used to measure sound absorptivity. Sonic tubes are used to carry out standing wave and transfer function approaches, and normal incident sound absorption coefficients are determined. For this small sample sizes are needed, and the procedures are inexpensive and simple to carry out. The absorption coefficients of a random incident are determined using the reverberation room technique. The results are fairly accurate despite not an easy measurement and large sample size.

Fire-resistant properties: Buildings can experience accidental or purposeful fire hazards that endanger human life, cause property damage, and weaken the structure. To minimize the negative social and economic effects of fire dangers, fire safety measures must be implemented from the building's preliminary design phase. According to the European standard BS EN 13,501-1, the materials are categorized from *A* (incombustible) to *F* (highly flammable) based on their combustibility.

3.2 *Multi-Criteria Decision-Making (MCDM) Technique.*

When choosing amongst options that are accessible and each described by several different (and sometimes competing) qualities, MCDM approaches are utilized. Some of the most popular MCDM techniques include TOPSIS, Analytical Hierarchy Process (AHP), and fuzzy logic-based methods, based on the problem; decision-maker has to choose the suitable MCDM. The majority of MCDM are discrete, with a constrained set of predetermined possibilities. A matrix can be used to represent a MCDM scenario with finite possibilities. It contains criteria (material characteristics) C_i, various alternatives (materials) A_i, which decision-makers must pick, based on relative relevance of criteria (or weightings), and the elements of x_{ij}, is the evaluation of alternative *i* in relation to criterion *j*. A typical case of MCDM is shown in Table 3.

In the given table A1 to A16 are the different material alternatives mentioned in Table 2, x_{11} to x_{nm} could be qualitative and quantitative values of various alternatives. Depending on the needs of the user, the parameters under each of above-mentioned criteria have to be chosen. Based on the weightage obtained, appropriate material is identified and can be applied for a building in that particular zone.

Table 3 A typical decision-making matrix of MCDM challenge for building-based insulation material

	w_1	w_2	w_3	w_4
	Thermal property (C_1)	Performance requirements (C_2)	Energy (C_3)	Environment (C_4)
A_1	x_{11}	x_{12}	x_{13}	x_{14}
A_2	x_{21}	x_{22}	x_{23}	x_{24}
A_3	x_{31}	x_{32}	x_{33}	x_{34}
A_n	x_{n1}	x_{n2}	x_{n3}	x_{n4}

4 Results and Discussions

Installing thermal insulation in buildings has been accepted as an efficient way for achieving productive energy utilization. Selection of appropriate insulation material can lower the initial cost and operating cost throughout the building's life. Building insulation is a technique for both energy conservation and lowering the greenhouse gas emissions that buildings produce. It has been noted that applying insulating material depends on the climatic conditions. In hot climates, exterior wall insulation is preferred as it prevents the entry of heat into the building envelope, whereas in cold climates, insulation is provided for interior walls so that heat does not escape from the interior.

Installing of regular insulating materials with low thermal conductivity improves the comfort inside a building but a heat island is created. Latent heat storage materials, which is an alternative by which thermal mass of building material, can be increased. They possess a special capacity for storing latent heat through phase transitions and releasing heat during the transition at a practically constant temperature.

In warmer climates, higher thermal capacity built structures with higher levels of comfort may be preferable to conventional buildings. On the other hand, structures with a lot of thermal mass in colder regions need more energy.

Based on the building envelope's heating and cooling load, this thickness can be determined. When the cost savings start to decline as insulation thickness is increased, the ideal thickness has been reached. Based on the building envelope's heating and cooling load, this thickness can be determined. The choice of installing PCM/insulation material is heavily influenced by the kind of building, location site, climate patterns, ventilation potentials, and properties of the site-specific nature of PCM or insulation. Software like as DesignBuilder, TRNSYS, and EnergyPlus may be used to evaluate PCM/insulation performance in buildings.

5 Conclusions

The study undertaken the interaction of energy, environment, performance, and thermal characteristics of materials and utilized the MCDM techniques for the selection of best alternative by considering various attributes of insulation material.

The following specific conclusions were observed to choose the materials required.

The state-of-the-art materials have the lowest thermal conductivity value compared to other types. Reduced interior peak temperatures and reduced risk of overheating during the hot summer may be achieved with the help of sustainable insulating materials. Additionally, insulation at the inside of a structure should be used in temperate and cold regions, whereas insulation outdoors should be used in hot, dry, and warm-humid climates.

To control interior relative humidness in a humid environment, to maintain a low noise level in a zone, and to offer safety in the event of flames, insulating materials must meet performance requirements such as hygroscopicity, acoustical, and firefighting properties.

Installation of PCMs can mitigate the effects of airtight, highly insulated homes that raise the danger of overheating and in the location where heating is predominate, highest need for cooling will rise during the peak summer months. The phase change materials (PCM) in the middle or inner layer of building envelopes produced better outcomes for cold and temperate regions; it is often not advisable for warm-humid climates. The thickness and volume of the insulating material, the price of the insulating material, and reduction of energy consumption and greenhouse gas emissions are the important elements that are to be carefully chosen in order to make the building's thermal insulation a cost-effective operation.

Thus, the developed framework helps the designers/decision-makers in choosing the finest insulating products for a climate zone according to the preferred requirements.

6 Scope of Future Work

1. Due to the low embedded energy utilization of sustainable insulating materials, they may be an excellent replacement for traditional insulation materials. However, because they are so highly flammable, more study is needed to enhance its fire-resistant qualities.
2. Although hygroscopic materials successfully control indoor relative humidity, moisture build-up increases material heat conductivity, which might have an influence on prospective energy savings. As a result, it is necessary to analyse how hygroscopic materials perform thermally while taking indoor relative humidity and heat transmission through the envelope into account.
3. For the developed framework, a case study can be performed to determine the given energy efficiency.

References

1. Molina AM, Ausina IT (2016) Energy efficiency and thermal comfort in historic buildings: a review. Renew Sustain Energy Rev 61:70–85. https://doi.org/10.1016/j.rser.2016.03.018
2. Aditya L, Mahlia TMI, Rismanchi B, Ng HM, Hasan MH, Metselaar HSC, Muraza O, Aditiya HB (2017) A review on insulation materials for energy conservation in buildings. Renew Sustain Energy Rev 73:1352–1365. https://doi.org/10.1016/j.rser.2017.02.034
3. Comakli K, Yuksel B (2004) Environmental impact of thermal insulation thickness in buildings. Appl Therm Energy 24(5–6):933–940. https://doi.org/10.1016/j.applthermaleng.2003.10.020

4. Asdrubali F, Alessandro FD, Schiavoni S (2015) A review of unconventional sustainable building insulation materials. Sustain Mat Technol 4:1–17. https://doi.org/10.1016/j.susmat.2015.05.002. 2015/07/01/
5. Jahan A, Edwards KL, Bahraminasa M (2016) Multi-criteria decision-making for materials selection. Multi-criteria decision analysis for supporting the selection of engineering materials in product design, pp 63–80. https://doi.org/10.1016/b978-0-08-100536-1.00004-7
6. Energy Conservation and Building Code (2009) User guide. https://beeindia.gov.in/sites/default/files/ECBC%20User%20Guide%20V-0.2%20%28Public%29.pdf
7. Hill C, Norton A, Dibdiakova J (2018) A comparison of the environmental impacts of different categories of insulation materials. Energy Build 162:12–20. https://doi.org/10.1016/j.enbuild.2017.12.009
8. Asdrubali F, Schiavoni S, Horoshenkov KV (2012) A review of sustainable materials for acoustic applications. Build Acoust 19(4):283–312
9. Yuan J (2018) Impact of insulation type and thickness on the dynamic thermal characteristics of an external wall structure. Sustainability 10:2835–2848. https://doi.org/10.3390/su10082835
10. Schiavoni S, Alessandro FD', Bianchi F, Asdrubali F (2016) Insulation materials for the building sector: a review and comparative analysis. Renew Sustain Energy Rev 62:988–1011. https://doi.org/10.1016/j.rser.2016.05.045
11. Li TT, Chuang YC, Huang CH, Lou CW, Lin JH (2015) Applying vermiculite and perlite fillers to sound-absorbing/thermal-insulating resilient PU foam composites. Fibers Polym 16(3):691–698. https://doi.org/10.1007/s12221-015-0691-8
12. Ibrahim MA, Melik RW (2003) Optimized sound absorption of a rigid polyurethane foam. Arch Acoust Q 28(4):305–312
13. Berardi U, Madzarevic U (2020) Microstructural analysis and blowing agent concentration in aged polyurethane and polyisocyanurate foams. Appl Therm Eng 164:114440. https://doi.org/10.1016/j.applthermaleng.2019.114440.2020/01/05/
14. Konig J et al (2019) Evaluation of the contributions to the effective thermal conductivity of an open-porous-type foamed glass. Construct Build Mater 214:337–343. https://doi.org/10.1016/j.conbuildmat.2019.04.109.2019/07/30/
15. Owoeye SS, Matthew GO, Ovienmhanda FO, Tunmilayo SO (2020) Preparation and characterization of foam glass from waste container glasses and water glass for application in thermal insulations. Ceram Int 46(8):11770–11775. https://doi.org/10.1016/j.ceramint.2020.01.211
16. Tingley DD, Hathway A, Davison B, Allwood D (2017) The environmental impact of phenolic foam insulation boards. Proc Inst Civil Eng Const Mat 170(2):91–103. https://doi.org/10.1680/coma.14.00022
17. Tutikian B, Nunes M, Leal L, Marquetto L (2012) Impact sound insulation of lightweight concrete floor with EVA waste. Build Acoust 19(2):75–88. https://doi.org/10.1260/1351-010x.19.2.75
18. Yao R, Yao Z, Zhou J, Liu P, Lei Y (2017) Mechanical, thermal and acoustic properties of open-pore phenolic multi-structured cryogel. IOP Conf Ser Mater Sci Eng 229. https://doi.org/10.1088/1757-899x/229/1/012034
19. Silvestre JD, Pargana N, de Brito J, Pinheiro MD, Durao V (2016) Insulation cork boards environmental life cycle assessment of an organic construction material. Materials 9(5). https://doi.org/10.3390/ma9050394
20. Sierra Perez J, Boschmonart Rives J, Dias AC, Gabarrell X (2016) Environmental implications of the use of agglomerated cork as thermal insulation in buildings. J Clean Prod 126:97–107. https://doi.org/10.1016/j.jclepro.2016.02.146
21. Robin O, Berry A, Doutres O, Atalla N (2014) Measurement of the absorption coefficient of sound absorbing materials under a synthesized diffuse acoustic field. J Acoust Soc Am 136(1):EL13–E19. https://doi.org/10.1121/1.4881321
22. Merli F, Anderson AM, Carroll MK, Buratti C (2018) Acoustic measurements on monolithic aerogel samples and application of the selected solutions to standard window systems. Appl Acoust 142:123–131. https://doi.org/10.1016/j.apacoust.2018.08.008

23. Moretti E, Merli F, Cuce E, Buratti C (2017) Thermal and acoustic properties of aerogels: preliminary investigation of the influence of granule size. Energy Procedia 111:472–80. https://doi.org/10.1016/j.egypro.2017.03.209
24. Pedroso M, Flores Colen I, Silvestre JD, Gomes MG, Silva L, Ilharco L (2020) Physical, mechanical, and microstructural characterisation of an innovative thermal insulating render incorporating silica aerogel. Energy Build. https://doi.org/10.1016/j.enbuild.2020.109793
25. Riffat S et al (2020) Sound absorption characteristics of KGM-based aerogel. Int J Low Carbon Technol. https://doi.org/10.1093/ijlct/ctaa005
26. Alotaibi SS, Riffat S (2014) Vacuum insulated panels for sustainable buildings: a review of research and applications. Int J Energy Res 38(1):1–19. https://doi.org/10.1002/er.3101
27. Baetens R, Jelle BP, Gustavsen A, GrynningS, (2010) Gas-filled panels for building applications: a state-of-the-art review. Energy Build 42(11):1969–1975. https://doi.org/10.1016/j.enbuild.2010.06.019.2010/11/01/
28. Su S et al (2020) Acoustic spectra of a gas-filled rotating spheroid. Eur J Mech B Fluids. https://doi.org/10.1016/j.euromechflu.2020.03.003.2020/04/07/
29. Del Rey R, Uris A, Alba J, Candelas P (2017) Characterization of sheep wool as a sustainable material for acoustic applications. Materials 10(11). https://doi.org/10.3390/ma10111277
30. Ahn E, Yeom D, Lee KI (2018) Experimental research on the indoor environment performance of complex natural insulation material: carbonized rice hull and rice hull. J Asian Architect Build Eng 16(1):239–246. https://doi.org/10.3130/jaabe.16.239
31. Buratti C, Belloni E, Lascaro E, Merli F, Ricciardi P (2018) Rice husk panels for building applications: thermal, acoustic and environmental characterization and comparison with other innovative recycled waste materials. Constr Build Mater 171:338–349. https://doi.org/10.1016/j.conbuildmat.2018.03.089
32. Muthuraj R, Lacoste C, Lacroix P, Bergeret A (2019) Sustainable thermal insulation biocomposites from rice husk, wheat husk, wood fibers and textile waste fibers: elaboration and performances evaluation. Ind Crop Prod 135:238–245. https://doi.org/10.1016/j.indcrop.2019.04.053
33. Flury M, Mathison JB, Wu JQ, Schillinger WF, Stockle CO (2009) Water vapor diffusion through wheat straw residue. Soil Sci Soc Am J 73(1). https://doi.org/10.2136/sssaj2008.0077
34. Ali M et al (2020) Thermal and acoustic characteristics of novel thermal insulating materials made of Eucalyptus Globulus leaves and wheat straw fibers. J Build Eng. https://doi.org/10.1016/j.jobe.2020.101452
35. Amber KP, Aslam MW, Ikram F, Kousar A, Ali HM, Akram N, Afzal K, Mushtaq H (2018) Heating and cooling degree-days maps of Pakistan. Energies 11:94. https://doi.org/10.3390/en11010094
36. Kaynakli O (2012) A review of the economical and optimum thermal insulation thickness for building applications. Renew Sustain Energy Rev 16:415–425. https://doi.org/10.1016/j.rser.2011.08.006
37. Mahlia TMI, Taufiq BN, Ismail HHM (2007) Correlation between thermal conductivity and the thickness of selected insulation materials for building wall. Energy Build 39:182–187. https://doi.org/10.1016/j.enbuild.2006.06.002
38. Bureau of Energy Efficiency (BEE) http://beeindia.gov.in
39. Arumugam P, Ramalingam V, Vellaichamy P (2022) Effective PCM, insulation, natural and/or night ventilation techniques to enhance the thermal performance of buildings located in various climates—a review energy and buildings. Energy Build 258:111840. https://doi.org/10.1016/j.enbuild.2022.111840

Designing a Virtual Twin for Structural Health Monitoring by Integrating BIM and Digital Twin Framework

Karthik Dasari and Aaditya Dogra

1 Introduction

Maintaining complex structures requires to gather the parametric data in a scientific manner to assess the damage patterns. An efficient monitoring system by adopting emerging technologies which involve autonomous decision-making technique would aid the structures to fully function by continuously tracking their strength and integrity. Structure health monitoring (SHM) system needs to be taken seriously along with the integration of various smart techniques and methods that are essential for the maintenance and supervision of the functional value of the structure [1, 2]. Being successfully implemented over the decades, SHM have large amount of data that can be used to aid autonomous decision systems if the artificial intelligence (AI) can be introduced in monitoring the construction projects [3, 4]. Industry 4.0 aims to replace manual intervention in decision management by developing concepts such as digital twin (DT) which can create a digitalized maintenance system by assembling all the data from a physical system [5–7]. Therefore, the main focus of the present work is to develop a virtual twin (VT) concept by combining the DT framework with virtual 3D representation of the structural performance to visually manage the structure without being physically present at site.

1.1 Virtual Twin

Sensor technologies combined with IoT formed a basis for AI systems. The data gathered from the sensors particularly represent an essential structural element of

K. Dasari (✉) · A. Dogra
CED, National Institute of Technology Srinagar, Srinagar, India
e-mail: dasari.karthik@nitsri.ac.in

R. K. Pancharathi et al. (eds.), *Low Carbon Materials and Technologies for a Sustainable and Resilient Infrastructure*, Lecture Notes in Civil Engineering 440,
https://doi.org/10.1007/978-981-99-7464-1_19

a DT [8]. The physical world can be directly observed through a physical sensor, or it can fuse the data indirectly with more than one physical sensor, i.e. by virtual sensors. By processing the input data from multiple physical sensors, virtual sensors can assess and process the variables that may not be possible to measure physically themselves [9]. The VT of SHM refers to the method of data processing and extracting using BIM (building information modelling) combining with DT concept to obtain real-time visual monitoring and maintenance of civil infrastructure. DT enabled the opportunity for the integration of the physical world to the digital world; thus, it is an updated and precise digital representation of the physical structure's data values. The visual data are stored in the computer for interactive display [10]. In SHM, these data are obtained by the sensors which are attached to the physical system to monitor the real-time information and condition. Three components make up a DT: (i) an entity that exists within the physical atmosphere (the real-world entity); (ii) an entity set in a virtual environment (the virtual counterpart); and (iii) providing a two-way connection system, namely a virtual-to-physical connection as well as a physical-to-virtual connection. VT is a 3D BIM model that involves DT framework and creates virtual interactions using BIM model, thereby creating a virtual communication between the physical entity (PE) and virtual entity (VE). A VT is the interactive virtual representation (virtual 3D model) of a physical object, whereas DT is a technical tool, which helps to find what is happening inside the structure. The changes in a PE get updated in its VE. Serving as a management tool, VT extends to aid in attending the maintenance of a structure to function for its designed life period and helps in predicting the future damages.

2 Background

Several digital tools were available that can frequently update the dynamic data but are not visually coordinated with spatial information. BIM is one such tool that can describe the attributes of hidden components and characteristics of digitization that could lead to effective solutions in SHM practices [11]. Digitally integrated SHM with BIM can be universally applicable to all construction projects. Damage detection and monitoring of structures are having a huge demand in terms of cost and time [12]. Besides conventional SHM practices, utilizing the BIM concepts is essential which focuses on entire life cycle of a construction project. Another emerging technology involving sensors has been active in the construction industry for continuous monitoring of structures by overcoming human-induced risks. BIM 3D models accurately monitor the structures with the help of sensor technology, thereby avoiding risks related to risky operations in the construction projects [13]. During SHM process, system generates large amount of monitoring data that can be utilized to predict the structural behaviour during the operation and avoid future incidents. Wireless communication technologies transferring the data over the internet using Internet of Things (IoT) along with real-time early warnings have become a fundamental field of research in SHM [14].

3 VT in SHM

The concept of VT will form a visual data agility for SHM practices to develop a virtually manageable system which can be achieved by integrating BIM and IoT concepts. VT controls both a physical and virtual representation of a physical system and its related processes. By exchanging the data between PE and VE, the information will be continuously updated. This information is exchanged within an online connection system (CS) that comprises of sensors and actuators. PE refers to structural components and subsystems of structural assets in SHM. The functioning of such systems is compromised when they are subjected to external changes, whether they are environmental or artificial. It is these processes that are responsible for the external actions like loads, temperature, humidity, etc. that affect the structure as a whole. In contrast, the VE uses mathematical models to create a digital representation or idealization based on a set of mathematical functions or a data structure. As the PE behaves over the design life of the system, the VE is updated accordingly. To access the health monitoring of a structure, VE offers a stage to perform risk assessment, predict future damages and examine remaining useful life which autonomously enables in-time decision-making. The conceptualization of VT in SHM is shown in Fig. 1.

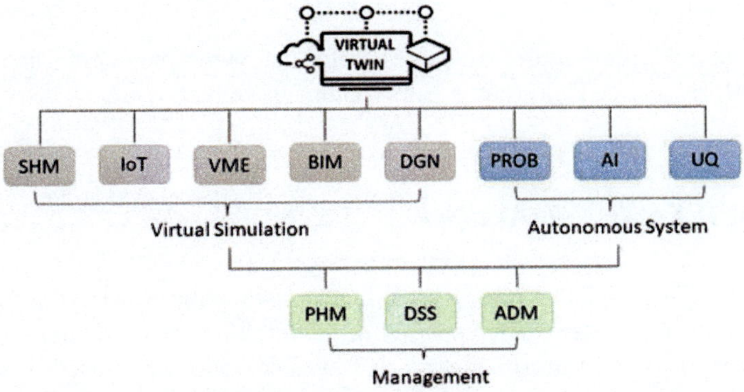

Fig. 1 Conceptualization of the virtual twin. Virtual twin is a concept of management of virtual simulation and some autonomous system. Abbreviations used; SHM: structural health monitoring, IoT: Internet of Things, VME: virtual management entity, BIM: building information modelling, DGN: application of diagnosis, PROB: probability (Bayesian learning inference) between PE and VE, AI: use of artificial intelligence, UQ: provides uncertainty qualification, PHM: application of prognostics, DSS: decision support system, ADM: autonomous decision-making

4 VT Framework for SHM

4.1 Mathematical Representation

The performance of a physical structure represented in the form of a mathematical understanding is signified by available dimensional vector $s(t) \in D \subset \mathbb{R}^{n_s}$ at time t, and environmental vector $e(t) \in \mathbb{R}^{n_e}$, where n_s and n_e represent the data values. To measure the state $s(t)$ at a time t, measurement through sensors $s(t) = d(e, w)$, where $d : \mathbb{R}^{n_e} \times \mathbb{R}^{n_w} \to$ D is expressed. Here, n_w is input vector for measuring the error. On the other hand, the virtual representation of the performance of a structure is referred as $\hat{s} \in \mu \subset \mathbb{R}^{n_s}$. This representation is shown by a model m, and therefore, $\hat{s} = m(\theta, u, e)$, where $m : \mathbb{R}^{n_\theta} \times \mathbb{R}^{n_u} \times \mathbb{R}^{n_e} \to \mathbb{R}^{n_s}$. This means that model m depends on the uncertain model parameter $n_{\theta,}$ including model input parameters u and e where $\theta \in \mathbb{R}^{n_\theta}$ and $u \in \mathbb{R}^{n_u}$. Finally, the model m is based on physical laws where it can be built from the PE data or can also be built from a combination of both PE and VE data sources. From the above concept, DT of a structure can be mathematically depicted by $\mathfrak{C}$, whereas $\mathfrak{C} \subset \mu \times D$. Mathematical representation of DT is shown as

$$\underbrace{m(\theta, u, e)}_{\hat{s}} \overset{CS}{\leftrightarrow} \underbrace{d(e, w)}_{s}, \tag{1}$$

where $\leftrightarrow$ symbol represents the exchange of information between virtual and physical entities performance $\mathfrak{s}$. Therefore, $\mathfrak{s} \in \mathfrak{C}$ which is modelled by CS.

4.2 Model Prediction Analysis

The VE model $m(\theta, u, e)$ requires a periodical update with regard to the PE states $\mathfrak{s}$ through monitoring system. The updating of model will vary based upon either the physical performance or historical data. A Bayesian approach is used in analysing the model. There will be an issue by updating the model with the data received from sensors, since there exists some parameters cannot be concluded. With the physically observed data D, the uncertain parameters θ and μ will be updated using Bayes' theorem as follows,

$$p(\theta|D, \mu) = \frac{p(D|\theta, \mu)p(\theta|\mu)}{p(D|\mu)}, \tag{2}$$

where $p(\theta|D,\mu)$ represents the posterior probability of θ and μ. The information $p(D|\theta,\mu)$ is the posterior probability function using Bayesian approach, and $p(D|\mu)$ is actual proof of physically gathered data. In the DT framework, the virtual model μ is given by $m(\theta, u, e)$, and a possibility model chosen for the w determines the posterior probability function. It can be noted that from the model-based method of autonomous context, the prior probability not only gives the uncertain parameters but also aids in decision-making amongst PE and VE. In fact, the prior probability approach shows the acceptable modelled states $s \in \mu$ of VE to forecast the observed output $s \in D$ in PE to make better decision-making with greater acceptance of the data.

4.3 Petri Nets for Autonomous Decision-Making

Petri nets (PNs) are basically a type of graphical programming language, and it is a bipartite graph used to model the functionality or behaviour of a system. The concept of the PNs in the present work has been limited to derive the workflow for DT part. PN structure is comprised of nodes, places and transitions. The direction of connection between places and transitions is shown by arcs. Data or values are stored in a place for the purpose of archiving them. As a circle in PN, a place represents a modelled state. This is similar to a structural member that has been damaged or a routine activity that has been inspected. Lines, box or rectangle represents the PNs transitions in which lines denote event/process and permit the model move to another state from its previous state. Places are temporarily visited by token; tokens are the values that can circulate through the graph. During execution at a particular time, tokens circulate over PN are known as marking. Arcs can be assigned with weights (multiplicity), a non-negative integer. If weight is not specified for an arc, it defaults to 1. The firing rule dictates when a token is added to output by removing a token from input place. In some cases, like whenever there are at least one token in all input places of transition, it can fire to represent changes in the marking. The mathematical representation of a PN is as follows,

$$\mathfrak{R} \triangleq (\mathrm{P}, \mathrm{T}, \mathrm{I}, \mathrm{O}, \mathrm{M}_0, \mathrm{W}), \tag{3}$$

where P indicates the set of n_p places, and T indicates n_t transition. I and O are the input and output places matrix ($n_p \times n_t$). M_0 is a marking vector which is derived by the following equation:

$$M_{k+1} = M_k + A^{\mathrm{T}} u_k. \tag{4}$$

Here, u_k denotes firing vector during execution at time k. $A \in \mathbb{N}^{\mathrm{n_t} \times \mathrm{n_p}}$ indicates the matrix occurred from the graph. In order to obtain the elements of the graph, the backward matrix must be subtracted from the forward matrix. Therefore,

$$A = A^{+} - A^{-} = a_{ij}^{+} - a_{ij}^{-},$$

where $i = 1$ to $n_t, j = 1$ to n_p. At time k, if a transition t_i is activated and according to the firing rule $u_{i,k} = \mathbb{I}_{\mathrm{i}}$, where $\mathbb{I}_{\mathrm{i}} = 1$, if $M_k(j) \geq a_{ij}^{-} \forall p_j \in P_{t_i}$, else $\mathbb{I}_{\mathrm{i}} = 0$. In the

Eq. 4, $M_k(j) \in \mathbb{N}$ is a marking vector at k, for P_j and P_{t_i} indicates the places that are related to t_i, $i = 1$ to n_t. There also exist other PNs, i.e. high-level Petri nets (HLPN), where we can understand the tokens and arcs in more flexible manner. HLPN applies extra algebraic conditions in firing transitions with regard to certain DT parameters such as input, environmental, model and data values. In HLPN, the rule for firing is stated as follows,

$$s.u_{i,k} = \mathbb{I}_{\mathrm{i}}.\mathbb{I}_{C_{\mathrm{i}}},\ \text{where,}$$
$$\mathbb{I}_{C_{\mathrm{i}}} = \begin{cases} 1\ if & C_{\mathrm{i}} = \text{true} \\ 0, & \text{otherwise} \end{cases} \tag{5}$$

Since $C_i = C_i(\theta, D)$ is a true/false variable, a transition t_i can only be fired upon activation, i.e. $\mathbb{I}_{\mathrm{i}} = 1$, and therefore, C_i will become true. There are some rubrics that summarizes HLPN processes for the present study. During firing of transition, all the tokens will be taken from their respective places and the same number of tokens will be produced in the output place. Transition would only activate when there is at least one token in all input places of transition. The transition t_i fires only when the transition condition $C_i =$ true. Then, the transition t_i eliminates a_{ij}^- from p_j by adding a_{ij}^+ for each jth place.

4.4 Integrating Sensing and Communication Technologies

As stated earlier, VT is a combination of sensor information appraised through DT model and visual presentation using BIM model. For this, combination of various tools and techniques such as smart sensors, communication modules and software was involved to develop an autonomous decision process. The autonomous system is achieved by integrating IoT technology that could enable a direct communication amongst the machines by evading the human intervention. For this to achieve, the physical system which is attached to the structure requires to be monitored for its performance states and thereby generating huge data through the sensing network. There are various protocols and standards implement for the hassle-free data exchange amongst several devices involved in the process. The overall design of the VT is split into four key parts as shown in Fig. 2: Microcontrollers and microprocessors, control sensors and actuators, communication channels (wired or wireless media, protocols and standards) and computing services, which include an API web interface on the front end and data analysis, data storage and IoT applications on the back end. Lastly, a data visualization system is modelled (e.g. BIM).

Fig. 2 Representation of VT process in four key parts concerning the technology integration

4.5 *Sensing Network*

In sensing network, the parameter data of a PE are gathered by smart sensors converting the physical state into a signal form. These smart sensors include actuators which turns the received signals into real physical affects, and the actions are triggered as per the command received. Microcontrollers/microprocessors are used to process the digital signals to convert them into useful information. Type and number of sensors to be used depend on what structural parameters to be analysed for monitoring. A wide variety of sensors are available which can be embedded inside or placed outside of the structures in the SHM practices.

4.6 *Data Transmission and Communication*

Data communication amongst the smart devices in the present work is done by implementing IoT sensors to perform the data exchange in the VT process. The present VT process utilises Wi-fi network for encrypted data communication amongst PE and VE. Further, this process uses google cloud for allowing the data to exchange between PE and VE, and HTTP is used for responding to the data requests from VE. These both transmission protocols run over TCP/IP connection standards. Smart sensors were tagged with regard to their IP address through HTTP for facilitating their communication with an unique identification. The Python-based kernel hosts these services, which communicate with the database as shown in Fig. 3.

4.7 *Integrating BIM Concept to Develop a Virtual Management Tool*

BIM encompasses a variety of software such as AutoCAD, Revit and Navisworks. In this experiment, we used Autodesk Revit for simulation purposes. In order to create a digital version of the structural member in accordance with BIM standards, it was decided to use a generic model from the software families and modify it according to the physical model in terms of its properties and dimensions.

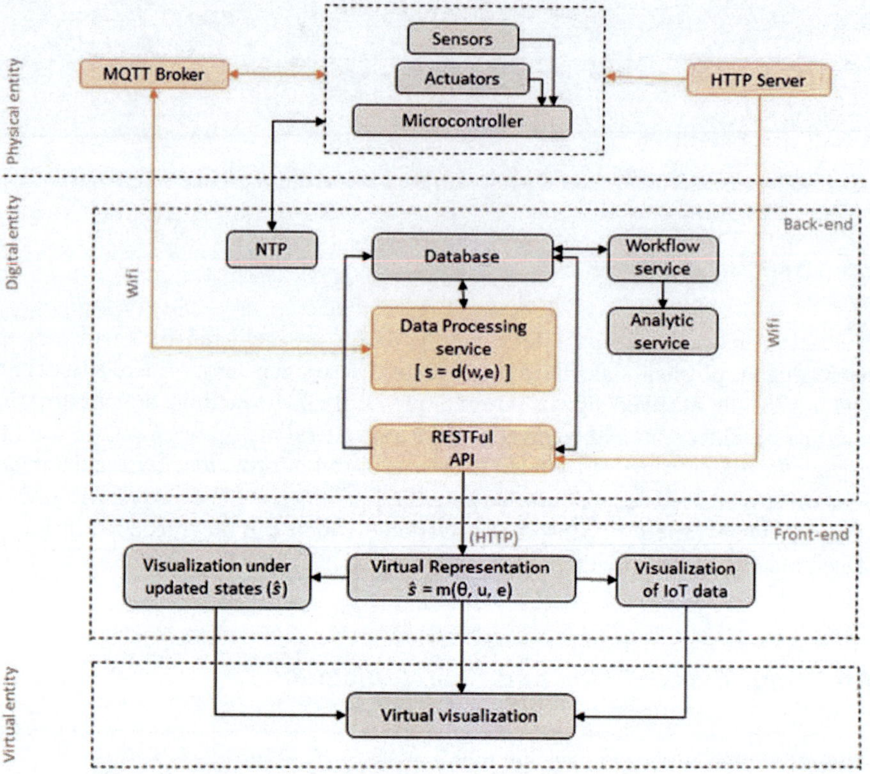

Fig. 3 Proposed VT integration process in which communication features were marked in pink

A 3D model of the project is created with Revit software, which is then used to create a virtual sensor. Using Dynamo, the virtual sensor is then linked to the physical sensor data, which is analysed and made decisions in the DT framework using CAD data. The response to the DT gathered data was added to the object categories in Dynamo as the input data for creating the virtual sensor. In Revit, Dynamo is a graphical programming tool that can be used to create a design. Information related to DT from Google Sheets linked to Dynamo through BIMOne. The Dynamo tool operates on the basis of connecting the required nodes. Nodes perform operations when they are linked to each other according to the function parameters. The programming of dynamo is done in two stages: In the first stage of the programming process, data input is connected to dynamo, and in the second stage, the data are connected to the BIM model.

5 Case Study

Laboratory experiment on a small-scale model is prepared to visualize the deformation of a structure in the VT environment. As part of the test model, a two-storey portal frame with fixed supports and rigid joints was used. A horizontal force (F_{5x}) is applied to node 5 of the frame as shown in Fig. 4. Structure integrity scenario is idealized depending upon the force applied, which activates the alarm when the applied force F_{5x} is higher than the predefined threshold value. Structures' elastic displacements are measured and modelled to determine the DT model. Mathematically, $s(e, w) = [s_{ix}, s_{iy}, s_{iz}]$, whereas $s(\theta, u, e) = [s_{ix}, s_{iy}, s_{iz}]$, where i indicates 1–6. In the present case study, joint 4 of the physical structure is considered for measuring the displacement by an ultrasound sensor. Therefore, $s = [0, 0, -, (s_{4x}, -, -), -, -]$, and the error vector is set to $[10^{-4}]$ sensitivity relating to distance measurement in metres. Based on these values, Bayesian inference classifies the likelihood function as Gaussian PDF. There are a total of $N_s = 5 \times 10^4$ simulations in this Bayesian inference module, which implements Metropolis–Hastings on a Bayesian basis. Additionally, a Gaussian PDF distribution is used to calculate the acceptance rate whose standard deviation falls within the recommended range [0.2, 0.4]. According to the beam element frame model, the VE joint displacement $\hat{s}$ is calculated as follows:

$$\hat{s}(\theta, u, e) = [K]^{-1}[F(\theta)]^{\mathrm{T}}, \tag{6}$$

where $F(\theta) = \left[F_1, F_2, 0, 0, \underbrace{\left\{ F_{5x}, 0, 0 \right\}}_{\theta}, 0 \right].$

$F(\theta)$ is a vector of applied forces and moments to the physical frame model. In this case, this is initially an unknown vector and is represented by a θ (a model parameter). An interval of 0.1–5 obtained by uniform PDF is used to determine the model parameter θ. This interval represents possible value of the referred force in

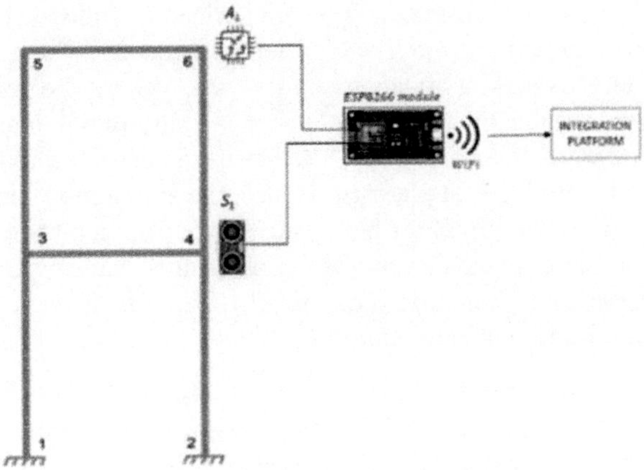

Fig. 4 Portal frame (laboratory model) connected with smart devices

$[N]$, namely $U[0.1, 5]$ $[N]$. In this case, $F1$ represents the forces, whilst $F2$ represents the moments both applied to joints 1 and 2. Since these joints are fixed, their values are not necessary. The displacements and rotations of these objects are therefore zero. From the Eq. 6, the elastic matrix of a portal frame with fixed joints is represented by the stiffness matrix $[K]$ of size 18×18, where $[K] = [K_{ij}]$, $i, j = 1$–6.

K_{ij} values depend upon the input variables [both material (E, A, I) and geometrical $(u = L_1, L_2, L_3)$]. These input variables are shown in Fig. 4. Moment of inertia and Young's modulus of the material is given as $I = 10^{-9}$ [m^4] and $E = 9 \times 10^8$ [N/m^2], respectively. Lastly, in this case study, environmental variables such as temperature and humidity were not considered in relation to structural response, thus $e = \emptyset = 0$.

In the case study, a test frame is set up to monitor its physical parameters using smart devices, one ultrasonic sensor S_1 attached to the frame at node 4, and one relay A_1 as shown in Fig. For the transformation of data, Arduino UNO (microcontroller) and IoT module ESP-8266 are connected to sensor. The data acquired from the sensor are sent to a cloud through the internet, and in turn, the data are linked to Dynamo BIM creating a VT environment. Thus, a VT communication platform is developed by linking the DT and virtual visualization though IoT.

Figure 5 is an illustration of the workflow process to perform autonomous decisions regarding integrity of structure. For the purpose of data collection, analysis and system rearm, the concept of HLPN was adopted which comprises of eight places (p_1 to p_8), seven transitions (t_1 to t_7) and two cold transitions ($\mathcal{E}$). The places represent system states for visual interpretation, such as 'new data arrival', 'adjusted/updated' and 'indicates warning' within the HLPN graphs that are shown in Fig with different colour text labels. The blue text labels represent the information about the place p_6 and one of the cold transitions. As transitions t_1 to t_7 fire, states of the DT system change, resulting in the number of automated actions. Some of these transitions are conditional transitions, i.e. provided with some conditions for firing. Each transition is described, and the actions associated with it are shown in Table 1. Transitions t_1, t_5 and t_6 are based on the conditions C_1, C_5 and C_6, respectively, and the third column of the below-mentioned table lists its algebraic predicates. The transitions will only activate if respective conditions C_i, $i = 1, 5, 7$ are fulfilled.

HLPN processes can be summarized as follows. Initially, the system is considered to start at time $k = 0$, where the new data from sensor S_1 should be received. Here, it was assumed to update the VE by the data gathered from PE and is represented by tokens p_1 and p_4. Thus, initial marking of the PN is given as $M_0 = (1, 0, 0, 1, 0, 0, 0, 0)^{\mathrm{T}}$. When t_4 is fired, the token is eliminated from place p_1 to be produced at place p_5. Hence, the autonomous decision system (DT) is able to identify the PE that is being subjected to an unknown force and needs to determine the force to update the VE. As a result of transitions t_1 and t_5, a decision is made based on the conditions C_1 and C_5. The mismatch evaluation provides the following criteria for enabling them:

$$J = \frac{\|s_{4x} - \tilde{s}_{4x}\|}{\tilde{s}_{4x}}, \tag{7}$$

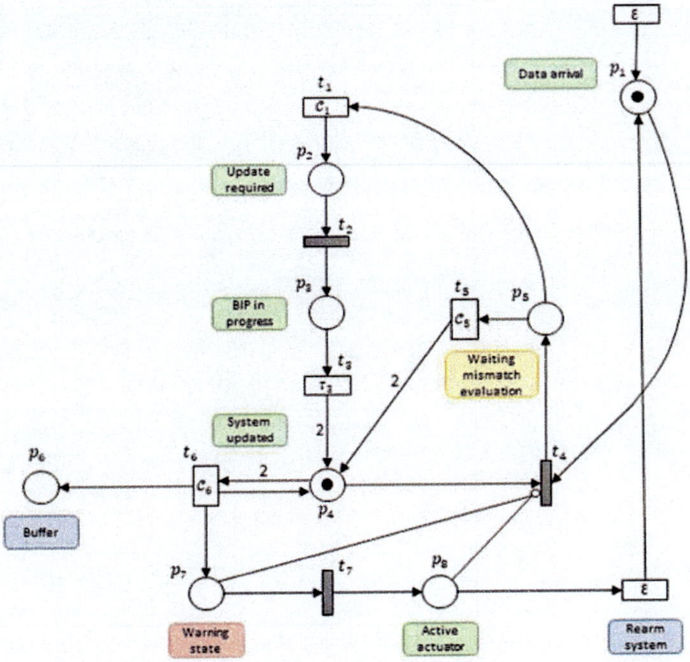

Fig. 5 HLPN model (dark rectangular shapes represent transitions, and circles represent places)

Table 1 HLPN transition. T3 shows time for the VE to process the Bayesian model

Transition	Type	Conditions	Description
t_1	Conditional	C_1: $\{j \geq 0.1\}$	Checks for mismatches
t_2	Thematic	–	Activate BIP
t_3	Event timed	Enable after τ_3	BIP execution
t_4	Thematic	–	Establish VE–PE mismatch
t_5	Conditional	C_5: $\{j < 0.1\}$	Checks for mismatches
t_6	Conditional	C_6: $\{\text{mean}\ (F_{5x}) \geq F)\}$	Analyses the inferred force
t_7	Thematic	–	Actuator is activated

where s_{4x} is horizontal displacement measured from PE at node 4 and $\tilde{s}_{4x}$ is earlier noticed value upon Bayesian inference, being at $k = 0$, $\tilde{s}_{4x} = 0$. Therefore, t_1 is fired when measured force value varies by greater than 10% as compared to previous recorded value and the DT performs structure inference updating. The workflow sequence is represented as $\{p_2, t_2, p_3, t_3\}$, which generate one token at p_4 and the system get in 'adjusted/updated' else t_5 would directly fire and produce one token to p_4. Here, PE to VE Bayesian updates are not required for the DT system. Despite the system being 'adjusted/updated', it remains as it was previously.

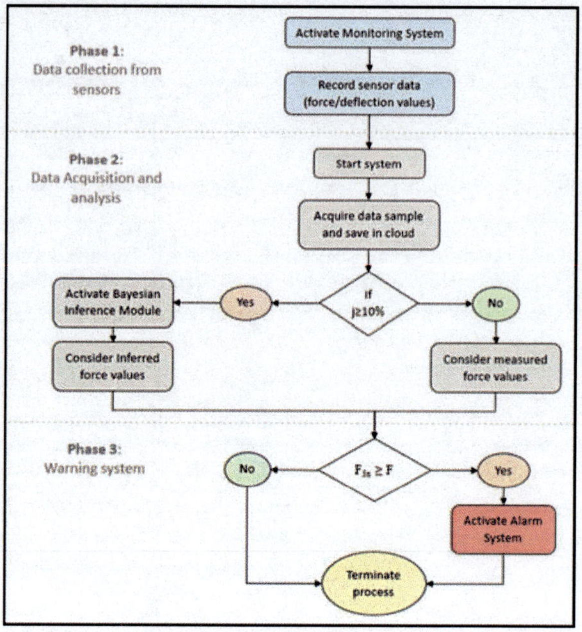

Fig. 6 Workflow cycle for the proposed twin model

According to the HLPN graph, the system is in an 'adjusted/updated' state, indicating that P_4 has been marked. In order to determine whether the inferred force exceeds threshold F, a comparison is made between the values. If so, transition t_6 is triggered. When the inferred force is greater as compared to threshold force F, the structure integrity is compromised. A series of warning states and actions by nodes $\{p_7, t_7, p_8\}$ are then activated by this transition. The system is set to 'indicate warning', and a visual alarm is activated. As the cold transition ($\mathcal{E}$) takes place, the system is rearmed to receive new data. As a result, the warning indication is discarded until it can be re-evaluated or discarded. As soon as the system enters into a warning indication, a token is generated in P_6 that acts as a buffer for information which can be analysed and monitored. Using this method, we can obtain valuable information when a structure having above threshold force F.

For the case study, the whole workflow cycle is categorized into three phases as shown in Fig. 6. This is the continuous process of the monitoring system, and this cycle continues for every new data generated.

5.1 Results and Discussion

In the test frame, node 5 of the smart device is loaded with a series of loads after the smart devices are activated. This results in a number of structural integrity scenarios being analysed. The workflow model was tested with 15 load cases, and DT behaviour was controlled with HLPN as explained in Fig. 5. Based on the applied and inferred forces, Fig. 6 illustrates the behaviour of a DT. A force/pressure sensor is fixed to the test frame to measure the applied force at node 5. In this manner, only visual validation is possible from the inferred values. A description of how DT responded to the test loads is given in Table 2. Furthermore, Tables 3 and 4 provide results of the behaviour of HLPN for test cases 1 and 12.

Table 2 Test structure at node 4: measured and inferred DT variables for 15 load cases

LCs	MF (N)	INF (N)	S_{4x} (mm)	$j \geq 10\%$	Warning
1	2.30	2.08	46	Yes	*Yes*
2	2.12	2.08	44	No	*Yes*
3	1.14	0.98	23	Yes	**No**
4	0.64	0.85	14	Yes	**No**
5	1.04	1.07	21	Yes	**No**
6	1.13	1.07	23	No	**No**
7	1.02	0.87	20	Yes	**No**
8	1.21	0.9	26	Yes	**No**
9	1.00	0.9	18	Yes	**No**
10	0.94	1.14	18	No	**No**
11	1.13	1.04	24	Yes	**No**
12	1.10	1.04	20	Yes	**No**
13	1.89	1.51	33	Yes	**No**
14	0.90	1.31	17	Yes	**No**
15	0.80	0.86	16	No	**No**

Table 3 PN workflow model events and actions under LC 1

State (k)	Marking vector (M_k)	Firing vector (u_k)	Events	PN interpretation
0	$[1, 0, 0, 1, 0, 0, 0, 0]^T$	0, 0, 0, 1, 0, 0, 0	New value observed	PN workflow begins
1	$[0, 0, 0, 0, 1, 0, 0, 0]^T$	1, 0, 0, 0, 0, 0, 0	C_1 (true)	Update check for VE and PE (true)
2	$[0, 1, 0, 0, 0, 0, 0, 0]^T$	0, 1, 0, 0, 0, 0, 0	Update is required	BIP inferred
3	$[0, 0, 1, 0, 0, 0, 0, 0]^T$	0, 0, 1, 0, 0, 0, 0	Run BIP	VE updating from PE
4	$[0, 0, 0, 2, 0, 0, 0, 0]^T$	0, 0, 0, 0, 0, 1, 0	C_6 (true)	$F_{5x} \geq F \rightarrow$ (true) (adjusted/updated)
5	$[0, 0, 0, 1, 0, 1, 1, 0]^T$	0, 0, 0, 0, 0, 0, 1	Action required	Activates actuator A_1
6	$[0, 0, 0, 1, 0, 1, 0, 1]^T$	0, 0, 0, 0, 0, 0, 0	Indicates warning	New data arrival (rearm)

Table 4 PN workflow model events and actions under LC 12

State (k)	Marking vector (M_k)	Firing vector (u_k)	Events	PN interpretation
0	$[1, 0, 0, 1, 0, 2, 0, 0]^T$	0, 0, 0, 1, 0, 0, 0	New value observed	A PN is initiated
1	$[0, 0, 0, 0, 2, 2, 0, 0]^T$	1, 0, 0, 0, 0, 0, 0	C_1 (true)	Update check for VE and PE (true)
2	$[0, 1, 0, 0, 0, 2, 0, 0]^T$	0, 1, 0, 0, 0, 0, 0	Requires updating	Invoke BIP
3	$[0, 0, 1, 0, 0, 2, 0, 0]^T$	0, 0, 1, 0, 0, 0, 0	Run BIP	Update of VE based on PE
4	$[0, 0, 0, 2, 0, 2, 0, 0]^T$	0, 0, 0, 0, 0, 0, 0	–	System rearm; awaiting new data

Table 3 shows observations for load case 1. Since $k = 1$ and condition C_1 are true, the transition t_1 is fired. Following the firing, one token is produced at position p_2, and Bayesian inference is performed. Once this is completed, the DT system will have two tokens placed in place p_4. Updated system produces higher inferred forces than threshold force $F = 1.8$ N.

After updating the system, the inferred force turns out to be higher than the threshold force $F = 1.8$ N. At time $k = 4$, transition t_6 occurs, and the warning sequence along with actuators is activated. For load case 12, the deflection at node 4 exceeds 10% of the previous measurement. Hence, DT needs to update VE by the inferred force in relation to PE. Workflow model waits for new data before continuing when the inferred force is less than the threshold value.

From these results, it is evident that the proposed DT model is capable of responding to sensor data in an efficient and effective manner. In this way, VE is updated according to PE's behaviour. This study has shown that actuators can affect the PE based on the VE's information. In addition to this, the case study studied also showed that HLPN acts as a workflow model for DT systems in order to provide effective management of interactions between the VE and PE. Finally, this DT is linked to Dynamo through cloud to visually update the structural parameters in the 3D BIM model.

6 Conclusion

The VT concept has been designed in a way that will assist in monitoring the current state of structures in the future. It was through the implementation of a proof of concept of a laboratory model using BIM, and IoT sensors that we get a first-hand glimpse of how to use a VT to remotely control SHM practices. In this way, appropriate measures can be taken in time to avoid accidents occurring in the future. By utilizing IoT and BIM data, VT makes it easier for the management to identify any critical condition for users and structures before it becomes a problem. This will allow them to take action earlier before any damage occurs. As part of this research, three technologies are integrated: BIM, SHM, and IoT, in order to present a set of structural health warning systems based on BIM model. The system can be summed up in five main functions that it provides: data visualization for monitoring, visual representation of structural components, location of risks and disasters, detection and control of on-site alarms automatically. It is important to assure that data are always up to date and stable by automating data acquisition and transmission. In addition, data mining should be enhanced with more intelligent algorithms and improve intelligence management. BIM model and monitoring data should be uploaded to a large cloud database, and background computations can also be done in the cloud server, which can save you a lot of time and help you better guide engineering applications.

References

1. Giurgiutiu V (2014) Structural health monitoring with piezoelectric wafer active sensors, 2nd edn.https://doi.org/10.1016/c2013-0-00155-7
2. Farrar CR, Worden K (2007) An introduction to structural health monitoring. Phil Trans R Soc A 365(1851). https://doi.org/10.1098/rsta.2006.1928
3. Flah M, Nunez I, Chaabene WB, Nehdi ML (2021) Machine learning algorithms in civil structural health monitoring: a systematic review. Arch Computat Methods Eng 28(4). https://doi.org/10.1007/s11831-020-09471-9
4. Farrar CR, Lieven NAJ (2007) Damage prognosis: the future of structural health monitoring. Phil Trans R Soc A 365(1851). https://doi.org/10.1098/rsta.2006.1927
5. Tygesen UT, Jepsen MS, Vestermark J, Dollerup N, Pedersen A (2018) The true digital twin concept for fatigue re-assessment of marine structures. In: Proceedings of the international conference on offshore mechanics and arctic engineering—OMAE, vol 1. https://doi.org/10.1115/OMAE2018-77915
6. Love PED, Matthews J (2019) The 'how' of benefits management for digital technology: from engineering to asset management. Autom Constr 107.https://doi.org/10.1016/j.autcon.2019.102930
7. Negri E, Fumagalli L, Macchi M (2020) A review of the roles of digital twin in CPS-based production systems. Value based and intelligent asset management.https://doi.org/10.1007/978-3-030-20704-5_13
8. van der Aalst WMP et al (2018) Views on the past, present, and future of business and information systems engineering. Bus Inf Syst Eng 60(6). https://doi.org/10.1007/s12599-018-0561-1
9. Kabadayi S, Pridgen A, Julien C (2006) Virtual sensors: abstracting data from physical sensors. In: Proceedings of the 2006 international symposium on the world of wireless, mobile, and multimedia networks. pp 587–592. https://doi.org/10.1109/WOWMOM.2006.115
10. Deng L et al (2021) Visualization and monitoring information management of bridge structure health and safety early warning based on BIM. J Asian Archit Build Eng. https://doi.org/10.1080/13467581.2020.1869013
11. Volk R, Stengel J, Schultmann F (2014) Building information modeling (BIM) for existing buildings—literature review and future needs. Autom Constr 38:109–127
12. Mishra M, Lourenço PB, Ramana GV (2022) Structural health monitoring of civil engineering structures by using the internet of things: a review. J Build Eng 48.https://doi.org/10.1016/j.jobe.2021.103954

13. Hou G, Li L, Xu Z, Chen Q, Liu Y, Qiu B (2021) A BIM-based visual warning management system for structural health monitoring integrated with LSTM network. KSCE J Civ Eng 25(8). https://doi.org/10.1007/s12205-021-0565-0
14. Lin Y-B et al (2021) The artificial intelligence of things sensing system of real-time bridge scour monitoring for early warning during floods. Sensors 21(14). https://doi.org/10.3390/s21144942

Impact of Inhibiting Factors on the Efficiency of Precast Construction Projects Using Kendall's Concordance Method

Sloka Gampa and Sri Kalyana Rama Jyosyula

1 Introduction

Concrete is considered an integral material in construction throughout the world, with it being used in a variety of projects. This includes projects for roads, buildings, or even canals, proving its versatility. The reason for this is because of its many pros, ranging from its ability to withstand different severe weather to general deterioration. Specifically, we can explore precast concrete, which has the advantage of lower costs and overall better quality, especially its durability. The present study seeks to analyze the impact of the inhibiting factors of precast construction projects. A survey was carried out by collecting the responses from the reputed precast construction industries in the southern part of India. The process of precast construction involves the creation of standardized structural components in a factory setting, which are later transported to the construction site for assembly. These elements, including walls, beams, floors, columns, and slabs, are commonly produced off-site. In a survey conducted by Skrzypczak [1], investors, architects, and contractors were asked about their opinions on precast structures. The results showed that investors focused primarily on implementation time, while architects placed greater importance on quality and technical specifications. In another study, Asamoah et al. [2] compared the cost of cast-in-place and precast concrete structures, with construction professionals expressing a preference for precast construction due to its lower life cycle costs and material wastage. Meanwhile, Tushar et al. [3] evaluated the environmental and economic impact of precast buildings and found that precast technology had better results in terms of greenhouse gas emissions and thermal performance during the production phase. To achieve a sustainable solution in construction,

S. Gampa · S. K. R. Jyosyula (✉)
Department of Civil Engineering, Ecole Centrale School of Engineering, Mahindra University, Hyderabad, India
e-mail: srikalyanarama.j@mahindrauniversity.edu.in

R. K. Pancharathi et al. (eds.), *Low Carbon Materials and Technologies for a Sustainable and Resilient Infrastructure*, Lecture Notes in Civil Engineering 440,
https://doi.org/10.1007/978-981-99-7464-1_20

the use of precast technology will achieve high-quality construction with reduced resource consumption such as cost and time during construction Wu et al. [4]. It was found that, on an average, the profitability of the structure of around 5–10% was achieved by replacing the conventional method with precast concrete method. Peng et al. [5] evaluated the thermodynamic and acoustic performance of precast composite wall panels using principles of heat transmission and acoustics. They applied the thermal conductivity equivalence method to determine the panels' thermal conductivity formula, concluding that it could be accurately calculated through this approach. Agrawal et al. [6] presented the differences between conventional construction and precast construction subjected to gravity and lateral loads. It is observed that the precast construction has shown better performance. Zhang et al. [7] proposed a novel approach to constructing large-span underground vaults using a combination of precast and cast-in-situ techniques. Through engineering practices and computational analysis, it was found that this combined method offers several significant advantages over the traditional cast-in-situ method alone. These benefits include a reduction in construction time by 50%, as well as the production of a high-quality fair-faced concrete surface and superior finishing. Albero et al. [8] aimed to develop an economical slab structure design that incorporated fire-resistant requirements and considered all available manufacturing technologies. The findings indicated that conventional designs were insufficient when factoring in the fire-resistant constraint, while the precast hollow core slab design was shown to be significantly impacted by this requirement. Waroonkun and Koojaroenpaisan [9] presented an article on the analysis of a conceptual model that considers the multiple factors that affect the efficiency of the precast process. It was done by statistical evaluation and regression analysis. Pradeep et al. (August 2014) concluded from their study that the precast construction is superior due to its quality pre-designing and prefabrication, lightweight, and economical construction. Li et al. [10] calculated the carbon footprint for precast concrete pile construction. It was concluded that the LCA of the overall project increased with the use of precast concrete for construction.

2 Methodology

The study gathered data on inhibiting factors through the literature review, interpersonal interviews, and on-site visits. The research team analyzed multiple factors to compile Table 1, which presents a list of 23 factors that were evaluated to determine their impact on precast construction projects.

Using the 23 factors identified, the research team created a questionnaire to assess professionals' level of agreement regarding the benefits of using precast concrete. These questions were distributed to several companies who work on precast projects. The survey was taken by civil engineers, design engineers, quality engineers, structural engineers, and other employees in the precast construction project. These professionals had top rate each of the factors alongside precast construction, 1–5; 1 being the least impactful and 5 the most impactful. A total of 115 responses were collected.

Table 1 Factors chosen to analyze the efficiency of precast construction

Factors
High quality
Fast construction
Less wastage
Large floor slabs possible
Improves handling ability
Dimensional accuracy
Extremely good fair-faced concrete finish
Mechanization and mass production
Walls with insulation possible
Ease of construction
Better health and safety standards
Economical for large and repetitive projects
Longer life and durability
Lower life cycle cost
Reduction of on-site labor
Reduction of on-site noise, disturbance, and activity
Acoustic insulation
Enables use of pre-tensioning techniques
Strength
Fire resistance
UV sensitivity
Chemical resistance
Weather resistance

The study employed the relative importance index (RII) to rank the advantages of precast concrete based on feedback from professionals involved in the selected projects. RII values were utilized to determine the relative significance of each factor compared to others within the same category. The formula used to calculate RII was given below.

Relative important index (RII) $= \sum W/AN$, wherein W represents the weights given by the respondents to each variable ranging from 1 to 5, A represents the maximum weight with 5 being the highest in this study, and N represents the total number of samples.

3 Results and Discussion

Table 1, which was ranked using the relative importance index, indicated that rapid construction and high-quality fair-faced concrete finishes were the primary advantages of precast concrete products, while space for acoustic insulation was deemed least important by professionals. The study also used Kendall's concordance analysis to assess the level of agreement among building professionals regarding the advantages of precast concrete, based on the significant inhibiting factors identified. The analysis produced Kendall's concordance coefficient, denoted as k, and Table 2 presents the degree of agreement among different professions on the benefits of precast concrete.

(We assumed that 4–5 ranking as very important and took those under consideration.)

From Table 3, grand mean $\breve{R} = \sum R_I = 11.74$, mean of ranking $= R_I$, number of professionals $= 4$, $n =$ number of factors ranked $= 23$, $[(R_I - \breve{R})^2]$ average $=$ 96.8814, $W = 96.8814/[23(23\text{^}2–1)/12]$

96.8814/1012 = 0.095
$W = 0.095$.

Irrespective of the number of rankings, the value of W falls between 0 and 1. A value closer to 1 indicates a strong level of agreement among the ranking sets.

To test the significance of W at a 95% confidence level, a hypothesis test was conducted. The null hypothesis (H_0) states that the set of rankings by civil engineers, quality engineers, design engineers, and structural engineers were unrelated. The alternative hypothesis (H_1) suggests that the set of rankings were related. A chi-square distribution was used to determine the significance of W, and the observed chi-square value was calculated using $\lambda^2 = k(n - 1)W$, where k is the number of groups being compared, which was 4 in this case. The critical chi-square value was obtained from a statistical table at $(k - 1)$ degrees of freedom. If the calculated chi-square value exceeded the critical value from the table, then the null hypothesis would be rejected, and the alternative hypothesis would be accepted.

If the value of W is significantly high, it indicates that the kth respondents have used similar standards in rating the nth aspect of the problem being studied. To interpret the data, the ratings were classified as very important (1–3), important (4–7), marginally important (8–11), and not important (12–14).

$\lambda^2 = 4(23–1)0.095 = 8.36.$

From the chi-square distribution tables, critical value = 13.28 for DOF(k) = 4.

The calculated λ^2 value of 13.28 is greater than the critical value of 8.36, leading to the rejection of the null hypothesis H_0 and acceptance of the alternative hypothesis H_1, indicating a strong agreement among civil engineers, quality engineers, design engineers, and structural engineers regarding the benefits of using precast concrete suspended slabs and columns.

Table 2 Ranking for various factors

Questions	1	2	3	4	5	Total	*W*	Mean	*AN*	*W*/*AN*	Ranking
High quality	0	1	7	28	67	103	470	4.5631	115	4.0869	5
Fast construction	0	0	0	24	79	103	491	4.7669	115	4.2695	1
Less wastage	0	2	8	25	68	103	468	4.5436	115	4.0695	6
Large floor spans possible	0	0	14	19	70	103	468	4.5436	115	4.0695	6
Improves handling ability	0	5	24	37	37	103	415	4.0291	115	3.6086	17
Dimensional accuracy	0	0	8	63	32	103	436	4.2330	115	3.7913	11
Extremely good fair-faced concrete finish	0	0	0	24	79	103	491	4.7669	115	4.2695	1
Mechanization and mass production	0	0	3	26	74	103	483	4.6893	115	4.2	3
Walls with insulation possible	0	21	4	49	29	103	395	3.8349	115	3.4347	21
Ease of construction	0	3	0	23	77	103	483	4.6893	115	4.2	3
Better health and safety standards	0	0	27	30	46	103	431	4.1844	115	3.7478	12
Economical for large and repetition projects	0	3	14	23	63	103	455	4.4174	115	3.9565	9
Longer life and durability	0	2	25	46	30	103	413	4.0097	115	3.5913	18
Low life cycle cost	0	1	12	66	24	103	422	4.0970	115	3.6695	14
Reduction of on-site labor	0	2	12	20	69	103	465	4.5145	115	4.0434	8
Reduction of on-site noise, disturbance, and activity	0	12	13	33	45	103	420	4.0776	115	3.6521	15
Acoustic insulation	0	32	17	26	28	103	354	3.4368	115	3.0782	23
Enables use of pre-tensioning techniques	0	11	13	52	27	103	404	3.9223	115	3.5130	19
Strength	0	0	16	37	50	103	446	4.3300	115	3.8782	10
Fire resistance	0	1	23	41	38	103	425	4.1262	115	3.6956	13

(continued)

Table 2 (continued)

Questions	1	2	3	4	5	Total	*W*	Mean	*AN*	*W*/*AN*	Ranking
UV sensitivity	0	1	40	34	28	103	398	3.8640	115	3.4608	20
Chemical resistance	0	2	44	40	17	103	381	3.6990	115	3.3130	22
Weather resistance	0	10	8	52	33	103	417	4.0485	115	3.6260	16

$W = [(R_I - R)^2]/ [n\,(n^2 - 1)/12] \rightarrow$ Kendall's coefficient of concordance

According to Kendall's concordance analysis, fast construction, extremely good fair-faced concrete finish, mechanization and mass production, ease of construction, and high quality were identified as the key advantages of using precast concrete.

4 Conclusion

The research findings suggest that civil engineers, quality engineers, design engineers, and structural engineers agree on the benefits of using precast concrete structural elements, with fast construction, extremely good fair-faced concrete finish, mechanization and mass production, ease of construction, and high quality being the key advantages. Continual study on the advantages and process of precast construction in south India will aid in the establishment of precast construction projects. The value of *W*, which measures the degree of agreement between the set of rankings, is 0.095 approx. to 0.01, indicating good agreement. The inhibiting factors listed and sent as a questionnaire were highly agreed upon and were found to be effective factors in precast construction projects. The major advantage of precast construction is its high real quality, which is ensured by standardized production methods and advanced quality control. Consistency throughout the production cycle enables development and customization according to customers' needs, emphasizing the interdependence of production and consumption. The quality of the final product depends primarily on the design method, quality of input materials, workmanship, and production technology, which can be easily achieved with precast construction.

Table 3 Overall ranking for various factors based on their total score

Advantages	CE	DE	QE	SE	SoR	NoF	MoR ($R1$)	R^	$R1 - R$^	$[(R1 - R\wedge)]^2$	OR
High quality	39	9	10	37	95	4	23.75	11.74	12.01	144.2401	5
Fast construction	42	9	10	42	103	4	25.75	11.74	14.01	196.2801	1
Less wastage	37	7	10	39	93	4	23.25	11.74	11.51	132.4801	6
Large floor spans possible	36	8	9	36	89	4	22.25	11.74	10.51	110.4601	8
Improves handling ability	31	7	7	30	75	4	18.75	11.74	7.01	49.1401	19
Dimensional accuracy	38	9	10	38	95	4	23.75	11.74	12.01	144.2401	3
Extremely good fair-faced concrete finish	42	9	10	42	103	4	25.75	11.74	14.01	196.2801	1
Mechanization and mass production	41	9	10	40	100	4	25	11.74	13.26	175.8276	3
Walls with insulation possible	32	7	7	32	78	4	19.5	11.74	7.76	60.2176	15
Ease of construction	41	9	10	40	100	4	25	11.74	13.26	175.8276	3
Better health and safety standards	32	7	6	31	76	4	19	11.74	7.26	52.7076	17
Economical for large and repetition projects	35	8	9	34	86	4	21.5	11.74	9.76	95.2576	11
Longer life and durability	31	7	7	31	76	4	19	11.74	7.26	52.7076	17
Low life cycle cost	36	9	10	35	90	4	22.5	11.74	10.76	115.7776	7
Reduction of on-site labor	34	8	9	38	89	4	22.25	11.74	10.51	110.4601	8
Reduction of on-site noise, disturbance, and activity	32	7	7	32	78	4	19.5	11.74	7.76	60.2176	15
Acoustic insulation	23	5	4	24	56	4	14	11.74	2.26	5.1076	22
Enables use of pre-tensioning techniques	33	7	7	32	79	4	19.75	11.74	8.01	64.1601	14

(continued)

Table 3 (continued)

Advantages	CE	DE	QE	SE	SoR	NoF	MoR ($R1$)	R^	$R1 - R$^	[($R1 - R$^)]^2	OR
Strength	36	8	9	34	87	4	21.75	11.74	10.01	100.2001	10
Fire resistance	33	8	8	31	80	4	20	11.74	8.26	68.2276	13
UV sensitivity	26	6	5	25	62	4	15.5	11.74	3.76	14.1376	20
Chemical resistance	26	6	5	22	59	4	14.75	11.74	3.01	9.0601	21
Weather resistance	36	8	9	33	86	4	21.5	11.74	9.76	95.2576	11

CE civil engineer, *DE* design engineer, *QE* quality engineer, *SE* structural engineers, *SoR* sum of ranking, *NoF* number of factors, *MoR* mean of ranking, *OR* overall ranking

References

1. Skrzypczak I (2023) Statistical quality inspection methodology in production of precast concrete elements. Materials 16(1):431
2. Asamoah RO, Ankrah JS, Offei-Nyako K, Tutu EO (2016) Cost analysis of precast and cast-in-place concrete construction for selected public buildings in Ghana. J Constr Eng 2016
3. Tushar Q, Zhang G, Bhuiyan MA, Navaratnam S, Giustozzi F, Hou L (2022) Retrofit of building façade using precast sandwich panel: an integrated thermal and environmental assessment on BIM-based LCA. Buildings 12(12):2098
4. Wu Z, Luo L, Li H, Wang Y, Bi G, Antwi-Afari MF (2021) An analysis on promoting prefabrication implementation in construction industry towards sustainability. Int J Environ Res Public Health 18(21):11493
5. Peng L, Xiaoyong L, Ying C, Zhiwu Y, Dayou Y (2020, December) Thermodynamic and acoustic behaviors of prefabricated composite wall panel. Structures 28:1301–1313
6. Agrawal A, Sanghai SS, Dabhekar K (2021, November) Comparative studies between precast and conventional cast-in-situ structural systems. IOP Conf Ser: Mater Sci Eng 1197(1):012062
7. Zhang JL, Liu X, Zhao JB, Yuan Y, Mang H (2021) Application of a combined precast and in-situ-cast construction method for large-span underground vaults. Tunn Undergr Space Technol 111:103795
8. Albero V, Saura H, Hospitaler A, Montalvà JM, Romero ML (2018) Optimal design of prestressed concrete hollow core slabs taking into account its fire resistance. Adv Eng Softw 122:81–92
9. Waroonkun T, Koojaroenpaisan R (2011) Modelling of factors impacting adoption of precast concrete systems
10. Li XJ, Zheng YD (2020) Using LCA to research carbon footprint for precast concrete piles during the building construction stage: a China study. J Clean Prod 245:118754

Evaluating the Performance of Precast Concrete Components for the Application in Low-Cost Housing

N. R. Dakshina Murthy, T. Vasudeva Rao, K. J. Rao, B. Sridhar, and Simhadri Raju

1 Introduction

The three basic needs of mankind are food, clothing, and housing. India has been fruitful in fulfilling the food and clothing requirements for the massive population but has yet to fulfil the housing need effectively. Regarding population, India stands second in the world, and the demand for housing is increasing continuously. The scenario of people moving to urban India is expanding at a rapid rate due to industrialization. The failure to create adequate housing and basic services to fulfil the needs of metropolitan areas has led in the growth of slums and unlawful settlements. According to the report of the Ministry of Housing and Urban Poverty's technical group on Urban Housing Shortage (2012–17), there is a shortage of approximately 19 million house units in India, with 96% belonging to the economically weaker sections (EWS) and the lower income group (LIG). The Housing for All programme, launched by the Indian government in 2015, aimed to provide the basic housing needs of disadvantaged urban residents by constructing 22 million dwellings by 2022.

N. R. Dakshina Murthy (✉) · T. Vasudeva Rao · K. J. Rao
Department of Civil Engineering, CBIT, Hyderabad 500075, India
e-mail: nrdmurthy@yahoo.com

T. Vasudeva Rao
e-mail: vasudeavarao_civil@cbit.ac.in

K. J. Rao
e-mail: jagannadharao_civil@cbit.ac.in

B. Sridhar
Department of Civil Engineering, Vasavi College of Engineering (A), Hyderabad, India
e-mail: b.sridhar@staff.vce.ac.in

S. Raju
Smart Build Prefab Pvt Ltd, Hyderabad, India

R. K. Pancharathi et al. (eds.), *Low Carbon Materials and Technologies for a Sustainable and Resilient Infrastructure*, Lecture Notes in Civil Engineering 440,
https://doi.org/10.1007/978-981-99-7464-1_21

To fulfil this, there is a need to develop large-scale affordable housing in India. The workforce shortage and increased price of building materials have raised the pressure on developers to provide an affordable buildings for all. To take advantage of mass housing, new construction technologies must deliver homes quickly with minimal labour and zero waste of materials, which conventional construction methods failed to do.

Utilizing innovative building practices, such as precast concrete construction, helped alleviate the housing crisis during the Second World War, particularly in Europe. Precast concrete construction is a building technology in which components are built off-site on an assembly line using sophisticated tools and equipment and then delivered to the construction site and assembled using building cranes. This kind of building is known as "precast concrete construction". Precast concrete construction has evolved into a more versatile building technology for the past half century. Today, this form of construction is utilized in a wide variety of construction and infrastructure projects worldwide. Although precast construction has substantially risen in the last decade, it forms only a small percentage of the total construction market. This is mainly because of the general reluctance of stakeholders to adopt non-conventional construction methods. Over the last five decades, precast concrete construction has evolved into flexible building techniques employed in numerous construction projects and building projects worldwide. Precast concrete offers several potential advantages over an on-site building. Precast concrete manufacturing is done on the ground level, which helps in project safety. A precast factory controls material quality and workmanship more than a cast-in-situ construction.

Precast concrete buildings have emerged as a viable alternative to traditional building methods. The construction industry has successfully combined precast concrete components with cast-in-situ concrete parts for ten years. A proper combination of precast and cast-in-situ structural parts is developed to fulfil quality and cost-effectiveness criteria. Compared to cast-in-situ construction, the combination allows infrastructure projects to reach better efficiency whilst maintaining quality control. By relying heavily on precast construction, on-site activities will be drastically reduced, resulting in a safer and cleaner working environment. However, it is important to have a good appreciation and encouragement from the client as it comes under cleaner production in management compared to conventional construction. The important criteria influencing the effective adoption of precast construction techniques include planning and understanding the close links between precast concrete component design, construction, detailing, execution, and manufacture.

The landscape of India's infrastructure business has shifted dramatically in recent years. Large initiatives such as townships, mass housing, and IT parks are now more frequent than two decades ago. They will only increase dramatically in the following years. The majority of such projects are still constructed utilizing traditional methods. As a result, the inherent benefits of these projects in terms of repetition and high-volume turnover must yet be explored. Seema et al. [1] concluded from their study that precast concrete technology, now known as an unconventional method in the Indian context, can provide both speed and quality of construction whilst also being beneficial for large-scale projects in terms of volume turnover and recurrences. The benefits of this precast technology may be fully realized in mass housing projects that include small, compact dwellings in significant quantities.

Rinkesh and Neeraj [2] discussed in detail the advantages and disadvantages of conventional methods of construction and precast construction and concluded that there was a cost difference between these methods, where the conventional method was economical and suitable for individual houses and precast construction will be apt for a large group of houses, where there is a labour shortage. Giovanni and Gennaro [3] evaluated the seismic capacity of the precast building by using the pushover analysis method.

They concluded that the available ductility of such industrial structures was more than the required one in a medium seismicity zone. In some circumstances, when the beam–column connection strength must be evaluated first, the proper evaluation is needed to determine seismic capacity. Yuka et al. [4] summarized the parameters, such as the type of concrete and the thickness of the lateral groove, which plays an important role in the strength of the precast structural elements. Chiara et al. [5] discussed the development of fragility functions using fundamental mathematics of RC precast industrial buildings in Italy. These are useful in estimating the loss of strength in seismic performance evaluation. The beam–column joint was explicitly included in creating the fragility curves because it plays an essential role (i.e. weak beam–column connection), which has been the primary cause of the collapse in previous seismic occurrences. Sára et al. [6] concluded that precast structures are superior to conventional ones from an environmental point of view.

All over the globe, the demand for precast is continuously increasing as it is eco-friendly, time-saving, and superior in quality control. When combined with prestressing methods, the precast techniques will yield good results. They can be adapted for a wide range of applications in civil engineering, such as buildings, parking structures, and bridges. In this experimental investigation, the performance of precast concrete elements was evaluated for the effective utilization of precast construction for low-cost housing.

2 Materials and Experimental Program

The smart-built prefab factory provided the materials used for manufacturing the precast elements, and the details of the same are given in the below sections.

2.1 Material Properties

Cement. Ordinary Portland cement of 53 grade confirming to IS: 269-2015 was used, and the physical properties of the cement are given in Table 1.

Aggregates. Crushed granite, which was locally available, was used as coarse aggregate, and river sand from the nearby river source was used as fine aggregate.

Table 1 Physical properties of cement

S. No	Properties	Test results
1	Standard consistency (%)	34
2	Specific gravity	3.13
3	Initial setting time (Min.)	145
4	Fineness of cement (%)	5

Table 2 Physical properties of coarse aggregates

Properties	Test results
Specific gravity	2.65
Bulk density (kg/m^3)	1430
Fineness modulus	6.7

Table 3 Physical properties of fine aggregates

Properties	Test Results
Specific gravity	2.6
Bulk density (kg/m^3)	1453
Fineness modulus	2.59

The physical properties of coarse and fine aggregates are given in Tables 2 and 3, respectively.

2.2 *Experimental Program*

The different precast elements which were supplied by the smart-built prefab factory that were considered in the experimental program were precast wall panels, precast beams, precast columns, precast column posts, and precast soak pit.

Three precast wall panel elements were tested under flexure to evaluate the flexural strength. The precast wall panels of size 2 m in length, 300 mm in depth, and 50 mm in thickness were considered. The grade of the concrete was M25. Uniformly distributed loads were applied on the wall panels, roof elements, and the deflections were recorded at various levels. The test set-up for the same is shown in Fig. 1.

The beam specimens were tested under flexure using uniformly distributed load to evaluate the flexural capacity, which was shown in Fig. 2. The precast beam elements of 2 m length with a cross section of 150 mm × 150 mm were considered, and the grade of the concrete was M25 grade. The beam was provided with 4 bars of 8 mm diameter and stirrups of 4 mm diameter.

The identical precast beams were also tested as columns in the vertical position in UTM of capacity 100 tons. For testing the column under axial compression, the

Fig. 1 Flexural testing of precast roofing sheets

Fig. 2 Flexural testing of precast beam elements

height of the column was considered as 1 m so that it could fit in the UTM machine, as shown in Fig. 3. The axial load-carrying capacity was calculated.

The precast members were also used as column posts for compound walls. These posts were generally subjected to wind loads. Typically, 1 m of the column will be buried into the ground, and 2 m length will act as a post above the ground level. The total moment developed due to wind loads was calculated. The beam was tested as a cantilever beam using a CTM of 200 tons capacity as shown in Fig. 4. One metre length of the beam was inserted between the two cross heads and fixed tightly. The load was applied with a hydraulic jack at the free end, which is 2 m in length. The deflections were recorded for different loads.

Fig. 3 Uniaxial compression of precast column elements

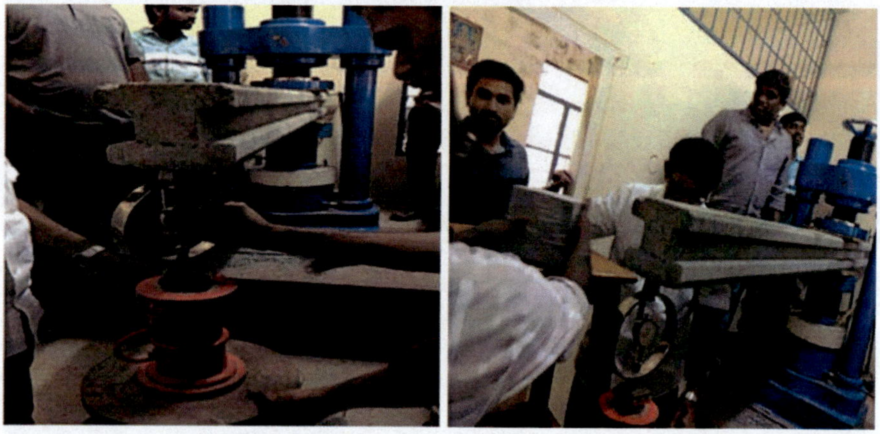

Fig. 4 Moment on a precast column

Another precast element, the precast soak pit, was tested for stability. The soak pits were fabricated using curved precast concrete panels connected by riveting, as shown in Fig. 5. Four small panels were connected circularly and riveted together to form a soak pit. The diameter of the soak pit was 1 and 1.2 m in height. Weep holes were provided at the bottom to release the pressure. The soak pits were buried in the ground and generally filled with soil. Wastewater pipelines will be connected to the soak pit. Wastewater coming out from the houses will directly enter the soak pit under

Fig. 5 Testing of soak pit

the ground, solid wastes will be distributed into the soil, and water percolates into soil layers through weep holes. Thus, the soak pit will be subjected to hoop tension from inside and compressive stresses from outside soil surrounding the periphery. For testing purposes, the soak pit was kept on the ground by tightening the rivets, as shown in Fig. 5. The soak pit was filled with sand and stone dust from the inside to create hoop tension. The rivets were observed for intactness. Then, the soak pit with sand and stone dust was filled with water for saturation. The water was continuously filled till water escaped from weep holes.

3 Results and Discussion

This section deals with the results of this experimental study, followed by a discussion.

3.1 Precast Wall Elements

The flexural strength of precast wall panel elements was calculated by subjecting the elements under flexure. Three precast wall panel elements were tested under flexure to evaluate the flexural strength. The ultimate and working stress of the three panels is given in Table 4. From the table, it can be understood that the precast wall panels exhibit better resistance under flexure.

Table 4 Flexural strength of precast wall panels

S. No	Panel designation	Size of the panel (m.)	Self wt. (kN/m)	Ultimate load (kN/m)	Ultimate stress (kN/m^2)	Service load (kN/m)	Working stress (kN/m^2)	Deflection at ultimate load (mm)
1	A1	2.13 × 0.3 × 0.05	0.345	3.26	10.8	2.17	7.23	16
2	A2	2.13 × 0.3 × 0.05	0.345	3.42	11.4	2.28	7.6	15
3	A3	2.13 × 0.3 × 0.05	0.345	3.36	11.2	2.24	7.46	15

3.2 *Precast Beam and Column Elements*

The flexural strength of precast beam elements was calculated by subjecting the elements under flexure. The deflections were within the allowable limit. The weight, dimension, area, failure load, and compressive strength of the precast column elements of different sections are given in Table 5. Amongst all the sectional elements, section I exhibited better compressive strength, and it further increased with an increase in the grade of the concrete. The foam concrete performed satisfactorily for the strength-to-weight ratio.

Table 5 Compressive load test on a column

S. No	Designation of the specimen	Wt. of the specimen (kg)	Dimensions $L \times B \times D$ (m)	Size of the groove in the column (mm)	Area of the column face (mm^2)	Load at failure (T)	Compressive strength (N/mm^2)
1	M20 I section	26.2	0.83 × 0.15 × 0.15	55 × 35 × 2 sides	18,650	27.1	14.25
2	M20 L section	30.84	0.83 × 0.15 × 0.15	55 × 35 × 2 sides	18,650	26.0	13.67
3	M30 I section	35.12	0.81 × 0.15 × 0.15	55 × 35 × 2 sides	18,650	27.4	14.40
4	Form concrete	7.62	0.82 × 0.15 × 0.15	55 × 35 × 2 sides	18,650	8.20	4.30

3.3 Precast Column Post

The precast column posts were generally subjected to wind loads (See Fig. 6). Typically, 1 m of a column will be buried into the ground, and 2 m length will act as a post above the ground level. The portion above the ground level will serve as a cantilever portion subjected to a uniformly distributed load. When wind acts on the wall panel, which are directly inserted in the grooves of columns, the effect was considered, and the same was transferred onto the column post as point loads at levels all over the height at regular interval of 300 mm. The moment was calculated by multiplying the load by the free length of the cantilever beam, which is 2 m.

Calculation of moment on the column post
Wind speed
$V_z = V_b * K_1 * K_2 * K_3$
$V_b = 44$ m/s
Considering category:4 highly spaced obstruction (Hyd)
$K_1 = 1.0$, $K_2 = 0.8$ and $K_3 = 1.0$
$V_z = 44 * 1 * 1 * 0.8 = 35.2$ m/s
$P_z =$ Design wind pressure
$P_z = 0.6 * V_z^2 = 0.6 * 35.2 * 35.2$

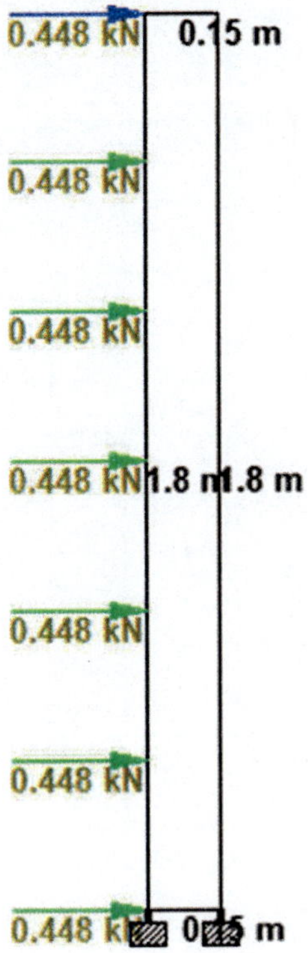

Fig. 6 Representation of wind loads on precast column post

Table 6 Moment carrying capacity of precast concrete column post

S. No	Designation of the specimen	Wt. of the specimen (kg)	Dimensions $L \times B \times D$ (m)	Cantilever portion length (m)	Designed moment (kN m)	Load at failure (kN)	Ultimate moment (kN m)
1	Column 1	26.2	2.43 × 0.15 × 0.15	1.82	2.8	2.81	5.14

$P_z = 743.42\ \text{N/m}^2 = 0.743\ \text{kN/m}^2$
Load on each panel @ 0.3 m = 2 * 0.3 * 0.743 = 0.448 kN
Moment = 0.448 * (1.8 + 1.5 + 1.2 + 0.9 + 0.6 + 0.3) = 2.8 kN m
Ultimate moment = 2.81 * 1.82 = 5.14 kN m > 2.8 kN m.

From Table 6, the moment carrying capacity of the precast column post was more than the moment estimated against wind loads. Therefore, the precast column post was safe.

3.4 Precast Roofing Elements

The experimental results of precast roofing elements 8 mm thick and 12 mm thick were given in Tables 7 and 8, respectively.

Table 7 Experimental results for roofing element (8 mm thick)

S. No	Load (kg)	Dial gauge readings				Load (kN)	UDL/Area (kN/m²)
		1	2	Average	Deflection (mm)		
1	42	11	3	7	0.07	0.411	0.122
2	81	19	5	12	0.12	0.803	0.239
3	122.2	21	6	13.5	0.135	1.212	0.361
4	159.2	25	9	17	0.17	1.577	0.470
5	198.8	26	10	18	0.18	1.969	0.587
6	237.2	31	12	21.5	0.215	2.326	0.694
7	271.2	40	12	26	0.26	2.65	0.791
8	311.3	40	15	27.5	0.275	3.05	0.910
9	346.7	42	16	29	0.29	3.397	1.014
10	411.5	46	26	36	0.36	4.036	1.204
11	475	52	30	41	0.41	4.659	1.390
12	523	59	40	49.5	0.495	5.013	1.496

Table 8 Experimental results for roofing element (12 mm thick)

S. No	Load (kg)	Dial gauge readings				Load (kN)	UDL/Area (kN/m^2)
		1	2	Average	Deflection (mm)		
1	52	15	9	12	0.12	0.51	0.152
2	97.4	19.5	15	17.25	0.1725	0.95	0.283
3	125	22	18	20	0.20	1.22	0.364
4	170.25	27	23	25	0.25	1.67	0.498
5	200.25	30	26	28	0.28	1.96	0.585
6	238	36	30	33	0.33	2.33	0.695
7	275.1	39	33	36	0.36	2.69	0.802
8	315.75	45	37	41	0.41	3.09	0.922
9	358.25	48	42	45	0.45	3.51	1.047
10	430.6	58	50	54	0.54	4.22	1.259
11	507.1	67	58	62.5	0.625	4.97	1.483
12	577.9	76	66	71	0.71	5.66	1.689
13	650.05	85	72	78.5	0.785	6.37	1.901
14	723.05	93	80	86.5	0.865	7.09	2.116
15	796.4	99	86	92.5	0.925	7.81	2.331
16	872.7	107	95	101	1.01	8.56	2.555
17	946.7	115	105	110	1.10	9.27	2.767

3.5 *Precast Soak Pit (Results)*

The riveted connection between the precast panels was observed, and any rivets were loosened. It is observed that the rivets were intact even under saturated condition subjecting the soak pit to hoop stress indicating precast panel elements were fit to use as a soak pit.

4 Conclusions

1. Precast construction is very suitable for rural housing with cost-effectiveness.
2. Roofing elements with textile reinforcement have shown better performance.
3. The precast element for column post is effective in resisting wind loads and, therefore, can be used for long-run compound walls.
4. Instead of going for cast-in-situ conventional concrete, precast elements like soak pits and column posts are convenient and suitable for handling and usage.
5. For a developing country like India, precast technology is one of the best practical construction methods.

Acknowledgements The authors place their sincere thanks to Sri. Simhadri Raju, Director, Smart Build Prefab, Pvt. Ltd. Hyderabad, for his support in providing the precast test specimen and facilitating to conduct some of the tests at their premises. The authors wish to extend their thanks to the Head of the Civil Engineering Department and Principal, Chaitanya Bharathi Institute of Technology (A), Hyderabad, Telangana, for, according to them, permission to use the laboratory facilities and the encouragement during the phase research.

References

1. Seema A, Vijay K, Dhawale AW (2016) Acceptance criteria for precast concrete construction technology in housing construction industry of Maharashtra state. Int J Appl Innovation Eng Manage 5(2):52–55
2. Rinkesh P, Neeraj S (2017) Study of prefabrication in India. Int J Adv Res Eng Sci Manage 1–6
3. Giovanni F, Gennaro M (2004) Seismic vulnerability of existing industrial precast structures. In: 13th world conference on earthquake engineering seismic vulnerability of existing industrial precast structures. Vancouver BC, Canada, pp 1–6
4. Yuka O, Norimichi S, Toshiyuki K (2008) Study on structural performance of precast concrete Balcony with HPFRCC. In: The 14th world conference on earthquake engineering, Beijing, China, pp 12–17
5. Chiara C, Vitor S, Helen C, Rui P, Roberto N (2014) Scenario damage analysis of RC precast industrial Structures in Tuscany, Italy. In: 2nd European conference on earthquake engineering and seismology. pp 1–12. https://doi.org/10.13140/2.1.1725.5683
6. Sára HH, Zsuzsa S (2014) Environmental assessment of a precast concrete building stock in a time perspective. Environ Eng Manag J 13(11):2797–2804. https://doi.org/10.30638/eemj.2014.312

Implementation of Base-Isolation Technique in the Design of Mega Hospital Building

P. Sivasubramanian, M. Manimaran, and U. P. Vijay

1 Introduction

In recent times, seismic design of hospital building structures has become very critical for structural damage control. Traditional design primarily focuses on the energy dissipation from yielding of structural members. This causes severe localized damage in few areas and prevents the structure from being used for immediate operation after earthquake.

Increasing the stiffness of the building by increasing member size or by adding shear walls will be typically necessary for strength and often is done to regulate the building's response to seismic events. In high seismic zones, increasing the member sizes has an additional drawback since it increases the building mass and leads to higher base shear. In such a situation, increasing damping rather than stiffness is a better approach. The purpose of using isolator is to increase the damping of the structure when it is subjected to lateral loads and to decrease the total structural response.

P. Sivasubramanian (✉) · M. Manimaran · U. P. Vijay
Larsen & Toubro Limited, Chennai, India
e-mail: sivasubramanian.p1@lntecc.com

M. Manimaran
e-mail: manimaranm@lntecc.com

U. P. Vijay
e-mail: upvijay@lntecc.com

R. K. Pancharathi et al. (eds.), *Low Carbon Materials and Technologies for a Sustainable and Resilient Infrastructure*, Lecture Notes in Civil Engineering 440,
https://doi.org/10.1007/978-981-99-7464-1_22

2 Building Profile

The building considered for this study is a hospital building, located in Zone 5, consisting of basement, ground + 9 floors. The building plan size at ground level is 175 × 85 m. The total height of the building is 41 m above ground level. Columns are spaced at 8.5 × 7.35 m center to center as shown in Fig. 1 and the building 3D view is shown in Fig. 2. Pile foundation system is used as per the geotechnical recommendations. Hospital buildings are considered with ordinary moment resisting frame with '*R*' value of 1 as per ASCE recommendation.

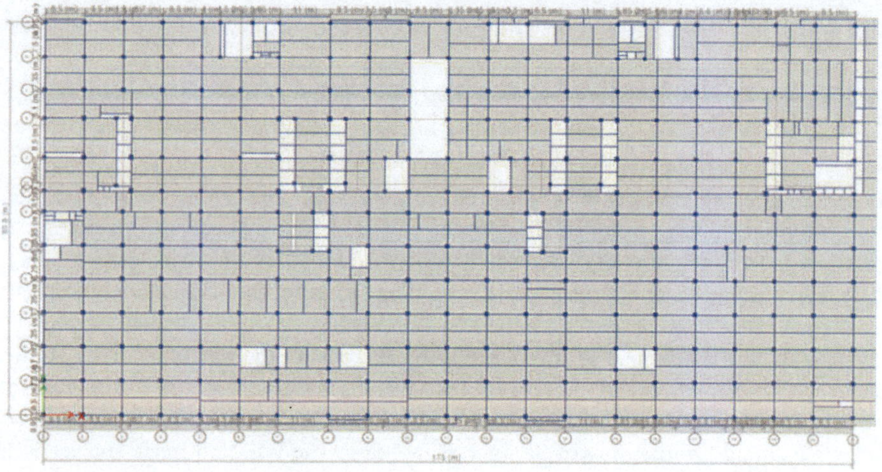

Fig. 1 Floor plan of building at ground level

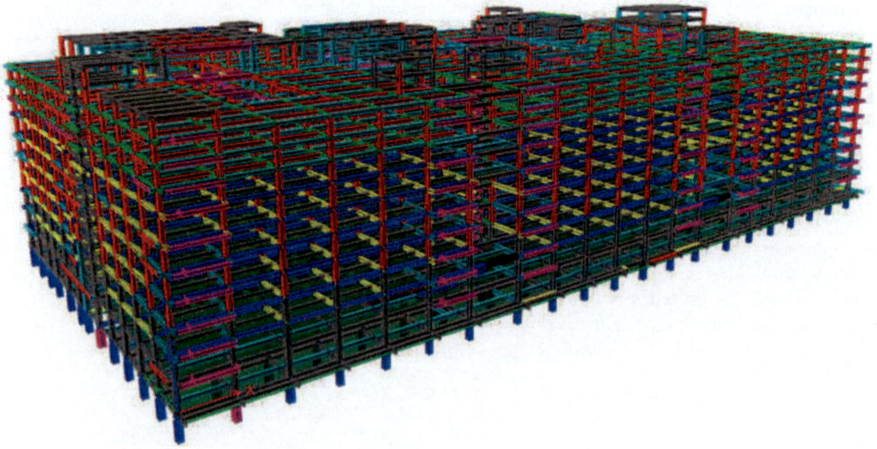

Fig. 2 Three-dimensional analytical model from ETABS

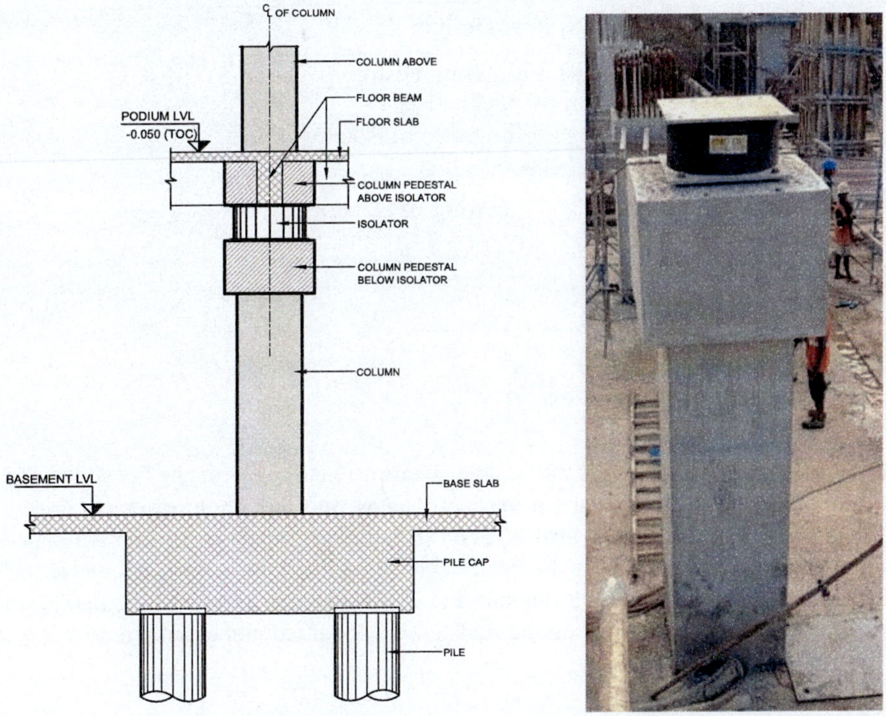

Fig. 3 LRB isolator in pedestal (left image) and site view (right image)

Beams and columns are modeled as frame elements, and above-ground RC walls are modeled as shell components. At each floor level, the slab is assigned as a rigid diaphragm. With the aid of the column pedestals, isolators are installed at the bottom of the ground floor beams as shown in Fig. 3.

3 Analysis of Structure with Isolated Building

The design and analysis of the base-isolated building procedure are explained through flow chart as given below.

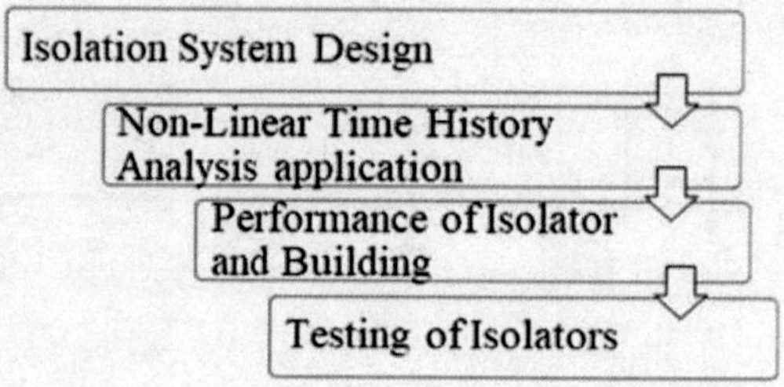

3.1 Isolation System Design

Isolation system combines of Lead Rubber Bearing (LRB) isolator and Pot slider. The system is designed to take care of vertical load as well as lateral movement during earthquake. Lead Rubber Bearing (LRB) isolators are made of rubber laminated shim plates with lead core at the center. The Pot sliders have an elastomeric pad with sliding plate on the top. Yield stiffness (K_d) and characteristic strength (Q_d) of isolators are calculated based on the ASCE 7-16 codal recommendation.

$$K_d = (A_e G)/T_r, \; Q_d = \pi/4 D_p^2$$

where A_e–Net bonded area of rubber, G-Shear modulus, T_r–Total height of rubber, D_p–Diameter of lead. The isolation system shall be designed and checked for 3 different LRB stiffness property cases. i.e., Upper bound, Nominal bound, and Lower bound cases. These stiffness properties are governed by Lambda (λ) factor and it is calculated based on the data of large set from previous projects over the year. Based on the seismic weight, lateral movement, and restoring forces, number of isolators/sliders for each group is finalized in the hospital building. The properties of isolator/slider are calculated based on the group effect and it is given as input in ETABS for nonlinear analysis. Figure 4 shows the input given in ETABS for LRB isolator.

The hysteresis loop (Force–Displacement curve) is plotted based on the formula given below.

$$F = Q_d + K_d \Delta$$

where Δ is the displacement of the system (Fig. 5; Table 1).

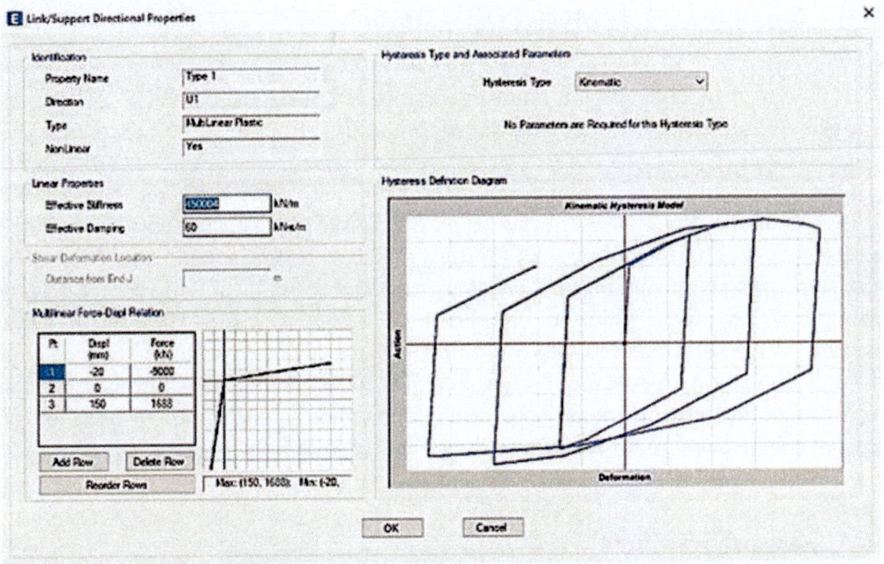

Fig. 4 LRB properties in ETABS

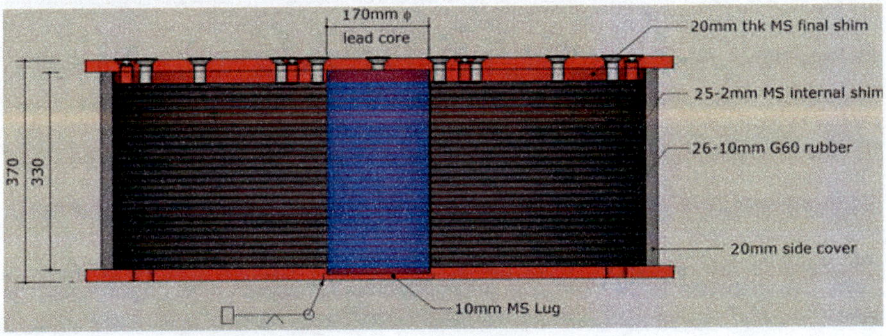

Fig. 5 Sectional view of 770LRB170

Table 1 Geometric properties of LRB

Name	Dia (mm)	Height (mm)	Rubber layer thk (mm)	Lead plug Dia (mm)
770LRB170	770	370	10	170
920LRB170	920	370	10	170
1020LRB170	1020	370	10	170

3.2 *Time History Analysis Application*

Seismic design parameters are considered from IS1983:2016 & ASCE 7-2016 for analysis and design of the isolation system and the structure. The parameters are given in Table 2.

Characterization of ten ground motion records compatible with the 'RESPONSE SPECTRUM-MCE-IS 1893' for soil type Medium-Stiff. These records are scaled so that they meet the requirements given in ASCE 7-16 for isolated structures. The first step consists of obtaining a set of as-recorded ground motion records in two orthogonal directions, which are similar to the target spectrum. The seismogenic features of the selected records should be similar to those expected for the area of study. The method proposed by Vargas et al. [1] for obtaining sets of ground motion records that fit a target response spectrum under a mean square metric is used for the present study. The target spectrum is shown in Fig. 6.

3.2.1 Record Selection Procedure

The approach for choosing ground motion records aligns with the methodology proposed by Vargas et al. [1]. This method involves assessing the mean squared error between the average response spectra of a set of records and the target spectrum function. Subsequently, the records are organized in ascending order based on these error values. This sequence ranks the record that closely matches the target spectrum at the top position, followed by others in succession. It is important to highlight that this selection process effectively mitigates the influence of individual records. The

Table 2 Seismic design parameters

Seismic zone	*V* with MCE (IS1893)
Zone factor, *Z*	0.36 (IS1893-2016 Part-1)
Importance factor, *I*	1.0 from ASCE 7 Cl 17.5.4
Response reduction factor, *R*	Concrete OMRF—ASCE 7-Cl 17.5.4
Soil category	Medium soil (IS1893)
Structural irregularity	None
Short period transition	0.55 s (IS1893)
Long period transition	4.0 s (IS1893)
Redundancy factor	1.0 from ASCE 7-Cl.17.2.3
MCE short period spectral acceleration	$S_{MS} = 0.36 * 2.5 = 0.9$ (IS1893)
MCE spectral acceleration at 1 s	$S_{M1} = 0.36 * 1.36/1 = 0.4896$ (IS1893)
Response spectrum curve	As per IS1893 with above parameters
Grade of concrete	M30 for super structure and foundation M50 for pedestals above and below isolator
Grade of steel	Fe-550D

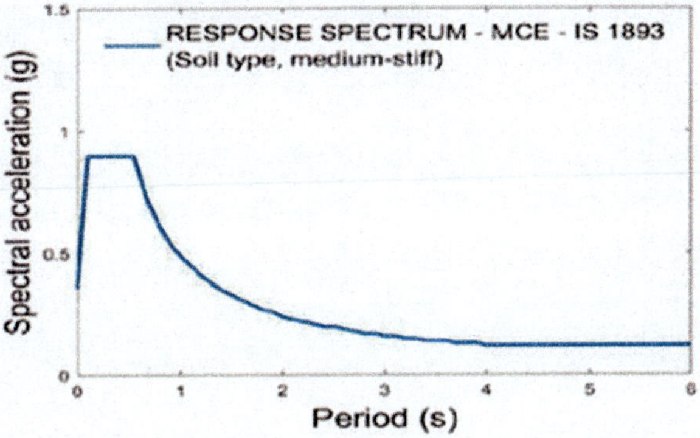

Fig. 6 Target spectrum

sorting process itself excludes records that could introduce substantial deviations in dynamic analyses and estimates of expected demand. Utilizing this method, a collection of ten specific ground motion records has been identified (see Fig. 7). A comparison of these spectra with the target spectrum in Fig. 6 is undertaken. For clarity, the primary components are denoted as '*X*' (Records on the right in Fig. 7) and '*Y*' (Records on the left in Fig. 7).

Note that, in addition to finding ground motion records compatible with a target spectrum, the method proposed by Vargas et al. [1] also could find signals with similar seismogenic features, including depth, magnitude, and epicentral distance (Table 3).

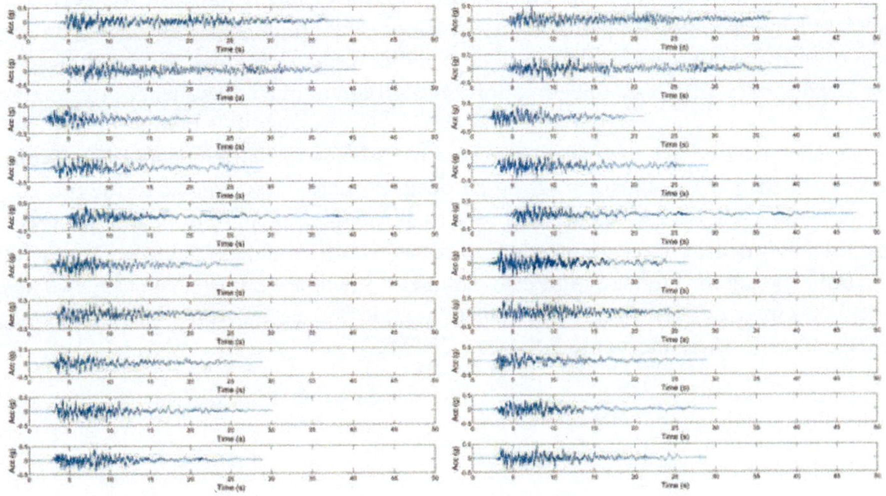

Fig. 7 As-recorded ground motion records; right: *X* direction; left: *Y* direction

Table 3 Seismological parameters of selected ground motion

Epicentral distance (km)	Depth (km)	Hypocentral distance (km)	Magnitude
06.8	3.8	07.8	6.9
22.0	3.8	22.3	6.9
18.6	9.2	20.8	6.5
05.0	8.3	09.7	6.1
01.8	8.3	08.5	6.1
04.9	8.3	09.6	6.1
07.8	9.2	12.1	6.5
15.6	8.1	17.6	6.0
04.6	9.2	10.3	6.5
10.5	9.2	14.0	6.5

3.2.2 Scaling of Ground Motions

ASCE 7-16 (section 17.3.3) sets out two key criteria that matched spectral records must satisfy. Firstly, for each pair of horizontal ground motion components, a square root of the sum of the squares (SRSS) spectrum is computed by determining the SRSS of the 5%-damped response spectra for the scaled components. Scaling is applied to each motion pair to ensure that, within the period range spanning from 0.75TMU (based on upper bound isolation system properties) to 1.25TML (determined from lower bound isolation system properties), the average SRSS spectra of all horizontal component pairs maintain values equal to or exceeding those on the response spectrum used for design (RESPONSE SPECTRUM-MCE-IS 1893).

Furthermore, in at least one direction, a minimum of one component of the record must undergo scaling to guarantee that none of its values fall below 90% of the target spectrum within the period range of [0.2TMU–1.25TML]. These conditions are established while considering the BI design sheet characteristics of the project, specifically TMU = 3.53 s and TML = 4.13 s. As illustrated in Fig. 8, the spectra of the selected records, once adjusted to meet the criteria, are depicted to comply with the 90% threshold. Figure 8 demonstrates that none of the adjusted spectra's values (depicted by grey lines) go below the 90% limit prescribed by ASCE regulations within the specified interval [0.2TMU–1.25TML].

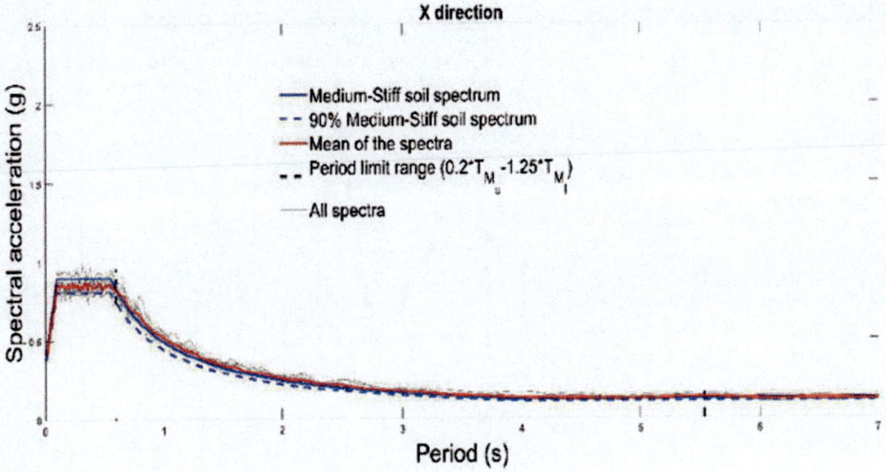

Fig. 8 Components selected to meet the 90% limit

4 Overview of Analysis Results

4.1 Story Drift

Story drift means the difference between two adjacent story lateral displacements. This drift will affect the structural and nonstructural elements during earthquake by the large displacements of structural and nonstructural elements. Hence limiting the story drift will not the hospital building adversely. And the maximum story drift is limited in both *X* and *Y* directions with ASCE 7-16 guidance of 1.5%. Refer Fig. 9.

4.2 Base Shear

Based on static analysis, target base shear of 7.0% seismic weight is determined for the hospital building. In time history analysis, by scaling according to ASCE 7-16, the desired base shear is attained. Furthermore, structural components of the hospital are designed to withstand a target base shear.

4.3 Restoring Force

The isolation system shall be configured to produce restoring lateral force for the hospital building as per ASCE 7-16. According to the codal clause of 17.2.4.4, the

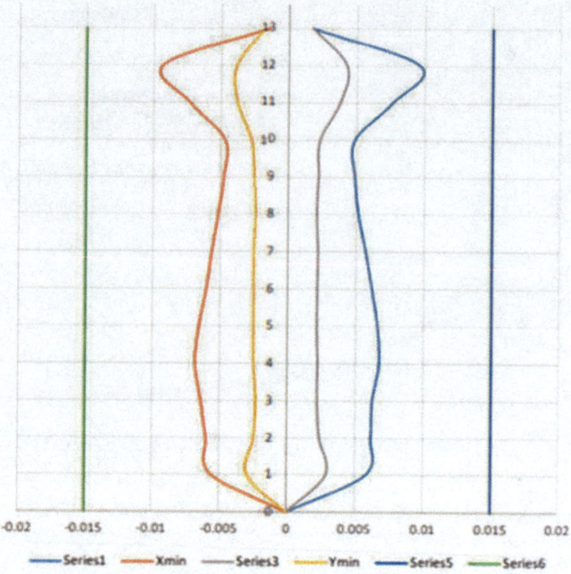

Fig. 9 Story drift of hospital

system's lateral force difference between the MCE condition and 50% of the MCE condition to the ratio of its seismic weight must be greater than 0.25 times.

5 Special Detailing in Base-Isolated Building

This hospital building concept is finalized by treating continuous access from upper floor to basement in staircase and lift areas. Special detailing is worked out for staircase flight and is separated in isolation plane as shown in Fig. 10.

For lift rails, two options are worked out. The first option is a cantilever wall hanging from ground beam and the rails are fixed over the walls and the wall is separated from basement by providing gap as shown in Fig. 11. In another option, wall lift supporting guide rail and block wall is separated from basement isolator plane by providing a joint as shown in Fig. 12.

The ground story diaphragm displacement value of 350 mm (maximum) from analysis is used as seismic separation joint in ground floor perimeter of the building as shown in Fig. 13. Retaining wall is designed as a cantilever wall to allow the lateral movement during seismic events and the joint treated with sealant.

MEP pipelines are flexible and sufficient length in the plane of isolator, will not break or damage during seismic movement of the building as shown in Fig. 14.

Fig. 10 Staircase flight joint between basement and ground

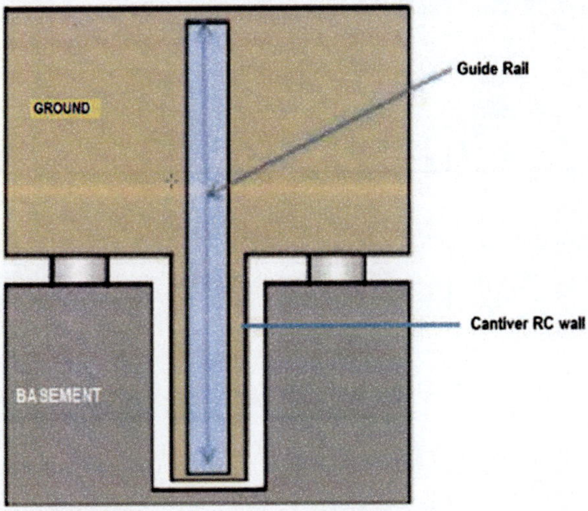

Fig. 11 Cantilever RC wall hanging from ground

6 Isolator Installation at Construction Site

The plan area of the pedestal is first considered, and a bearing template made of a steel plate at least 20 mm thick is used, with the position of the bolt template following the drawing in Fig. 15. Through the use of 6 mm hooked shape wire and free flow, non-shrinking cement-based grout, mock grouting is completed with the fewest possible air spaces and bubbles. Minimum 95% full contact area grout needs to be achieved inside the diameter of isolator location. After completion of the mockup, isolators are installed. The erection of isolator should be done in bottom pedestal based on

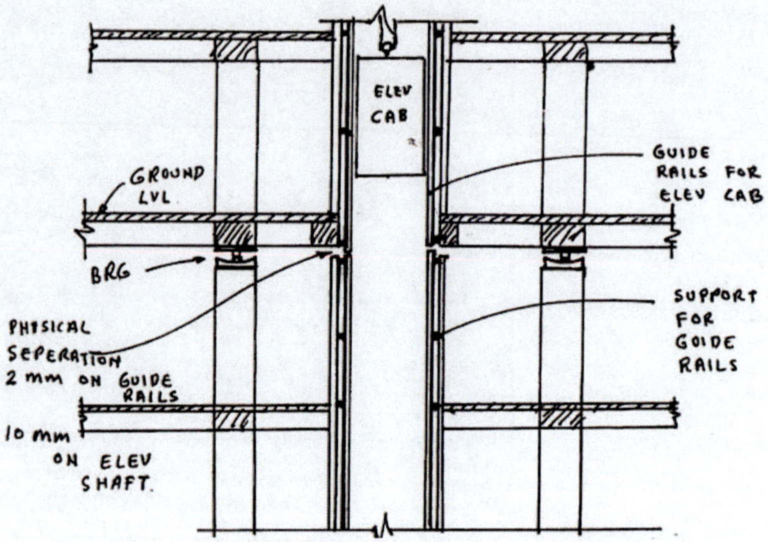

Fig. 12 Gap in guide rail between ground and basement

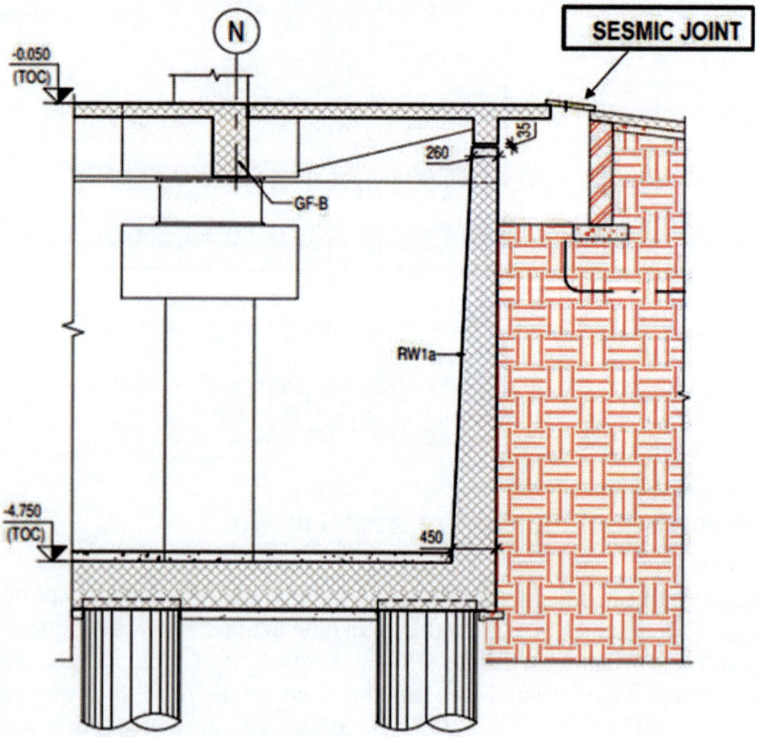

Fig. 13 Seismic joint in building perimeter

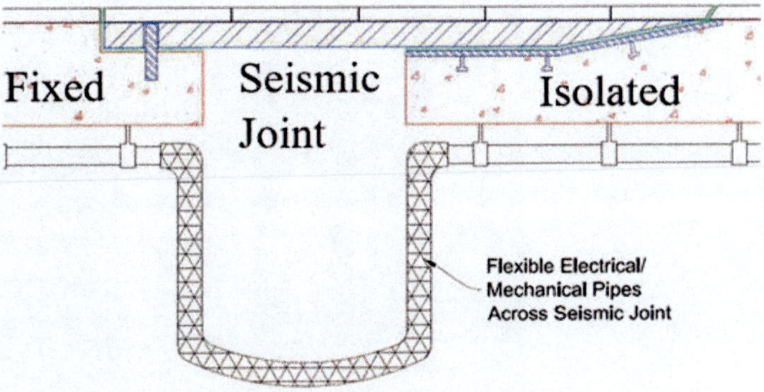

Fig.14 MEP cables in between fixed and isolated building

Fig. 15 Standard template in site to cast the pedestal

standard operating procedure. The same construction team shall be used to do the installation of isolator in remaining location of the hospital building.

7 Testing of Isolator

Prototype testing has been done for various types of LRB isolators as per ASCE 7-16 guidelines at Seismic Robinson's laboratory. The test results are verified with analytical model properties and found to be similar.

The test setup is shown in Fig. 16. The basic concept of the Prototype is such that the lateral force is applied to achieve target displacement at compressive load. The readings are noted by digital sensor at very small time intervals (in the range of ms) to check the lateral force required to achieve particular displacement, those values

Fig. 16 Testing rig

are plotted in the graph to get the hysteresis loop. From that loop, the value of k_{eff} (Load/Disp) and value of damping can be arrived. The arrived values of the k_{eff} and damping should be within prescribed limits from the target values.

770LRB170 Isolator Prototype testing parameters like axial load and lateral displacements which arrived from analytical model are given as input through computer. The test results are plotted in Fig. 17.

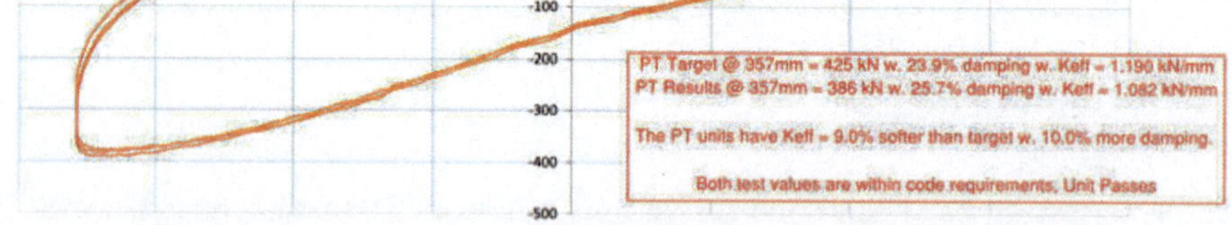

Cycle	Disp Min (mm)	Disp Max (mm)	Force Min (kN)	Force Max (kN)	EDC (kNm)	k effective (kN/mm)	Damping %
1	-357.2	358.0	-397.5	394.2	225.3	1.107	25.3
2	-357.2	358.0	-389.1	380.4	225.0	1.076	26.0
3	-357.2	358.0	-385.0	376.1	222.4	1.064	25.9

Fig. 17 Load versus displacement chart

8 Conclusion

National Disaster Management Authority has stated that hospitals need to function even post-earthquake. Base-Isolated hospitals have significantly contributed to function even after earthquake. The importance of the Base-Isolation system in a hospital building is addressed and Base-Isolation system is implemented successfully at the construction site. The design and analysis procedure of the isolated building are briefed in the above sections.

The Hospital building drift is verified with the limit of 1.5% and found within the codal allowance of ASCE 7-16.

The Isolation system restoring performance is verified and found that the system's lateral force difference between the MCE condition and 50% of the MCE condition to the ratio of its seismic weight is greater than 0.25 times.

To serve as a handy reference for upcoming projects, the challenges associated with maintaining the isolator on the basement roof are discussed along with a workable solution.

Isolator properties used in analytical model are verified through prototype test results and the results are found to be within the limiting value of 10% as per ASCE 7-16 codal references.

References

1. Vargas YF, Pujades LG, Barbat AH, Hurtado JE (2013) Capacity, fragility, and damage in reinforced concrete buildings: a probabilistic approach. Bull Earthquake Eng 11(6)
2. ASCE 7-16 Minimum design loads and associated criteria for building and other structures
3. IS1983-2002 Part 1 criteria for earthquake resistant design of structures
4. Seismic Robinson Limited. https://www.robinsonseismic.com
5. Anbazhagan P, Bajaj K, Patel S (2015) Seismic hazard maps and spectrum for Patna considering region-specific seismotectonic parameters. Nat Hazards 78(2):1163–1195

Mass Concreting Techniques Used in Various Hospital Projects for LINAC Block

J. Praveen, Harine Thangaraj, and V. G. Lavanya

1 Introduction

LINAC Bunkers house the "LINEAR ACCELERATOR" which is used extensively for radiation therapy as shown in Fig. 1. Proton Therapy facility for advance cancer treatment houses a cyclotron, energy selection system, beam transport system and rotating gantry. The bunkers are designed to contain harmful radiation generated during the treatment and prevent its leakage to the outside environment. As per the guidelines of Atomic Energy Regulatory Board (AERB), massive thick walls and slabs are constructed to avoid radiation leakage as shown in Fig. 2. Such huge volume of concrete develops significant internal temperature which must be controlled to prevent the occurrence of cracks. Use of concrete in these massive elements requires innovative, unique and special precautionary measures to be followed. Critical project data where LINAC block is constructed is given in Table 1 [1].

J. Praveen (✉) · H. Thangaraj · V. G. Lavanya
Larsen & Toubro Limited, Chennai 600089, India
e-mail: PRAVEENJAYANTHI@lntecc.com

H. Thangaraj
e-mail: harine@lntecc.com

V. G. Lavanya
e-mail: vgoutami@lntecc.com

R. K. Pancharathi et al. (eds.), *Low Carbon Materials and Technologies for a Sustainable and Resilient Infrastructure*, Lecture Notes in Civil Engineering 440,
https://doi.org/10.1007/978-981-99-7464-1_23

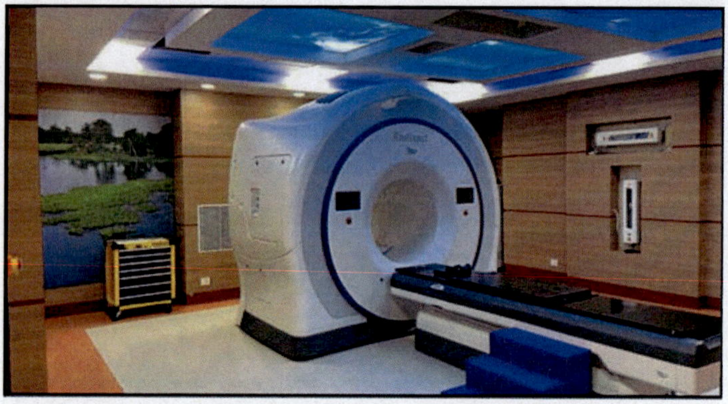

Fig. 1 LINAC treatment unit after installation

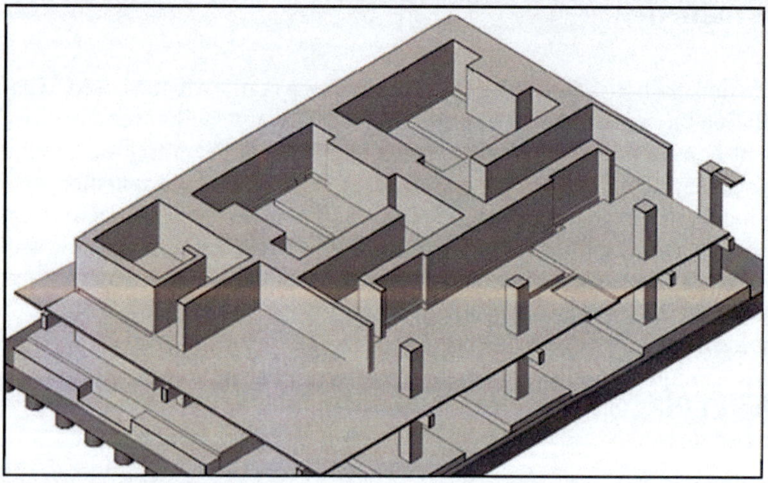

Fig. 2 LINAC block with mass concrete elements

Table 1 Critical project data

Description	Bunker size $L \times B$	Height	RC wall thickness	RC roof slab depth	Volume of concrete
SCI cancer centre	12.5 m × 29 m	5.70 m	0.60–2.40 m	0.7–2.40 m	2258 m^3
L2 prototype (3 sites)	25 m × 14.5 m	5.70–10.80 m	0.60–2.40 m	0.7–2.40 m	4164 m^3
L3 prototype (6 sites)	25 m × 14.5 m	5.70–10.80 m	0.60–2.40 m	0.7–2.40 m	7114 m^3

2 Special Techniques and Quality Control Measures

2.1 Formwork Systems

Special insulated formwork arrangement is used to prevent the structure from getting exposed to the atmosphere. Wall concrete surface is insulated by keeping specially constructed PUF foam boards at both faces. For the slab, the bottom shuttering is fixed with PUF boards, while the top surface is coated with curing compound and covered with tarpaulin sheet followed by Thermocol panels to ensure minimal temperature variation between inside and outside.

PUF boards consist of a pair of 12 mm thick plywood as outer layer with 50 mm thick high-density Thermocol (25 kg/m) in between. Formworks of approximate size 2.4 m × 3.6 m (with size variation as per site requirement) are constructed and erected in place at site by cranes. 2″ × 3″ wooden runners at 300 mm spacing are used to maintain the stability of the panels which is further backed by H-16 aluminium beams. Form works on each side of the walls are anchored together with 16 mm consumable tie rods running across the wall thickness. As an alternative to the sandwiched PUF board, 18 mm thick monolithic plywood board panels (without Thermocol) as shown in Fig. 3 are also used based on availability at site.

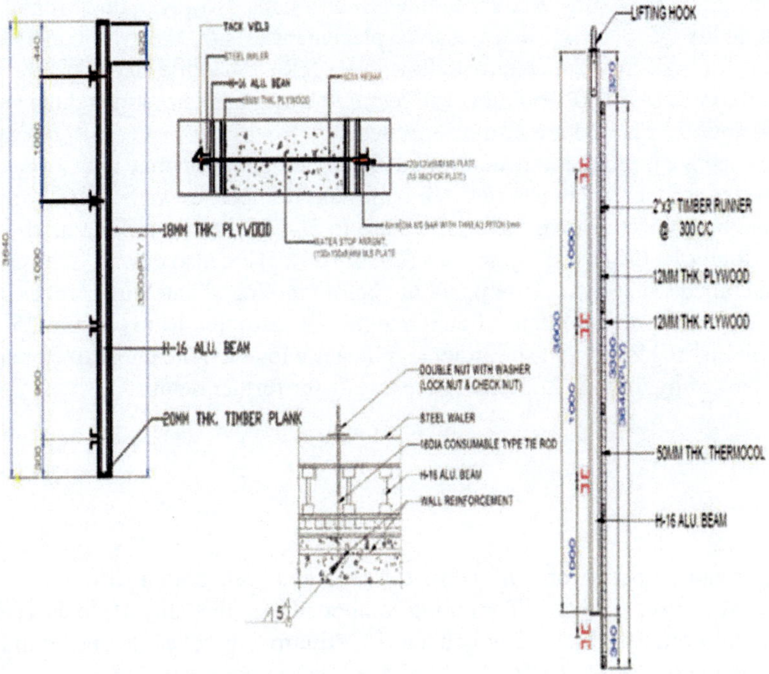

Fig. 3 Formwork arrangement

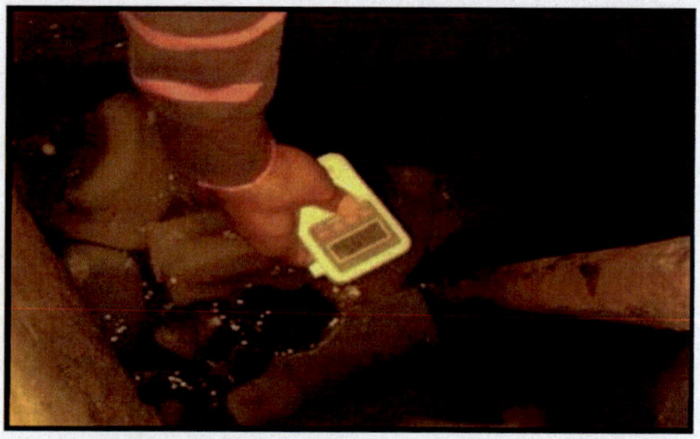

Fig. 4 Lowering water temperature with ice to < 4 °C

De-shuttering of wall is done only when the internal peak temperature is less than 400 °C, so that variation of temperature between the core temperature and ambient temperature is less than 190 °C.

2.2 *Concrete Mix Design*

Concrete grade of M30 is used for both walls and slabs. However, due to challenges in availability of materials and concrete placement at site, the wall concrete mix design is for high-flow concrete with flow index 520–550 while slab concrete mix is with slump value of 150–160 mm. The required water-cement ratio and the amount of cementitious material are also determined.

The concrete is temperature controlled by preparing the concrete with low temperature water and coarse aggregates. Ice cubes are dumped in water, based on exact calculations, to bring down the temperature to 3–4 °C before mixing as shown in Fig. 4. Similarly, the coarse aggregates (20 mm down) are also dowsed with crushed ice and chilled water to lower temperature before mixing in batching plant as shown in Fig. 5. Placing temperature of concrete mix is restricted to maximum 25 °C as shown in Fig. 6. Use of fly ash further significantly lowered the internal temperature due to heat of hydration [2]. Refer Annexure–1 for further details.

2.3 *Curing*

Curing is done on the top surface of the slab and wall bottom lifts with curing compound as shown in Fig. 7. The curing compound is applied directly to the concrete surface followed by a layer of polythene (LDPE/tarpaulin) sheet as shown in Fig. 8 to prevent the moisture loss from the surface. Thermocol sheets are placed over the polythene membrane to provide necessary insulation as shown in Fig. 9.

Fig. 5 Lowering aggregate temperature with ice

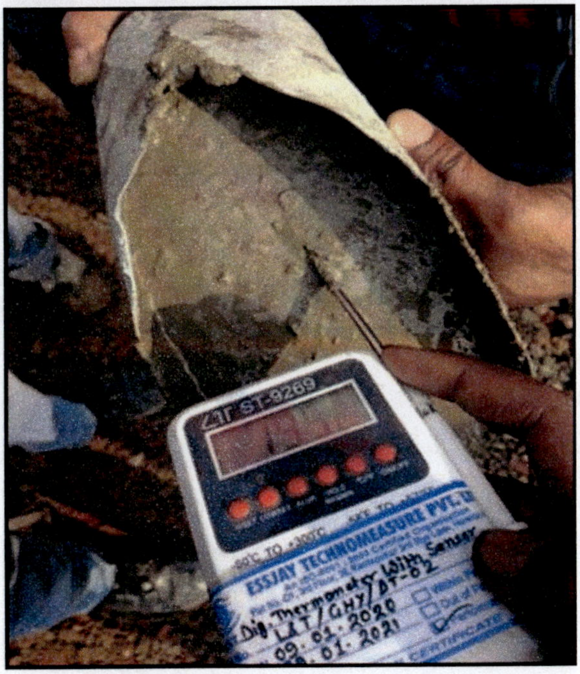

Fig. 6 Concrete temperature at pouring point < 25 °C

Fig. 7 Application of curing compound over concrete surface

Fig. 8 Application of thermocol over LINAC slab

3 Temperature Monitoring

Temperature observations are made in the concrete by using thermocouples (Fig. 10) as per GFC (good for construction) drawing. The placing of thermocouples in a wall/ slab is done in the following manner.

Fig. 9 Application of polyethene sheet over thermocol

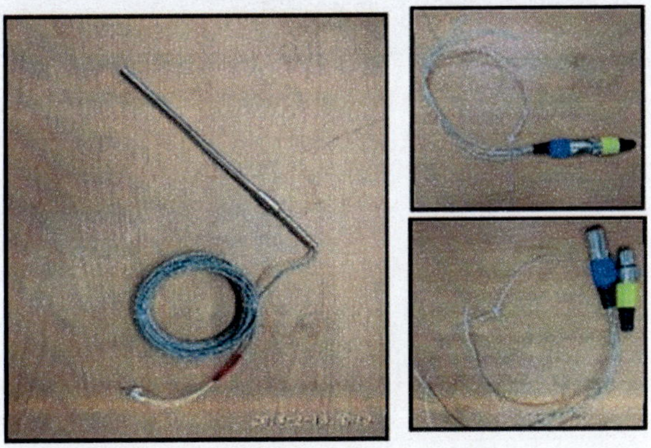

Fig. 10 Thermocouple sensor and connectors

1. Thermocouples are fixed to a rebar as shown in Fig. 11.
2. At the centre point to the thickness of the raft/wall, three thermocouples are placed with its monitoring pins placed at the top, middle and bottom points of the wall as shown in Fig. 12.
3. The top point and the bottom point are 200 mm below and above the level of casting respectively and the middle point is at the centre to the entire depth of the level of casting.
4. In walls, thermocouples of one number are placed in the left and right sides of the wall, with its pin positioned at the centre, lying parallel to the centre point of the thermocouple placed at the middle of the wall. The thermocouple readings are taken at regular intervals as shown in Fig. 13.

Fig. 11 Thermocouples fixed to rebar

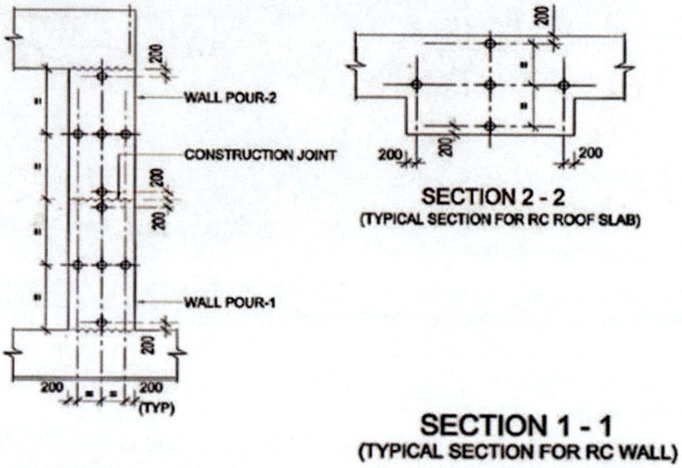

Fig. 12 Thermocouple placement in walls and slab

4 Calculations

Thermal properties of mass concrete in heating zone are determined as per Clause 4.2.8 of IS 14591:1999 [2] with following inputs.

1	Cement content in concrete	530 kg/m^3
2	Total coarse aggregate in concrete	944 kg/m^3

3	Total fine aggregate in concrete	714 kg/m^3
4	Total water content in concrete	186 kg/m^3
5	Density of concrete (ρ)	2610 kg/m^3
6	Mean annual air temperature	26 °C
7	Mean summer temperature	32 °C
8	Mean winter temperature	20 °C
9	Mean annual river water temperature	20 °C
10	Specific heat of concrete (C)	857.33 J/kg°C
11	Thermal conductivity of concrete (K)	2.235 5 J/m s °C
12	Ultimate adiabatic rise in temperature (T_o)	68.32 °C
13	Factor m for the concrete	0.025/h
14	Exposure period of the lift (t)	72 h
15	Lift height (D)	3.00 m
16	Desired placement temperature of concrete	7 °C
17	Temperature of ingredients of concrete at the time of mixing	
a	Cement	30 °C
b	Coarse aggregate	20 °C
c	Fine aggregate	28 °C
d	Batched mixing water	5 °C
18	Ice per cent of mixing water	60 per cent
19	Permissible tensile stress at 10% of the compressive strength of concrete	3.0 MPa
20	Coefficient of thermal expansion of concrete (α)	4×10^{-6}/°C
21	Sustained modulus of elasticity of concrete ($E\ \rho$)	2×10^4 MPa
22	Restraint factor (R)	1.00
23	Compressive strength of concrete	40 MPa

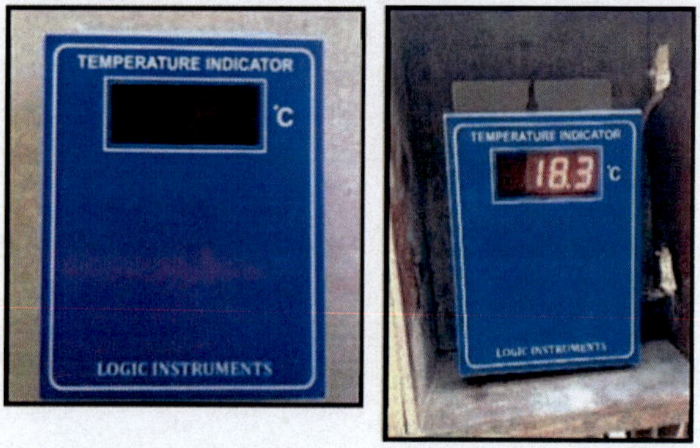

Fig. 13 Digital temperature monitor and concrete placing temperature

4.1 Computations

A **Verification of Desired Placement Temperature of Concrete**

Placement temperature of concrete is computed by the formula given in Clause 4.2.6.2 of IS 14591:1999,

$$T_p = \frac{0.22\left(T_{ca}W_{ca} + T_{fa}W_{fa} + T_cW_c\right) + T_fW_f + T_wW_w - 79.6W_i}{0.22\left(W_{ca} + W_{fa} + W_c\right) + W_f + W_w + W_i} = 5.3\,°\mathrm{C}$$

The placement temperature at the site of work depending on conditions is expected to be about 7 °C as desired.

B **Determination of Loss of Heat from the Surface of the Lift**

(i) As per Clause 4.2.7.2 of IS 14591:1999, drop in temperature of the concrete due to heat loss q_3 is obtained by the formula:

$$T_3 = \eta T_o$$

The value of η is obtained from Fig. 1 after calculating mt and $\frac{D}{\sqrt{4h_d^2 t}}$
$m = 0.025/\mathrm{h}$ and $t = 72$ h

$$mt = 0.025 \times 72 = 1.80$$

and $\frac{D}{\sqrt{4h_d^2 t}} = \frac{3.00}{\sqrt{4 \times h_d^2 \times 72}}$

The diffusivity h_d^2 of concrete is calculated by the formula:

$$h_d^2 = \frac{K}{C\rho} = 9.994 \times 10^{-7}\,\mathrm{m}^2/\mathrm{s}$$

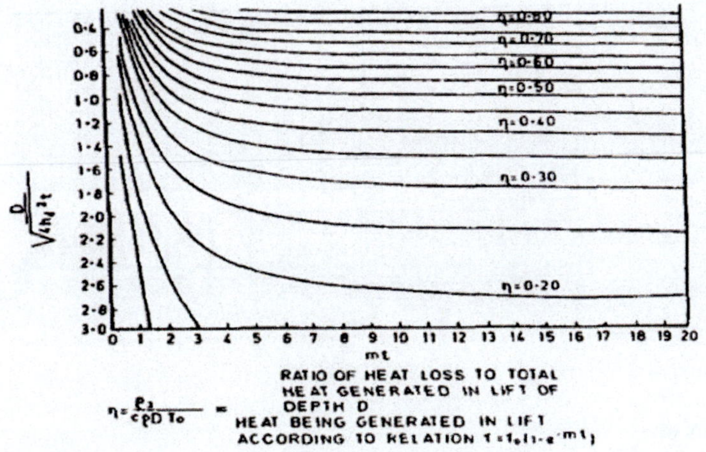

Fig. 14 Part of the total heat lost from the surface of the lift

$$h_d^2 = 3.598 \times 10^{-3}\,\text{m}^2/\text{h}.$$

Hence $\frac{D}{\sqrt{4h_d^2 t}} = 2.95$ Now, corresponding to the values of $mt = 1.80$ and $\frac{D}{\sqrt{4h_d^2 t}} = 2.95$, the value of η as obtained from Fig. 14 is 0.10.
Thus, $T_3 = \eta T_o = 6.832\,°\text{C}$.

(ii) As per Clause 4.2.7.3 of IS 14591:1999, temperature drop due to heat loss q_2 is calculated by the formula:

$$T_2 = \frac{2h_d^2\theta_o}{D\sqrt{\pi}}\left[\sqrt{\frac{t}{h_d^2}} - I\right]$$

Now,

$$\theta_o = \text{Placement temperature} - \text{Atmospheric temperature} = -25\,°\text{C}$$

From Fig. 15,

$$\frac{I}{\frac{D}{2h_d^2}} = 0.0001.$$

$$I = 0.041$$
$$T_2 = -4.78\,°\text{C}$$

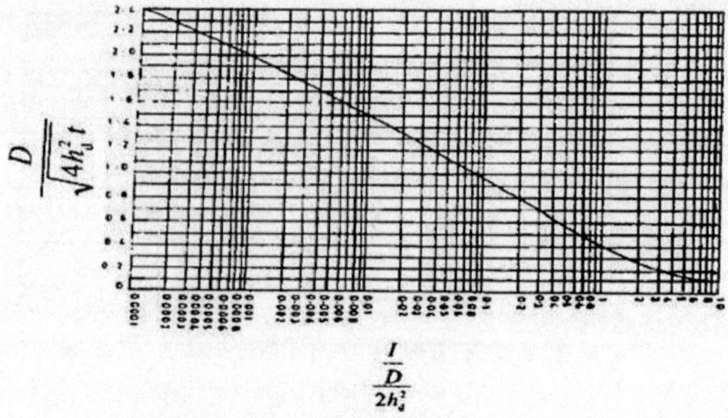

Fig. 15 Plot of $\frac{I}{\frac{D}{2h_d^2}}$ versus $\frac{D}{\sqrt{4h_d^2 t}}$

(iii) As per Annex A of IS 14591:1999, temperature drop due to heat loss q_1 is calculated by the formula:

$$T_1 = \frac{2h_d^2\theta_2}{D\sqrt{\pi}} \times I$$

Here,

$$\theta_2 = \left(T_p + T_o - T_3 - T_2\right) - \text{Atmospheric Temperature} = 41.27\,^\circ\text{C}$$

Hence,

$$T_1 = 0.003\,^\circ\text{C}$$

Therefore, the total temperature drop due to loss of heat from the exposed surface of the lift,

$$T_L = T_3 + T_2 + T_1 = 2.055\,^\circ\text{C}$$

C **Evaluation of Thermal Stresses**

As per Annex A of IS 14591:1999, the maximum temperature attained by concrete $= T_p + T_o - T_L = 73.265\,^\circ\text{C}$.

Where, $T_o = \frac{HW}{\rho C} = 68.32\,^\circ\text{C}$ [$H = 343{,}527$ J/kg and $W = 445$ kg/m^3].

Final stable temperature of concrete is $T_s = 26\,^\circ\text{C}$ which is the mean annual air temperature.

Thermal Stress is $F = \alpha E_p R\left(T_p + T_o - T_L - T_s\right) = 3.78\,\text{MPa} < 4.3\,\text{MPa}$, hence acceptable.

5 Results and Discussions

Graph between time from concrete placing vs temperature is shown in Fig. 16. Following observations are made from the graphs.

- Maximum core temperature readings obtained from thermocouples is 70 °C typically after 48 h of concrete placing and the ambient temperature is 32 °C.
- Maximum temperature difference is 34 °C to 39 °C.

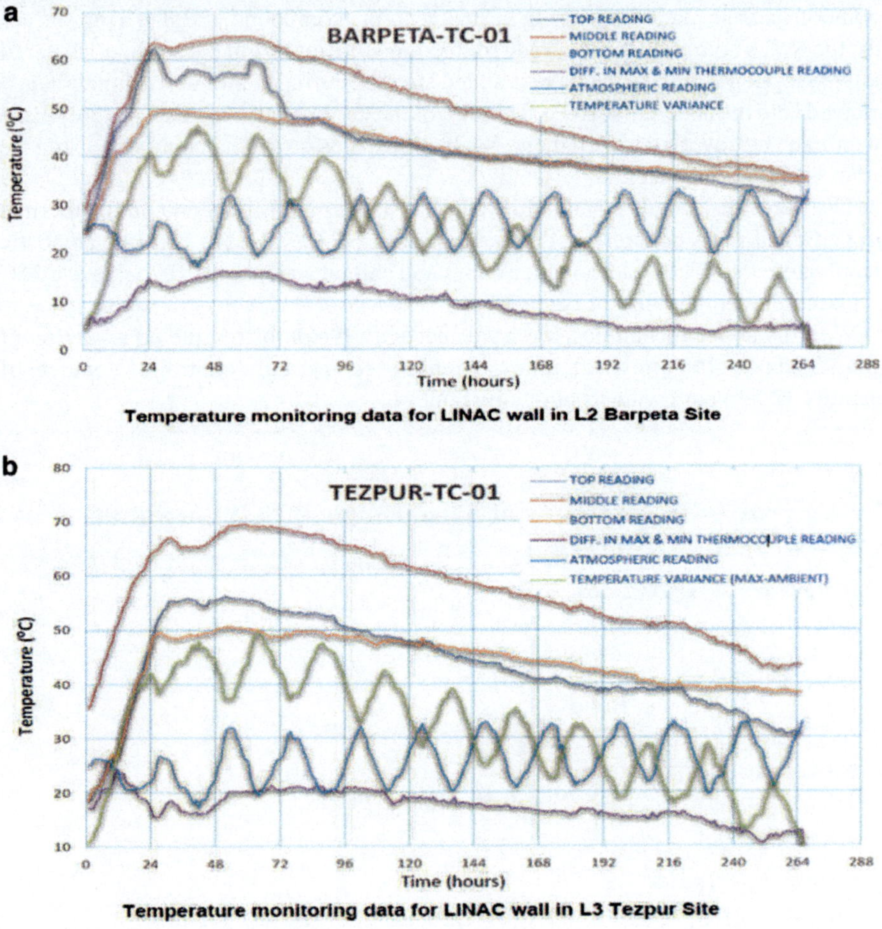

Fig. 16 **a** Temperature monitoring data for LINAC wall Barpeta site. **b** Temperature monitoring data for LINAC wall Tezpur site

- Proper thermal insulation techniques as mentioned above must be followed till the temperature difference between core temperature and ambient temperature is 20 °C.
- Typical time taken to maintain the temperature difference of 20 °C is 7 to 14 days.

6 Conclusion

Several procedures were used throughout the project to prevent thermal cracks and radiation leakage, including curing i.e. using curing compound, the slab's top surface and the wall's bottom lifts are cured. To stop moisture loss from the surface, a layer of polythene (LDPE/tarpaulin) sheet is added after the curing compound is immediately sprayed onto the concrete surface. Thermocol sheets are used to insulate the polythene membrane by covering it with them. We have also performed batch mixing using ice cubes and ice-cold water.

The usage of ice cubes and cold water has had a major influence on the thermal properties of mass concrete in the heating zone. 68.32 °C is the ultimate adiabatic temperature rise. The addition of ice cubes and chilled water helped to achieve a 7 °C of placement temperature of concrete.

With the above techniques, we have successfully completed the construction of LINAC bunker structures with innovative use of temperature-controlled concrete of quantity 13,536 cum with strict quality control.

7 Annexure–1: Summary of Mix Design at SCI Guwahati

Properties of Aggregates
Source:

(a) Coarse Aggregate: Shillong
(b) Fine Aggregate: Boko
(c) Cement: Dalmia OPC 43 Grade
(d) Fly Ash: Bongaigaon
(e) Admixture: 4111 NS (Sika)–Mid Range PC

Coarse Aggregate
Impact value: – 22.92
Crushing value: – 23.47
Specific Gravity: – 2.717
Water Absorption: – 0.32
Flakiness (20 mm):15–18%
Flakiness (10 mm):14–19%.

Fine Aggregate
Specific Gravity: – 2.612
Water Absorption: – 1.23
FM: 2.4–2.6
Zone–3.

A **Design Mix content per cum**

(a) 20 mm: 330 kg
(b) 10 mm: 614 kg
(c) Sand: 714 kg
(d) Cement: 360 kg
(e) Fly Ash: 170 kg
(f) Water Cement Ratio: 0.35
(g) Admixture: 2.39 kg (Dosing 0.45%)
(h) Concrete Flow: 640 mm (Initially), 550 mm (After 60 min)

B **Concrete Strength at the time of Design mix**

(a) 7 days Strength–33.45 Mpa
(b) 28 days Strength–43.44 Mpa

C **Concrete Strength after pouring.**

(a) 7 days Strength–28.54 Mpa (Avg)
(b) 28 days Strength–38.64 Mpa (Avg)

Ambient Temperature During Concrete

(a) Winter Season–Min-11 °C Max-28 °C
(b) Summer–Min: 21 °C Max: 34 °C

Pouring Temperature—< 25 °C
Maximum Temperature difference—34–39 °C (In both Summer and Winter)
Time is taken to maintain temperature difference between core temperatures: 20 °C: 7–14 days.

References

1. Regulatory requirements and guidelines for new radiation therapy facility. Atomic Energy Regulatory Board
2. IS 14951 (1999) Temperature control of mass concrete for dams—guidelines. BIS

Recycled and Renewable Materials for Sustainable Construction

Image Processing Techniques for Assessment of Deterioration in Recycled Aggregate Concrete Due to Chemical Attack

Madhavi Latha Kasulanati and Rathish Kumar Pancharathi

1 Introduction

Recycling of concrete practised in present-day construction overcomes the problem of depletion of natural sources and reduces the wastage of land required for the disposal of demolished material [1]. It is now made mandatory to utilise recycled material in construction [2]. The natural aggregate used in construction of a structure can be replaced by 20% with recycled aggregate for reinforced concrete and 25% for plain concrete as per IS 383:2016 [3]. The mandatory guidelines according to the report in 2016 by construction and demolition waste management necessitate utilisation of 10–20% of materials from construction and demolition waste [1]. The aggregates are classified based on the attainment from masonry rubble, concrete rubble and a blend of natural and recycled aggregate according to RILEM TC121-DRG (1994) [4]. The Recycled aggregate can be replaced up to 35 and 25% for strength class C25/30 and C30/37 according to German standards DIN 4226-100 [5] in conventional concrete.

The durability of a structure can be estimated partially by visual examination, destructive tests on samples and non-destructive tests conducted according to the requirement [6]. The deterioration of concrete can be due to internal effects like poor quality materials, faulty workmanship etc., and due to external effects like acid attack, sulphate attack [7], chloride attack, etc. The structures constructed using Recycled Aggregate Concrete (RAC) have to be further analysed for their microstructure, as the interfacial transition zone of concrete affects the quality [8]. The images of the

M. L. Kasulanati (✉) · R. K. Pancharathi
Department of Civil Engineering, National Institute of Technology Warangal, Warangal, Telangana 506004, India
e-mail: madhavi@student.nitw.ac.in

R. K. Pancharathi
e-mail: rateeshp@nitw.ac.in

R. K. Pancharathi et al. (eds.), *Low Carbon Materials and Technologies for a Sustainable and Resilient Infrastructure*, Lecture Notes in Civil Engineering 440,
https://doi.org/10.1007/978-981-99-7464-1_24

affected surface captured using high-resolution camera can be analysed for the shape, texture and colour variations for different chemical attacks. The results obtained can further be correlated with the evaluation based on other non-destructive tests.

2 Methodology

The conventional concrete and Recycled Aggregate Concrete (RAC) cubes are cast using Particle Packing Methods (PPM) [9, 10] and IS 10262:2019 for strengths of 30 and 50 MPa. The proportions of concrete mix for natural and Recycled Aggregate Concrete (RAC) using IS (IS: 10262:2019) method and PPM are presented in Tables 1 and 2.

It can be observed that PPM can ensure a decrease in the cement content and better packing, resulting in higher strengths in both 30 and 50 MPa concretes for both Natural and Recycled Aggregate Concrete (RAC).

The concrete specimens cast were cured for 28 days. The images of specimens before exposure to chemical attack are captured. The specimens are then subjected to attack by four chemical species viz., HCl, H_2SO_4, $MgSO_4$ and NaCl for 28 days. After the exposure period, the specimens are again surface dried and the images of the surfaces are captured using a high-definition camera. All six surfaces of the cube are captured and the images captured before and after exposure are analysed using image J software. Each image is given a unique value of visible colour in the form of Red, Green and Blue values (RGB). The average RGB values of the six surfaces of each cube are determined and are converted to '*H*', '*S*' and '*I*' colour space values using the following equations [11].

$$H = \cos^{-1}\left\{\frac{0.5 * [(R - G\} + (R - B)]}{\sqrt{(R - B)^2 + (R - B) * (G - B)}}\right\} \quad (1)$$

Table 1 Mix proportions of concrete using IS 10262:2019 mix design method

Constituents in Kg/m^3	IS 10262:2019			
	30 MPa		50 MPa	
	NAC	RAC	NAC	RAC
Cement	296.88	296.88	381.70	381.70
Fly ash	74.22	74.22	95.42	95.42
CA	1338.40	1290.20	1018.80	1013.60
FA	592.70	588.90	758.10	694.20
Super plasticizer	–	2.37	3.05	3.81
Water	151.81	151.81	151.81	151.81
W/C	0.41	0.41	0.32	0.32

CA Coarse Aggregate; *FA* Fine Aggregate; *W/C* Water-cement ratio

Table 2 Mix proportions of concrete using particle packing method (PPM)

Constituents in Kg/m^3	Particle packing method of mix design				
	30 MPa			50 MPa	
	NAC		RAC	NAC	RAC
Cement	268.83		279.24	329.26	355.39
Fly ash	67.20		69.81	65.85	63.97
CA	Sieve size (mm)	385.31	423.36	529.34	403.01
	16–12.5				
	12.5–10	363.91	322.56	350.17	307.06
	10–4.75	299.69	80.64	260.24	76.76
FA	4.75–2.36	278.28	322.56	465.77	307.06
	2.36–1.18	256.88	342.72	209.87	326.25
	1.18–0.6	278.28	282.24	86.03	268.67
	0.6–0.15	278.28	262.08	99.74	249.48
Super plasticizer	2.68		2.79	3.29	3.55
Water	134.69		156.71	151.65	160.44
W/C	0.45		0.40	0.34	0.33

CA Coarse Aggregate; *FA* Fine Aggregate; *W/C* Water-cement ratio

$$S = 1 - \frac{\min\{R, G, B\}}{I} \tag{2}$$

$$I = \frac{R + G + B}{3} \tag{3}$$

'*H*' indicates "Hue" and describes the pure colour; '*S*' indicates "saturation" which represents contamination of a colour with white colour. '*I*' indicates "Intensity" which represents the brightness intensity of a particular colour [11]. The HSI values determined for conventional and Recycled Aggregate Concrete (RAC) are compared for different chemicals and the results are analysed. The colour variation for different chemicals in HSI colour space can be used as a measure of deterioration of a structure when exposed to these chemicals [12].

The Conventional and Recycled Aggregate Concrete (RAC) specimens are also tested for durability loss when exposed to different chemicals and quantified according to the standard stipulations of ASTM C 666-97 [13]. The durability of test specimens after 28 days of exposure to chemical attack is determined from the variation of diagonal length, mass loss and strength loss using Durability Loss Factor (DLF) determined by using the following equations [14].

$$\text{DLF} = \text{MLF} \times \text{DiF} \times \text{SLF} \tag{4}$$

DLF = Durability Loss Factor; MLF = Mass Loss Factor; DiF = Distortion Factor and SLF = Strength Loss Factor.

Mass Loss Factor considers the mass loss with respect to initial mass of specimen affecting the durability of the specimens.

$$\text{MLF} = \frac{M_1 - M_2}{M_1} \times 100 \tag{5}$$

M_1 = Mass of specimen before exposure to chemicals; M_2 = Mass of specimen after exposure to chemical attack.

Distortion Factor considers the change in diagonal length with respect to initial length of specimen influencing the durability.

$$\text{DiF} = \frac{D_1 - D_2}{D_1} \times 100 \tag{6}$$

D_1 = Initial diagonal length before exposure; D_2 = Diagonal length of the specimen after exposure to chemical attack.

Strength Loss Factor considers the loss of strength with respect to the initial strength causing the deterioration of concrete.

$$\text{SLF} = \frac{S_1 - S_2}{S_1} \times \frac{N}{M} \tag{7}$$

S_1 = Strength of control specimen without exposure to chemical attack; S_2 = Final strength of specimen after exposure to the attack of chemical; N = Number of days at which the durability factor is to be determined. M = Number of days at which the exposure is to be terminated.

3 Results and Discussion

The HSI colour space values are determined from the average RGB values for natural specimens of 30 and 50 MPa strength designed using IS 10262:2019 [15] and PPM [10] methods. The properties of materials used for making concrete satisfy the stipulations in the codes [16–18]. The values of '*H*', '*S*' and '*I*' for NAC specimens exposed to HCl, H_2SO_4, $MgSO_4$ and NaCl compared to the values before exposure are shown in Table 3.

Similar to Natural Concrete (NA), Recycled Aggregate Concrete (RAC) specimens are also evaluated for RGB values and the corresponding HSI colour space values are determined for 30 and 50 MPa strength and the results are shown in Table 4.

From Tables 3 and 4, it can be observed that the values of Hue, '*H*' decreased for HCl exposure as compared to the control specimen. This is true both for NA and RA specimens of 30 and 50 MPa strength. Exposure of concrete to HCl results in the

Table 3 HSI values of conventional concrete (NAC) before and after exposure to chemicals

Chemical	30 MPa			50 MPa		
	H	*S*	*I*	*H*	*S*	*I*
IS						
CS	64.22	9.16	49.59	62.20	8.24	56.22
HCl	61.20	15.58	67.99	60.11	14.28	72.13
H_2SO_4	64.50	14.85	69.65	63.21	13.08	76.28
NaCl	58.60	9.83	53.28	56.80	9.74	57.83
$MgSO_4$	60.59	12.48	57.30	58.71	11.58	69.47
PPM						
CS	61.10	7.68	43.22	60.40	6.47	61.71
HCl	58.64	13.27	76.25	57.21	12.18	78.84
H_2SO_4	61.40	12.87	77.82	60.80	11.27	78.90
NaCl	55.32	8.87	60.48	54.21	7.29	62.37
$MgSO_4$	56.22	10.57	70.21	55.40	9.47	72.58

CS Control Specimen

Table 4 HSI values of recycled aggregate concrete (RAC) before and after exposure to chemicals

Chemical	30 MPa			50 MPa		
	H	*S*	*I*	*H*	*S*	*I*
IS						
CS	65.6	11.28	51.48	64.32	10.54	58.17
HCl	64.11	17.37	69.24	63.28	16.48	74.32
H_2SO_4	66.21	16.58	71.42	65.55	15.13	78.37
NaCl	62.28	11.36	55.76	60.12	11.08	59.27
$MgSO_4$	63.45	14.37	59.47	62.44	13.84	71.24
PPM						
CS	62.24	9.81	45.28	60.8	8.43	63.16
HCl	61.42	15.37	78.51	58.81	14.66	81.11
H_2SO_4	64.27	14.67	79.46	62.17	13.22	80.45
NaCl	59.27	10.76	62.22	58.46	9.18	64.18
$MgSO_4$	61.65	12.74	72.38	60.27	11.63	74.63

CS Control Specimen (No exposure to chemicals)

formation of red colour initially and further causes the exposure of aggregate on the surface, causing the reduction of pure colour on the surface [19]. The value of '*H*' for H_2SO_4 after 28 days of exposure is slightly higher compared to control specimen as the reaction with concrete caused formation of a white residue along with exposure of aggregate on the surface [20]. White salt depositions are found on the specimen surface exposed to NaCl [19] and so the value of '*H*' is reduced compared to the control specimen. Due to the sulphate attack, the surface turned to dark grey and so the value of '*H*' is reduced for specimens exposed to $MgSO_4$. The maximum variation of '*H*' value is observed for the specimen exposed to H_2SO_4, followed by HCl. The variation is minimum for specimens exposed to NaCl and $MgSO_4$.

The values of Saturation, '*S*' indicating the colour contamination with white colour varied according to the extent of deterioration caused by several chemicals. It can be observed that the value of '*S*' for specimens exposed to HCl is higher compared to the value of control specimen as the initial grey surface of the concrete specimen changes to bright red colour and then to greyish white colour exposing the aggregates. The specimens exposed to H_2SO_4 showed a slight increase in the value of '*S*' compared to control specimens due to the formation of white residue with aggregate exposure [21]. The variation of value of '*S*' for specimens subjected to the attack of NaCl and $MgSO_4$ is low due to reduced deterioration. The value of '*S*' is higher for specimens exposed to HCl compared to other chemicals.

The values of Intensity, '*I*' indicating the brightness intensity specimens for HCl increased compared to the control specimen because of the removal of cement from the surface of the specimens. Similarly, the value of '*I*' increased for specimens exposed to H_2SO_4 visibly evident with fraction of white powder and aggregate protruding after 28 days of exposure. The values of '*I*' also increased for the concrete units exposed to NaCl and $MgSO_4$, however, the variation is slightly less due to lower deterioration of exposed surfaces [22]. The value of '*I*' is higher for specimens exposed to H_2SO_4 indicating higher variation in the surface brightness, in turn, indicating greater deterioration., followed by HCl, $MgSO_4$ and NaCl respectively. Also, the HIS values decrease with increase in the grade of concrete due to improvement in the quality of concrete. The HSI values are lower for concrete specimens cast based on PPM method compared to concrete cast using IS 10262:2019 indicating the decrease in the deterioration using PPM. The values are however higher in Recycled Aggregate Concrete (RAC) due to lower quality of recycled aggregate because of the adhered mortar content on surface of aggregate.

The mass loss, diagonal loss and strength loss of conventional concrete designed using IS and PPM methods for 30 and 50 MPa strength are shown in Tables 5 and 6.

From Tables 5 and 6, it can be observed that the mass loss, diagonal loss and strength loss are maximum for specimens exposed to H_2SO_4 followed by HCl, $MgSO_4$ and NaCl respectively. This is attributed to higher deterioration of concrete exposed to H_2SO_4 compared to other chemicals. The Mass loss, Diagonal loss and Strength loss of RAC designed using IS and PPM methods for 30 and 50 MPa strength are shown in Tables 7 and 8.

From Tables 7 and 8, it can be observed that the mass loss, diagonal loss and strength loss are maximum in RAC specimens exposed to H_2SO_4 followed by HCl, $MgSO_4$ and NaCl respectively. The variation is similar as in NAC, however, the losses are more in RAC compared to NAC due to inferior quality of concrete. Also, the losses are lower for concrete specimens cast using Particle Packing Methods (PPM) compared to IS 10262:2019 mix design method. The variation of percentage

Table 5 Initial and final average values of mass, diagonal and strength of NAC at 28 days of exposure in chemicals (IS method)

Mix method	IS					
Concrete strength	30 MPa			50 MPa		
Mass loss	M_1	M_2	$(M_1 - M_2)$	M_1	M_2	$(M_1 - M_2)$
HCl	2.60	2.43	0.17	2.62	2.42	0.20
H_2SO_4	2.66	2.28	0.38	2.63	2.43	0.20
NaCl	2.67	2.66	0.01	2.58	2.56	0.02
$MgSO_4$	2.68	2.65	0.03	2.62	2.60	0.02
Diagonal loss	D_1	D_2	$(D_1 - D_2)$	D_1	D_2	$(D_1 - D_2)$
HCl	17.12	16.90	0.22	17.23	16.80	0.43
H_2SO_4	17.25	16.10	1.15	17.21	16.24	0.97
NaCl	17.21	17.10	0.11	17.25	16.95	0.30
$MgSO_4$	17.10	16.95	0.15	17.32	17.30	0.02
Strength loss	S_1	S_2	$(S_1 - S_2)$	S_1	S_2	$(S_1 - S_2)$
HCl	37.60	28.00	9.60	37.60	35.00	15.10
H_2SO_4	37.60	22.00	15.60	37.60	23.00	14.60
NaCl	37.60	35.00	2.60	37.60	48.00	2.10
$MgSO_4$	37.60	30.00	7.60	37.60	42.00	8.10

Table 6 Initial and final average values of mass, diagonal and strength of NAC at 28 days of exposure in chemicals (PPM method)

Mix method	PPM					
Concrete strength	30 MPa			50 MPa		
Mass loss	M_1	M_2	$(M_1 - M_2)$	M_1	M_2	$(M_1 - M_2)$
HCl	2.54	2.37	0.17	2.55	2.43	0.12
H_2SO_4	2.59	2.30	0.29	2.61	2.45	0.16
NaCl	2.57	2.53	0.04	2.62	2.59	0.03
$MgSO_4$	2.59	2.52	0.07	2.51	2.49	0.02
Diagonal loss	D_1	D_2	$(D_1 - D_2)$	D_1	D_2	$(D_1 - D_2)$
HCl	17.20	16.35	0.85	17.21	16.60	0.61
H_2SO_4	17.20	16.25	0.95	17.21	16.30	0.91
NaCl	17.22	17.05	0.17	17.30	17.25	0.05
$MgSO_4$	17.30	17.25	0.05	17.20	17.10	0.10
Strength loss	S_1	S_2	$(S_1 - S_2)$	S_1	S_2	$(S_1 - S_2)$
HCl	37.60	31.00	6.60	37.60	32.00	5.60
H_2SO_4	37.60	25.00	12.60	37.60	33.00	4.60
NaCl	37.60	38.00	2.80	37.60	50.00	2.80
$MgSO_4$	37.60	40.00	0.80	37.60	44.00	8.80

Table 7 Initial and final average values of mass, diagonal length and strength of RAC (28 days of exposure in chemicals-IS method)

Mix method	IS					
Concrete strength	30 MPa			50 MPa		
Mass loss	M_1	M_2	$(M_1 - M_2)$	M_1	M_2	$(M_1 - M_2)$
HCl	2.40	2.01	0.39	2.43	2.20	0.23
H_2SO_4	2.42	1.63	0.79	2.48	1.72	0.76
NaCl	2.48	2.30	0.18	2.40	2.28	0.12
$MgSO_4$	2.50	2.27	0.23	2.40	2.24	0.16
Diagonal loss	D_1	D_2	$(D_1 - D_2)$	D_1	D_2	$(D_1 - D_2)$
HCl	17.00	16.30	0.70	17.10	16.60	0.50
H_2SO_4	17.20	15.40	1.80	17.00	15.50	1.50
NaCl	17.00	16.80	0.20	17.30	17.20	0.10
$MgSO_4$	17.30	16.90	0.40	17.32	16.88	0.44
Strength loss	S_1	S_2	$(S_1 - S_2)$	S_1	S_2	$(S_1 - S_2)$
HCl	35.02	24.00	11.02	48.12	36.24	11.88
H_2SO_4	35.02	19.00	16.02	48.12	26.00	22.12
NaCl	35.02	32.00	3.02	48.12	44.12	4.00
$MgSO_4$	35.02	26.84	8.18	48.12	39.12	9.00

Table 8 Initial and final average values of mass, diagonal length and strength of RAC (28 days of exposure in chemicals-PPM method)

Mix method	PPM					
Concrete Strength	30 MPa			50 MPa		
Mass loss	M_1	M_2	$(M_1 - M_2)$	M_1	M_2	$(M_1 - M_2)$
HCl	2.50	2.15	0.35	2.50	2.32	0.18
H_2SO_4	2.51	1.78	0.73	2.56	2.02	0.54
NaCl	2.52	2.40	0.12	2.55	2.48	0.07
$MgSO_4$	2.52	2.45	0.07	2.50	2.40	0.10
Diagonal loss	D_1	D_2	$(D_1 - D_2)$	D_1	D_2	$(D_1 - D_2)$
HCl	17.20	17.00	0.20	17.20	17.01	0.19
H_2SO_4	17.15	15.90	1.25	17.10	16.24	0.86
NaCl	17.20	17.10	0.10	17.32	17.30	0.02
$MgSO_4$	17.20	17.00	0.20	17.30	17.18	0.12
Strength loss	S_1	S_2	$(S_1 - S_2)$	S_1	S_2	$(S_1 - S_2)$
HCl	38.20	28.00	10.20	51.10	42.10	9.00
H_2SO_4	38.20	23.00	15.20	51.10	28.84	22.26
NaCl	38.20	36.00	2.20	51.10	49.21	1.89
$MgSO_4$	38.20	31.00	7.20	51.10	44.17	6.93

loss of mass, diagonal length and strength of concrete are shown in Figs. 1, 2 and 3 respectively.

From Figs. 1, 2 and 3, it is evident that, with increase in grade of concrete, there is a decrease in mass loss, diagonal loss and strength loss and this is true with both methods of mix design. However, in case of Particle Packing Method of concrete mix design, these values are lower. It can also be seen that Recycled Aggregate Concretes are inferior to Natural Aggregate based Concrete for both grades and for concretes designed using IS and PPM methods. However, it is less in case of PPM method. It can hence be confirmed from this that Particle Packing Method of mix design is the optimum method for both conventional and Recycled Aggregate Concrete (RAC).

The mass loss, diagonal loss and strength loss are further used for calculating Mass Loss Factor (MLF), Distortion Factor (DiF) and Strength Loss Factor (SLF) and ultimately, the Durability Loss Factor (DLF) is determined from Eqs. 4, 5, 6 and

Fig. 1 Percentage mass loss in NAC and RAC

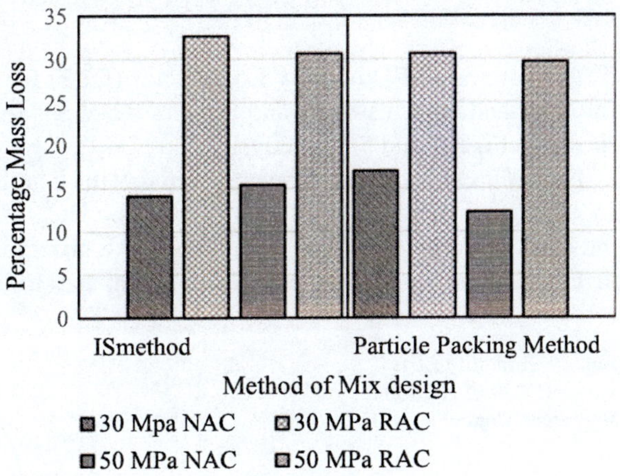

Fig. 2 Percentage diagonal loss in NAC and RAC

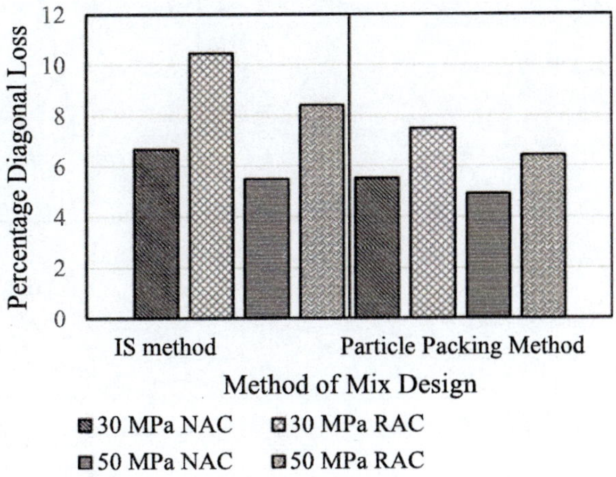

Fig. 3 Percentage strength loss in NAC and RAC

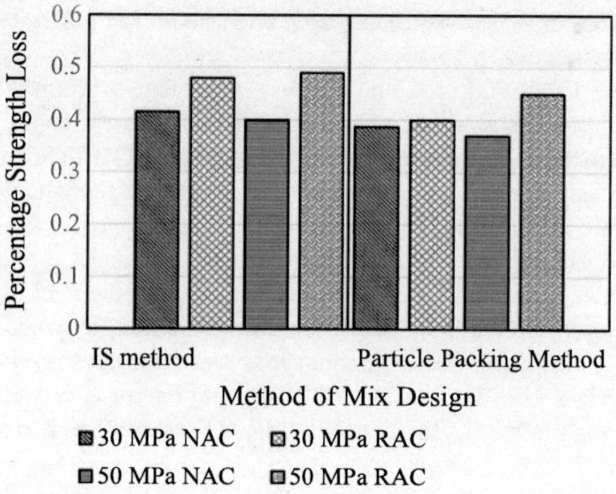

7. The variation of Durability Loss Factor (DLF) for different chemical exposures for NAC and RAC using IS and PPM mix designs for 30 and 50 MPa strengths is shown in Figs. 4 and 5 respectively.

From Figs. 4 and 5, it can be observed that the Durability Loss Factor is maximum for specimens exposed to H_2SO_4 followed by HCl, $MgSO_4$ and NaCl. Higher DLF indicates more deterioration. It can hence be observed from Figs. 4 and 5 that the deterioration is the highest with H_2SO_4 compared to other chemical agents.

Fig. 4 Durability Loss Factor (DLF) of Natural Aggregate Concrete (NAC)

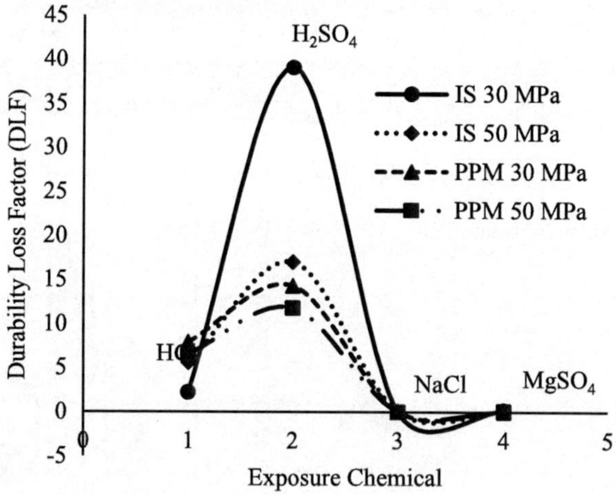

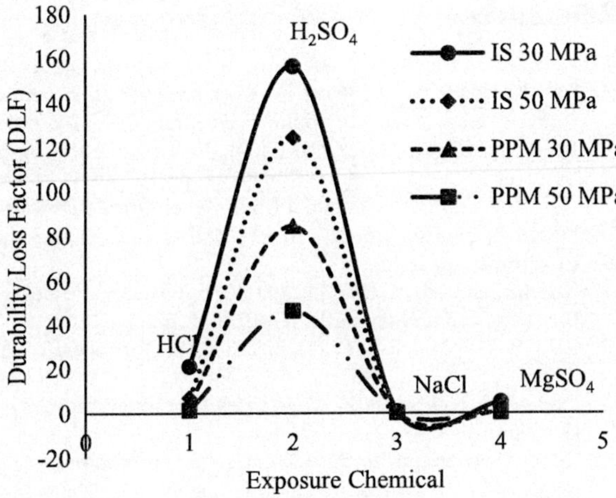

Fig. 5 Durability loss factor (DLF) of recycled aggregate concrete (RAC)

4 Conclusions

- The RGB values of specimens obtained from image analysis before and after exposure are converted to *H*, *S* and *I* colour space values to quantify the deterioration in concrete, due to different chemical species.
- The colour variation is more pronounced in case of specimens exposed to HCl and H_2SO_4 compared to NaCl and $MgSO_4$, indicating, greater deterioration in concrete, due to acid attack.
- The variation of Hue (*H*) indicating pure colour is more for H_2SO_4, after 28 days of exposure which is evident from the colour changes from dark grey before exposure to white with aggregates exposed on the surface.
- The saturation (*S*) values are higher for HCl exposure after 28 days, due to the drastic colour variation of specimen from dark grey to bright red and further leads to formation of white residue with exposure of aggregates on the surface.
- The Intensity (*I*) of colours reflecting white is maximum for specimens exposed to H_2SO_4 for 28 days followed by HCl. This variation is very less in specimens exposed to NaCl and $MgSO_4$.
- The Mass Loss Factor (MLF), Distortion Factor (DiF), Strength Loss Factor (SLF) are the maximum for concrete exposed to H_2SO_4 compared to HCl. These values are further lower for exposure to NaCl and $MgSO_4$
- A unified parameter Durability Loss Factor (DLF), combining MLF, DiF and SLF are evaluated. This value is higher in concretes exposed to H_2SO_4 for 28 days period indicating greater deterioration. This can further be correlated with strength of concrete determined through destructive tests.
- The variation of primary colours of surfaces of concrete specimen and the corresponding *H*, *S* and *I* colour space values and the durability loss factors are correlated and it can be used to identify the influence of chemical attack on a real-life structure.

References

1. Calixto J, Esenarro D, Rojas W, Rodriguez C (2022, January) Influence of the use of recycled coarse aggregates on the compressive strength for the production of concrete of 350 kg/cm^2. In: Proceedings of 2021 4th international conference on civil engineering and architecture. Springer Nature Singapore, Singapore, pp 193–206
2. Guidelines on Environmental Management of C&D Waste Management in India by Central Pollution control Board (2017) Ministry of Environment, Forest and Climate change, Government of India
3. Indian Standards (2016) IS: 383:2016, coarse and fine aggregate for concrete specification. Bureau of Indian Standards, New Delhi, India
4. RILEM TC 121-DRG (1994) Specifications for concrete with recycled aggregates. Mater Struct 27:557–559
5. DIN 4226–100-2 (2002) Aggregates for mortar and concrete, part 100: recycled aggregates, Germany
6. Costa A, Appleton J (1999) Chloride penetration into concrete in marine environment part I: main parameters affecting chloride penetration. Mater Struct 32(4):252–259. https://doi.org/10.1007/BF0247369594
7. Nadir HM, Ahmed A (2022) The mechanisms of sulphate attack in concrete–a review. Mod Approaches Mater Sci 5(2):658–670
8. Khatib JM (2005) Properties of concrete incorporating fine recycled aggregate. Cem Concr Res 35(4):763–769. https://doi.org/10.1016/j.cemconres.2004.06.017
9. Fennis S, Walraven J (2012) Using particle packing technology for sustainable concrete mixture design. Heron 57:73–101
10. Pradhan S, Kumar S, Barai SV (2017) Recycled aggregate concrete: Particle Packing Method (PPM) of mix design approach. Constr Build Mater 152:269–284. https://doi.org/10.1016/j.conbuildmat.2017.06.171
11. Reddy VGP, Tadepalli T, Pancharathi RK (2022) Surface imaging based non-destructive assessment of concrete deterioration using hue-saturation-intensity colour space. Measurement 197:111311
12. Banerjee A, Reddy VGP, Pancharathi RK, Tadepalli T (2022) Colour-stability analysis for estimation of deterioration in concrete due to chemical attack. Constr Build Mater 321:126288
13. ASTM C 666–97, Standard test method for resistance of concrete to rapid freezing and thawing. ASTM International, West Conshohocken, PA, USA
14. Venkateswara Rao S, Seshagiri Rao MV, Ramaseshu D, Rathish Kumar P (2012) Durability performance of self-compacting concrete. Mag Concr Res 64(11):1005–1013
15. Indian Standards: 10262-2019 Concrete mix proportioning-guidelines. Bureau of Indian Standards, New Delhi, India
16. Indian Standards:12269-2013 Specifications for 53 grade ordinary Portland cement. Bureau of Indian Standards, New Delhi, India
17. Indian Standards: 383-2016 Coarse and fine aggregate for concrete-specification. Bureau of Indian Standards, New Delhi, India.
18. Indian Standards: 10500-2012 Drinking water-specification. Bureau of Indian Standards, New Delhi, India
19. Zhang F, Hu Z, Wei F, Wen X, Li X, Dai L, Liu L (2021) Study on concrete deterioration in different NaCl-Na_2SO_4 solutions and the mechanism of cl-diffusion. Materials (Basel) 14:1–23. https://doi.org/10.3390/ma14175054
20. Ashish K, Singh B, Verma SK (2016) The effect of attack of chloride and sulphate on ground granulated blast furnace slag concrete. Adv Concr Constr 4(2):107–121. https://doi.org/10.12989/acc.2016.4.2.107
21. Megid WA, Khayat KH (2020) Methodology to evaluate variations in concrete color caused by white cement substitutions and forming materials. Mater Struct 53:1–15
22. Reddy VGP, Krishna BM, Tadepalli T, Kumar PR (2020) Image-based deterioration assessment of concrete. Mater Today: Proc 32

Performance of Basalt Fibre Reinforced High Density Concrete Using Hematite Ore Aggregates

T. Harini, B. Sneha, S. Swathy, and R. Vidjeapriya

1 Introduction

In general, the radiations are classified into two types. They are Electro Magnetic waves and Nuclear Particles. Among all electromagnetic waves, X and Gamma rays with high energy at high frequency are to be shielded to protect human health. These two are similar to light rays but they have high energy penetrating power and Gamma rays that can be adequately absorbed by an appropriate thickness of concrete shield. The absorption of Gamma rays is directly proportional to the density of material which is used as a shielding barrier. If density is more, the absorbing character of Gamma rays is increased, if density is less absorbing character decreases. The Biological hazard of radiation upsets the chemical balances in the tissue and causes the death of cells. If enough cells are affected, then the various organs of human will fail to function. Hence radiation should be avoided. High density of concrete helps in shielding and it has the attenuation character of Gamma and Neutron rays. Standard heavy density concrete, using natural iron ore aggregate, can achieve densities up to 3850 kg/m^3 [1]. Fibres are generally used to increase the strength. Here we specifically use basalt fibre as it is a natural fibre and does not require any treatment to use it along with high density concrete, economical and for strength gain. There are many aggregates, whose specific gravity more than 3.5 is available around the world for making a heavy density concrete. Out of these, commercial barite, magnetite, limonite, magnetite, hematite aggregates (Fig. 1) are employed. Among the above, the Hematite Ore in Fig. 2 is selected as a heavy aggregate and is quite suitable for preparing structural/shielding elements for Gamma rays attenuation.

T. Harini (✉) · B. Sneha · S. Swathy · R. Vidjeapriya
College of Engineering, Guindy, Anna University, Chennai, Tamil Nadu, India
e-mail: harinithirumaran197@gmail.com

R. Vidjeapriya
e-mail: vidjeapriya@annauniv.edu

R. K. Pancharathi et al. (eds.), *Low Carbon Materials and Technologies for a Sustainable and Resilient Infrastructure*, Lecture Notes in Civil Engineering 440,
https://doi.org/10.1007/978-981-99-7464-1_25

Fig. 1 Hematite aggregate

Fig. 2 Hematite ore

Initial research on high density concrete investigated the special aggregates needed to produce concrete shields capable of attenuating high and almost equal intensities of Gamma and neutron radiation at a temperature of 100 °C. The basic approach to high density concrete mix design is revived to show that theoretical calculations for concrete mixes do give results that are reliable but laboratory tests on high density concrete were lacking [2]. Later investigative work was carried out on a high density concrete aggregates. Properties of the high Density aggregates and the criteria for their use in concrete was stated. Study of Thermal properties of high density concrete was carried out compared with a normal-density dolomite-limestone based concrete [3]. High density concrete was examined for its radiation shielding of nuclear reactors and other structures that require radiation impermeability. In general hematite ores increase the density of concrete hence wall thickness of shielding wall is reduced to provide radiation shielding. Other physical and mechanical properties were also studied [4]. Studies in recent years stated concrete shields are structural elements, which should meet certain requirements in order to retain and absorb various kinds of radiation and to protect the staff and environment in a reliable way. Various difficulties in quality control and production of high density aggregates and high density concrete were discussed [5]. Further to reduce neutron and Gamma radiation, the shielding capability of high density concretes mixed with Ferro boron in different proportions (25, 50 and 75%) replacing granite aggregate was examined. When the percentage of Ferro boron increases, the radiation shielding property increases but it should be added with caution. If an excess amount of ferro boron is present in the mix, it adversely affects the strength of concrete [6].

Early studies in fibre reinforced concrete explained that the concrete is moderate in the compressive properties but lacks tensile properties. In order to compensate for the lack of the tensile strength properties, fibres can be inserted into the plain cement concrete. The test results indicated that when fibres are incorporated into the concrete, it enhances the strength properties and further increases the volume fraction of basalt fibres has a decreasing trend [7]. Basalt fibres were used in combination of other fibres. One of the studies investigated original basalt fibre (BF), and the sizing-removed BF through the heat treatment were pneumatically dispersed by using an air compressor and by incorporating into normal strength concrete to basalt fibre reinforced concrete was produced. The concrete's impact resistance performance increases when different lengths of basalt fibres are mixed than a single length basalt fibre [8]. Basalt fibre is used in combination with polypropylene. The increase in shear strength of reinforced concrete (RC) beams due to addition of basalt and polypropylene fibre was investigated. About 2.5% of basalt fibres and 0.6% of polypropylene fibres produced the best results. The inclusion of polypropylene or basalt fibres considerably enhanced the ductility of the beams. The inclusion of 0.6% polypropylene fibres enhanced the shear strength, peak deflection, ductility and initial stiffness [9]. In this research study, various percentages of basalt fibres are introduced into high density concrete made of Hematite aggregates to find the optimum percentage of basalt fibres that give Superior strength properties. The maximum compressive strength and flexural strength of basalt fibre reinforced high density concrete are compared with ordinary high density concrete.

2 Experiment

2.1 Characterization Materials

In this research study, materials such as Grade 53 OPC cement, hematite ore aggregates (Fine and Coarse) as in Fig. 1, superplasticizer and basalt fibres were used. The high density aggregate which contains high concentration of hematite ore was brought. Superplasticizer Master Glenium SKY 8631 was procured. The basalt fibre of 6 mm length was also procured.

2.2 Materials and Methods

The 53 Grade ordinary Portland cement is used for casting and its chemical composition and physical performance are used. Penna OPC 53 grade cement is used. The well-graded crushed hematite boulder or stone with a particle size of 12.5 and 20 mm obtained from the quarry site is used as the coarse aggregate. The well-graded hematite sand obtained from crushed stones with a fineness modulus of 4.89 is used as the fine aggregate. Both hematite fine and coarse aggregate are obtained from the quarry site "hematite boulders and stone crushed for construction". Cluster basalt fibres (CBF) of size 30s 6 mm length are used as additive in the concrete mix. Basalt fibres are produced from basalt rocks by melting them and converting the melt into fibres. Basalts are rocks of igneous origin. These fibre types are manufactured from the raw basalt stones. Thus, the mechanical properties of these fibres are close tensile strength ranges from 3000 to 4800 MPa, elastic modulus ranges from 95 to 110 GPa, and density is about 2.6 g/cm^3. Admixture used for conventional concrete was Rheobuild 1125 and high density concrete was Master Glenium SKY 8631 of high range water reducer cum retarding plasticizer. Le-Chattier method is followed to determine specific gravity as per IS 4031-part 11 1996. Specific gravity of cement obtained by testing is 3.15 and it is in the range of 3.10–3.16 as per IS code. Sieve analysis is performed by sieving the aggregates of known mass through 90 μm sieve to determine the fineness of cement as per IS code. Fineness of cement obtained by sieving is 4%. As per IS, it should be less than 10%. Initial setting time of cement is 162 min and final setting time of cement is 302 min. For an Ordinary Portland Cement of any grade, the initial setting time is 30 min. The Final Setting Time is 600 min (10 h). So, it satisfies as per IS code. Pycnometer is used to find specific gravity of aggregates. Specific gravity of hematite fine aggregate obtained is 3.53 and that of coarse aggregate is 3.9 and 3.83 for 20 and 12.5 mm size. As per IS 2386, the specific gravity should be in between 3.6 to 4.0 for coarse and 3.5 to 3.8 for fine aggregate. The aggregates are graded as per IS code 383. The aggregate gradation is determined by the fineness modulus method where aggregates are passing through the sieve as per IS standard to classify its size. Gradation of both fine and coarse hematite aggregate are well graded as it provides an S shape curve.

2.3 *Procedure for Mixing*

Initially, batching of each component of concrete was performed in a weighing balance and it was prepared for mixing. Now, the cement and fine aggregate are mixed in the mixer until the mixture is thoroughly blended and is of uniform colour. Add the coarse aggregate and mix with cement and fine aggregate mixture until the coarse aggregate is uniformly distributed throughout the batch. Adequate quantity of superplasticizer was mixed with water. Add water and mix it until the concrete appears to be homogenous and of the desired consistency. Slump test was carried out before filling the moulds. Well mixed concrete is filled in oil moulds and it is filled in layers with tamping of concrete in between each layer. The filled moulds are air dried for 24 h until the concrete is set. It is transferred to a water pond for curing.

2.4 *Tests*

After curing process, all the specimens were stored at room temperature until the tests were conducted. At 7 and 28 days, the concrete cube specimens were tested for compressive strength and flexural strength. The average results of three identical specimens for each group were taken.

2.5 *Mix Design*

Mix proportion of M30 basalt fibre reinforced high density concrete is shown in Table 1. The water/cement (w/c) ratio is 0.43 and the volume content of basalt fibre is kept at 0.5, 1, 1.5 and 2%.

Table 1 Mix proportion of basalt fibre reinforced high density concrete (M30)

Concrete ingredients	Water	Cement	Fine aggregate	Coarse aggregate		Density (Kg/m^3)
				20 mm	12.5 mm	
				50%	50%	
Weight (kg/m^3)	159	370	950.67	895.50	868.29	3244
Mix proportion	0.43	1	2.57	4.77		

3 Results and Discussion

3.1 *Fresh Concrete Properties*

Concrete is called Fresh concrete when it has not been set. By checking the fresh property of concrete, properties such as workability can be checked. The relative quantities of binding material, aggregates (fine and coarse), water and admixture are controlled and mixed together to get concrete in the wet state. In fresh concrete, the following properties are tested for satisfying its property.

3.1.1 Workability of Concrete

The ease, at which concrete is being mixed, transported, placed and vibrated without segregation is called workability. It is measured in terms of slump cone test. Slump is essentially a measure of consistency of rich mix. It is tested as per Indian standards and practice. For the ease of performing the test in field and due to high workability of the mix, slump cone test was opted. The results of it are tabulated in Table 2.

From the table, the average slump value of high density concrete reinforced with varied percentage of basalt fibre is 140 mm. From the Results, it is evident that the mix has high workability (Ranging from 100 to 150 mm as per IS 456 2000) of 140 mm. Workability of Concrete is influenced by various external factors such as site's weather condition, mode of compaction, etc. Quality of raw materials used for producing concrete also affects workability. These factors include water/cement ratio, size and shape of aggregates, admixtures, etc. Hence adequate care has been taken while grading of aggregates and mix has been designed to provide high workability. This satisfies the properties of fresh concrete and design parameters are satisfied.

3.2 *Hardened Properties of Concrete*

In order to assess physical and mechanical properties of conventional and heavy density concrete suitable size of specimens were casted and tested at the appropriate ages as per guidelines of Indian standards. Concrete Hardening is a continuous process. Initially, chemical bonds are created between cement and water when they come in contact and they adhere to the aggregates to form the concrete. Reworking of concrete should be avoided to facilitate for full achievement of strength. Hardening

Table 2 Slump value of basalt fibre reinforced high density concrete

Conventional concrete ($M1$)	130 mm
High density concrete by using Hematite aggregate ($M2$)	140 mm

can only be carried out in concrete in presence of moisture. Curing is an important process that aids in the development of strength of concrete. With the help of curing process, concrete gains 65% in 7 days and 100% in 28 days. In hot weather, absence of curing results in development of shrinkage cracks. Water Ponding method of curing is used as it is very simple.

3.2.1 Compressive Strength Test

Compressive strength is the ratio between load applied till get failure and contact area of the cube. It represents the concrete characteristics. With the results from this test, the quality of concreting and accuracy of mix design can be ascertained. Compressive strength test of concrete can be carried out in a cube or cylinder. IS 456 2000 provides a procedure for concrete cube compressive strength test. ASTM C39/C39M provide guidelines for Cylinder Concrete Compressive Strength Test. Here the test has been carried out as per regulations in IS 456 2000. Concrete cube mould of 100 mm × 100 mm × 100 mm has been used. The test results of early compressive Strengths are tabulated in Tables 3 and 4.

From the above mentioned Tables 3 and 4, it can be clearly witnessed that the basalt fibre reinforcement of about 1% increases the compressive strength of high density concrete (with Fibres) compared with Non fibre reinforced high density concrete. The basalt Fibre has increased the compressive strength due to the bridging effect of fibres. Concrete when subjected to compressive load fails under shear with a 45° crack. Due to the introduction of basalt fibre in concrete, some shear is carried by the basalt fibres. Hence this improves the load carrying capacity of concrete in a considerable amount. It could be clearly seen from the results target 7 day strength has been achieved by the concrete. The seven day compressive strength is a good indicator of strength gaining of concrete. If satisfactory results are not obtained, the mix needs to be redesigned. The following Fig. 3 shows a graph plotted between Average Compressive Strength of Concrete versus Percentage of Basalt Fibres.

Table 3 7 day average compressive strength of high density concrete

S. No	Age at testing (days)	Average compressive strength in MPa
1	7	25.77

Table 4 7 day average compressive strength of basalt fibre reinforced high density concrete

S. No	Fibre %	Age at testing (days)	Average compressive strength in MPa
1	0.5	7	26.4
2	1.0	7	29.3
3	1.5	7	24.6
4	2.0	7	20.4

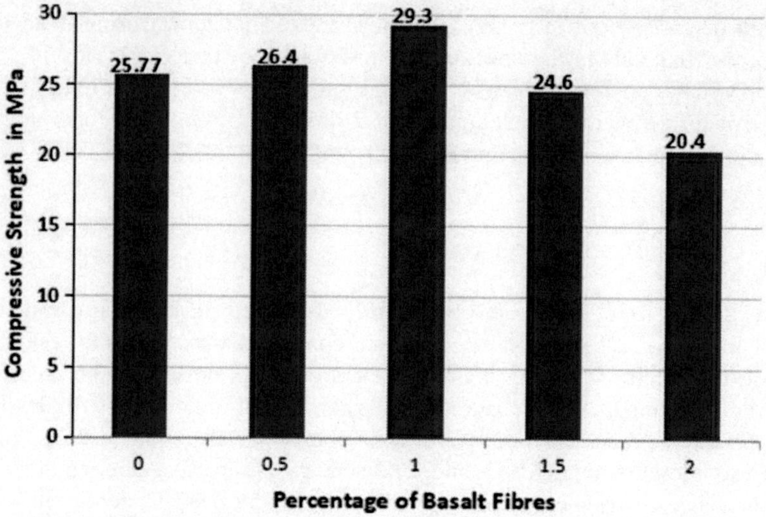

Fig. 3 Graph between average compressive strength of BFRP concrete versus percentage of basalt fibres (for 7 days)

A concrete cube gains its 100% target compressive strength at 28 days of curing. It is an important parameter that has to be tested. The concrete gains compressive strength very quickly till 14 days (About 90%) and it gains about 99–100% of compressive strength in 28 days and continues to gain compressive strength in future. Hence 28 day compressive strength criteria allows us to check any discrepancies in strength from 7 to 28 days. Following Tables 5 and 6 represent the 28 days Average Compressive Strength of High Density Concrete and Basalt fibre Reinforced High density Concrete with varied Percentages of basalt fibre.

From the tabulated results, it could be clearly seen that the optimum quantity of basalt fibre is about 1% of the weight of cement. The compressive strength increases

Table 5 28 days average compressive strength of high density concrete

S. No	Age at testing (days)	Average compressive strength in MPa
1	28	37.49

Table 6 28 days average compressive strength of basalt fibre reinforced high density concrete

S. No	Fibre %	Age at testing (days)	Average compressive strength in MPa
1	0.5	28	38.20
2	1.0	28	42.86
3	1.5	28	35.25
4	2.0	28	32.72

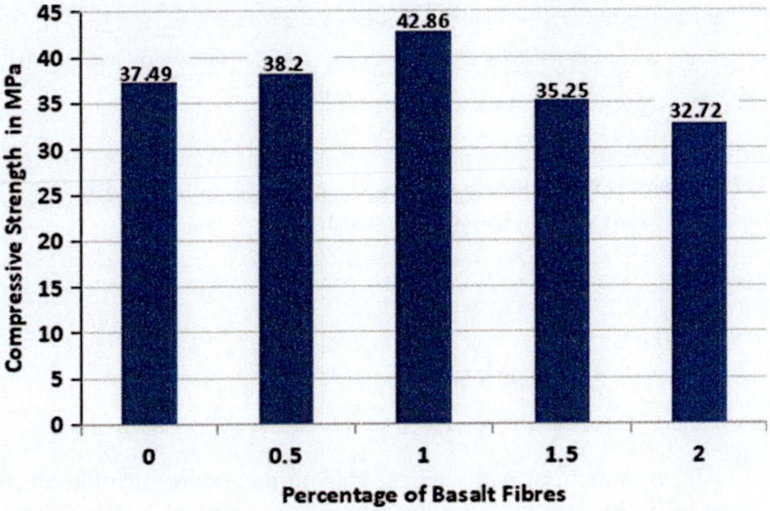

Fig. 4 Graph between average compressive strength of BFRP concrete and percentage of basalt fibres (for 28 days)

from 0.5 to 1% and it falls down from 1 to 1.5%. This shows that Peak Compressive strength of basalt fibre is achieved at about addition of 1% of basalt fibres. Similar results have notes in seven days compressive strength. The target strength of concrete has been achieved. Hence in our mix design, selection of materials constituting concrete and quality control process is adequate and accurate. Following Fig. 4 represents the graph between Average Compressive Strength of BFR Concrete and Percentage of Basalt Fibres in Concrete.

3.2.2 Flexural Strength Test

The tensile properties of concrete are measured in terms of flexural strength of concrete. When the flexural strength of concrete is high, it can be used for Stress-Bearing Restoration. Flexural strength increases with hike in compressive strength and age of concrete as well. The dimensions of the beam used for the test are 100 mm × 100 mm in cross section and 50 mm in span. Flexural strength of concrete is roughly about 10–20% of compressive strength of concrete. In order to verify the flexural strength, the sample is tested at lab. Tables 7 and 8 represent the average flexural strength of high density concrete and average flexural strength of basalt fibre reinforced concrete in varied percentage of basalt fibres.

From the table, it can be inferred that the 7 day compressive strength of basalt fibre reinforced high density concrete at 1% of basalt fibre is the highest. It is also higher than that of flexural strength of high density concrete. It could be seen that the compressive strength and flexural strength follow a trend of being highest at the

Table 7 7 day average flexural strength of high density concrete

S. No	Age at testing (days)	Average compressive strength in MPa
1	7	3.63

Table 8 7 day average flexural strength of basalt fibre reinforced high density concrete

S. No	Age at testing (days)	Percentage of basalt fibre	Average compressive strength in MPa
1	7	0.5	3.93
2	7	1.0	4.31
3	7	1.5	3.42
4	7	2	3.15

addition of 1% by weight of basalt fibres. This further ensures that introduction of about 1% of basalt fibres is the optimum quantity for improved hardened properties of concrete. Fibres are bound to improve the tensile property of concrete. It is one of the cheapest method available to improve tensile characteristics of concrete when compared to reinforcing with steel. Various research work has been going on in the field of fibre reinforced concrete due to it being economical in nature and improving the tensile properties of concrete. The following Fig. 5 represents a graph between average flexural strength of basalt fibre reinforced concrete versus percentage of basalt fibres for 7 days.

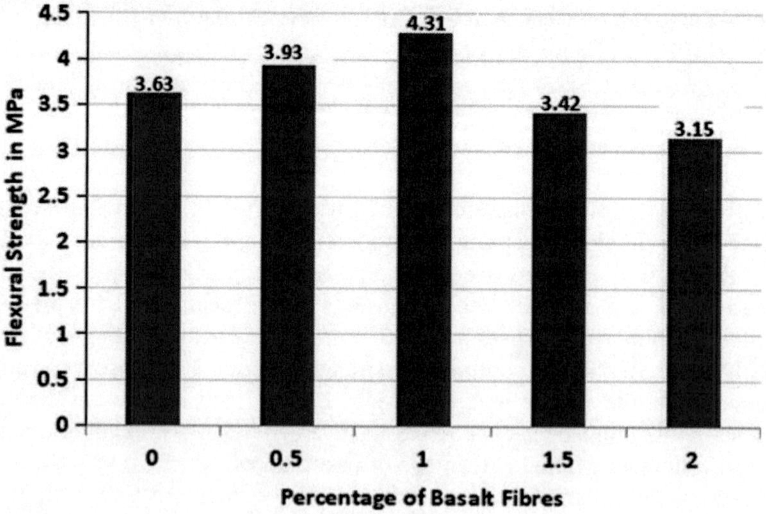

Fig. 5 Graph between flexural strength of BFRP concrete versus percentage of basalt fibres (for 7 days)

Table 9 28 day average flexural strength of high density concrete

S. No	Age at testing (days)	Average compressive strength in MPa
1	28	6.53

Table 10 28 day average flexural strength of basalt fibre reinforced high density concrete

S. No	Age at testing (days)	Percentage of basalt fibre	Average compressive strength in MPa
1	28	0.5	6.81
2	28	1.0	7.31
3	28	1.5	6.60
4	28	2	5.40

The concrete beam was also subjected to flexure test after 28 days of curing. The beam was supported on rollers and the central load was distributed between 2 points and then it was transmitted to the beam. The loading was continued until the failure of the beam. Depending upon the distance between crack and the nearest support in the tension region of the beam, formula for estimation of flexural strength of concrete was chosen. Following Tables 9 and 10 represent the 28 days average flexure strength of high density concrete and basalt fibre reinforced high density concrete with varied Percentages of basalt fibre.

From the table, it is clear that the 28 days flexure strength of basalt fibre reinforced high density concrete is significantly higher than that of flexure strength of high density concrete. These results also follow the trend of other test results by providing maximum flexural strength at 1% by weight of basalt fibres. It can be ascertained that addition of 1% of basalt fibre is optimum for producing maximum compressive and flexural strength for high density concrete. The following Fig. 6 represents a graph between average flexural strength of basalt fibre reinforced concrete versus percentage of basalt fibres for 28 days.

4 Conclusions

Experimental study of replacement of Hematite aggregate (High Density Aggregate) for normal aggregate was attempted for preparation of high density concrete along with reinforcing the concrete with basalt fibres. Following are the conclusions arrived.

- By varying the percentage of basalt fibre in concrete in the order of 0.5, 1, 1.5 and 2%, it could be found that maximum strength properties are achieved at 1% of basalt fibres in high density concrete. At lower percentages such as 5 and 1%, basalt fibres improve the strength characteristics by absorbing the energy released during cracking of concrete and improving ductility of concrete. At higher

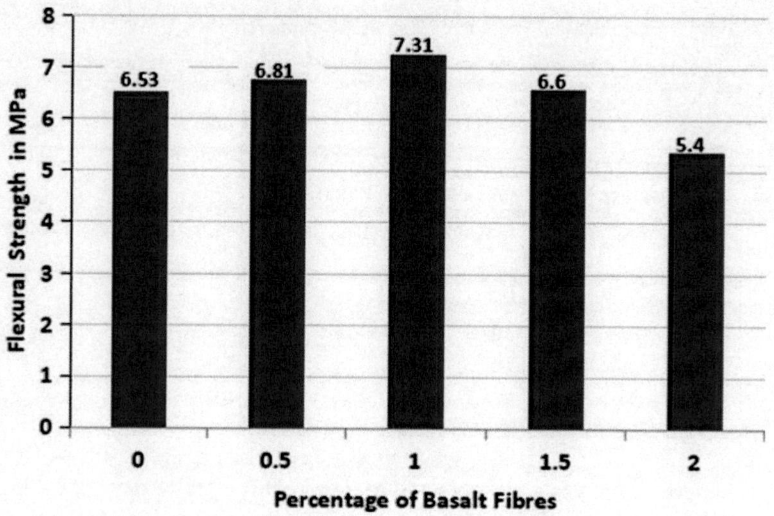

Fig. 6 Graph between flexural strength of BFRP concrete versus percentage of basalt fibres (for 28 days)

percentages, due to higher quantity of fibres, it doesn't get incorporated in the mix causing a balling effect.

- As per the test results, it could be clearly seen that combination of basalt fibres at 1% (optimum per cent of basalt fibre) and high density concrete demonstrates increase of compressive and flexural strength than those of unreinforced high density concrete. Compressive strength increases because a part of the load causing shear failure in direct compression is taken up by the fibres. Flexural strength of high density concrete increases due to the improved tensile properties of fibre reinforced high density concrete.
- The characteristic compressive strength of fibre reinforced high density concrete was 15% higher than high density concrete without fibre reinforcement concrete. The compressive strength increases from 0.5% of basalt fibres to 1% (Maximum Compressive strength). The strength decreases from 1 to 1.5% and a decreasing pattern follows on. The decrease in compressive strength is due to the balling effect of concrete.
- Flexural strength of fibre reinforced high density concrete was 12% higher than that of high density concrete without fibre reinforcement. Tensile properties of concrete have been improved. Flexural strength follows similar pattern to that of compression. Due to the balling effect in the mix (at 1.5 and 2% of basalt fibres), the strength of basalt fibre reinforced high density concrete is lower than that of normal high density concrete. With increased flexural strength, beam dimensions can be reduced during the design. High density BFR concrete resists bending better than high density concrete.

Acknowledgements The authors would like to express gratitude to those who were involved in the research and the institutional support of this work, especially to technicians of Structural Engineering Laboratory, Faculty of Civil Engineering, CEG Campus, Anna University, Chennai-600025.

References

1. Volkman DE (2006) Concrete for radiation shielding. Significance of tests and properties of concrete and concrete making materials. ASTM Int. https://doi.org/10.1520/stp37766s
2. Northup TE (1965) High density concrete for gamma and neutron attenuation. Office of Scientific and Technical Information.https://doi.org/10.2172/4640326
3. Mukherjee PK (1992) Properties of high-density concrete. J Test Eval 20(1). https://doi.org/10.1520/jte11901j
4. Gencel O, Brostow W, Ozel C, Filiz M (2010) Concrete containing hematite ore used as shielding barriers. Mater Sci 16(3):249–256. https://doi.org/10.1016/j.anucene.2011.08.010
5. Neidzweidzka DJ, Lessing PA (2020) High density and radiation shielding concrete, 2nd edn. In: Developments in the formulation and reinforcement of concrete. Elsevier. https://doi.org/10.1016/b978-0-08-102616-8.00009-5
6. Roslan MKA, Ismail M, Kueh ABH, Zin MRM (2019) High-density concrete: exploring Ferro boron effects in neutron and gamma radiation shielding. Constr Build Mater 215:718–725. https://doi.org/10.1016/j.conbuildmat.2019.04.105
7. Srividhya S, Vidjeapriya R, Neelamegam M (2021) Improving mechanical and durability properties of hypo sludge concrete with basalt fibres and SBR latex. Adv Concr Constr 12(4):327–337. https://doi.org/10.12989/ACC.2021.12.4.327
8. Li YF, Hung JF, Syu JY, Chang SM, Kou WS (2022) Influence of sizing of basalt fiber on the mechanical behaviour of basalt fiber reinforced concrete. J Market Res 21:295–307. https://doi.org/10.1016/j.jmrt.2022.09.045
9. Murad Y, Jabbar HA (2022) Shear behavior of RC beams prepared with basalt and polypropylene fibers. Case Stud Constr Mater 16:e00835. https://doi.org/10.1016/j.cscm.2021.e00835

Production of Artificial Aggregates and Their Impact on Properties of Concrete

Gopal Bharamappa Bekkeri, Kiran K. Shetty, and Gopinatha Nayak

1 Introduction

The most widely used material in today's infrastructure is concrete, which uses natural resources like river sand and natural gravel for its aggregate requirements. The majority of concrete is made up of coarse aggregate [1], which causes natural rock beds to be depleted. The government implements strict guidelines for acquiring natural aggregates as part of its efforts to safeguard these resources. As a result, researchers concentrate to a great extent on developing artificial aggregates manufactured from other materials, such as industrial waste, to replace natural aggregate [2]. On the other side, thermal coal-fired power stations generate a huge quantity of fly ash (FA), but only a small fraction of it is put to good use. FA causes significant air and water pollution, creating disposal issues and environmental damage. In this instance, FA is employed to form an artificial coarse aggregate that is tested for usage as a replacement material in concrete. The production of fly ash aggregates (FAA) is an acceptable action to significantly expand the use of FA, which is currently underutilized in enormous quantities in most of the world's nations [1, 3–6].

Manufacturing aggregate has adopted agglomeration, an extensively utilized method, in powder metallurgy for size expansion. In this work, cold bonding, an environmentally friendly and energy-efficient hardening technique, has been adopted to produce aggregates from industrial waste [7]. Cold bonding necessitates the use of cementitious binder and/or relies on the pozzolanic reactivity of the FA, ultimately leading to a type of matrix bonding. Until now, low-calcium bottom ash, class-F fly

G. B. Bekkeri · K. K. Shetty (✉) · G. Nayak
Department of Civil Engineering, Manipal Institute of Technology, Manipal Academy of Higher Education, Manipal, Karnataka 576104, India
e-mail: kiran.shetty@manipal.edu

G. Nayak
e-mail: nayak.gopinath@manipal.edu

R. K. Pancharathi et al. (eds.), *Low Carbon Materials and Technologies for a Sustainable and Resilient Infrastructure*, Lecture Notes in Civil Engineering 440,
https://doi.org/10.1007/978-981-99-7464-1_26

ash, and class-C fly ash have been used to manufacture FAA [8]. There are numerous benefits to using lightweight aggregate in concrete. These include (i) reducing the dead load, which may lead to smaller, lighter high-rise structures and smaller footing sizes. This could lead to less cement and perhaps less reinforcement. (ii) Pre-cast elements that are lighter and smaller require more affordable and minor handling and transportation tools. (iii) Column, slab, and beam size reductions increase the available space. (iv) Excellent thermal insulation. (v) Increased fire resistance [9, 10].

The current study deals with the production of FAA for different proportions of FA and cement. The physicomechanical characteristics of manufactured artificial aggregates were determined and compared with natural coarse aggregates (NCA). Concrete was produced by replacing NCA with FAA in different proportions, and their fresh and hardened properties were investigated.

2 Material and Methodology

The study makes use of the following materials.

2.1 Cement

Cement is the binding material that holds other materials together to create a mass. In a study, cement of 43 grade was used and tested as per IS: 269-2015 and IS 4031-1996. It has shown a specific gravity of 3.12, fineness of 93%, and standard consistency of 32%, with initial and final setting times of 147 and 352 min, respectively.

2.2 Fine Aggregates

The fine aggregate used in this study was natural river sand and locally sourced. Its properties were checked as per IS: 383-2016 and IS: 2386-1963. It exhibited a specific gravity of 2.61, water absorption of 1.2%, and loose and compacted bulk density of 1580 kg/m^3 and 1602 kg/m^3, respectively. According to gradation, it drops under zone II; hence, it can produce plain and reinforced concrete.

2.3 Coarse Aggregates

In a study, commercially available coarse aggregates of angular shape and size range from 20 to 4.75 mm were used and tested according to IS: 383-2016. The test results have shown a specific gravity of 2.8, water absorption of 0.67%, crushing value of 30.3%, and loose and compacted bulk density of 1550 kg/m^3 and 1680 kg/m^3, respectively.

2.4 Fly Ash

FA (class F) was obtained for a study from the Udupi power plant and used in different proportions with cement to produce artificial aggregates. It is tested as per IS 3812 (Part 1)-2003. The test results showed a specific gravity of 2.19, fineness of 355 m^2/kg, and particle retained on 45 microns in percentage by mass is 30%.

2.5 Production of Artificial Aggregates

A granulation or agglomeration process employing a disc-type granulator machine is a step-in manufacturing fly ash aggregates. A standard disc-type granulator was set up with a speed of 55 rpm and a constant angle of 42° throughout the entire production of aggregate. Figure 1 illustrates the disc's dimensions, which were 0.5 m in diameter and 0.10 m in depth. The mix proportions used to form different types of artificial aggregates are listed in Table 1. The artificial aggregates were prepared using the cold-bonding method and maintained duration agglomeration for 20 min. The fresh aggregates were air-dried for 1 day to attain a minimum strength, followed by water curing for 28 days.

Fig. 1 Production of fly ash aggregates. **a** disc-type pelletizer, **b** green pellets, and **c** FAA after curing

Table 1 Mix proportions of FAA

Mix ID	Fly ash (%)	Cement (%)	Angle	Speed (rpm)	Duration (min)
F90C10	90	10	42°	55	20
F85C15	85	15	42°	55	20
F80C20	80	20	42°	55	20
F75C25	75	25	42°	55	20

3 Characterization of Fly Ash Aggregates

The gradation of aggregates was done as per IS-383:2016. The produced artificial FAA were tested for specific gravity, bulk density, and water absorption as per IS-2386:1963 part 3. The crushing strength of aggregates was examined as per IS-2386:1963 Part 4.

4 Surface Treatment of Artificial Aggregates

The FAA were subjected to a surface treatment that involved soaking the aggregate in cement slurry and saturated calcium hydroxide solution separately to enhance the performance of aggregates. The treated solution incorporates the aggregates' surface pores and enhances the aggregates' properties. FAA with particle sizes larger than 4.75 mm were chosen, cleaned, and dried for treatment with the cement slurry. Then, a container filled with cement slurry with a w/c ratio of 0.5. The aggregate was placed in the container, shaken for 3 min, screened out, and dried for 28 days. FAA, with a more than 4.75 mm particle size, were chosen for treatment with the calcium hydroxide solution.

The aggregate was then thoroughly rinsed, immersed in a saturated calcium hydroxide solution for 2 days, filtered out, and dried for 28 days. Figure 2 shows the surface treatment of aggregates with cement slurry and calcium hydroxide solution.

5 Production of Fly Ash Aggregate Concrete

The concrete mix was designed for M30 grade by replacing natural coarse aggregates with FAA by 30, 50, and 70%. Table 5 shows the mix proportions obtained for concrete of different replacements. According to the mix design, the ingredients were mixed and cast into 0.15 × 0.15 × 0.15 m cubes, cylinders of 0.15 m diameter and 0.30 m height, and beams of 0.70 × 0.15 × 0.15 m. All the casted specimens were demoulded after 24 h, followed by water curing for 7, 14, and 28 days. The concrete produced was checked for fresh properties as per IS: 1199-1959 and hardened properties as per IS 516-1959.

Fig. 2 Surface treatment of aggregates with **a** cement slurry, **b** calcium hydroxide solution

6 Properties of Artificial Fly Ash Aggregates

6.1 Particle Size Distribution

The FAA are rounded in shape with a smooth texture, while conventional aggregates are angular with a rough texture. The gradation of conventional and artificial aggregate is represented in Fig. 3. Both aggregates were comprised of particles of all sizes and were well-graded.

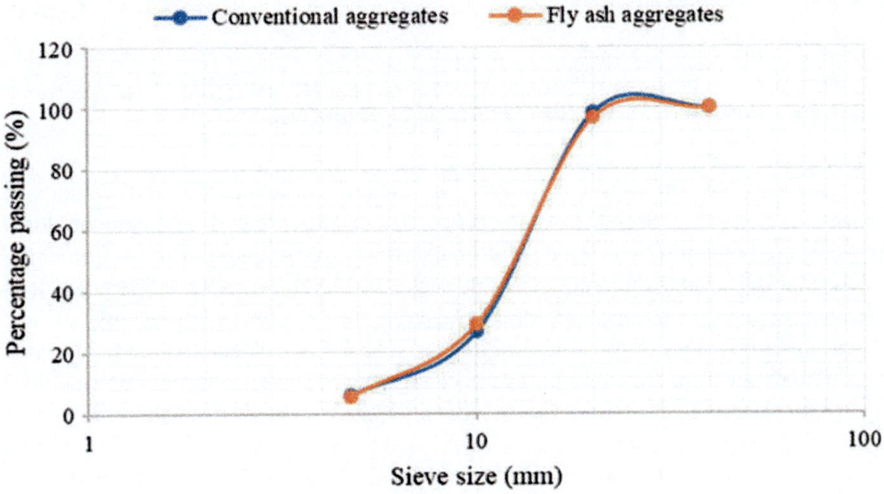

Fig. 3 Gradation of FAA and conventional aggregates

Table 2 Physical and mechanical properties of FAA

FA:cement	Specific gravity	Bulk density (kg/m^3)	Water absorption (%)	Crushing value (%)
90:10	1.57	949	19.37	38.63
85:15	1.62	960	17.19	38.07
80:20	1.64	987	14.21	33.12
75:25	1.71	1009	12.38	32.01

6.2 Physical and Mechanical Properties

Table 2 and Fig. 4 show the physicomechanical characteristics of FAA prepared for various proportions of FA and cement. FAA's specific gravity and bulk density ranged from 1.57–1.71 to 949–1009 kg/m^3, respectively. Hence, they can be considered lightweight aggregates. The obtained results show that the specific gravity and density of aggregates increased with the increased cement content, mainly due to cement of higher specific gravity and the formation of a denser paste matrix. The aggregates' water absorption and crushing strength improved with the increased percentage addition of cement due to the formation of an increased amount of C–S–H, which in turn results in denser microstructure and makes the aggregates less permeable by reducing voids. The aggregates' water absorption and crushing value lie between 19.37–12.38% and 38.63–32.01, respectively. It is worth noting that aggregates containing a higher percentage of fly ash raise water absorption due to the fly ash of higher water absorption. It makes the aggregates porous resulting in lightweight with less strength.

6.3 Properties of Surface-Treated Aggregates

The characteristics of coarse aggregate significantly influence the performance of the mixture. Tables 3 and 4 show the properties of FAA treated with cement slurry and saturated calcium hydroxide solution exhibiting significant improvisation in the water absorption and crushing strength compared to non-treated aggregates. It is because there may be impregnation of treated solution into the aggregates' surface pores, making it less permeable and denser. The FAA treated with cement slurry exhibited slightly enhanced properties than aggregates treated with saturated calcium hydroxide. Hence from the obtained results presented in Fig. 5, it can be revealed that by treating aggregate surfaces with cement slurry and saturated calcium hydroxide, water absorption can be reduced in the range of 34.5–41.92% and crushing strength of 7.92–12.67%.

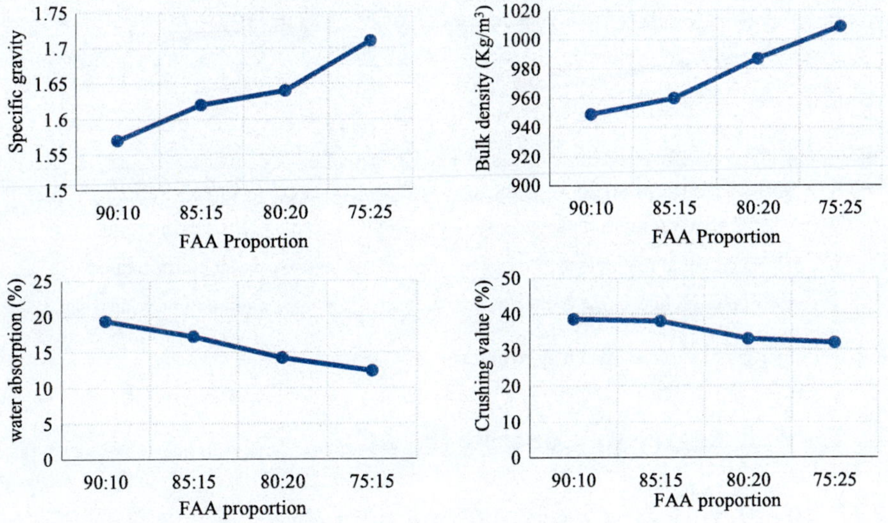

Fig. 4 Properties of FAA

Table 3 Properties of cement slurry treated aggregates

FA:cement	Water absorption (%)	Crushing value (%)
90:10	10.93	34.21
85:15	10.07	34.68
80:20	8.33	31.79
75:25	7.19	28.41

Table 4 Properties of calcium hydroxide treated aggregates

FA:cement	Water absorption (%)	Crushing value (%)
90:10	12.26	36.72
85:15	10.86	36.01
80:20	10.43	31.52
75:25	8.11	29.66

Table 5 Concrete mix proportions

FAA replacement (%)	OPC (kg/m^3)	Sand (kg/m^3)	NCA (kg/m^3)	FAA (kg/m^3)	Water (l)
Normal concrete	384	670.45	1173.0	–	192
30%	384	670.45	821.5	130.765	192
50%	384	670.45	586.768	217.942	192
70%	384	670.45	410.73	305.63	192

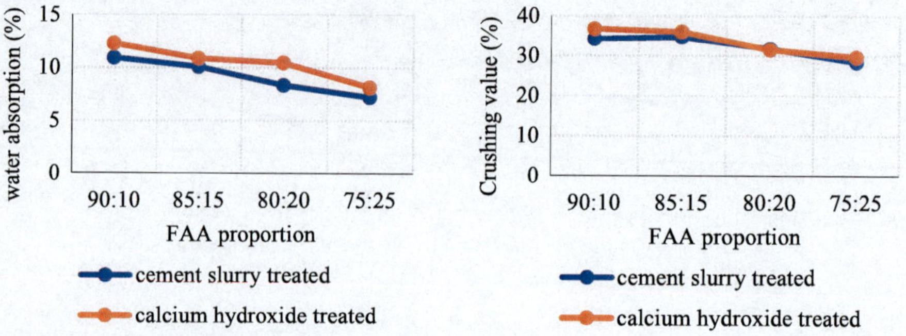

Fig. 5 Properties of surface-treated aggregates

7 Fly Ash Aggregate Concrete (FAAC)

7.1 Mix Design for Concrete Containing FAA

The materials adopted for this study satisfy the requirements for using them in concrete. In accordance with IS-10262-2000, a mix design without admixtures or superplasticizers has been developed for the M-30 grade. Table 5 displays details about the various blends and their composition.

7.2 Workability of Concrete

Figure 6 shows the concrete workability measured by the slump test for a given water-cement ratio. The concrete incorporated with FAA exhibited good workability compared to conventional concrete, mainly due to aggregates of rounded shape. The results show that the concrete workability increased with the increased replacement level of FAA because rounded shape FAA requires less water for friction between them and makes more water available to fluidify the mix.

7.3 Concrete's Density

The density of the concrete containing FAA ranged between 1906 and 2263 kg/m^3, whereas for the same w/c ratio, the density of conventional concrete was 2413 kg/m^3. It has been observed that the concrete's density was reduced with the increased replacement level of FAA. By incorporating FAA, the density of the concrete can be reduced up to 21% for 70% replacement compared to conventional concrete.

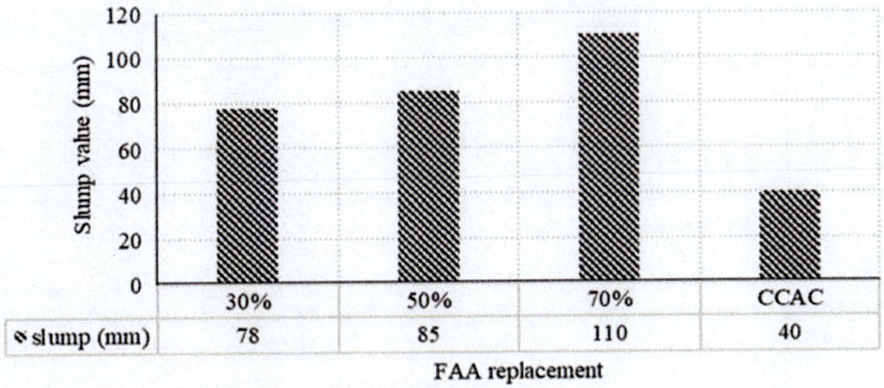

Fig. 6 Workability of concrete

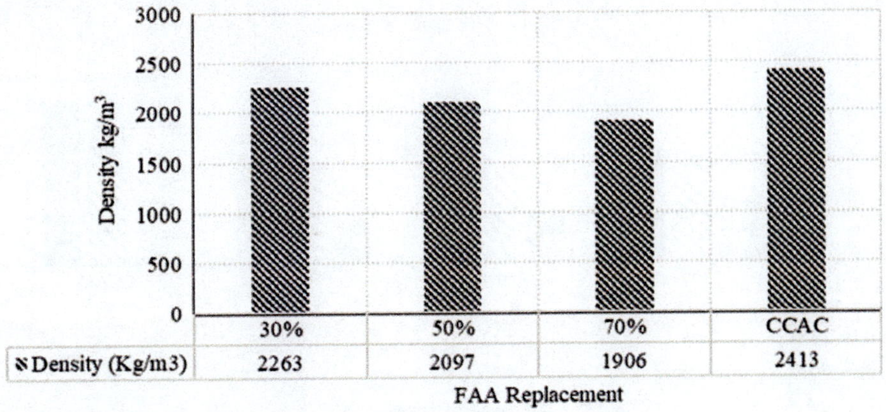

Fig. 7 Density of concrete

Figure 7 presents the concrete density variations as the FAA replacement percentage increases.

7.4 Mechanical Properties

The concrete incorporated with cement slurry treated FAA of proportion 75:25, showing the best physical and mechanical properties compared to other proportions, employed in the concrete mix of M-30 grade for 30, 50, and 70% replacements to natural coarse aggregates. Figures 8 and 9 show characteristic strength and tensile strength for 7, 14, and 28 days of curing, and Fig. 10 shows flexural strength for 28 days of curing only. The results show that the concrete strength decreased with the

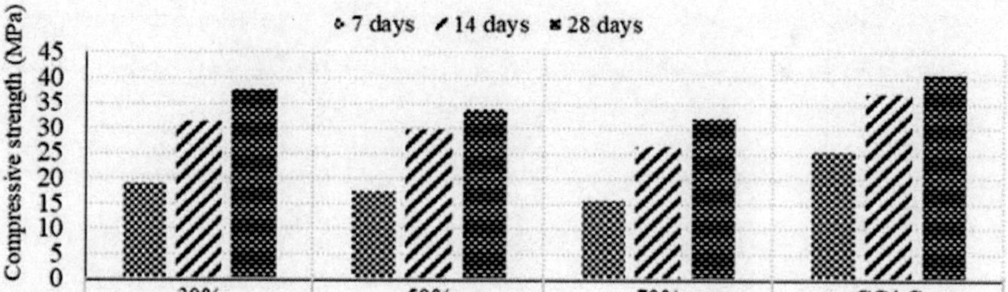

	30%	50%	70%	CCAC
7 days	19.3	17.7	15.9	25.7
14 days	31.3	29.8	26.3	36.9
28 days	37.6	34.2	32.3	41.1

FAA replacement

Fig. 8 Characteristic strength of FAAC

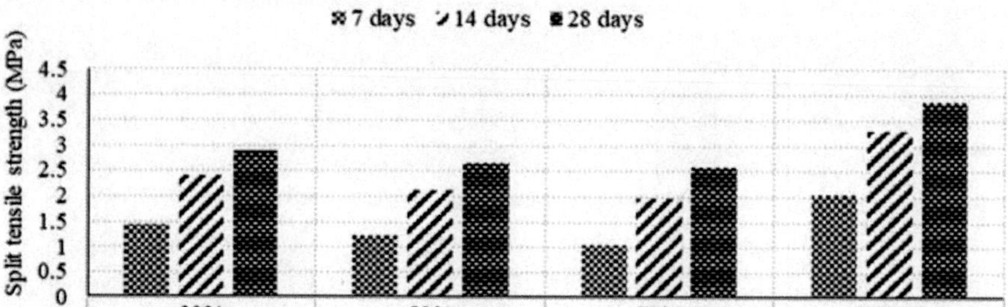

	30%	50%	70%	CCAC
7 days	1.45	1.24	1.07	2.06
14 days	2.4	2.12	1.98	3.29
28 days	2.89	2.67	2.59	3.88

FAA replacement

Fig. 9 Split tensile strength of FAAC

increased replacement level of FAA due to the increased volume of FAA of lower strength. The strength properties of FAAC are inferior compared to the strength properties of conventional concrete. However, they satisfied the code requirements to use it as a structural construction material.

8 Failure Mode

The failure mode of concrete can be identified as bond failure and aggregate failure. In concrete incorporated with FAA, failure on aggregates has been observed, as shown in Fig. 11, which is attributed to fly ash aggregates of weaker strength than natural aggregates.

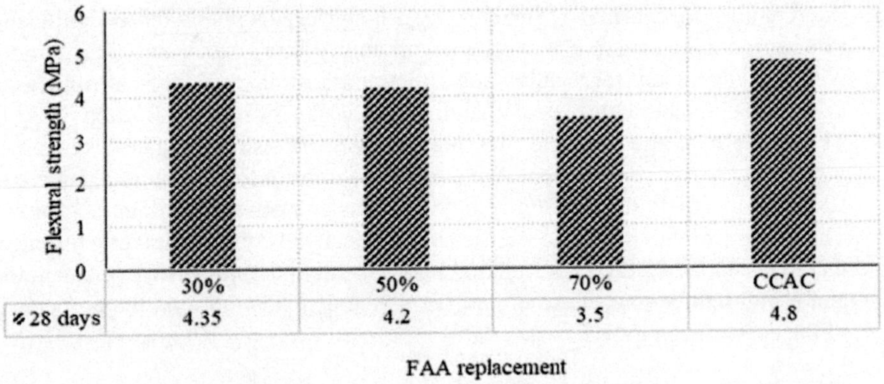

Fig. 10 Flexural strength of FAAC

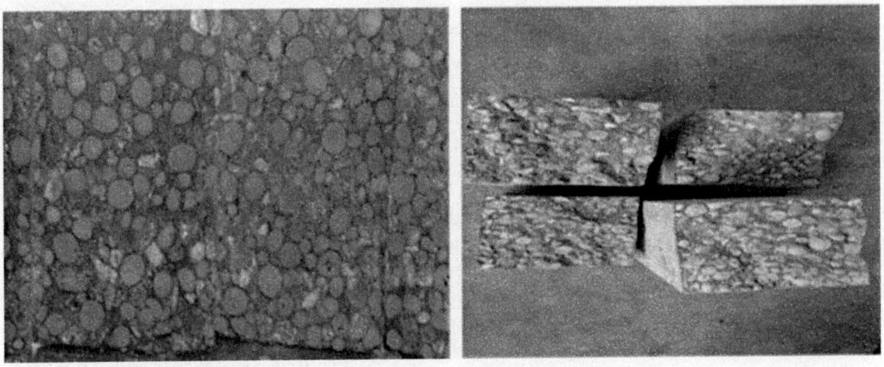

Fig. 11 Failure mode of FAAC

From the above experimental studies, it can be said that the FAA aggregates can be an effective replacement for conventional aggregates under all the above cases, as all the specimen has reached the characteristics strength of M30 Concrete.

9 Conclusion

1. Fly ash aggregates' specific gravity and bulk density are low compared to conventional normal aggregates and can also be considered lightweight aggregates.
2. The fly ash aggregates exhibit high water absorption and lower strength than conventional normal aggregates. However, these can be improved to a certain extent by employing surface treatment with cement slurry or saturated calcium hydroxide solution.

3. In comparison to normal aggregates, fly ash aggregates play a more significant role in producing concrete of excellent workability for a given water-cement ratio without causing any segregation and affecting the strength. This is attributed to aggregates of spherical shape demanding less water for internal friction.
4. The strength of the concrete incorporated with fly ash aggregates decreased with the increased replacement levels. They are inferior to conventional concrete strength but satisfy the requirements to use it as a structural material.
5. The strength of the concrete decreased as the density of concrete decreased, hence there existence of a direct correlation between strength and density of concrete, and it highlighted that the strength of concrete depends more on the properties of the coarse aggregate.

Therefore, from the above-drawn conclusions, it can be suggested that developing fly ash aggregates artificially by agglomeration method leads to the effective utilization of industrial waste and overcoming the scarcity of natural coarse aggregates to produce concrete in the near future.

References

1. Sahoo S, Selvaraju AK, Suriya Prakash S (2020) Mechanical characterization of structural lightweight aggregate concrete made with sintered fly ash aggregates and synthetic fibres. Cem Concr Compos 113. https://doi.org/10.1016/j.cemconcomp.2020.103712
2. IEEE (2012) 2012 International conference on advances in engineering, science and management. IEEE
3. Kockal NU, Ozturan T (2010) Effects of lightweight fly ash aggregate properties on the behavior of lightweight concretes. J Hazard Mater 179:954–965. https://doi.org/10.1016/j.jhazmat.2010.03.098
4. Gomathi P, Sivakumar A (2014) Synthesis of geopolymer based class-F fly ash aggregates and its composite properties in concrete. Arch Civ Eng 60:55–75. https://doi.org/10.2478/ace-2014-0003
5. Dinakar P (2013) Properties of fly-ash lightweight aggregate concretes. Proc Inst Civil Eng Constr Mater 166:133–140. https://doi.org/10.1680/coma.11.00046
6. Terzić A, Pezo L, Mitić V, Radojević Z (2015) Artificial fly ash based aggregates properties influence on lightweight concrete performances. Ceram Int 41:2714–2726. https://doi.org/10.1016/j.ceramint.2014.10.086
7. González-Corrochano B, Alonso-Azcárate J, Rodas M (2009) Characterization of lightweight aggregates manufactured from washing aggregate sludge and fly ash. Resour Conserv Recycl 53:571–581. https://doi.org/10.1016/j.resconrec.2009.04.008
8. Vasugi V, Ramamurthy K (2014) Identification of design parameters influencing manufacture and properties of cold-bonded pond ash aggregate. Mater Des 54:264–278. https://doi.org/10.1016/j.matdes.2013.08.019
9. Kayali O (2008) Fly ash lightweight aggregates in high performance concrete. Constr Build Mater 22:2393–2399. https://doi.org/10.1016/j.conbuildmat.2007.09.001
10. Abbas W, Khalil W, Nasser I (2018) Production of lightweight geopolymer concrete using artificial local lightweight aggregate. In: MATEC web of conferences. EDP sciences

Properties of Concrete Produced Using Waste Polypropylene Fiber

Lakshmi Vara Prasad Meesaraganda and Md. Athar Kazmi

1 Introduction

The disposal of discarded face masks has become a critical global concern [1]. The disposal of such waste has a high environmental impact, while its use in manufacturing or construction has a lower economic impact [2, 3]. This garbage has the benefit of being readily available in huge quantities and at a low cost. The market for construction materials is dominated by cementing materials due to their vital role [4]. Combining concrete technology with environmental technology, incorporating polymeric waste into concrete is an innovative step toward sustainability [5–7]. In order to reduce the dependency on natural resources and to make use of waste, many experiments have been carried out by incorporating such wastes in civil engineering. By doing so, the exhaustion of natural resources is minimized, and the CO_2 emission while manufacturing cement is reduced to a great extent [8, 9].

The coronavirus (COVID-19) epidemic has boosted the use of personal protective equipment (PPE), such as face masks and plastic gloves [3]. The general public's use of face masks helps reduce COVID-19 transmission but has significant environmental repercussions. According to estimates from June 2020, 129 billion face masks were disposed of each month [10]. These calculations were done before some countries enacted laws requiring people to wear face masks worldwide. Face masks left in the street, parking lots, or neighborhood parks are becoming common due to increased littering. The masks' light weight makes them quickly pulled into rivers, lakes, and oceans, where they can break down into tiny pieces of plastic [11]. As a result, discarding masks or disposing of used personal protection equipment improperly

L. V. P. Meesaraganda · Md. A. Kazmi (✉)
Department of Civil Engineering, National Institute of Technology, Silchar, Assam 788 010, India
e-mail: mdatharkazmi108@gmail.com

L. V. P. Meesaraganda
e-mail: mlvprasad@civil.nits.ac.in

R. K. Pancharathi et al. (eds.), *Low Carbon Materials and Technologies for a Sustainable and Resilient Infrastructure*, Lecture Notes in Civil Engineering 440,
https://doi.org/10.1007/978-981-99-7464-1_27

could result in issues with wildlife and aquatic life or even the extinction of animals and marine life. Additionally, non-biodegradable polymers, which take hundreds of years to decompose in the environment, are used to make single-use masks [12]. As a result, quick action is needed to combat the COVID-19 epidemic and reduce the environmental impact of PPE disposal.

The utilization of waste items such as PPE has become vital in order to decrease the negative environmental implications caused by the epidemic [13]. Due to the pandemic, the usage of PPE has increased drastically. This PPE is made of polypropylene (PP), which will take more than 400 years to degrade in the environment. So, this PP is treated as non-biodegradable material. Because of the increased consumption, the load of discarding this PP has become numerous [7].

This research explores the advantages of PP fibers in the form of shredded masks, a PPE category. The shredded masks are added to the conventional M40-grade concrete at a calculated proportion. The test results have shown an increase in concrete's compressive, tensile, and flexural strength. When these PP fibers are added to concrete, the elongation problem of the material is controlled.

2 Materials and Methods

2.1 *Materials*

53-grade commercial Portland cement confirming to IS 12269-2013 [14] is used in this project. Coal-fired power stations' by-product Class F fly ash is utilized. Locally available fine aggregate of grade II and 10 and 20 mm nominal maximum size of coarse aggregate was utilized. A superplasticizer named Sunanda Polytancrete NGT super plasticizing admixture from Sunanda specialty coating PVT.LTD has been used in this present experimental work, and water drinking quality standard was used in the project. Waste polypropylene fiber (WPF) obtained from waste shredded masks is utilized; the face masks were shredded into 2 cm × 0.5 cm sizes for this study, as shown in Fig. 1.

2.2 *Methods*

To get effective properties, comparative analysis and testing of concrete specimens are carried out before and after fiber inclusion. The concrete mixing was carried out as per IS 10262:2019 [15] and IS 456:2000 [16]. One without WPF and three varying amounts of WPF were applied in the concrete mix considered in this study, conventional concrete having zero WPF (CC), concrete with 0.1% WPF by weight of cement used (FC-I), concrete with 0.2% WPF by weight of cement used (FC-II),

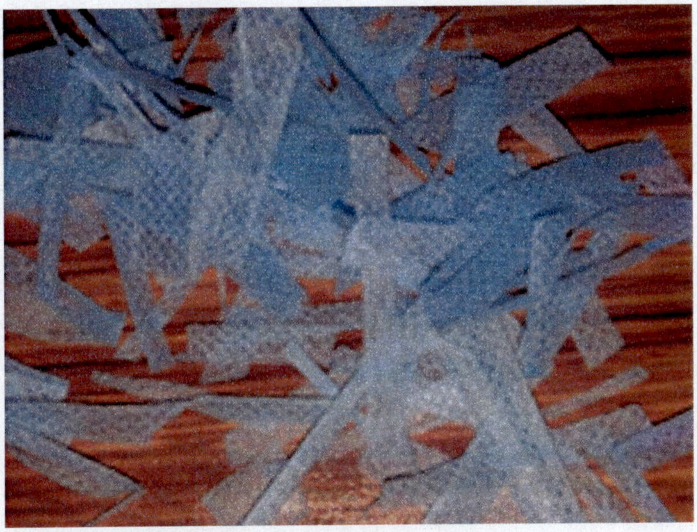

Fig. 1 Single-use face mask shredded into 2 cm × 0.5 cm

concrete with 0.3% WPF by weight of cement used (FC-III). The weight of each ingredient per unit volume of concrete for each mix ID is presented in Table 1.

In order to evaluate the concrete fresh qualities, slump value test was performed; testing view of it is shown in Fig. 2. A test for compressive strength was performed on 150 mm cube specimens at 7, 28, and 56 days after water curing; the test setup is shown in Fig. 3. On a cylindrical specimen with a 150 mm diameter and 300 mm height, a splitting tensile strength test was performed at 28 and 56 days of water curing, Fig. 4 shows the splitting tensile testing. At 28 and 56 days of water curing, a flexural strength test was conducted on a beam specimen with a length of 500 mm and a cross-section of 100 × 100, and the test setup for the flexural strength test is shown in Fig. 5. The test for drying shrinkage was carried out on specimens having dimensions of 75 × 75 × 300 mm. The specimens were kept in moist air

Table 1 Weight of ingredients per cubic meter of concrete

Ingredients	CC	FC-I	FC-II	FC-III
Cement (kg)	392	392	392	392
Fly ash (kg)	97	97	97	97
Water (kg)	178	178	178	178
Superplasticizer	3.07	3.07	3.07	3.07
Water/binder	0.36	0.36	0.36	0.36
Coarse aggregate (kg)	1046	1046	1046	1046
Fine aggregate (kg)	704	704	704	704
WPF (g)	–	390	780	1170

Fig. 2 Slump test

for 3 days followed by 28 and 90 days of water curing. Figure 6 shows the drying shrinkage testing. The details of the entire test conducted and, the method of testing are presented in Table 2.

3 Results and Discussion

3.1 Slump Value

Figure 7 shows the result of the control and WPF concrete slump values. The slump value decreases as the dose of WPF increases, the mix FC-I and FC-II having 4.8% and FC-III with 9.2% lesser slump value than CC. To improve the workability of concrete, a water-reducing additive is employed with lower water-cementitious ratios. All of the mixes fall under the heading of excellent workability. Similar findings were made by Chen et al. [20], who discovered that slump decreases as polypropylene content rises.

Fig. 3 Compressive strength testing

3.2 *Compressive Strength*

The results of compressive strength at 7, 28, and 56 days are shown in Fig. 8; three specimens are tested on each testing day from each mix ID, and the average of these compressive strengths obtained is reported here. With the addition of WPF up to a certain optimum dose into concrete, the increase in the compressive strength was observed; beyond that, it tended to decrease on all testing days, and the mix FC-II showed the highest compressive strength. A 2.67, 1.85, and 1.3% higher compressive strength was found in FC-II than in CC at 7 days, 28 days, and 56 days of testing, respectively. The beautifully packed gaps between the coarse aggregate, sand, and cements components, which led to maximum compressive strength, cause the rise in compressive strength brought on by the incorporation of WPF. Another factor contributing to maximum strength may be the WPF's strong bond with cement paste. However, using more WPF than the recommended 0.2% may result in improper compaction and workability, preventing future compressive strength increases at higher WPF doses. The published literature was found to be consistent with the higher dose of PP fiber than the optimum, decreasing the compressive strength [21].

Fig. 4 Splitting tensile strength testing

3.3 *Splitting Tensile Strength*

The splitting tensile strength test was done at 28 and 56 days, and the result obtained is shown in Fig. 9; three specimens are tested on each testing day from each mix ID, and the average of this compressive strength obtained is reported here. With the addition of WPF up to a certain optimum dose into concrete, the increase in the splitting tensile strength was observed; beyond that, it tends to decrease. The highest splitting tensile strength was found in the mix FC-II on both testing days. An 11% and 7.33% higher splitting tensile strength was a noticed in FC-II than in CC at 28 days and 56 days of testing, respectively. This show that the WPF acts as a minute, uniformly distributed reinforcement throughout the concrete matrix. However, a higher dose of WPF than 0.2% may cause improper bonding, compaction, and workability and hence forbid further increment in splitting tensile strength at a higher dose of WPF. A similar observation was noticed by Liu et al. [22] that a higher dose of PP fiber than the optimum decreases the splitting tensile strength.

Fig. 5 Flexural strength testing

Fig. 6 Length comparator for shrinkage test

Table 2 Details of tests conducted

Sr. No	Test	Age on test day	Specimen size, mm (in.)	Test method
1	Slump value	–	–	1199 (part-2) [17]
2	Compressive strength	7, 28, 56	Cube—150 × 150 × 150(5.91 × 5.91 × 5.91)	IS 516 (part-1/sec-1) [18]
3	Splitting tensile strength	28, 56	Cylinder—Φ 150 × 300 (Φ 5.91 × 11.81)	IS 516 (part-1/sec-1) [18]
4	Flexural strength	28, 56	Prism—100 × 100 × 500 (3.94 × 3.94 × 19.69)	IS 516 (part-1/sec-1) [18]
5	Drying shrinkage	28, 90	Prism—75 × 75 × 300 (2.95 × 2.95 × 11.81)	ASTM C 157/C 157 M – 08 [19]

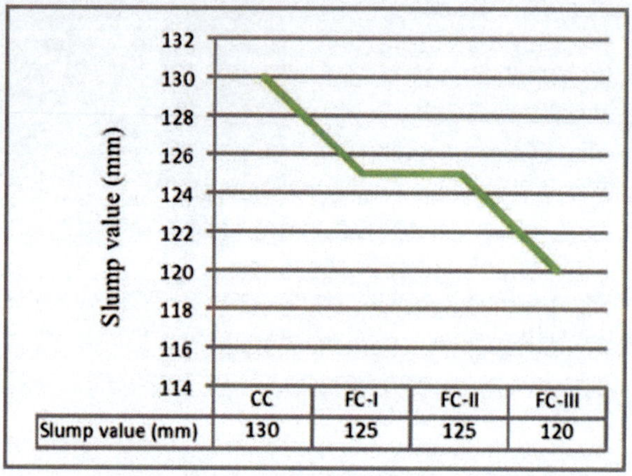

Fig. 7 Slump values for all mixes

3.4 Flexural Strength

The results of the flexural strength test, which was performed at 28 and 56 days, are presented in Fig. 10. Three specimens are tested on each testing day from each mix ID, and an average of this compressive strength obtained is reported here. With the addition of WPF up to a certain optimum dose into concrete, the increase in flexural strength was observed; beyond that, it tends to decrease. The highest flexural strength was observed in the mix FC-II on both testing days. A 13% and 8.66% higher flexural strength was found in FC-II than in CC at 28 days and 56 days of testing, respectively. This show that the WPF acts as a minute, uniformly distributed reinforcement throughout the concrete matrix. The increase in flexural strength due to the incorporation of WPF is due to excellently packed spaces in between the coarse aggregate, sand, and cement components, culminating in maximum flexural strength. Another reason for maximum strength might be the strong bonding of WPF

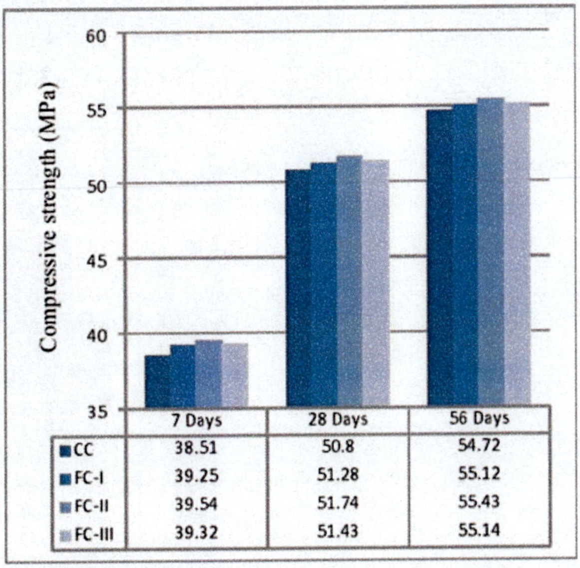

Fig. 8 Compressive strength value for all mixes

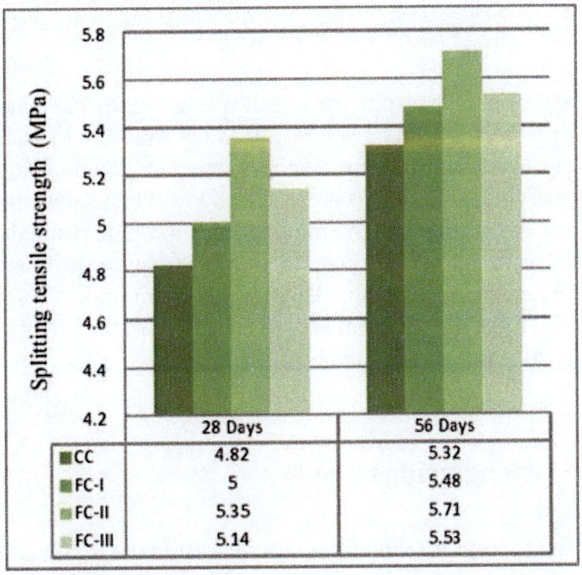

Fig. 9 Splitting tensile strength value for all mixes

with cement paste; however, higher dose of WPF than 0.2% may cause improper compaction and workability and hence forbid further increment in flexural strength at a higher dose of WPF. The published literature was found to be consistent with the higher dose of PP fiber than the optimum, decreasing the flexural strength [23].

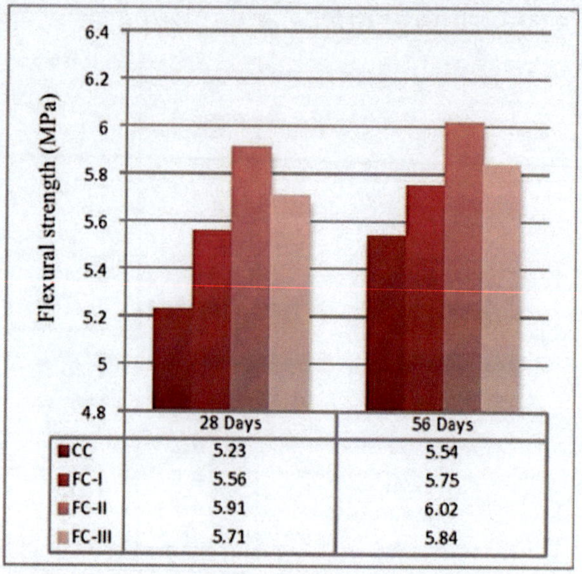

Fig. 10 Flexural strength value for all mixes

3.5 Drying Shrinkage

Experimental results show that all the mixes have very less shrinkage strain, even with the addition of WPF; it does not make any major changes in shrinkage strain. The test for drying shrinkage is done after 28 and 90 days of water curing. The result of drying shrinkage is presented in Fig. 11; it shows that at 28 and 90 days of testing day, a marginal increment in drying shrinkage strain due to the addition of WPF than CC is observed. The increment in the drying shrinkage strain is very small and can be deemed acceptable. The marginal increment in shrinkage strain in mixes FC-I, FC-II, and FC-III compared to CC is due to the lower specific gravity of WPF, which is soft in nature and tends to shrink.

4 Conclusions

The present research is on using waste face masks made up of polypropylene fiber as fibrous materials in concrete. The finding of the experimental investigation shows that the addition of WPF into concrete acts as a fiber. Due to the incorporation of WPF in concrete, compressive, splitting tensile, and flexural strength are greatly improved. A small optimum dose of WPF in concrete gives the best result in terms of mechanical properties. A 1.85%, 8%, and 13% higher compressive, splitting tensile, and flexural strength was observed in the mix FC-II than CC at 28 days of testing. A marginal reduction in slump value and increment in the drying shrinkage strain was observed due to the addition of WPF; however, the reduction in slump and increment

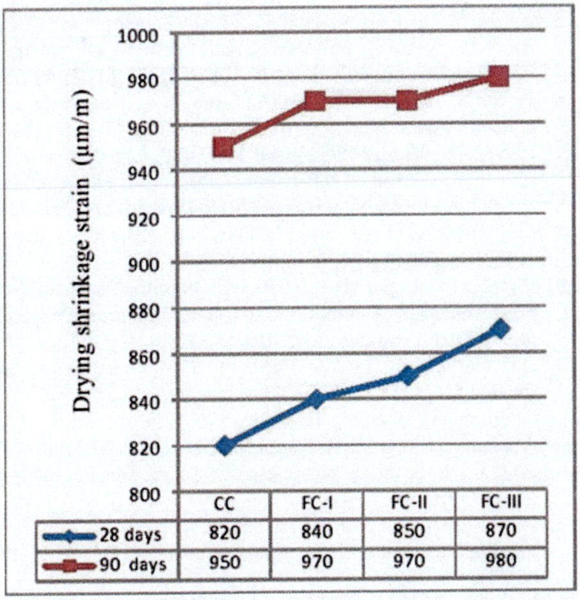

Fig. 11 Drying shrinkage value for all mixes

in drying shrinkage is small, which is deemed to be acceptable. Hence, the addition of WPF into concrete gives a double-fold benefit, one is improving concrete mechanical properties, and the other is solving the disposal issue of waste polypropylene face masks. Also, this practice is environmentally friendly and cost-effective.

Acknowledgements The authors are grateful to the director of the National Institute of Technology Silchar, Assam, India, for providing all the research facilities for carrying out the research.

Competing Interest The authors declare that they have no known competing financial interests or personal relationships that could have appeared to influence the work reported in this paper.

References

1. Hantoko D, Li X, Pariatamby A, Yoshikawa K, Horttanainen M, Yan M (2021) Challenges and practices on waste management and disposal during COVID-19 pandemic. J Environ Manage 286:112140. https://doi.org/10.1016/j.jenvman.2021.112140
2. Akinwumi II, Domo-Spiff AH, Salami A (2019) Marine plastic pollution and affordable housing challenge: Shredded waste plastic stabilized soil for producing compressed earth bricks. Case Stud Constr Mater 11:00241. https://doi.org/10.1016/j.cscm.2019.e00241
3. Nzediegwu C, Chang SX (2020) Improper solid waste management increases potential for COVID-19 spread in developing countries. Resour Conserv Recycl 161:104947. https://doi.org/10.1016/j.resconrec.2020.104947
4. Demirboga R, Kan A (2012) Thermal conductivity and shrinkage properties of modified waste polystyrene aggregate concretes. Constr Build Mater 35:730–734. https://doi.org/10.1016/j.conbuildmat.2012.04.105

5. Aslani F, Deghani A, Asif Z (2020) Development of lightweight rubberized geopolymer concrete by using polystyrene and recycled crumb-rubber aggregates. J Mater Civ Eng 32:04019345. https://doi.org/10.1061/(ASCE)MT.1943-5533.0003008
6. Prasad MLV, Kumar PR (2007) Strength studies on glass fiber reinforced recycled aggregate concrete. Asian J Civ Eng Build Hous 8(6):677–960
7. Rao MVK, Kumar PR, Prasad MLV (2015) Influence of type and size of fine aggregate on glass fibre reinforced self-compacting mortars (GFRSCM). Int J Conc Technol 1(2):13–21
8. Fang C, Zhang M, Yu R, Liu X (2015) Effect of preparation temperature on the aging properties of waste polyethylene modified Asphalt. J Mater Sci Technol 31:320–324. https://doi.org/10.1016/j.jmst.2014.04.019
9. El-Naga IA, Ragab M (2019) Benefits of utilization the recycle polyethylene terephthalate waste plastic materials as a modifier to asphalt mixtures. Constr Build Mater 219:81–90. https://doi.org/10.1016/j.conbuildmat.2019.05.172
10. Dharmaraj S, Kumar VA, Hariharan S, Manibharathi A, Show PL, Chong CT, Ngamcharussrivichai C (2021) The COVID-19 pandemic face mask waste: a blooming threat to the marine environment. Chemosphere 272:129601. https://doi.org/10.1016/j.chemosphere.2021.129601
11. Ahmadinia E, Zargar M, Karim MR, Abdelaziz M, Shafigh P (2011) Using waste plastic bottles as additive for stone mastic asphalt. Mater Des 32:4844–4849. https://doi.org/10.1016/j.matdes.2011.06.016
12. Hake SL, Damgir RM, Awsarmal PR (2020) Utilization of Plastic waste in Bitumen Mixes for Flexible Pavement. Transp Res Procedia 48:3779–3785. https://doi.org/10.1016/j.trpro.2020.08.041
13. Prasad MLV, Kumar PR, Oshima T (2009) Development of analytical stress-strain model for glass fiber reinforced self compacting concrete. Int J Mech Solids 4(1):25–37
14. IS 12269:2013 (2015) Ordinaraly Portland cement 53 Grade. Bureau Indian Standard, New Delhi, p 1
15. IS 10262 (2019) Concrete mix proportioning—guidelines. Bureau Indian Standard, New Delhi, p 2
16. IS 456:2000 (2000) Plain and reinforced concrete. Bureau Indian Standard, New Delhi, p 4
17. 1199 (Part-2) Fresh concrete—methods of sampling, testing and analysis-determination of consistency of fresh concrete. Bureau Indian Standard, New Delhi, p 1
18. IS 516: Part I (Section 1). Hardened concrete—methods of test—testing of strength of hardened concrete. Bureau Indian Standard, New Delhi, p 1
19. ASTM C 157/C 157M-08 Standard test method for length change of hardened hydraulic-cement mortar and concrete. ASTM International, 100 Barr Harbor Drive, PO Box C700, West Conshohocken, PA 19428-2959. United States
20. Chen Y, Zou H, Liang M, Liu P (2013) Rheological, thermal, and morphological properties of low-density polyethylene/ultra-high-molecular-weight polyethylene and linear low-density polyethylene/ultra-high-molecular-weight polyethylene blends. J Appl Polym Sci 129:945–953. https://doi.org/10.1002/app.38374
21. Prasad MLV, Saha P, Kumar PR (2016) Self compacting reinforced concrete beams strengthened with natural fiber under cyclic loading. Comput Concr 17(5):597–612. https://doi.org/10.12989/cac.2016.17.5.597
22. Abukhettala M, Fall M (2021) Geotechnical characterization of plastic waste materials in pavement subgrade applications. Transp Geotech 27:100472. https://doi.org/10.1016/j.trgeo.2020.100472
23. Kazmi MA, Prasad MLV (2022) Investigation on mechanical, durability and time-dependent properties of standard grade recycled aggregate concrete with pozzolanic material. Indian J Eng Mater Sci 29(1):62–70. https://doi.org/10.56042/ijems.v29i1.45448

Performance Evaluation of Ternary Blended Alkali-Activated Mortars Incorporated with Industrial Waste Byproducts—A Step Toward Sustainability

Sunil Nandipati, G. V. R. Srinivasa Rao, and G. Mallikarjuna Rao

1 Introduction

Concrete is universally accepted material for construction and infrastructure development. It is used in various projects such as building construction. Annually, around 30 billion tons of concrete are consumed globally [1]. Unfortunately, the production of concrete using OPC can lead to the release of large amounts of greenhouse gases into the atmosphere. For instance, if the 1 ton of concrete that is produced requires 1.5 tons of raw materials, then the production of this material directly releases around 0.5 tons of CO_2. On the other hand, the auxiliary processes of the construction industry can also contribute to the emission of around 0.8–1.0 tons of CO_2 per ton of OPC [2, 3]. Hence, there is an urgent need for alternative binders emphasizing sustainable construction practices to minimize the carbon footprint.

Alkali-activated binders (AAB) have been extensively explored for the last few decades for sustainable construction materials to provide an alternative to OPC-based but CO_2-intensive materials while providing beneficial mechanical properties and durability. One of the main advantages of AAB is the valorization of waste that is usually disposed of in landfills, occupying an increasing volume every year,

S. Nandipati (✉) · G. V. R. Srinivasa Rao
Department of Civil Engineering, AU College of Engineering (A), Andhra University, Visakhapatnam, Andhra Pradesh, India
e-mail: snandipa@gitam.edu

S. Nandipati
Department of Civil Engineering, GST, GITAM Deemed to Be University, Visakhapatnam, Andhra Pradesh, India

G. Mallikarjuna Rao
Department of Civil Engineering, Vardhaman College of Engineering (Autonomous), Hyderabad, Telangana, India

R. K. Pancharathi et al. (eds.), *Low Carbon Materials and Technologies for a Sustainable and Resilient Infrastructure*, Lecture Notes in Civil Engineering 440,
https://doi.org/10.1007/978-981-99-7464-1_28

a significant sustainability concern in India. AAB is produced from the alkaline activation of an amorphous precursor that can be calcareous, siliceous, or silico aluminous. Some precursors are suitable for alkaline activation, such as fly ash (FA) [4, 5], ground granulated blast-furnace slag (GGBS) [6, 7], metakaolin (MK) [8], copper slag (CS) [9], red mud (RM) [10], ceramic tile waste (CTW) [11], rice husk ash (RHA) [12], silica fume (SF) [13], glass powder (GP) [4], bottom ash (BA) [14] that are mostly dumped as a waste material every year in India. Binary AAB can be assembled to further improve the mechanical and durability properties of concrete. Also, the most common activators are sodium-based compounds such as sodium hydroxide (SH) and sodium silicate (SS).

Several studies have been conducted in choosing the precursors and activators based on the AAB's potential strength, eco-friendliness, durability, and lower cost. However, recent studies have indicated that calcium-content-based precursors could significantly enhance AAMs' hardening, strength, and setting properties [6, 7]. In addition to calcium, other important factors such as the curing environment and the alkali-activated source material type can also contribute to the development of concrete that is environmentally friendly.

CTWD is a fine powder generated during the finishing processing of ceramic tile products. It was found that the daily waste produced by ceramic factories is approximately 30% [15]. These ceramic waste products significantly impact health, air, and disposal challenges. Disposing of ceramic waste in landfills causes environmental issues such as groundwater contamination and air pollution. Industrial solid waste disposal is a global issue today era.

This study mainly studied on utilization of CTWD as an alternative precursor, which is rich in alumina and silica; it can be effectively used for the development of AAMs. The literature review on the topic revealed that not many studies had been studied on CTWD-based AAMs about the high concentration of molarity, higher binder-to-aggregate ratio, fluid-to-binder ratio, and durability [16]. Therefore, this research focuses on determining the optimum mix proportion based on mechanical and water transport properties. This study also examines the influence of specific parameters for developing AAMs using CTWD and the combination of FA and GGBS.

2 Experimental Program

2.1 Materials

Preparation of AAMs includes CTWD, FA, and GGBS as precursors and Na_2SiO_3 and NaOH as alkali activators. All the materials used in this research were carefully collected and stored in containers to avoid contamination. For the present study, GGBS was collected from Sri Vishnu Sai Saravana Enterprises, Visakhapatnam. Low calcium FA from NTPC-Simhadri thermal power plant, Visakhapatnam. The ceramic

Table 1 Physical properties

Particulars		IS method	CTWD	FA	GGBS
Grain size analysis	Sand (%)	[17]	15.1	24.8	12.7
	Silt and clay (%)		84.9	75.2	87.3
Consistency (%)		[18]	33.7	29.8	25.5
Specific gravity			2.51	2.15	2.85
Autoclave expansion (%)			0.08	NA	0.045

Table 2 Chemical characterization

Sample	CTWD (%)	FA (%)	GGBS (%)
SiO_2	59.62	61.34	34.25
TiO_2	0.99	1.376	0.37
Al_2O_3	16.62	28.47	10.72
MnO	0.108	0.38	0.33
Fe_2O_3	3.52	4.10	0.55
CaO	5.72	1.11	34.88
MgO	4.533	0.40	10.98
Na_2O	0.79	0.62	0.21
K_2O	2.11	0.34	0.31
P_2O_5	0.123	0.622	0.171
SO_3	3.806	0.26	7.139

wastes were collected from sirazee tiles private limited, a ceramic manufacturer in Tadepalligudem, (AP) India. All precursors were sieved through 90 μm, where crusher dust (CD) was used as filler material. Properties of precursors used are presented in Tables 1 and 2. Commercial grade (SH and SS) are acquired from Sai Chemicals, Visakhapatnam. SH was purchased as flakes (99.58% purity), and anhydrous SS (ratio of Na_2O: SiO_2 is 1:0.2.2) was used in this research work.

2.2 *Variation of the Binder in the Mix Proportion*

Preparation of AAMs by a blend of CTWD, FA, and GGBS at varying proportions. NaOH of 10 molarity (10M), Alkaline solution to binder (AS/B) ratio of 0.55, and SS to SH ratio of 2.5 were kept constant in the trail mixes. The AAMs Mix Design is represented in Table 3.

Table 3 AAMs mix design

Mix No	Binder (% by weight)			AS: B	B: CD	SS: SH
	CTWD	FA	GGBS			
AAMs$_{1}$	0	0	70	0.55	2.33	2.5
AAMs$_{2}$	10	10	50	0.55	2.33	2.5
AAMs$_{3}$		20	40			
AAMs$_{4}$		30	30			
AAMs$_{5}$		40	20			
AAMs$_{6}$		50	10			
AAMs$_{7}$	20	20	30	0.55	2.33	2.5
AAMs$_{8}$		30	20			
AAMs$_{9}$		40	10			
AAMs$_{10}$		10	40			
AAMs$_{11}$	30	10	30	0.55	2.33	2.5
AAMs$_{12}$		30	10			
AAMs$_{13}$		20	20			
AAMs$_{14}$	40	10	20	0.55	2.33	2.5
AAMs$_{15}$		20	10			
AAMs$_{16}$	50	10	10	0.55	2.33	2.5

2.3 *Preparation and Test Procedure of AAMs*

All precursors were weighted accurately. SS and SH as alkali activators were mixed and stored in an airtight container one day before the casting of cubes. The mixing process of AAMs for 2–5 min in a rotary mixer until a uniform mix is obtained. The consistency and homogeneity of the fresh concrete mixture are evaluated before it is placed into the production line. After it has been prepared, the flow rate and setting time of the mixture are calculated as per ASTM standards [19] and [20], respectively. The alkali-activated mix was taken to moulds of sizes $100 \times 100 \times 100\ mm^3$ and $140 \times 40 \times 40\ mm^3$ and compacted using the table vibrator. The moulds with an alkali-activated mix are left for outdoor curing for one day and de-molded after 24 h. All the samples were tested to evaluate the compressive strength at 3, 7, and 28 days, flexural strength, and water absorption test for 28 days in outdoor curing in accordance with ASTM standards [21], [22], and [23] respectively.

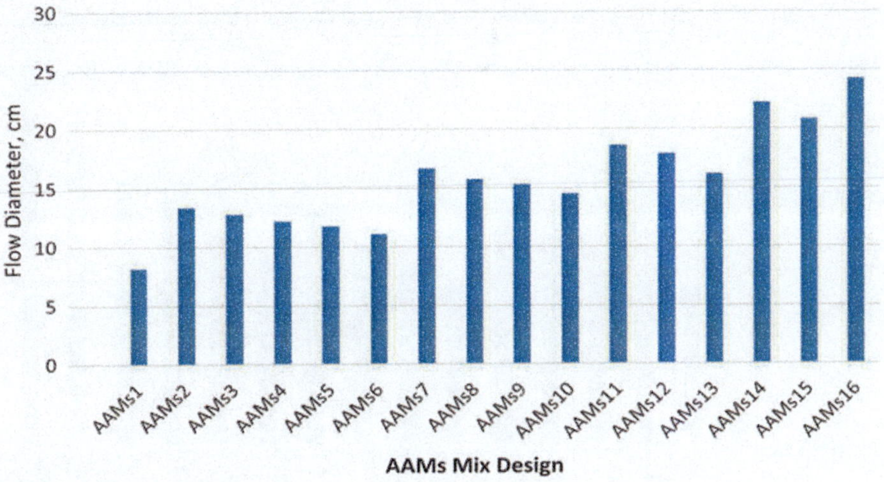

Fig. 1 Flow values of AAMs

3 Results and Discussions

3.1 Flow Value

Figure 1 represents flow values of AAMs with a range of 13.4 cm to 24.25 cm with an increase of 10%–50% CTWD, respectively, compared to the control mix ($AAMs_1$) of 8.2 cm. The flow value often decreased due to GGBS and FA contents. In mix $AAMs_6$, flow diameter was measured to be the lowest value. The increase in the FA content can significantly decrease the workability of mortar materials due to their porous structure. On the other hand, an increase in the ceramic content can improve the AAMs' workability [24].

3.2 Final Setting Time

The higher volume of CTWD longer will be the setting time of mortars (62.65 min) (Fig. 2). When the GBBS content increased, and CTWD content decreased, a significant reduction in setting time was observed. With the increased level of FA content, final setting times are increased, which is a longer setting time reported with $AAMs_{16}$ compared to the control mix ($AAMs_1$). Due to higher GGBS content and lower CTWD content, which will reduce the setting properties [25].

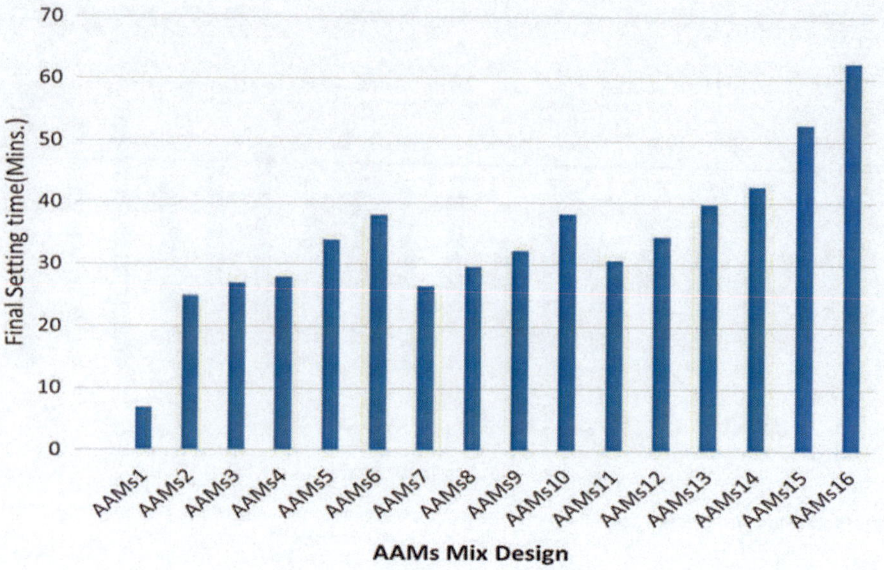

Fig. 2 Final setting times of AAMs

3.3 Hardened Density

Figure 3 summarizes the change in AAMs densities. Increased CTWD content affects the AAMs' density due to the specific gravity of ceramic particles than GGBS. In the comparison, AAM_2 (2.254 g/cm^3) with 10% CTWD, 10% FA, and 40% GGBS presented higher density than the control mix of AAM_1 (2.396 g/cm^3). Moreover, the AAM_{16} mix shows the lowest density of 1.951 g/cm^3. Generally, higher density is observed due to the increased content of CTWD, and slag compared to FA.

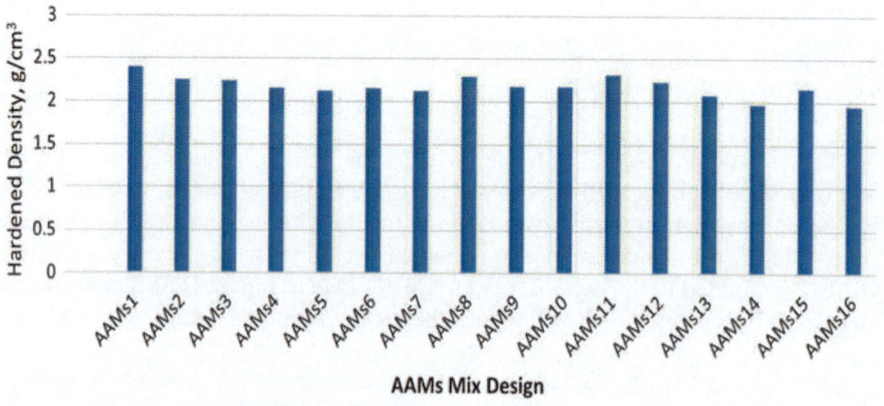

Fig. 3 Density of ternary blended AAMs at 28 days

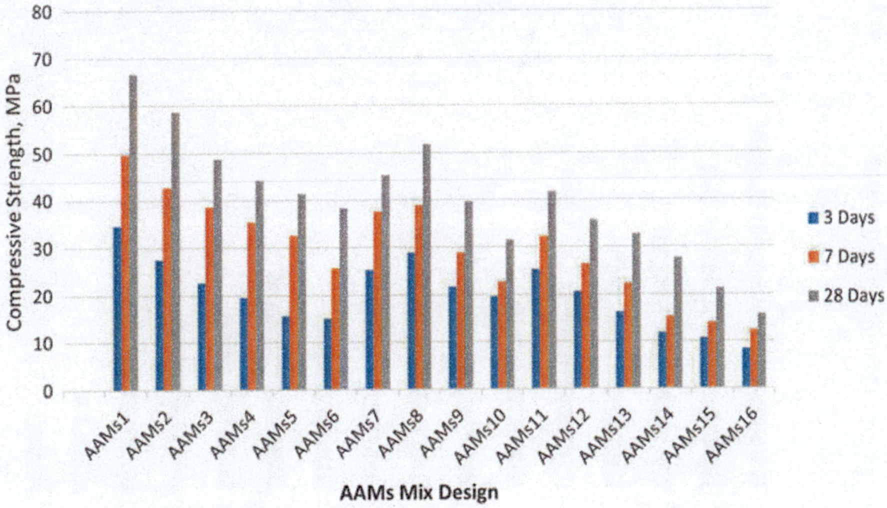

Fig. 4 Compressive strength of AAMs

3.4 Compressive Strength (CS)

Figure 4 highlights the effect of CTWD on CS of AAMs. From the experimental investigation, it is observed that with increase in CTWD content, the Compressive Strength was reduced from 58.72MPa to 15.66 MPa at the age 28 days. The significant strength reduction is due to a decrease in the low calcium content of CTWD and FA precursors. Likewise, the reduction in GBBS content from 70% ($AAMs_1$) to 10% ($AAMs_{16}$) led to a weakening in the CS from 66.67 to 15.66 MPa, respectively, for 28 days. Generally, the CS increases due to the process of hydration. The decrease in CS is due to several factors, such as alkali activation and low reactivity of CTWD and FA than GGBS [24].

3.5 Flexural Strength (FS)

Figure 5 presents the FS results of ternary blended AAMs. The flexural strength has been reduced with the effect of CTWD, which was reported to be 5.28–1.40 MPa at 28 days. In trail mixes, strength is significantly reduced due to FA and CTWD contents. At 50% CTWD, 10% FA, and 10% GGBS, lower strength of 1.40 MPa was observed. From this hypothesis, the method of exponential regression was applied to correlate the test results data given in Eq. (1), with R^2 values of 0.968, respectively (Fig. 6). These values are relatively insignificant, lies a higher confidence interval.

$$FS = 1.0288e^{0.0285x} \tag{1}$$

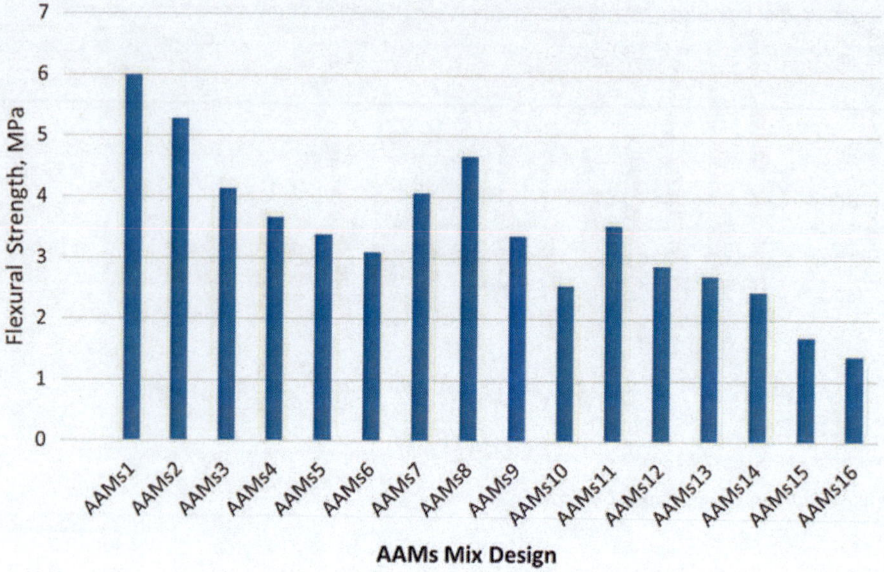

Fig. 5 Flexural strength of AAMs

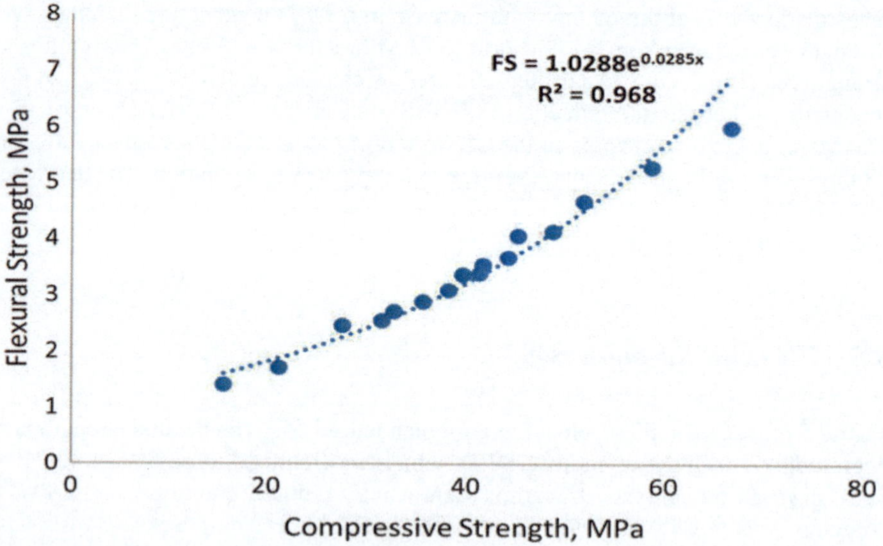

Fig. 6 Relationship between CS and FS of AAMs

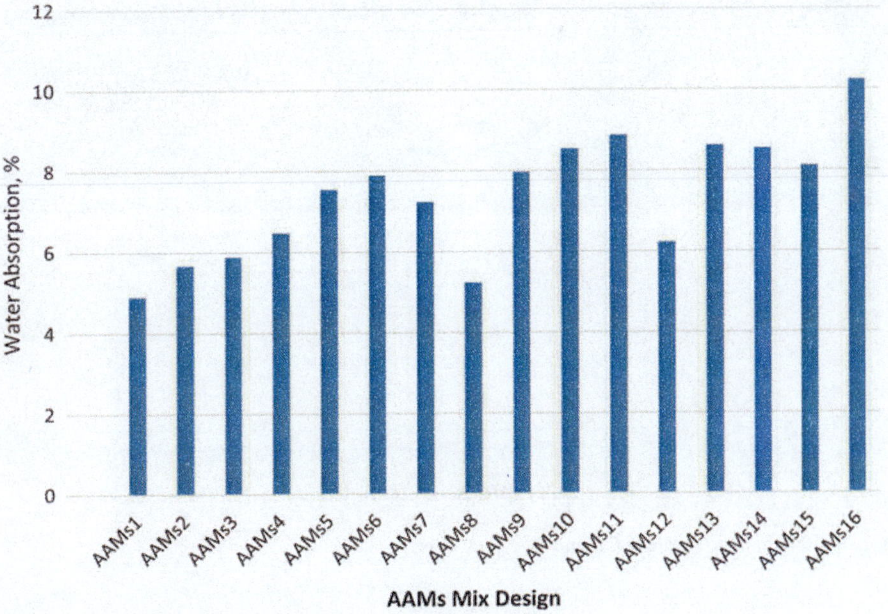

Fig. 7 Water absorption of ternary blended AAMs (28 days)

3.6 Water Absorption (WA)

From the study, the rate of increase in WA is due to effective change in the volume of CTWD content (Fig. 7). Variation in alternative precursors predicts a change in WA from 5.67 to 10.22% when it is compared to 4.89% of control mix. The amount of semi-partially reacted and non-reacted particle is due to CTWD content and FA, which would result in a more porous structure [24].

Figure 8 highlights the relationship between CS and WA of AAMs. The experimental study was correlated using the exponential regression approach given in Eq. 2.

$$\text{WA} = 12.57\text{e}^{-0.014x} \tag{2}$$

4 Conclusions

This research study investigates the evaluation of reusing CTWD to develop novel AAMs as sustainable alternative materials. From this current work, the following findings can be drawn are.

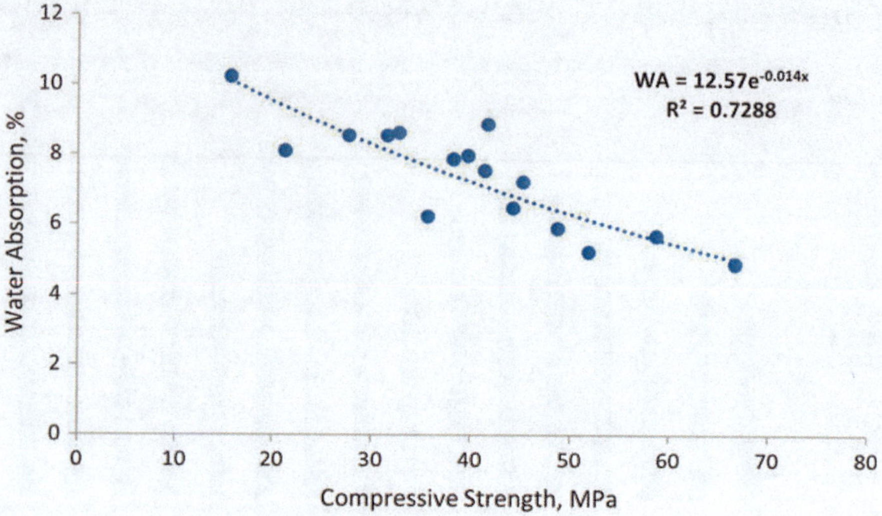

Fig. 8 Relationship between CS and WA of AAMs

- AAMs were made more workable by adding CTWD to GGBS and FA. Increasing the content of CTWD proportionally increases the setting times due to chemical composition and less rate of reactivity. At 40% of CTWD, significant change in setting time has been observed.
- Variation in volumes of precursors, shows the variation of compressive strength between 15MPa and 58 MPa at the age 28 days, it can be used for building applications. At $AAMs_2$, the Optimum compressive strength is 58 MPa for a mix proportion composition of 10% CTWD, 50% GGBS, and 10% FA in comparison to the higher volume of ceramic content with lower GGBS and FA.
- Lower percentage effect of water absorption was observed when compared to all AAMs with a volume of CTWD to control mix. This new type of AAM is are more environmentally friendly and affordable and emits fewer CO_2 emissions than traditional OPC mortars. It also helps reduce cost and energy as part of sustainability.
- Use of waste industrial byproducts in AAMs will help in addressing the challenges like, mineral resource scarcity, disposal problems, optimum usage, and environmental pollution. As a sustainability point of AAMs incorporated with industrial byproducts will make material ecofriendly.

References

1. Miller SA, Monteiro PJ, Horvath A (2017) Towards sustainable concrete. Nat Mater 16(7):698–699
2. Cen K, Wei J (2019) Empirical assessing cement CO_2 emissions based on China's economic and social development during 2001–2030. Sci Total Environ 653:200–211. https://doi.org/10.1016/j.scitotenv.2018.10.371
3. Collins FG, Turner LK (2013) Carbon dioxide equivalent (CO_2-e) emissions: a comparison between geopolymer and OPC cement concrete. Constr Build Mater 43:125–130. https://doi.org/10.1016/j.conbuildmat.2013.01.023
4. Singh RJ, Ashwin R, Murmu AL, Jameel M (2021) Influence of glass powder incorporated foamed geopolymer blocks on thermal and energy analysis of building envelope. J Build Eng 43:1–12. https://doi.org/10.1016/j.jobe.2021.102520
5. Singh NB (2018) Fly ash-based geopolymer binder: a future construction material. Minerals 8:299. https://doi.org/10.3390/min8070299
6. Thandiwe S, Tsotetsi N, Mashifana T (2021) Synthesis of ambient cured GGBFS based alkali activated binder using a sole alkaline activator: a feasibility study. Appl Sci 11:5587. https://doi.org/10.3390/app11135887
7. Guneet S, Uthej V (2020) Assessing properties of alkali activated GGBS based self-compacting geopolymer concrete using nano-silica. Case Stud Constr Mater 12:e00352. https://doi.org/10.1016/j.cscm.2020.e00352
8. Liang C, Zaiqin W, Yuanyi W, Jing F (2016) Preparation and properties of alkali activated metakaolin-based geopolymer. Materials 9:767. https://doi.org/10.3390/ma9090767
9. Asudullah Khan K, Raut A, Rama Chandrudu C, Sashidhar C (2021) Design and development of sustainable geopolymer using industrial copper byproduct. J Clean Prod 278:123565. https://doi.org/10.1016/j.jclepro.2020.123565
10. Singh S, Aswath MU, Ranganath RV (2020) Performance assessment of bricks and prisms: red mud based geopolymer composite. J Build Eng 32:101462. https://doi.org/10.1016/j.jobe.2020.101462
11. Mahmoodi O, Siad H, Lachemi M, Dadsetan S, Sahmaran M (2020) Development of ceramic tile waste geopolymer binders based on pre-targeted chemical ratios and ambient curing. Constr Build Mater 258:120297. https://doi.org/10.1016/j.conbuildmat.2020.120297
12. Chao-Lung H, Trong-Phuoc H (2015) Effect of alkali-activator and rice husk ash content on strength development of fly ash and residual rice husk ash-based geopolymers. Constr Build Mater 101:1–9. https://doi.org/10.1016/j.conbuildmat.2015.10.025
13. Jithendra C, Elavenil S (2020) Effects of silica fume on workability and compressive strength properties of aluminosilicate based flowable geopolymer mortar under ambient curing. SILICON 12:1965–1974. https://doi.org/10.1007/s12633-019-00308-0
14. Khan KA, Raut A, Chandrudu CR, Sashidhar C (2019) Mechanical and water transport performance of sustainable geopolymer composite using binary blends of pozzolanic. Mater J Tekno 81(5):139–147. https://doi.org/10.11113/jt.v81.13530
15. Senthamarai RM, Devadas Manoharan P (2005) Concrete with ceramic waste aggregate. Cem Concr Compos 27:910–913. https://doi.org/10.1016/j.cemconcomp.2005.04.003
16. Huseien GF, Mirza J, Ismail M, Ghoshal S, Hussein AA (2017) Geopolymer mortars as sustainable repair material: a comprehensive review. Renew Sustain Energy Rev 80:54–74. https://doi.org/10.1016/j.rser.2017.05.076
17. IS 2720:1985(Part 4). Methods of test for solids, p 3
18. IS 1727 (1967): Methods of test for pozzolanic materials
19. ASTM C1437-15 (2015) Standard test method for flow of hydraulic cement mortar
20. ASTM C191-08. Standard test methods for time of setting of hydraulic cement
21. ASTM C109 (2008) Standard test method for compressive strength of hydraulic cement mortars
22. ASTM C348 (2014) Standard test method for flexural strength of hydraulic-cement mortars

23. ASTM C20-00 (2015) Standard test methods for apparent porosity, water absorption, apparent specific gravity, and bulk density of burned refractory brick and shapes by boiling water. ASTM International. West Conshohocken. PA
24. Huseien GF, Sam ARM, Shah KW, Asaad MA, Tahir MM, Mirza J (2019) Properties of ceramic tile waste based alkali-activated mortars incorporating GBFS and fly ash. Constr Build Mater 214(30):355–368. https://doi.org/10.1016/j.conbuildmat.2019.04.154
25. Kumar S, Kumar R, Mehrotra SP (2010) Influence of granulated blast furnace slag on the reaction, structure and properties of fly ash based geopolymer. J Mater Sci 45:607–615. https://doi.org/10.1007/s10853-009-3934-5

Experimental Investigation on Mechanical Properties of Recycled Geopolymer Concrete Aggregates-Based Geopolymer Concrete (M40 Grade)

N. S. M. Ravi Kumar, Prince Kumar, and S. Venkateswara Rao

1 Introduction

Cement is considered as the second most used material after drinking water and it has been estimated that the world uses 30 billion tonnes of concrete each year [1]. Production of one tonne of cement emits around 1 tonne of CO_2 [1]. This situation may become critical if no remedial steps are taken to optimize the carbon emission from cement industries.

Many industrial by-products are used to partially or fully replace the Portland cement to reduce greenhouse gas emissions related to cement manufacture. The by-products commonly used are fly ash (FA) [2], ground granulated blast furnace slag (GGBFS) [3], stone waste [4], copper slag [2] etc. Agricultural residues can also be effective in replacing cement. Some of them are palm oil fuel ash [4], corn cob ash [5], rice husk ash [6, 7], wood waste ash [8] etc.

An alternative to conventional cement concrete in recent years is found to be geopolymer concrete. Geopolymer mainly contains silica and alumina in high percentage and it is way different from cement in production. Some activators are required to activate it, mainly in the form of alkalies. Higher pH value enhances the rate of geopolymerization, so alkalies are used to dissolve the solid alumino silicates by alkaline hydrolysis. As cement manufacturing produces a huge amount of carbon dioxide, these geopolymers have come with a great solution in this aspect of carbon-reduced environment [9].

Construction industries play a major role in the economic development of any nation. Due to the rapid growth in construction projects in developing nations, there is high demand to manage the disposable materials [1]. These disposable materials primarily come from either the construction or the demolition of various structures.

N. S. M. R. Kumar (✉) · P. Kumar · S. V. Rao
Civil Engineering Department, NIT-W, Hanamkonda 506004, India
e-mail: NS712014@student.nitw.ac.in

R. K. Pancharathi et al. (eds.), *Low Carbon Materials and Technologies for a Sustainable and Resilient Infrastructure*, Lecture Notes in Civil Engineering 440,
https://doi.org/10.1007/978-981-99-7464-1_29

Ministry of environment, forest and climate change (MoEFCC) has defined these "Construction and Demolition" (C&D) waste as disposable materials from repair and demolition work in the building sector [10]. In addition to this, the waste, which are generated from rehabilitation work of infrastructure projects, are also under C&D wastes [1]. Replacing natural aggregates with RCA partially or completely reduced the cost of disposal of C&D wastes and also helps in reusing the natural aggregates again as it still has potential left with it.

Studies have been done on the physical properties of the geopolymer concrete containing various fractions of RCA. A study reflected that the workability of the fly ash—GGBFS based geopolymer concrete moderately increased due to the increased proportions of recycled aggregate [11]. The slump value was observed to be reduced with the increment in the fraction of recycled aggregates [11, 12]. A small increase was seen in both the setting times of FA-GGBFS-based geopolymer concrete with the increasing fractions of recycled concrete aggregates over natural ones [11]. No much significant changes occurred up to 20% replacement level, but decreased when the replacement percentage was increased [13]. For various replacement (0, 50 and 100%) level of RCA, it was found that the young's modulus of concrete decreased significantly with increase in the RCA [12].

There are no reputable figures available regarding the percentage of the market that is occupied by GPC or the volume of C&D waste that is produced by GPC on an annual basis. But there has been a possibility of more people using GPC instead of PCC, which is supported by the literature [14]. When the first structures produced of GPC reach the end of their usable life in the next few decades, it is projected that the quantity of waste GPC will make up as a percentage of the total amount of rubbish concrete will rapidly rise. So, the construction industry is still figuring out how to use GPC and trying it out in small ways. It is important to pay close attention to the potential problems that could arise when GPC structures reach the end of their **usable** life. This work could be a key factor in deciding the recyclability of geopolymer concrete aggregates.

The present research contributes new information to our understanding of the application of recycled geopolymer concrete aggregates to M40 grade GPC based on FA and GGBFS. The characteristics of GPC made with various replacements of recycled geopolymer concrete aggregates (RCA) as coarse aggregates are experimentally investigated in this work. Mechanical properties such as compression, split tensile and flexural tensile strengths as well as workability, were studied experimentally in this work.

2 Experimental Program

In this work, experimental research is done on the characteristics of GPC (M40 grade) at different replacements of coarse aggregates (NCA) with RCA from geopolymer concrete as parent concrete at an interval of 25% till 100% replacement. The recycled geopolymer concrete aggregates are produced from the geopolymer concrete specimens after 120 days. Ambient curing was adopted for all the mixes.

2.1 Description of Raw Materials Used

GGBS and FA are used as blended binding materials in this study. FA (Class-F) that confirmed to IS: 3812-2013 was obtained from NTPC Ramagundam, India, GGBFS, according to IS:12089-1987, was acquired from AASTRA Chemicals, India. Table 1 shows the physical characteristics of the FA and GGBFS. Fine aggregates meeting the zone-II gradation requirements of IS:383-2016 were employed in this study. The coarse aggregates of maximum size 20 mm and the gradation for NCA and RCA were maintained the same confirming to IS:383-2016. According to IS:2386-1963 (Part I–V), the aggregates properties were examined, and the tested results are shown in Table 2.

A blend of liquid sodium silicate (SS) and 14M concentration of sodium hydroxide (SH) as alkaline solution were used at a proportion of 2.5:1 for developing M40 grade GPC. Sodium silicate (silica modulus of 2.19) and sodium hydroxide of purity 98% were procured from AASTRA Chemical Ltd. India. Conplast SP:430 Superplasticizer from Fosroc chemicals, India, was used at 4% (by weight) of binder content in

Table 1 Physical characteristics of GGBS and Fly ash

S. No	Property	Fly ash	GGBFS
1	Specific gravity	2.16	2.86
2	Blaine's fineness (m^2/kg)	381	412

Table 2 Mechanical characteristics of NCA, RCA and fine aggregates

Description of property	NCA	RCA	Fine aggregates
Specific gravity (G)	2.85	2.61	2.61
Aggregate crushing value (%)	18.5	28.55	–
Aggregate impact value (%)	9.16	29.87	–
Water absorption (%)	0.276	2.79	0.935
Los Angeles abrasion value (%)	24.5	35.44	–
Bulk density (kg/m^3)	1640	1390	1574
Attached old mortar content (%)	–	30.72	–

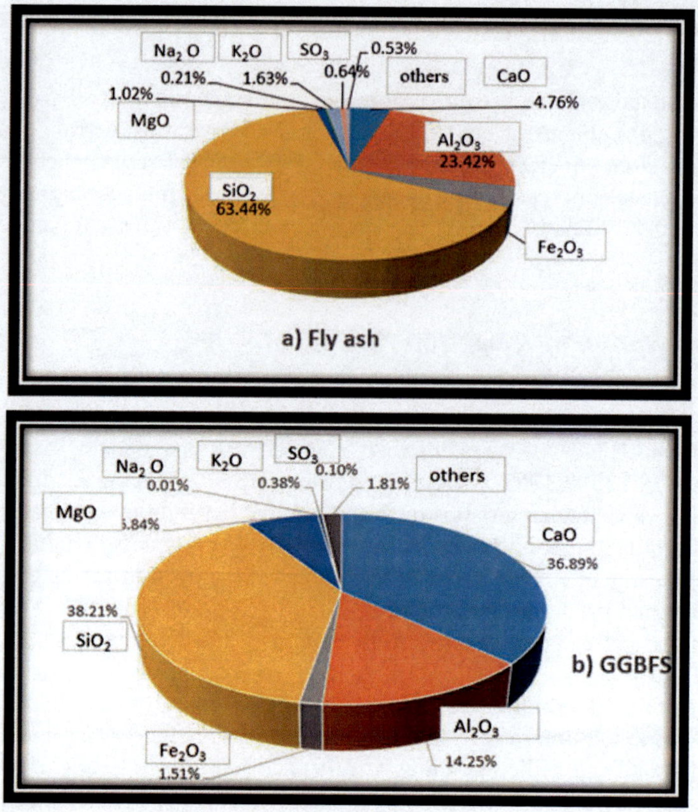

Fig. 1 **a** FA and **b** GGBFS chemical characteristics

all mixes. Figure 1 illustrates the chemical characteristics of the binding materials and Fig. 2 shows all the raw materials that were used for investigation in this work.

2.2 *Mix Proportions*

In this experimental investigation, mix design was prepared by referring to Llyod and Rangan approach [15]. The M40 grade GPC (with a target strength of 48.25 MPa at 28 days) was developed with a binder content of 450 kg/m^3 with equal proportions of FA and GGBFS and a solution–binder ratio of 0.45 at 14M NaOH. Extra water of 5% with binder content was used for workability point of view in all mixes. The behaviour of FA and GGBS-based GPC was studied at different replacements of NCA with RCA from GPC at an interval of 25% till 100% replacement. Table 3 shows the details of mix design particulars for NCA and RCA-based GPC. The conventional method of mixing procedure was adopted for M40 grade FA and GGBS-based GPC.

Fig. 2 Raw materials

Table 3 Details of Mix proportions in kg per cubic meter

MIX ID	GGBFS (kg/m^3)	FA (kg/m^3)	Alkaline solution/ Binder ratio	Alkaline solution (kg/m^3)	Fine Aggregates (kg/m^3)	NCA (kg/m^3)	RCA (kg/m^3)
RCA0	225	225	0.45	202.5	699	1049	0
RCA25	225	225	0.45	202.5	699	787	262
RCA50	225	225	0.45	202.5	699	524.5	524.5
RCA75	225	225	0.45	202.5	699	262	787
RCA100	225	225	0.45	202.5	699	0	1049

2.3 *Tests Conducted*

The fresh concrete was used to evaluate the workability and the hardened concrete specimens were used for mechanical characteristics. The codal provisions and the specimen sizes used for evaluating the influence of RCA on FA and GGBFS-based GPC were presented in Table 4.

Table 4 Specimens tested and their codal standards

Test conducted	Specimen size	Confirming to codal standards
Slump (workability)	–	IS:1199-1959
Compressive strength	15 × 15 × 15 cm cube	IS:516-1959
Split tensile strength	10 cm diameter × 20 cm height	IS:5816-1999
Flexural tensile strength	10 × 10 × 50 cm	IS:516-1959

3 Experimental Results and Discussions

The behaviour of FA and GGBS-based GPC of M40 grade with various replacements of recycled geopolymer concrete aggregates (RCA) to natural coarse aggregates (NCA) is experimentally investigated in this work. The experimental study was done on workability (slump) and mechanical properties, such as compressive strength, split tensile strength and flexural tensile strength and all test results are presented in Table 5.

3.1 Fresh Properties on GPC Concrete

To determine the workability of concrete, a slump test was performed according to IS:1199-1959. Figure 3 shows the effect of RCA replacement content on the workability of GPC. The GPC displayed a reduction in slump fall with increasing the content of RCA replacement.

Table 5 The combined results on effect of replacing NCA with RCA in GPC

Mix ID	Vertical slump (mm)	Dry density (kg/m^3)	Compressive strength (MPa)	Splitting tensile strength (MPa)	Flexural strength (MPa)
RCA0	33	2422	53.8	3.50	5.27
RCA25	31	2389	51.7	3.32	5.08
RCA50	29	2358	49.3	3.24	4.79
RCA75	27	2334	47.2	2.97	4.43
RCA100	23	2315	46.8	2.80	4.22

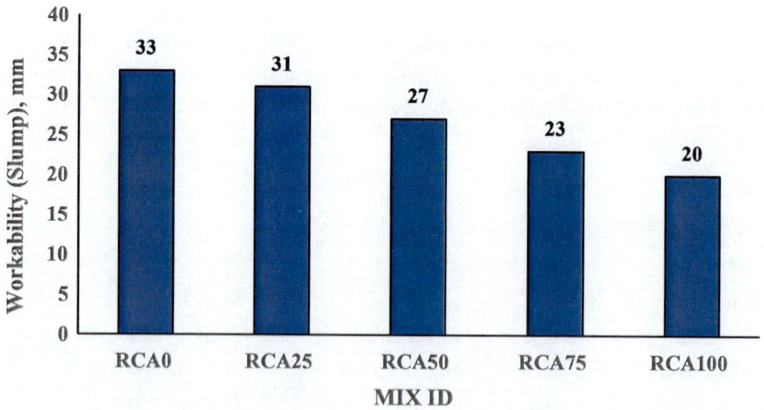

Fig. 3 Effect of RCA content on the slump of M40 Grade GPC

Fig. 4 Hardened concrete specimens used in this study

3.2 Hardened Properties of GPC

The mechanical behaviour was studied on hardened concrete samples. Figure 4 shows the specimens for investigating the mechanical behaviour.

Compressive Strength. The compressive strength was carried out using 3000 kN capacity of CTM as per IS: 516-1959. Figure 5 shows the specimens for compressive strength during testing in CTM at different RCA replacements. Figure 6 shows the influence of RCA replacement content on the compressive strength of M40 grade GPC. The target strength as per IS:10262-2019 for M40 grade of concrete is 48.25 MPa. From the results, the GPC compressive strength shown to be reduced with increasing the replacement of RCA, which was due to less crushing strength of RCA compared to NCA. The GPC mix with 50% RCA attained higher value than target strength i.e., 49.3 MPa. Hence, upto 50% replacement can be treated as feasible limit in FA and GGBS-based GPC.

Splitting Tensile Strength. After 28 days of ambient curing, the tensile strength of GPC concrete cylinders based on RCA was evaluated as per IS:5816-1999. Figure 7 shows the effects of substituting RCA for NCA in GPC on the tensile strength, and Fig. 8 shows the specimens after investigation of splitting tensile strength. The GPC splitting tensile strength shown to be reduced with increasing the replacement of RCA, with 100% RCA the reduction was about 20%.

Flexural Tensile Strength. After 28 days of ambient curing, the flexural tensile strength of GPC concrete prims was evaluated using 2-point loading as per IS:516-1959. The specimen after testing for rupture strength was shown in Fig. 9. The effect of replacing NCA with RCA on FA and GGBS-based M40 grade GPC was shown in Fig. 10.

The flexure tensile strength of GPC shown to be reduced with increasing the replacement of RCA, with 100% RCA the reduction was about 19%.

Fig. 5 Concrete specimens during compression test **a** RCA0, **b** RCA25, **c** RCA50, **d** RCA75 and **e** RCA100

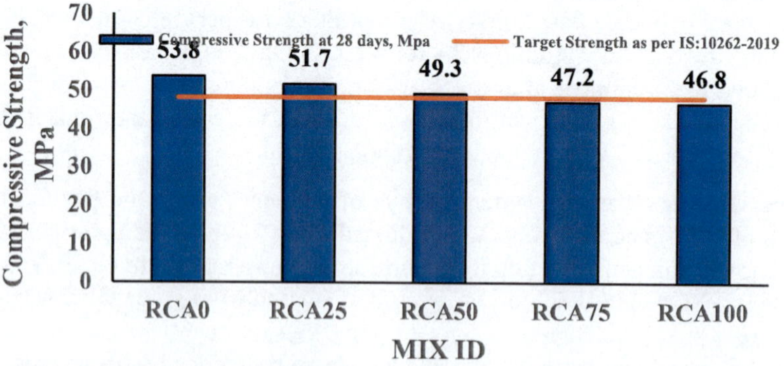

Fig. 6 Influence of RCA content on M40 Grade GPC

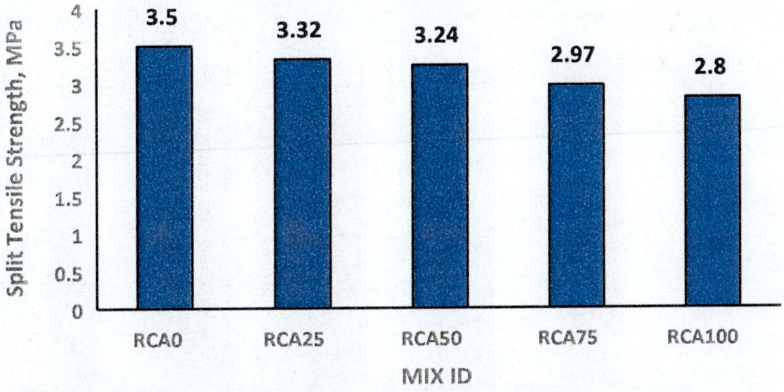

Fig. 7 Effect of RCA content on the split tensile strength of M40 Grade GPC

Fig. 8 Failed samples after splitting tensile strength test

Fig. 9 Flexure tensile testing of specimen

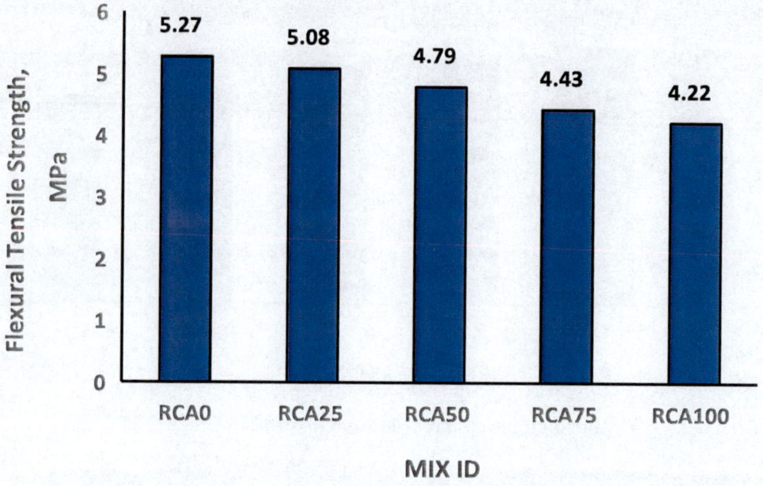

Fig. 10 Effect of RCA content on the flexure tensile strength of M40 Grade GPC

4 Conclusions

The experiments on GGBS-FA-based GPC mixes by integrating recycled aggregates are part of the current systematic investigations. At regular intervals of 25% till 100%, the recycled coarse concrete aggregates (RCA) were substituted for the conventional granite coarse aggregates (NCA). The precise findings that were reached in light of the workability and mechanical strength performance results were as follows:

- The workability of GPC was decrease as the replacement of NCA with RCA due to the additional water absorption by RCA.
- The workability reduced by 39%, upon replacing 100% natural coarse aggregate (NCA) with recycled concrete aggregate (RCA) in the GPC,
- Compressive and tensile strength of GPC shows that as the replacement level of RCA increases, the strength of the GPC mixes observed decreasing, due to lesser crushing strength of RCA.
- There is no much significant reduction was observed upon replacing NCA with 25% RCA on workability and mechanical properties for M40 grade GPC.
- The incorporation of RCA in GPC shows the satisfactory results up to 50% replacement content in GPC mixes.
- The splitting and flexural tensile strengths were reduced by 20% and 19%, respectively on strength of GPC upon replacing NCA with RCA completely.

These experimental findings, which primarily deal about the mechanical properties of the RCA-based GPC, demonstrate that a careful use of recycled geopolymer concrete aggregates, up to 50%, might be utilized for making the M40 grade GPC.

Acknowledgements The authors would like to thank the DST-SERB for financial assistance.

References

1. York IN, Europe I (2021) Concrete needs to lose its colossal carbon footprint. Nature 597:593–594. https://doi.org/10.1038/d41586-021-02612-5
2. Mehra P, Thomas BS, Kumar S, Gupta RC (2016) Jarosite added concrete along with fly ash: properties and characteristics in fresh state. Perspect Sci 8:69–71. https://doi.org/10.1016/j.pisc.2016.03.012
3. Zawrah MF, Gado RA, Feltin N, Ducourtieux S, Devoille L (2016) Recycling and utilization assessment of waste fired clay bricks (Grog) with granulated blastfurnace slag for geopolymer production. Process Saf Environ Prot 103:237–251. https://doi.org/10.1016/j.psep.2016.08.001
4. Ul Islam MM, Mo KH, Alengaram UJ, Jumaat MZ (2016) Durability properties of sustainable concrete containing high volume palm oil waste materials. J Clean Prod 137:167–177. https://doi.org/10.1016/j.jclepro.2016.07.01
5. Charitha V, Athira V, Jittin V, Bahurudeen A, Nanthagopalan P (2021) Use of different agro-waste ashes in concrete for effective upcycling of locally available resources. Constr Build Mater 285:122851. https://doi.org/10.1016/j.conbuildmat.2021.122851
6. Siddika A, Amin MR, Rayhan MA, Islam MS, Mamun MAA, Alyousef R, Amran YHM (2021) Performance of sustainable green concrete incorporated with fly ash rice husk ash and stone dust. Acta Polytech 61(1):279–291
7. Siddika A, Mamun MAA, Ali MH (2018) Study on concrete with rice husk ash. Innov Infrastr Solut 3(1). https://doi.org/10.1007/s41062-018-0127-6
8. Arunkumar K, Muthukannan M, Suresh A, Ganesh AC (2022) Hybrid fibre reinforced eco-friendly geopolymer concrete made with waste wood ash: a mechanical characterization study. Eng Appl Sci Res 49:235–247. https://doi.org/10.14456/easr.2022.26
9. Akbarnezhad A, Xiao JZ (2017) Estimation and minimization of embodied carbon of buildings: a review. Buildings 7(1):5. https://doi.org/10.3390/buildings7010005
10. Ministry of Environment, Forest and Climate Change (MoEFCC). http://www.moef.nic.in/sites/default/files/SWM20160
11. Hu Y, Tang Z, Li W, Li Y, Tam VWY (2019) Physical-mechanical properties of fly ash/GGBFS geopolymer composites with recycled aggregates. Constr Build Mater 226:139–151. https://doi.org/10.1016/j.conbuildmat.2019.07.211
12. Xie J, Wang J, Rao R, Wang C, Fang C (2019) Effects of combined usage of GGBS and fly ash on workability and mechanical properties of alkali activated geopolymer concrete with recycled aggregate. Compos Part B Eng 164. https://doi.org/10.1016/j.compositesb.2018.11.067
13. Shaikh FUA (2016) Mechanical and durability properties of fly ash geopolymer concrete containing recycled coarse aggregates. Int J Sustain Built Environ 5(2):277–287. https://doi.org/10.1016/j.ijsbe.2016.05.009
14. El-Naggar KAM, Amin ShK, El-Sherbiny SA, Abadir MF (2019) Preparation of geopolymer insulating bricks from waste raw materials. Constr Build Mater 222:699–705
15. Lloyd NA, Rangan BV (2010) Geopolymer concrete with fly ash. In: Second international conference sustainable construction materials technology, vol 7

Studies on Evaluation of Structural Properties of Mixed Multi-layered Recycled Plastic Lumber

Upasana Surya Kiran, M. N. Shariff, and U. M. Sulthana

1 Introduction

Plastic is one of the most useful manufactured materials to date. However, they are non-biodegradable and often lead to environmental pollution [1]. Up-scaling of plastic and reuse is not very common, especially with domestic plastic waste. To reduce the impact on landfills and use plastic waste in a useful manner, some up-scaling of plastic in the form of recycled plastic lumber (RPL) has been attempted in the past [2]. The use of RPL has been successfully demonstrated in manufacturing outdoor furniture which are subject to low stresses [3]. Recycled PET (Polyethylene Terephthalate) is used as geotextiles, roof insulations, and apparel [4]. Recycled plastic tiles have also been recently used in the construction of pavements in Bengaluru, India [5]. Recycled plastic aggregate has been extensively used in the manufacturing of sustainable concretes [6]. However, the quantity of recycled plastic consumed in concrete is only a fraction and hardly solves the problem at hand. One of the major drawbacks of RPL has been to maintain consistency in the material properties. Further, low compression and flexural modulus and low density seem to limit their applications in load-resisting structural forms. RPL inherently exhibits a visco-elastic behaviour leading to increasing deformations under sustained loads [7].

U. S. Kiran (✉) · M. N. Shariff
Department of Civil Engineering, Indian Institute of Technology Bombay, Powai, Mumbai, Maharashtra 400076, India
e-mail: upasana1.surya@gmail.com

M. N. Shariff
e-mail: shariff@iitb.ac.in

U. M. Sulthana
Department of Civil Engineering, National Institute of Technology Tiruchirappalli, Tiruchirappalli, Tamilnadu 610101, India
e-mail: smash@nitt.edu

R. K. Pancharathi et al. (eds.), *Low Carbon Materials and Technologies for a Sustainable and Resilient Infrastructure*, Lecture Notes in Civil Engineering 440,
https://doi.org/10.1007/978-981-99-7464-1_30

It is known that the properties of RPL are dependent on the thermal treatment they are subjected to. When RPL is subjected to higher temperatures, it has shown lesser strength in comparison to specimens subjected to cooler or sub-zero temperatures [8]. RPL is also susceptible to fire, as it has very low fire rating. Nevertheless, RPL has been used in marine pilings and corrosive environments and has shown durable performance [9].

Despite the drawbacks, RPL can potentially become an alternate structural construction material on a large scale through engineered solutions. The absence of standard codal provisions limits its use at a commercial scale. Recycled plastic lumber manufactured using multi-layered plastics (RPL-MLP) (collected through domestic waste) can be engineered as a sustainable material for developing structural members. To test the efficacy of this material, it is necessary to ascertain its mechanical properties.

In this present study, mixed multi-layered recycled plastic sourced from domestic waste is studied. The lumber elements are manufactured by commercial plastic manufacturers in Mumbai. The mechanical properties such as tensile strength, flexural strength, compression strength, flexural modulus, compression modulus, tensile modulus, shear strength, coefficient of thermal expansion, specific density, and exposure to high temperatures are studied through experiments.

2 Experimental Program

The main objective of this testing program is to understand the mechanical properties of RPL and ascertain its use in structural applications. The tests were carried out at Indian Institute of Technology, Bombay and the National Institute of Technology, Tiruchirappalli. All mechanical property tests were carried out in ambient conditions of 27 °C (± 3 °C).

2.1 Compression Tests

Using the RPL samples available, cubes of size 31 mm × 31 mm × 31 mm were cut and milled ensuring the specimen to be square and having straight edges, without any visible internal defects (cavities) as shown in Fig. 1. The samples were weighed and their densities evaluated. The compression tests were carried out in compliance with ASTM D695-15 standards [10] using a servo-controlled strain-based universal testing machine. The testing was conducted at a rate of loading corresponding to 0.5 mm/min. The test was carried out on three samples to verify the repeatability and consistency.

The failure pattern of the cube failed in compression due to crushing is shown in Fig. 2.

Fig. 1 Compression cube

Fig. 2 Failure pattern of cube in compression

The corresponding stress vs strain plots generated are shown in Fig. 3. The peak compressive stress is in the range of 22.5 to 28 MPa which can be attributed to the micro-voids present inside the sample. The strain at peak stress (ϵpk) lies in the range of about 0.12–0.25, whilst the strain at ultimate (ϵu) is 0.40–0.44. The initial stiffness (modulus of elasticity Ecomp) of the three samples is ranged between 280 and 353 MPa. Even though having a disadvantage in terms of low compression modulus, it can still be used in high-deformable structures with the help of appropriate investigations. Such materials find applications in high-seismic regions, where the drift demands can be significant. The addition of fillers and additives can further improve the compression modulus and density of the corresponding material used in the present study.

Fig. 3 Stress–strain curve of RPL under compression

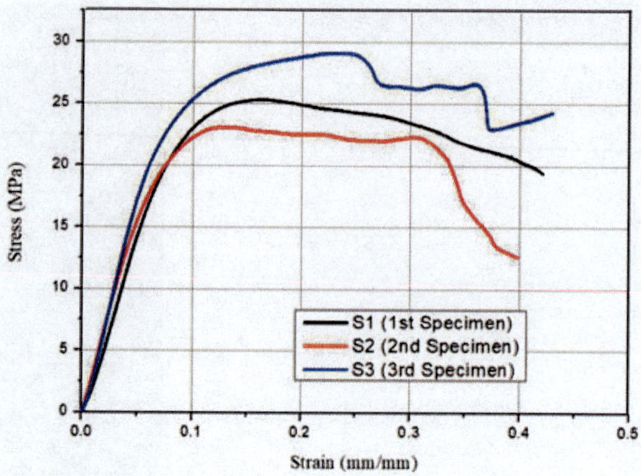

Fig. 4 Stress–strain curve of RPL under compression

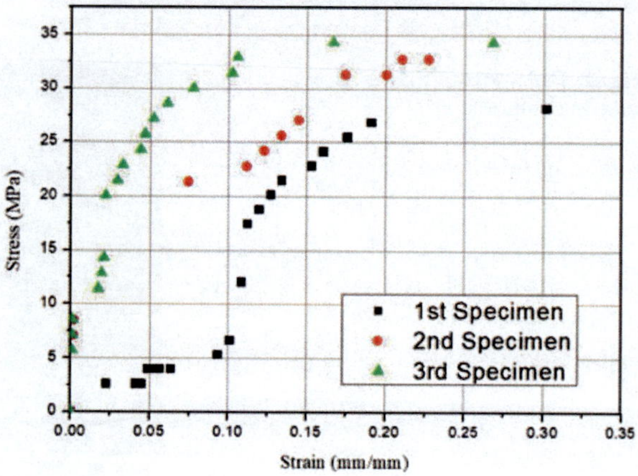

A parallel test was also carried out at NIT Trichy on a load-controlled UTM under the conditions of being thermally treated at 105 °C.

The corresponding stress vs strain plots generated are shown in Fig. 4. The peak compressive stress is in the range of 26–34 MPa, which can be attributed to micro-voids present inside the sample. The strain at peak stress (ϵpk) is in the range of about 0.17–0.22, whilst the strain at ultimate (ϵu) is about 0.22–0.3. This shows that the RPL member is highly deformable under compression.

2.2 *Tension Tests*

Efforts were done to conduct tension tests on dog-bone shaped samples cut out of RPL members following the relevant ASTM D638-22 standards [11]. The specimen had a cross-section size of 19 mm × 19 mm and gauge length of 50 mm as shown in Fig. 5. The tests were conducted using a displacement-controlled UTM at a rate of 5 mm/min as shown in Fig. 6.

There were challenges in carrying out the tension test on the UTM as the grips had to be prestressed to an optimal value without crushing the specimen. Rectangular mild steel plates were attached to the bulk heads of the specimen to avoid the crushing. After a few initial trials, the gripping pressure could be controlled, and the test was carried out.

The failure pattern of the dog-bone sample failed in tension is shown in Fig. 7.

The corresponding stress–strain plots generated are shown in Fig. 8. It was observed that the peak tensile stress is 3.0 MPa. The initial tensile stiffness (modulus of elasticity Eten) is about 210 MPa. The strain at peak tensile stress (ϵtpk) is about 0.025, whilst the tensile strain at ultimate (ϵtu) is 0.05. This shows that the RPL member is highly deformable under tension.

Fig. 5 Dog-bone shaped specimen for tensile test

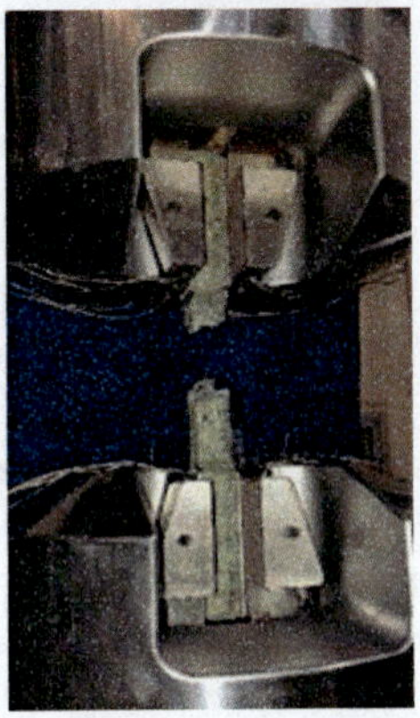

Fig. 6 Failure of dog-bone shaped specimen

Fig. 7 Failure pattern of dog-bone shaped Specimen in tensile test

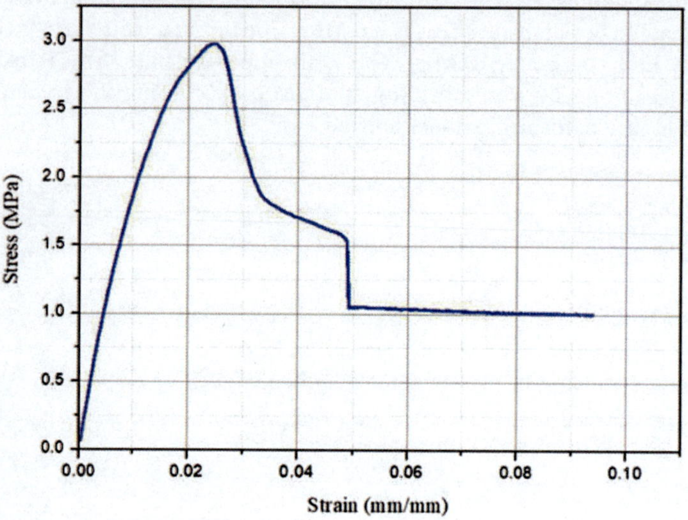

Fig. 8 Stress–strain graph of tension test

2.3 *Flexure Tests*

Flexure tests were conducted on RPL samples cut out of lumbers following the ASTM D6109-19 standards [12]. The specimens had a size of 600 mm × 120 mm × 30 mm as shown in Figs. 9a and b.

The specimens were tested in both joist and plank positions with joist position and plank position conduction of testing shown in Figs. 9a and b, respectively. The failure pattern of the flexure beam (joist position) is shown in Fig. 9c, whilst for the plank position it is shown in Fig. 9d. The tests were carried out in a load-controlled bending test machine with a rate of loading of 20 N/s.

The load–deflection plot of the specimen in joist and plank positions is shown in Figs. 10 and 11, respectively. The flexural strength for the corresponding material was determined to be about 11.08 MPa and 45.3 MPa for the specimen kept in plank and joist position, respectively. The flexural modulus Eflex for joist and plank positions ranges from 10.5 MPa to 12.9 MPa and 6.3 to 6.7 MPa, respectively.

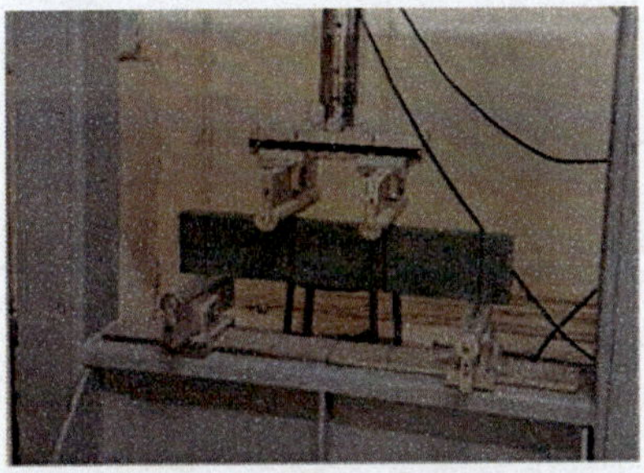

(a) Specimen kept in joist position

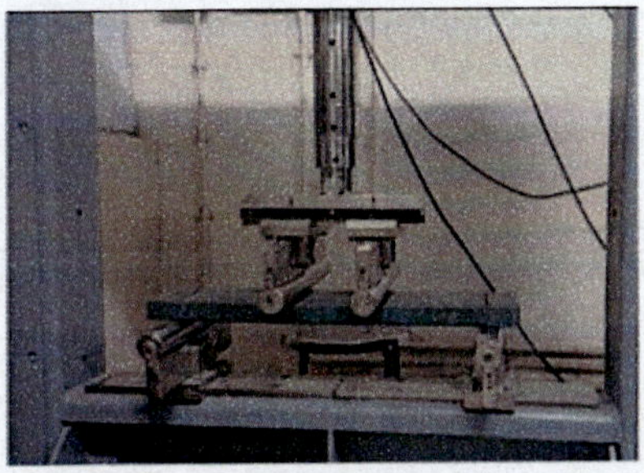

(b) Specimen kept in plank position

(c) Failure pattern of specimen in joist position

(d) Failure pattern of specimen in plank position

Fig. 9 Flexural tests

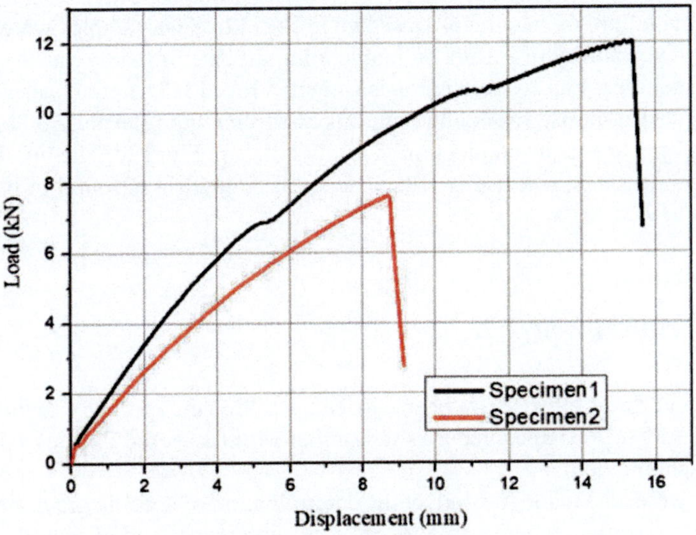

Fig. 10 Load–displacement graph in Joist position for 2 specimens

2.4 *Shear Tests*

Shear tests were conducted on RPL samples cut out of lumbers in accordance with ASTM D732-17 [13]. The specimens prepared were 50 mm × 50 mm (square shaped), having a thickness of 10 mm and a central hole of 11 mm diameter, to

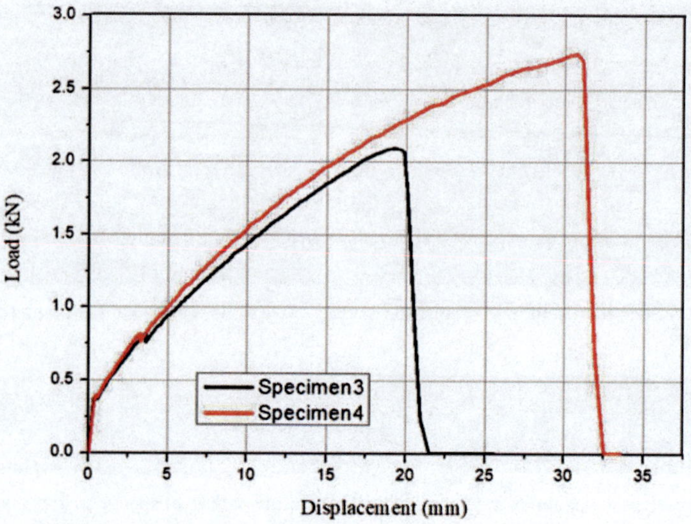

Fig. 11 Load–displacement graph in plank position for 2 specimens

facilitate the failure of specimens as shown in Fig. 12a. Five samples were tested to check the repeatability of results by maintaining the rate of loading as 0.5 mm/min and corresponding stress–strain curve is shown in Fig. 13. The punch shear toolbox was fabricated in-house for conducting this test following relevant ASTM code as shown in Fig. 12c and the conduction of test is shown in Fig. 12d. The failure pattern of the specimen is shown in Fig. 12b. The shear strength was found to be ranging from 7.47 to 9.31 MPa.

2.5 Specific Gravity

In the present study, efforts have been carried out to find the specific gravity and bulk density of RPL-MPL (Recycled Plastic Lumber-Multi-Layered Plastic) using water displacement method.

Efforts were also made to conduct the determination of specific gravity test using GP Thinner (specimens were able to sink fully in that instead of water) using the methodology described in ASTM D6111-19 standards [14], however, the results were not conclusive. Thus, results using that method were discarded.

The specimens used here were squared prisms of dimensions 60 mm × 30 mm × 30 mm as shown in Fig. 14.

A beaker was filled with water up to the mark of 700 ml. Then after sinking the RPL-MPL specimens in each case, the water mark increased to 750 ml. Thus, volume displacement in this case was found to be 50 ml for each specimen and accordingly,

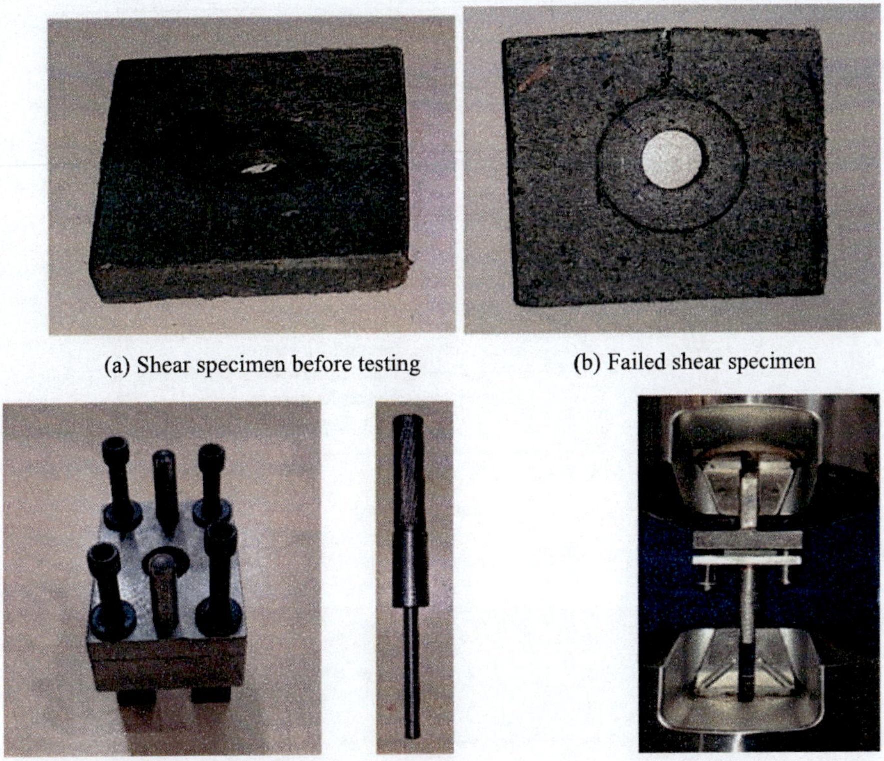

(a) Shear specimen before testing (b) Failed shear specimen

(c) Shear punch-tool setup box (d) Shear test in progress

Fig. 12 Shear tests

the weight of each specimen in water would be 50 g. The average specific gravity was found to be as 1.03 and detailed readings are tabulated as shown in Table 1.

2.6 *Coefficient of Thermal Expansion*

Linear coefficient of thermal expansion (α) was determined after keeping six recycled plastic lumber specimens having dimensions of 60 mm $\times$ 30 mm $\times$ 30 mm as shown in Fig. 14 at the temperatures prescribed in ASTM D6341-21 standards [15] for not less than 48 h. It was observed that α varied significantly between the specimens as shown in Table 2.

The results of all tests carried on RPL are summarised in Table 3.

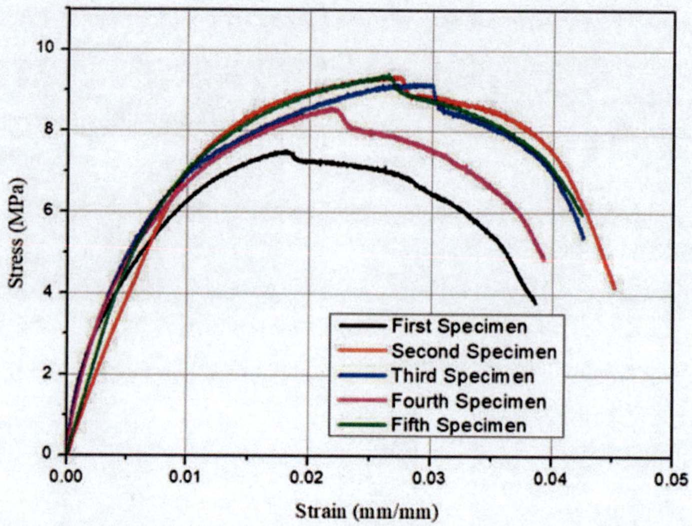

Fig. 13 Stress–strain graph for shear tests of 5 specimens

Fig. 14 Squared Prism shaped recycled plastic lumber specimens

Table 1 Determination of specific gravity

S. No. (specimens)	Weight of specimen in air (g)	Volume displacement = weight of specimen in water (g)	Specific gravity (weight of specimen in air/ weight of specimen in water)	Bulk density (kg/ m^3)
1	53.0	50	1.060	981
2	51.8	50	1.036	959
3	51.9	50	1.038	961
4	50.9	50	1.018	942
5	49.9	50	0.998	924

Table 2 Linear coefficient of thermal expansion

Specimen No	Initial length (mm)	Temperature (°C)	Changed length (mm)	Linear coefficient of thermal expansion $\frac{1}{L_0} \times \frac{(L_2-L_1)}{(T_2-T_1)}$
1	60.6	− 34	60	9.104×10^{-4}
2	62.7	24	63.2	4.124×10^{-4}
3	62	− 34	61.7	9.171×10^{-4}
4	61.5	60	61.7	2.258×10^{-4}
5	61.1	24	61.2	2.273×10^{-4}
6	60.1	60	60.3	4.068×10^{-4}

Table 3 Combined test results

Properties	Values
Compressive strength (MPa)	25.86
Compressive modulus of elasticity (MPa)	325.00
Tensile strength (MPa)	2.98
Tensile modulus of elasticity (MPa)	208.00
Shear strength (MPa)	8.68
Specific gravity	1.03
Linear coefficient of thermal expansion (α)	4.33×10^{-4}
Flexural modulus for joist position (MPa)	11.71
Flexural modulus for plank position (MPa)	6.53
Flexural strength for joist position (MPa)	45.38
Flexural strength for plank position (MPa)	11.08

3 Conclusion

The points listed below were drawn in the form of conclusions from the present study.

- The compressive strength of the RPL specimens used in this study was about 25 MPa, which is comparative to normal strength concrete. Hence, the material can be used to resist compressive forces when acted upon it. However, the initial compression modulus was found to be only about 325 MPa, which is significantly lesser than concrete having similar compressive strength. This implies that the material would undergo larger deflections when subjected to the same load.
- The tensile strength of the RPL specimens used in this study was about 3 MPa, which is comparative to normal strength concrete.
- The flexural and shear strength of the RPL specimens used in this study was about (45 MPa for joist position and 11 MPa for plank position) and 9 MPa respectively.

- The material can be used in constructing low stress structural elements, where deflections are not a concern. However, more tests are needed to confirm its fire resistance and bonding with reinforcement. This material can be then used in constructing structural members.

Acknowledgements The authors thank the lumber plastic material supplied by Shakti Plastics, Mumbai.

Conflict of Interest The authors declare no conflict of interest.

References

1. Kumar S, Panda AK, Singh RK (2011) A review on tertiary recycling of high-density polyethylene to fuel. Resour Conserv Recycling 55(11):893–910
2. Baishya P, Jain A, Bora M P, Goswami K (2022) Reduction of groundwater contamination by converting plastic waste to plastic lumber
3. Nosker TJ, Renfree R (2000) Recycled plastic lumber: from park benches to bridges. In: Approved for proceedings of R 2000 5th world congress, Toronto, Canada
4. Parkhill LD (1992), Viability of the current market for geotextiles manufactured with post-consumer recycled polyethylene terephthalate, the 1990–1999-Mines Theses & Dissertations
5. Shanker R, Khan D, Hossain R, Islam MT, Locock K, Ghose A, Sahajwalla V, Schandl H, Dhodapkar R (2022) Plastic waste recycling: existing Indian scenario and future opportunities. Int J Environ Sci Technol 1–18
6. Jin R, Chen Q (2013) An investigation of current status of green concrete in the construction industry. In: 49th ASC annual international conference proceedings, pp 1–8
7. Chen CW, Salim H, Bowders JJ, Loehr JE, Owen J (2007) Creep behaviour of recycled plastic lumber in slope stabilization applications. J Mater Civil Eng 19(2):130–138
8. Carroll DR, Stone RB, Sirignano AM, Saindon RM, Gose SC, Friedman MA (2001) Structural properties of recycled plastic/sawdust lumber decking planks. Resour Conserv Recycling 31(3):241–251
9. Guades E, Aravinthan T, Islam M, Manalo A (2012) A review on the driving performance of frp composite piles. Compos Struct 94(6):1932–1942
10. ASTM (2010) Test method for compressive properties of rigid plastics. Int ASTM 10
11. ASTM (2014) Test method for tensile properties of plastics. Int ASTM
12. ASTM (2013) Standard test methods for flexural properties of unreinforced and reinforced plastic lumber and related products. ASTM Int (2013)
13. ASTM (2017) Standard test method for shear strength of plastics by punch tool. Int ASTM
14. ASTM (2019) Standard test method for bulk density and specific gravity of plastic lumber and shapes by displacement. Int ASTM
15. ASTM (2016) Standard test method for determination of the linear coefficient of thermal expansion of plastic lumber and plastic lumber shapes between –30 and 140°f (–34.4 and 60°c). Int ASTM

Sustainable Technologies in Structural Design and Construction

Experimental and Numerical Investigation on Stress–Strain Relationship of Graphene Oxide and Fly Ash-Based Concrete Under Axial Compression

P. V. R. K. Reddy and D. Ravi Prasad

1 Introduction

Cement is the most significant and commonly used building material, which is a major component of both mortar and concrete. A significant method for enhancing the rheological, strength, durability, and other characteristics of cement composites is the introduction of admixtures [1] and fibres [2]. Cement paste performance has recently been improved due to the implementation of nanomaterials like carbon nanotubes (CNTs) and graphene [3, 4]. Several investigations on the impact of CNTs on hydration and strength characteristics of cementitious composites have been performed [3]. The flexural and compressive strengths were enhanced by the incorporation of small quantity of CNTs into cement composites [5]. The primary CNT-involved reactions in cement matrix include the nucleation effect, the filling of microscale capillary pores, and nanoscale gel pores. However, due to their intense van der Waals forces and high aspect ratios, CNTs frequently form CNT bundles [6]. Additionally, it has been stated that weak matrix bonding is indicated by CNTs sliding away from the cementitious matrix when under tension [7]. As a result of their low dispersion and weak interaction with the cementitious matrix, CNTs have only had a limited impact on the characteristics of cementitious materials.

Graphene oxide (GO), a two-dimensional graphene derivative, has a variety of oxygenated functionalities, such as carboxyl and carbonyl groups located at the sheet edges, and hydroxyl and epoxy groups on their basal planes, which make GO sheets

D. R. Prasad
Department of Civil Engineering, National Institute of Technology, Warangal, India

P. V. R. K. Reddy (✉)
Department of Civil Engineering, Anil Neerukonda Institute of Technology and Sciences, Visakhapatnam, India
e-mail: pvrkreddy20@gmail.com; pvrkreddy.ce@anits.edu.in

R. K. Pancharathi et al. (eds.), *Low Carbon Materials and Technologies for a Sustainable and Resilient Infrastructure*, Lecture Notes in Civil Engineering 440,
https://doi.org/10.1007/978-981-99-7464-1_31

hydrophilic [8]. Owing to the existence of these oxygenated functionalities, GO can easily produce stable dispersion in water that is generally composed of 1-nm-thick sheets, unlike CNTs. When it comes to mechanical characteristics, the modulus of elasticity and tensile strengths of GO are about 32 GPa and 130 MPa, respectively, values that are higher than that of cementitious materials [8]. Owing to its large surface area and excellent mechanical characteristics, GO is increasingly being used in a variety of applications [9]. The ability to produce massive quantities of GO from cheap graphite powder is very significant. Because GO offers an additional dimension of interface with cement matrix, it may be preferable to 0D nanoparticles and 1D nanotubes for changing a variety of matrix parameters such as mechanical, rheological, and permeability characteristics [10]. As a result, compared to cement composites reinforced with CNT, GO exhibits a higher growth in compressive strength at a small content. The addition of GO to cement can control hydration as well as enhances flexural, tensile, and compressive strengths of cement paste [4, 11, 12]. The cement composites showed a significant growth in flexural strength (61%), tensile strength (79%), and compressive strength (39%) when the dosage of GO was 0.03% [11]. However, the incorporation of GO into cement influences the workability and enhanced rheological characteristics, comparable to other nanomaterials like CNTs and nano-silica [4, 12]. Mini-slump test with the addition of GO at 0.20 weight percent revealed that the workability was reduced by about 19% [12].

However, few studies have examined how GO affects the behaviour of cement concrete and the majority of previous works have emphasized on microstructural and mechanical behaviour of cement paste and cement mortar with the incorporation of GO. The combined effect of fly ash and GO on compression behaviour of cement concrete is a vital question that has not yet been investigated. The combined effect of fly ash and GO on the compression behaviour of cement concrete has been investigated in this study. Additionally, the finite element-based ATENA-GiD software was used to create the nonlinear numerical model with the experimental results to simulate the uniaxial compression behaviour. This study aims to present a thorough understanding of the compression stress–strain behaviour of concrete with the combination of GO incorporation and fly ash replacement.

2 Experimental

2.1 *Materials*

The materials employed in the present experimental study are ordinary Portland cement (OPC) of 53grade conforming to IS:269, fly ash of Class F conforming to IS:3812(Part-1), Polycarboxylate ether-based superplasticizers (PCE) conforming to IS:9103, sand and coarse aggregate (CA) conforming to IS:383.

The technical characteristics of the GO, which has been utilized in this study as a nanomaterial, are presented in Table 1. The concentration of GO solution was maintained at 4 g/L, and ultrasonication was used to disperse it in water. In order to characterize GO, SEM, EDX, FTIR, and XRD studies were conducted and the findings are presented in Fig. 1. It can be seen in the scanning electron micrograph of GO in Fig. 1a, the surface morphology is folded and wrinkled. The composition of elements in GO is shown in Fig. 1b, and it consists of 75.28% carbon, 24.38% oxygen, and 0.34% sulfur. FTIR spectra shown in Fig. 1c provide evidence that the surface of GO possesses oxygenated functions. The XRD pattern shown in Fig. 1d illustrates that the diffraction peak appeared at an angle of $2\theta = 11.9°$ with the spacing between the layers was 0.74 nm.

Table 1 Technical characteristics of GO

Purity	Number of layers	Average lateral dimension	Thickness	Surface area	Bulk density
99%	1–4	5–10 μm	0.8–2 nm	100–240 m^2/g	0.124 g/cm^3

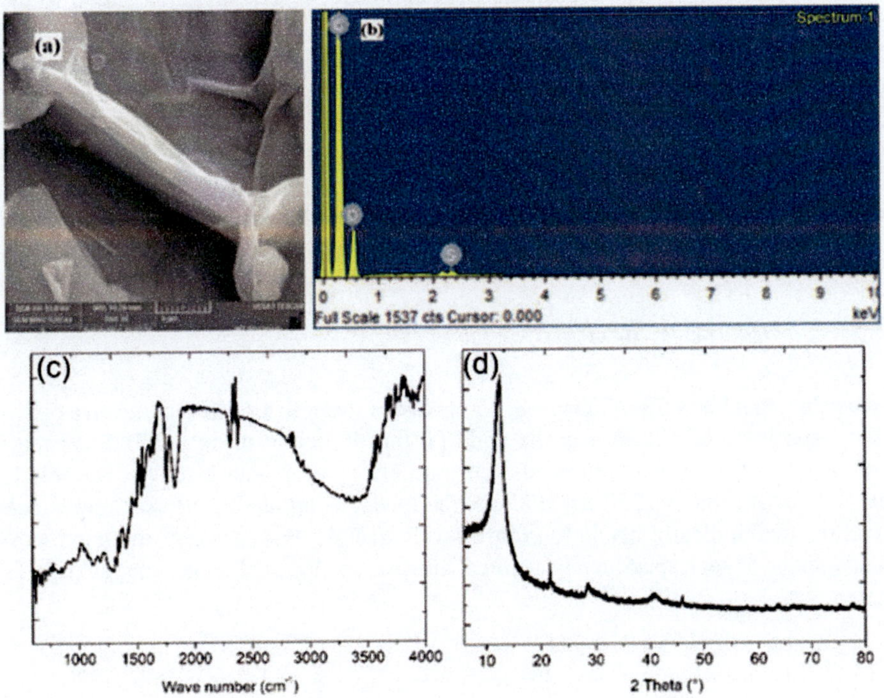

Fig. 1 Microstructure of GO **a** SEM, **b** EDX, **c** FTIR, **d** XRD

Table 2 Mix proportions of concrete per m^3

S. No	Mix	W/C ratio	PCE (%)	GO (kg)	Fly ash (kg)	OPC (kg)	Sand (kg)	CA (kg)
1	CC	0.45	0.6	0	0	375	840	1060
2	CG	0.45	0.6	0.56	0	375	840	1060
3	CGF10	0.45	0.6	0.56	37.5	337.5	840	1060
4	CGF20	0.45	0.6	0.56	75.0	300	840	1060
5	CGF30	0.45	0.6	0.56	112.5	262.5	840	1060

2.2 *Concrete Specimen Preparation*

The mix proportions of concrete considered in the present investigation are shown in Table 2 and are in compliance with IS: 10262–2019 standard. Water to cementitious material ratio (w/c) in all mixes was 0.45. Five concrete mixes were considered and designated as CC, CG0, CGF10, CGF20, and CGF30. The GO was added to the concrete at 0.15% and the fly ash at different percentage replacements of 0%, 10%, 20%, and 30% by cement weight. The overall water content, including the mixing water and GO dispersion in water, was 168.75 kg/m^3. To obtain a uniform mix, all the concrete mix constituents were mixed for about 4–6 min in a power-driven mixer. The fresh concrete mix is poured into moulds and vibrated using an electrical vibrating table. Concrete moulds were then wrapped and cured for 24 h. Subsequently, the specimens were taken out and cured at a standard condition until they were tested.

2.3 *Compressive Strength*

According to ASTM C39/C39M-18, compressive strength test is executed on a cylindrical specimen of 100 mm in dia and 200 mm in height using a 200 T capacity compression testing machine at the curing periods of 7 and 28 days. The test is carried out at a rate of 250 $\pm$ 50 kPa/s, the load was applied gently without shock until the load indicator began to progressively decline. For each test, the maximum load applied to each specimen was noted, and average load value of 3 specimens was noted.

2.4 *Stress–Strain Behaviour*

A uniaxial compression test was carried out on a cylindrical specimen with dimensions of 100 mm in dia and 200 mm in height to assess the stress–strain behaviours of concrete mixes. At a rate of loading 0.25 $\pm$ 0.05 MPa/s, the load has been applied monotonously without shock. A data acquisition system that has been coupled to a load cell and LVDTs was used to measure the specimen's longitudinal deformations in response to the load applied.

2.5 *Static Modulus of Elasticity*

Static modulus of elasticity was evaluated as per ASTM C469-14. This test procedure is developed to evaluate the elastic chord modulus of concrete cylinders that have received longitudinal compressive stress. The ratio of the change in compressive stress and longitudinal strain at two places on the stress–strain curves has been used to obtain elastic chord modulus. The first point is considered at 0.000050 mm/mm longitudinal strain, whilst the second point at a load of 40% of the ultimate load. The average of three identical specimens was taken into account.

2.6 *Numerical Modelling*

A finite element software was used to perform the nonlinear numerical modelling. In the current investigation, ATENA-GiD software was utilized to simulate the stress–strain relation of concrete with GO and fly ash. GiD is an interactive tool for geometric modelling that also acts as the data source for ATENA analysis. The material characteristics of a model were derived from the findings of experiments. The behaviour of concrete produced with GO and fly ash was modelled as nonlinear with softening portion under uniaxial stress.

3 Result and Discussions

3.1 *Compressive Strength*

Compressive strength of concrete samples CC, CG, CGF10, CGF20, and CGF30 at 7 and 28 days is shown in Figs. 2 and 3. Compressive strength of all samples of CC, CG, CGF10, CGF20, and CGF30 increased with the curing age. It can be revealed that the compressive strength of CG was greater than the CC mix at 7 and 28 days. Furthermore, growth rate in strength at 7 days was higher than 28 days. These findings are consistent with relevant research findings [9, 12, 13]. The compressive strength of CG mix was enhanced by 58.2% and 47.7% compared to the CC mix. The oxygenated functional groups that are connected to GO nanosheets remain unchanged due to the fact that GO functions as a catalyst and contributed to the acceleration of the cement hydration process. These oxygen functionalities serve as active locations for cement particle attraction. Thus, the presence of GO enhanced the cement composite at the nano-level, resulting in the compressive strength improvement of concrete composite incorporated with nano-GO compared to the CC mix [9, 12]. Replacing cement with fly ash was diminished compressive strength of the CGF10, CGF20, and CGF30 mixes compared to the CG mix. However, compressive strength in CGF10, CGF20, and CGF30 mixes was improved compared to that in CC mix at curing period of 7 and 28 days. The improvement of compressive strength in CGF10, CGF20, and CGF30 compared to CC was 51.6%, 45.4%, and 20.7% at 7 days, and 41.9%, 35.9%, and 13.8% at 28 days, respectively. The disadvantage of fly ash in delaying strength improvement at early stage was offset by GO [14].

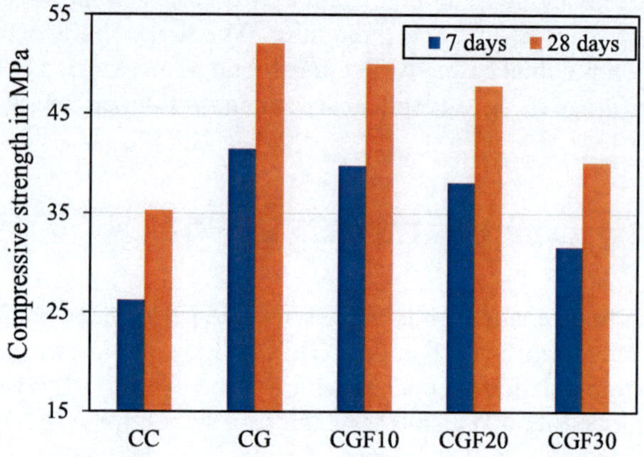

Fig. 2 Results of compressive strength

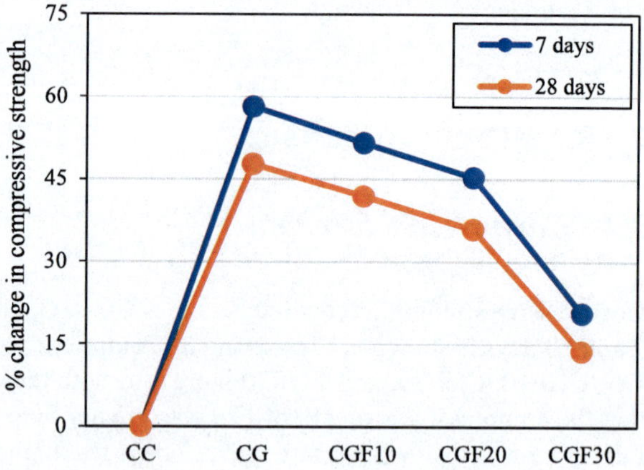

Fig. 3 Percentage change in compressive strength

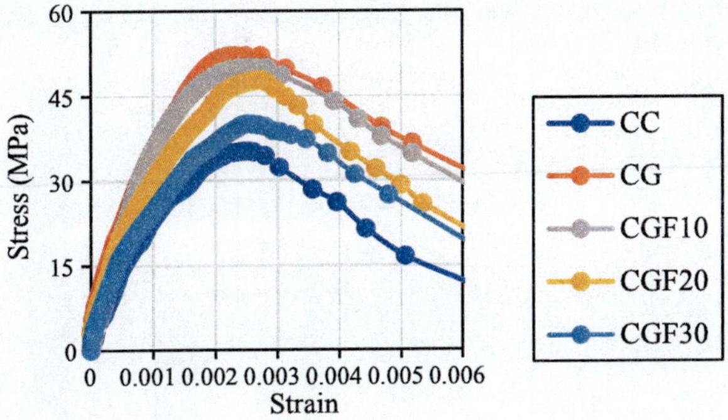

Fig. 4 Stress–strain relationship

3.2 *Stress–Strain Behaviour*

The stress–strain curves of all concrete mixes at a curing period of 28 days were shown in the Fig. 4. The stress–strain plots were plotted by considering the average values of the three similar specimens. The stress–strain plots show that the incorporation of GO has a significant influence on the stress–strain relationship. This can be due to the GO nanosheets, which have high surface area to weight ratio, the nucleation process for the production of hydration products is accelerated, and strong covalent connections are formed at the interface between the cement matrix and the dispersion [9, 12]. However, the CGF10, CGF20, and CGF30 mixes containing cement replacement with fly ash along with the incorporation of GO exhibited the improvement in stress–strain behaviour compared to the CC mix. The findings demonstrated that appropriate quantities of fly ash replacement and GO addition improved the stress–strain behaviour of concrete mixes [10, 14, 15].

3.3 *Static Modulus of Elasticity*

The elastic modulus of various concrete mixes was shown in Fig. 5. It can be seen that the CG mix showed highest improvement in static elastic modulus at 28 days curing compared to CC mix. Several investigations on the impact of GO addition in cementitious composite have shown that GO enhances static modulus of elasticity whilst improving mechanical interlocking of the cement matrix interface reinforced with GO [9, 12]. The improvement in static elastic modulus of CG, CGF10, CGF20, and CGF30 mixes was 28.7%, 26.1%, 21.2%, and 8.7%, respectively, compared to the CC mix. Because of its reinforcing effect on concrete strength and elastic modulus, GO is a potential reinforcement in concrete composites [9–14].

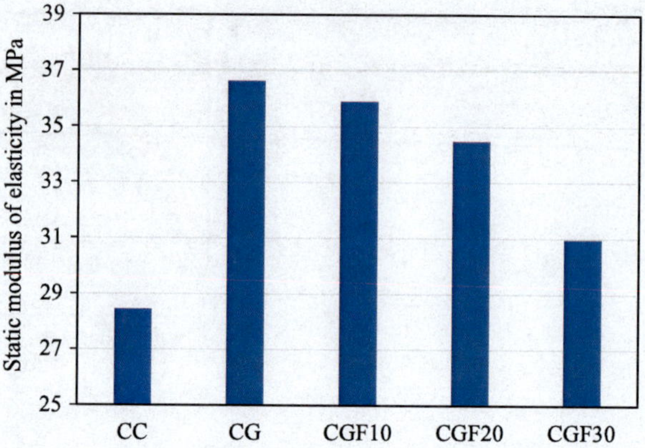

Fig. 5 Static modulus of elasticity

3.4 *Numerical Modelling*

Finite Element Method (FEM) is a powerful tool for simulating the behaviour of concrete. Experimental data was used to create and evaluate a finite element model of a CGF cylinder. The ATENA-Gid software tool was used to generate a finite element model of the specimen. Hexahedron and tetrahedron elements were used to develop the FEM model in ATENA-Gid. The concrete was represented by the hexahedron element, whilst the elastic support was represented by the tetrahedron element. The material properties obtained from the experimental results were used to simulate the cylinder. The details of the discretized model are shown in Fig. 6. Stress–strain behaviour of concrete mixes using numerical analysis is shown in Fig. 7. The strength values of experimental and numerical analysis are compared and presented in Table 3. The numerical analysis results are consistent with the experimental findings with less than 15% error.

4 Conclusions

The present study investigated the incorporation of GO at constant dosage with various percentage replacements of cement with fly ash in developing a concrete composite. The following are the conclusions of the present investigation.

1. The developed concrete composite has remarkably increased the compressive strength by 47.7–13.8% with constant GO at 0.15% and the fly ash replacements at 0–30%.
2. The combined effect of GO and fly ash has shown improvement in stress–strain behaviour concrete composite compared to control concrete mix.

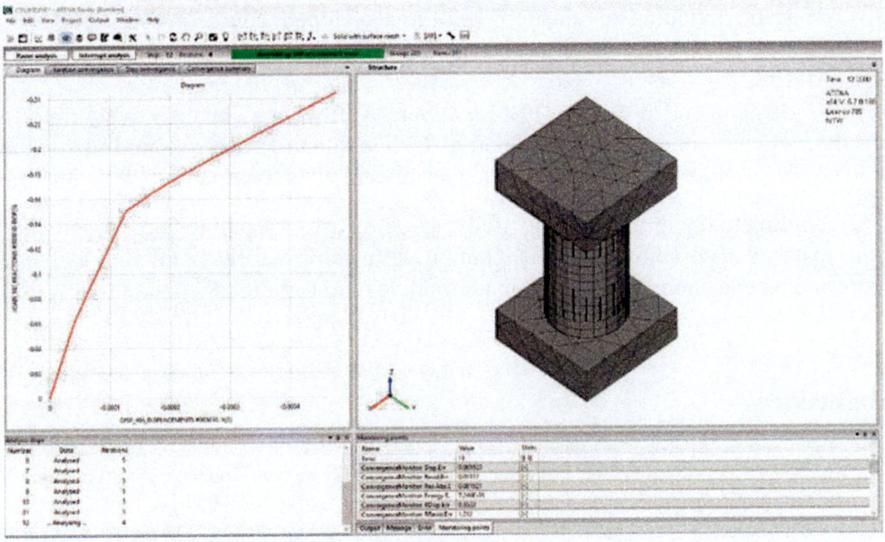

Fig. 6 The details of the discretized model

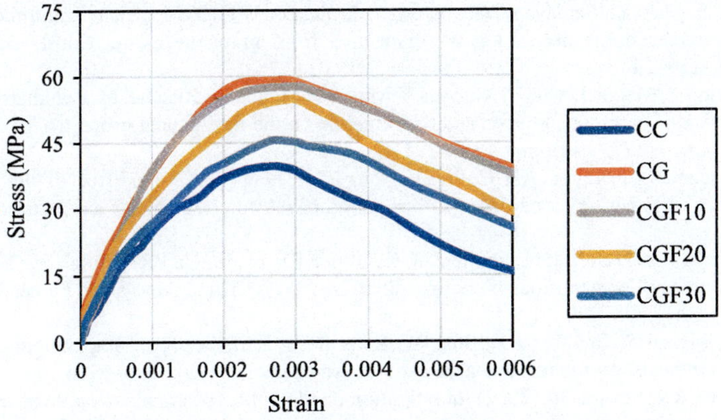

Fig. 7 Stress versus strain of all concrete mixes from ATENA

Table 3 Comparison strength values of experimental and FEM model

S. No	Mix	Experimental	ATENA	Error
1	CC	35.30	39.76	12.64
2	CG	52.10	59.09	13.42
3	CGF10	50.00	57.66	15.32
4	CGF20	47.89	54.92	14.68
5	CGF30	40.10	45.67	13.89

3. The static modulus of elasticity of the developed concrete composite was enhanced by 28.7–8.7% with constant GO at 0.15% and the fly ash replacements at 0–30%.
4. It is also concluded that the FEM software ATENA may be effectively used in the validation of experimental findings with less than 15% error, minimizing the resources and effort needed in the experimental investigation.

According to the findings of this study, using fly ash as a supplementary cementitious material in combination with GO improves mechanical characteristics and may therefore be considered a sustainable alternative to conventional concrete.

References

1. KumarVVP PrasadDR (2019) Influence of supplementary cementitious materials on strength and durability characteristics of concrete. Adv Concr Constr 7(2):75–85
2. SridharRK PR (2019) Experimental and numerical study on damage evaluation of hybrid fiber-reinforced concrete. Asian J Civ Eng 20(5):745–758
3. Chen SJ, Collins FG, Macleod AJN, Pan Z, Duan WH, Wang CM (2011) Carbon nanotube-cement composites: a retrospect. The IES J Civil Struct Eng 4(4):254–265
4. Chuah S, Pan Z, Sanjayan JG, Wang CM, Duan WH (2014) Nano reinforced cement and concrete composites and new perspective from graphene oxide. Constr Build Mater 73(30):113–124
5. Sobolkina A, Mechtcherine V, Khavrus V, Maier D, Mende M, Ritschel M, Leonhardt A (2012) Dispersion of carbon nanotubes and its influence on the mechanical properties of the cement matrix. Cement Concr Compos 34(10):1104–1113
6. Zou B, Chen SJ, Korayem AH, Collins F, Wang CM, Duan WH (2015) Effect of ultrasonication energy on engineering properties of carbon nanotube reinforced cement pastes. Carbon 85:212–220
7. Cwirzen A, Cwirzen KH, Nasibulin AG, Kaupinen EI, Mudimela PR, Penttala V (2009) SEM/AFM studies of cementitious binder modified by MWCNT and nano-sized Fe needles. Mater Charact 60(7):735–740
8. Zhu Y, Murali S, Cai W, Li X, Suk JW, Potts JR, RuoffRS (2010) Graphene and graphene oxide: synthesis, properties, and applications. Adv Mater 22(35):3906–3924
9. Reddy PVRK, Prasad DR (2022) Investigation on the impact of graphene oxide on microstructure and mechanical behaviour of concrete. J Build Pathol Rehabil 7(1):1–10
10. Reddy PVRK, Prasad DR (2022) A study on workability, strength and microstructure characteristics of graphene oxide and fly ash based concrete. Mater Today Proc 62:2919–2925
11. Lv SH, Ma YJ, Qiu CC, Sun T, Liu JJ, Zhou QF (2013) Effect of graphene oxide nanosheets of microstructure and mechanical properties of cement composites. Constr Build Mater 49:121–127
12. Reddy PVRK, Prasad DR (2023) The role of graphene oxide in the strength and vibration characteristics of standard and high-grade cement concrete. J Build Eng 63:105481
13. Reddy PVRK, Prasad DR (2023) Graphene oxide reinforced cement concrete—a study on mechanical, durability and microstructure characteristics. Fullerenes, Nanotubes, Carbon Nanostruct 31(3):255–265
14. Reddy PVRK, Prasad DR (2022) Synergetic effect of graphene oxide and fly ash on workability, mechanical and microstructural properties of high-strength concrete. Jordan J Civ Eng 16(3):507–517

15. Reddy PVRK, Prasad DR, Singh SB, Gopalarathnam M, Kodur VK, Matsagar VA (2023) Fiber reinforced polymeric materials and sustainable structures. Static and dynamic mechanical properties of graphene oxide and fly ash based concrete (pp 279–288). Springer Nature Singapore, Singapore

Flexural Behaviour of Hybrid and Graded Fibre Reinforced Concrete

P. Anuradha, D. Annapurna, and K. L. Radhika

1 Introduction

Concrete is the most widely used construction material in the world due to its ability to get cast in any form and shape. However, concrete has some deficiencies like, Low tensile strength, Low post cracking capacity, Brittleness and low ductility, Limited fatigue life etc.

The fundamental problem of normal cement concrete is caused by the micro-cracks that exist at the mortar-aggregate boundary. This condition can be overcome by incorporating fibres into the composition. Various fibre types, such as those found in conventional composite materials, can be added to the mix to make concrete more durable or capable of resisting the development of cracks. These fibres assist in load transmission at the interior micro-cracks. This form of concrete is known as fibre-reinforced concrete (FRC).

Fibre is a small piece of reinforcing material that possesses certain characteristics which make its uniqueness in arresting the cracks when it is incorporated into concrete. Fibre besides arresting cracks also improves the static as well as the dynamic properties of the concrete [1, 2]. Fibre Reinforced Concrete is a composite material consisting of mixtures of cement, coarse aggregate, fine aggregate, or concrete and discontinuous, discrete, uniformly dispersed fibres. Based on fibre used, Fibre reinforced concrete is of different types and also the properties of fibre reinforced concrete get enhanced. The hybridization of fibres provides improved specific or synergistic characteristics not obtainable by addition of a single fibres. The fibres

P. Anuradha (✉) · D. Annapurna · K. L. Radhika
Department of Civil Engineering, University College of Engineering (A), Osmania University, Hyderabad, Telangana, India
e-mail: anuradha.p@uceou.edu

D. Annapurna
e-mail: annapurna.d@uceou.edu

R. K. Pancharathi et al. (eds.), *Low Carbon Materials and Technologies for a Sustainable and Resilient Infrastructure*, Lecture Notes in Civil Engineering 440,
https://doi.org/10.1007/978-981-99-7464-1_32

in hybrid concrete increase the Flexural strength, Impact strength, and Toughness properties of concrete [3]. The small and soft fibres control crack initiation and propagation of small cracks [4]. The important factor which influences the properties and behaviour of fibre reinforced concrete is the aspect ratio of fibre. Hybrid Fibre Reinforced Concrete (HFRC) utilizes complementary fibres to improve the properties of concrete, and the performance of HFRC is better than that of a single FRC [5].

Basalt fibre is a new environmental protection fibre, hence, vast research work is being carried out. Basalt fibre is prepared by melting basalt stone at 1450–1500 °C through Platinum rhodium alloy bushing. Basalt fibre has a higher modulus of elasticity, better temperature resistance, impact resistance, and chemical stability than ordinary glass fibre. It is cost-effective when compared to glass fibre [6]. Therefore, an attempt has been made to investigate the Mechanical Properties, mainly focusing on the Flexural Behaviour of Steel and Basalt Fibres reinforced concrete beams, having fibres of different lengths and proportions. Investigation is carried on the effectiveness of Hybrid and Graded fibres in arresting the macro- and micro-cracks in concrete, which are responsible for the early failure.

2 Experimental Investigation

The Present experimental work is carried out in four stages.

2.1 Materials

The following ingredients were utilized in the current project: OPC 53 grade (Ordinary Portland Cement), Fly Ash (FA), Alcco fine, Polycarboxylate Ether (PCE) as super plasticizer, Fine Aggregate, Coarse Aggregate, Water, Crimped Steel fibres (50 and 12.5 mm) (Fig. 1), and chopped Basalt fibres (6 mm) (Fig. 2). Alccofine (Fig. 3) is a special class micro-fine calcium silicate concrete material of particle size much less than Cement, FA, and silica fume. It is normally used in high Strength Concrete which is important in respect of workability as well as strength. The specific gravity of Alccofine is observed as 2.83.

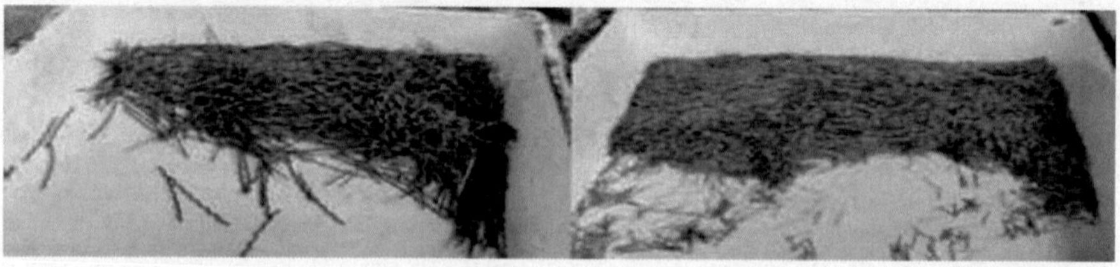

Fig. 1 Steel fibres 50 and 12.5 mm

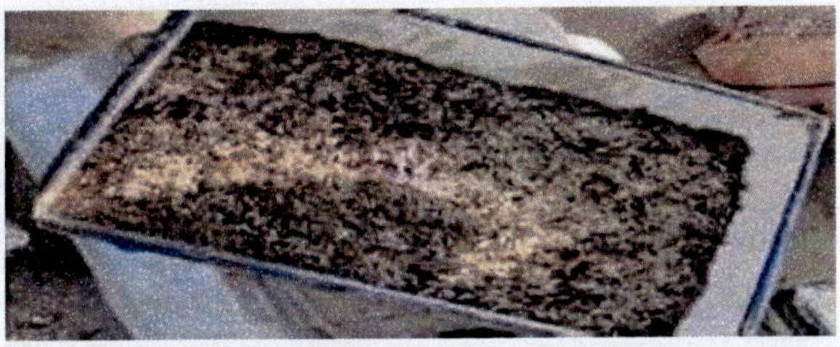

Fig. 2 Basalt fibre

Fig. 3 Alccofine

Based on the trial mixes, the compositions for M40 grade concrete were obtained as stated in Table 1.

In the first stage, Graded steel fibre reinforced concrete is prepared using crimped Steel fibres cold drawn type, from mild steel wire of length 12.5 mm and 50 mm. These steel fibres were added in the proportions of 75:25, 50:50, and 25:75 for the fibre volumes of 0.5%, 1.0%, 1.5%, and 2.0%. For each proportion of fibres in each volume of fibres, 6 cubes of standard size 150 mm were cast to identify the volume with the best proportion of steel fibres based on 28 days compressive strength. The total number of cubes casted in this stage is 72.

In the next stage, Graded steel fibres were mixed with chopped Basalt fibres making it a Hybrid Graded Fibre Reinforced Concrete (HGFRC). The steel and basalt fibres combinations tried here are 100:0, 80:20, 60:40, 40:60, 20:80, 0:100. To find out the best mix proportion from the different volumes selected based on the detailed literature survey [7–9] conducted on various grades of concrete, the fibre volume to be incorporated into the concrete has been fixed as 0.5%, 1.0%, 1.5%, and 2.0%. Based on compressive strength results achieved from 144 cubes, optimum HGFRC mix will be achieved.

Table 1 Mixture compositions of M40 concrete (kg/m^3)

OPC	Fly ash	Alccofine (kg)	Water	Fine aggregate	Coarse aggregate	Super plasticizer
337.5	67.5 (15%)	45 (10%)	131	709	1200	3.6

Fig. 4 Reinforcement bars

Fig. 5 Placing concrete in beam mould

In the succeeding stage, the mechanical properties of HGFRC are identified, including its strength in terms of compression, tension, and flexure. Further, in the last stage, flexural behaviour of Hybrid graded fibre reinforced concrete beams of size 1500 × 2300 × 150 mm is investigated. Based on balanced beam design, 2–12 mm ϕ and 1–16 mm ϕ bars as tension steel and 2–10 mm ϕ bars as compression steel zone are used in RC beam as displayed in Figs. 4 and 5.

3 Results and Discussions

The discussions of all the test results that were conducted are explained in the following sections.

3.1 *Workability of Concrete*

The workability of the control mix was collapsible at a slump value of 163 mm. After the addition of the Fibres (steel and basalt) into the concrete, the mix become a bit harsh and depicting true slump, as shown in Fig. 6, compared to collapse slump of control mix as shown in Fig. 7.

Fig. 6 Slump of HGFRC

Fig. 7 Slump of control mix

3.2 Compressive Strength of Graded Steel FRC

The compressive strength results shown in Fig. 8 depict that steel fibre proportion of 50:50 is giving maximum value in all the % fibre volume. Hence, Steel fibre grading of 50:50 is chosen for further stages of study.

3.3 Compressive Strength of Hybrid Graded FRC

Compressive strength results of Hybrid Graded FRC with different percentage variation of steel and basalt fibres with % volume variation are shown in Fig. 9.

When 0.5% of fibres are added, 80:20 proportion gave the highest value and the remaining proportions decreased gradually. The same trend is observed in 1% fibre volume mixes. But When 1.5% fibre content was added, 60:40 yielded the result which was the highest amongst all the fibre content and proportions. The basalt fibres packed the concrete's porous structure in different directions, which led to an enhancement in the values of properties, which is the cause of the strength enhancement of hybrid graded fibre reinforced concrete. The compressive strength decreases as the fibre content is raised over 1.5%. This occurred as a result of the fibre content's tendency to bunch, problems with compaction, and a lack of bonding between the fibres, and the elements of the concrete. These elements also result in a harsh concrete mixture that doesn't guarantee homogeneous mixing.

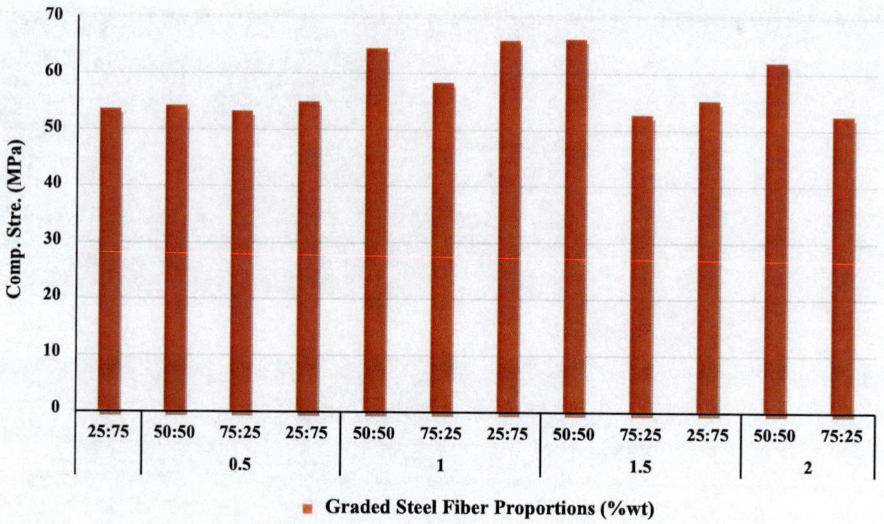

Fig. 8 Bar graph depicting the compressive strength results with graded steel fibre

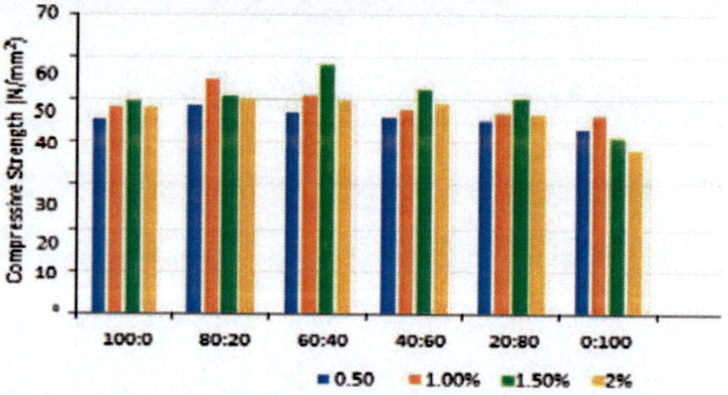

Fig. 9 Compressive strength results of hybrid graded FRC

3.4 Mechanical Properties of HGFRC

The following discussion is focused on the results in relation to the mechanical characteristics of both types of concrete.

3.4.1 Compressive Strength

The 28 days mean compressive strength of the Control mix is 56.67 N/mm^2, whilst the same for HGFRC specimens at the fibre content of 1.5% and proportion of 60:40

after the age of 28 days of curing is 62.02 N/mm^2, which is showing that there is an increase of 9.44% in the compressive strength of HGFRC specimen. This is because the pores in the concrete are reduced due to addition of hybrid fibres having different aspect ratio which increases the resistance of concrete against failure.

3.4.2 Split Tensile Strength

The Control mix cylinder's mean Tensile strength is 4.816 N/mm^2, whilst the HGFRC cylinder's mean Split tensile strength is found to be 6.583 N/mm^2 at the optimized 1.5% fibre content. It is clear from the empirical findings that the split tensile strength of the HGFRC specimen has an increase of 36.69%. There is a substantial enhancement in the split tensile strength as the steel and basalt fibres in a blended state have enhanced the bridging activity and reduced the propagation of cracks, thereby increasing its tensile strength.

3.4.3 Modulus of Rupture (MOR)

The MOR of the normal concrete mix was estimated as 4.986 MPa, whereas the Hybrid graded fibres concrete had an increase of MOR. The modulus of rupture based on flexural strength revealed for HGFRC with 1.5% fibre content was 7.594 MPa. The previously stated finding indicates that fibres have the ability to delay the emergence of micro-cracks in concrete. The largest improvement in MOR over normal concrete for basalt FRC with 1.5% fibre volume percentage was 52.3%.

3.5 *Flexural Behaviour of Beams*

The general flexural behaviour of the three different kinds of beams, including the distribution of cracks, the load–deflection reaction, and the distribution of strain in the concrete and reinforcement, is summarized in this section.

3.5.1 Development of Cracks

During the analysis, deflections at the initial crack and the ultimate load were discovered. At 58.86 kN, the control beam (CB) had its initial crack, which was followed by flexural cracks. It was evident that the beam's failure in flexure had little energy absorption prior to the failure. The control beam's mean ultimate load is 230.5 kN.

3.5.2 Load–Deflection Behaviour

To investigate the loss of brittle material character of concrete, the load–deflection curves for HGFRC were generated, from the data collected during the testing of beams. From the P- Δ graph, shown in Fig. 10, it is noted that initially for different concrete mixes, the curves were linear and then curved with the increasing load. When the load reached its peak value, it was found that the linear P-curve diverged from the linear response. This denotes the start of the crack initiation process.

The increase in fibre content causes an increase in the length of segment until reaching its peak. The addition of fibre content of 1.5% caused the increase in Pmax. The quantity of fibres in concrete also had an impact on the post-critical portion of the P-d curve. The evident force decrease and strain softening characteristics were seen for reference concrete. The inclusion of basalt fibre has an impact on both the pre-peak and post-peak portions of the load–deflection curve of the concrete beam. A typical load–deflection diagram of a specimen during three-point loading shows that the material behaves linearly up to the first crack stress, then through a strain hardening stage after the first crack up to the ultimate flexural load, and finally, a post-ultimate load stage.

The load–deflection behaviour of the HGFRC beam with reduced main reinforcement was very different from the other two types of beams. The beam had a decreased percent of main reinforcement and the fibre content of 1.5% was the same as that of another specimen.

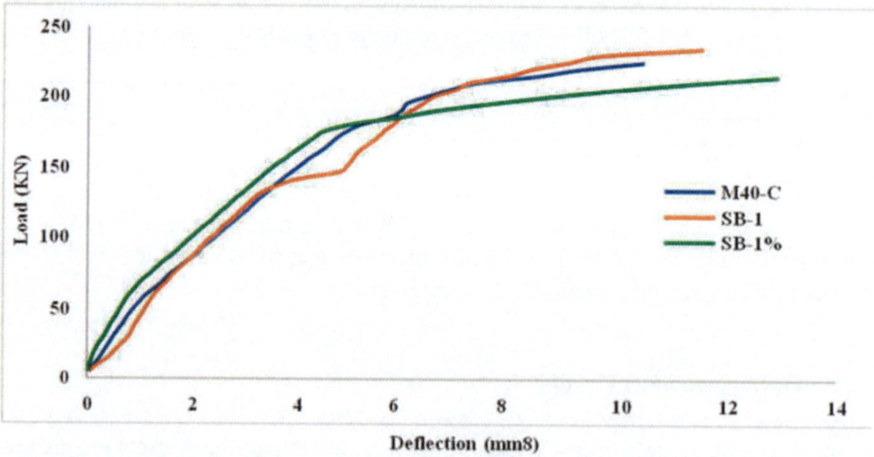

Fig. 10 Load–deflection behaviour of three type of beams

3.5.3 Crack Distribution

At the pure bending zones of all 3 kinds of reinforced concrete beams, vertical flexural fractures first appeared. Then, as the stress increased, inclined shear fractures were created. Each beam experienced flexure failure. A ductile mechanism of failure known as tension failure, longitudinal tension-steel yielded in all types of beams first before concrete crushing.

3.5.4 Crack Widths

In addition to the load-deformation behaviour of fibre reinforced concrete beams, the cracking characteristics have been also observed during the tests. Figure 11 shows the diagram of crack width for the different types of reinforced concrete beams. This means that the steel fibres act as crack arresters and give great benefit in controlling the crack propagation. The crack widths were not more than 3 mm. For the control beam, the crack width was less but the CB failed due to a long shear crack connecting the support point and the loading point. But for the HGFRC beams, no shear failure in shear was visible and the fibres could resist the crack formation and propagation.

The fibre-added reinforced concrete beam exhibits significantly reduced cracking. This data demonstrates the exceptional resilience to tensile cracking of the fibre reinforced concrete.

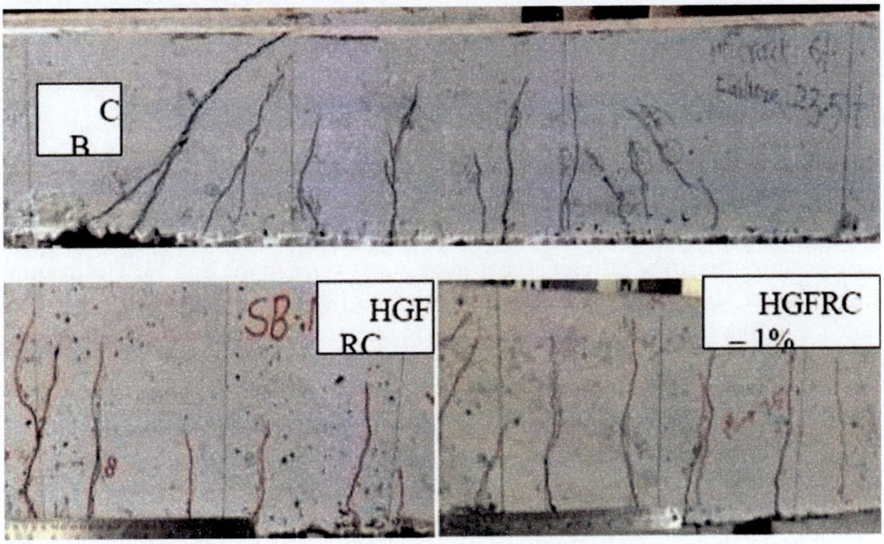

Fig. 11 Crack spacings of control beam (CB), HGFRC, HGFRC-1% reinforcement

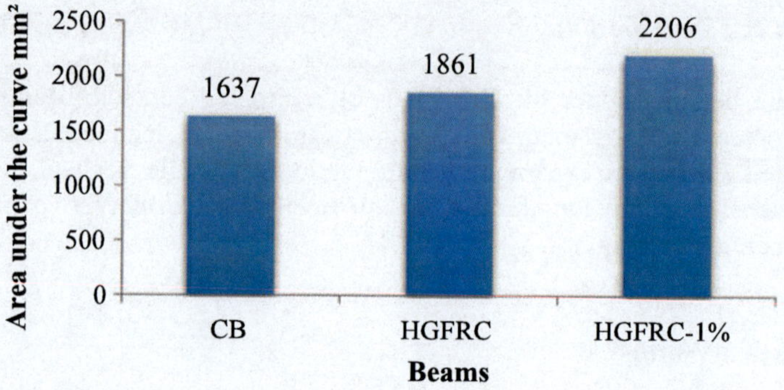

Fig. 12 Flexural toughness of beams up to ultimate deflection

3.6 Flexural Toughness

Flexural toughness is a criterion for concrete's ability to absorb energy and is also used to evaluate the material's ductility and crack resistance. It is the region underneath the concrete's load–deflection curve in flexure.

It is clear from the Fig. 12 that more area under the curve is possessed by the beam with reduced reinforcement. This study displays the improvement in composites' bearing capacity following cracking. This is explained by the fibres' ability to prevent and bridge fracture growth. The rise in peak load is primarily responsible for the improvement in flexural toughness for HGFRC, but the increase in bearing capacity after cracking is also responsible for the improvement in flexural toughness for HGFRC strengthened with less main steel %.

3.7 Flexural Strength

The Flexural strength of different types of beams is given in the Fig. 13. The flexural strength of the HGFRC beam has an increase in its flexural strength than the control beam due to the action of bridging fibres. And the steel fibres are proven to increase the flexural characteristics of the concrete when incorporated. There has been a 4.33% increase in the beam flexural strength. But the flexural strength of the HGFRC-1% beam has a clear drop in its flexural strength. This is clearly due to the reduction in its main reinforcement steel. As there was not enough steel to take the stress from concrete, resulting in the fibres taking the stress. But the short discrete fibres, which are good at arresting cracks and bridging the cracks, failed to take the stress from the concrete directly. So, the decrease in the flexural strength of the HGFRC-1% beam is 4.36%.

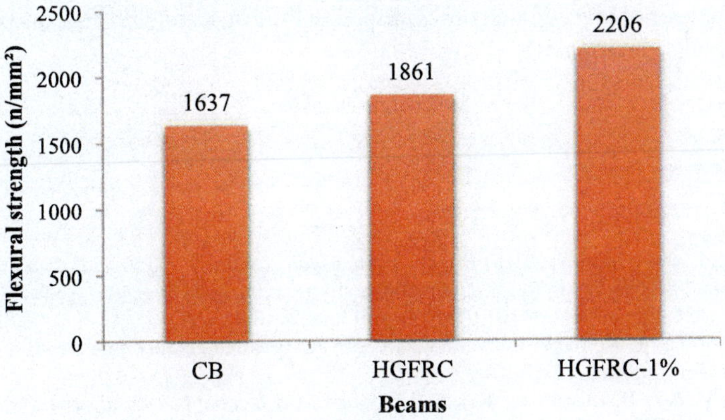

Fig. 13 Flexural strengths of CB, HGFRC, HGFRC-1%

4 Conclusions

The following results are derived from the experimental work and petrographic examination.

- Due to the ultra-fine physical property of Alcofine, the workability of concrete was maintained well throughout the work. Addition of fibres into concrete had drastically reduced the workability.
- The water-cement ratio could be lowered by using a superplasticizer built on polycarboxylate ether (PCE).
- The best volume of steel and basalt fibres was found at 1.5% with a proportion of 60:40 by performing a compression strength test as the basis.
- When the volume of fibres was increased more than 2%, the effect of balling of concrete was taking place and the workability was affected the most.
- Improvement of 9.44% in HGFRC compression strength was observed after 28 days of curing.
- Split tensile strength of HGFRC specimens was found to be increased by 36.69% when compared to control specimens. The flexural strength of HGFRC prism specimens was also found to have increased by 52.3% at the optimum fibre content of 1.5%.
- The flexural strength of beams that were incorporated with the optimum fibre content was found to increase by 4.33%, whilst the HGFRC-1% beam has a decrease in its flexural strength by 4.36% performance. The pre-peak behaviour of the three types of specimens was almost the same in the flexural strength test. But the post-peak characteristics were different from one another.
- The HGFRC beam had an Ultimate Peak-load value and enhanced deflection than the control beam. This indicates that the fibres' impact on the concrete's post-peak effectiveness was quite evident. The fracture behaviour of concrete is influenced

by both types of investigated fibres, although the effects on pre-peak and post-peak performance are distinct.

References

1. Li B, Xu L, Shi Y, Chi Y (2018) Effects of fiber type, volume fraction and aspect ratio on the flexural and acoustic emission behaviors of steel fiber reinforced concrete. Constr Build Mater 181:474–486. https://doi.org/10.1016/j.conbuildmat.2018.06.065
2. Joshi A, Reddy P, Kumar P, Pramodhatker (2016) Experimental work on steel fibre reinforced concrete. Int J Sci Eng Res 7(10):971–981
3. Sarmah M, Roy B, Mozumder RA et al (2018) Effect of chopped basalt fibers on the cyclic behavior of RCC beam-column subassemblies. Arab J Sci Eng 43:1865–1874. https://doi.org/10.1007/s13369-017-2801-y
4. Kao MKM, Ali M (2021) Effect of basalt fibers on mechanical properties of calcium carbonate whisker-steel fiber reinforced concrete. Materials 14(23):7455–7460
5. Wang D, Ju Y, Shen H, Xu L (2019) Mechanical properties of high- performance concrete reinforced with basalt fiber and polypropylene fiber. Constr Build Mater 197:464–473
6. Prathipati SRRT, Rao CBK (2021) A study on the uniaxial behavior of hybrid graded fiber reinforced concrete with glass and steel fibers. Innovative Infrastruct Solutions 6(2):64–73
7. Mobasher B, Yao Y, Chotesoranakom (2015) Analytical solutions for the flexural design of hybrid steel fiber reinforced concrete beams. Eng Struct 100:164–177
8. Krassowska J, Lapko A (2013) The influence of steel and basalt fibers on the shear and flexural capacity of reinforced concrete beams 7(7):789–795
9. Mohod MV (2012) Performance of steel fiber reinforced concrete. Int J Eng Sci 1(12):1–4

Torque and Twist Response of Under-Reinforced and Completely Over Reinforced Concrete Beams with 'U' Ferro Cement Wrap

G. C. Behera, T. D. G. Rao, and C. B. K. Rao

1 Introduction

Torsion is one of the basic loads. It acts on structural elements individually or with combination of axial loads, flexure or with shear. At early eighteenth centuries, torsion is not considered for design purpose due to its complex nature. Primary or secondary torsions may take place in structural elements. Structural elements due to ageing, updates in codal provisions or change of load path due to remodelling of structure need to be retrofitted rather than demolition. Concrete structures can be retrofitted with other materials like steel jacketing or with FRP. Time of repair, cost of retrofitting material, availability of retrofitting material and workmanship are controlling factors for repairing of distressed members [1]. Recently FRP is in driver's seat as retrofitting material. Due to torsion, shear is induced on a structural element which requires full wrap for retrofitting. Due to constructional difficulties, a 'U' wrap may be selected over full wrap torsion [2]. A significant research in FRP with full jacket for torsional retrofitting has been carried out [3]. With FRP 'U' wrap some investigation were carried and found there is significant improvement in strength [4] and [3]. Repair strategy should be economic, easy and practicable. FRP retrofitting is suitable for developed countries not for developing countries Behera [5], Behera et al. [6, 7]. Ferrocement is a good substitute for FRP. According [8], ferrocement is impermeable, durable, easily flowable and also provides higher strength. The micro cracks also can be arrested [9].

G. C. Behera (✉)
Government College of Engineering Kalahandi, Bhawanipatna, Bhawanipatna, India
e-mail: beheragb@gmail.com

T. D. G. Rao · C. B. K. Rao
NIT Warangal, Warangal, India

R. K. Pancharathi et al. (eds.), *Low Carbon Materials and Technologies for a Sustainable and Resilient Infrastructure*, Lecture Notes in Civil Engineering 440,
https://doi.org/10.1007/978-981-99-7464-1_33

1.1 Research Significance

For quantification of torsional parameters, the first and foremost method is the experimental one. A mathematical model may be developed to quantify the same with and the results of the model must be well in agreement with experimental values. A good deal of research was carried by the author to predict torsional parameters of ferrocement 'U' wrap (2014, 2018, 2022) by experimental investigation as well as by analytical model. It was observed that Experimental results cannot quantify torque and twist of a new beam with different dimensions and analytical method needs some computational programs to predict torque and twist. This challenge inspired to find a solution to predict torque and twist with easy methods. Here, with the help of soft computing, solution is found out. Prediction of torque and twist of torsionally under reinforced and completely over-reinforced beams with ferrocement 'U' wrap beams has been presented here as design of a torsionally balanced section may not be feasible always as it depends on many parameters. The under reinforced and completely over reinforced beams were taken for study as the later can resist the maximum torque while the former can undergo maximum twist.

The outcome of the present study is to predict torsional parameters such as torque and twist of ferrocement 'U' wrap RC under reinforced and completely over reinforced beams with different grades concrete at cracking stage and ultimate stage.

2 Different Methods to Predict Torque and Twist

2.1 Experimental Programme

Eight numbers of beams having cross sections 125 mm wide, 250 mm depth and lengths of the 2 m were tested during this investigation. Beam details, material and properties are presented in Table 1. Cross section of the beam and testing of beam in torsion in the torsion test rig is presented in Fig. 1.

2.2 Analytical Model

The model was developed with reference to Hsu's softened truss model. Modification in the material properties and asymmetry of material in faces were incorporated. Separate equations were applied for elastic, cracking and post cracking zones. Details of the model is available in Behera et al. [7, 10].

Table 1 Beam designation and material properties

Sl. No	Designation	Dimensions (mm)	Constituent material property						
			Ferrocement matrix (MPa)	Concrete (MPa)	Longitudinal Steel		Transverse steel		No. of mesh layers
					No. of bars, Diameter	Yield Strength (MPa)	Diameter, Spacing	Yield Strength (MPa)	
1	U3N	125 × 250	40	35	4 nos.,6 mm	350	2 legged 6 mm @ 100 mm c/c	350	3
2	U4N	125 × 250	40	35	4 nos.,6 mm	350	2 legged 6 mm @ 100 mm c/c	350	4
3	U5N	125 × 250	40	35	4 nos.,6 mm	350	2 legged 6 mm @ 100 mm c/c	350	5
4	Co3N	125 × 250	40	35	4 nos.,12 mm	440	2 legged 8 mm @ 100 mm c/c	465	3
5	Co4N	125 × 250	40	35	4 nos.,12 mm	440	2 legged 8 mm @ 100 mm c/c	465	4
6	Co5N	125 × 250	40	35	4 nos.,12 mm	440	2 legged 8 mm @ 100 mm c/c	465	5
7	U4H	125 × 250	55	60	6 nos.,6 mm	350	2legged,6 mm @ 70 mm c/c	350	4
8	Co4H	125 × 250	55	60	6 nos.,12 mm	440	2 legged,10 mm @ 70 mm c/c	445	4

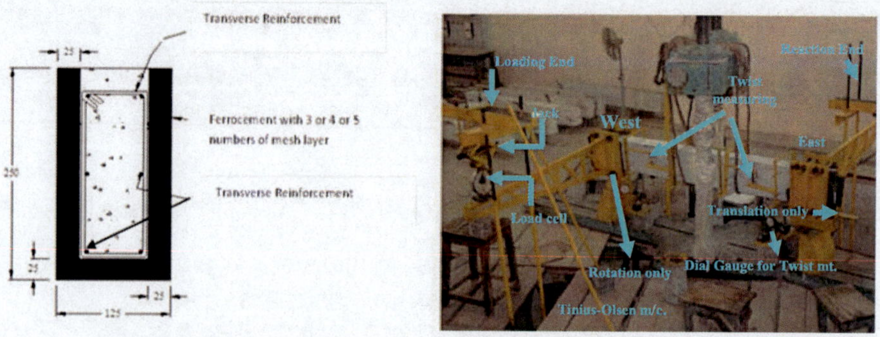

Fig. 1 Beam cross section and torsion testing

2.3 *Soft Computing Method*

For the quantification of torsional parameters by experimental method, proto type structures are to be demolished. Experimental method is time consuming as well as not cost effective. The prediction of torsional parameters needs programming part to solve many 22 unknowns by 22 equations. Both the methods do not provide any single equation to predict torsional parameters. To overcome these problems, soft computing method can play a vital role.

2.3.1 Multivariate Adaptive Regression Spline (MARS)

Simple equations were developed by Multivariate adaptive regression spline (MARS) to predict the desired parameters. This method was developed by Friedman [11]. Some of results of experimental value are used for fitting and others are used for testing. No such basic assumptions are required for this method. So, sometimes it is known as black box method. The performer is unable to find the relation between various parameters. This model is the outcome research work with the help of MARS, final equations for torque and twist at cracking and ultimate stage were found as mentioned below

$$\begin{aligned} T_{\text{cracking}}(\text{kNm}) = [&5.5324 + \max(0, \text{Fly} - 350) * 0.00154 \\ &- \max(0, 350 - \text{Fly}) * 0.0002689 + \max(0, \text{spacing}) \\ &* 0.0012240 + \max\ (0,\ \text{mortar strength} - 40) * 0.0680449 \\ &- \max\ (0, 40 - \text{mortar strength}) * 0.0206587] \end{aligned}$$

$$\begin{aligned} T_{\text{ultimate}} = [&6.752 - \text{maximum } 0, 0.32265 \\ &- \ \text{spacing of longitudinal reinforcement} * 2.7323 \\ &- \text{Maximum of } \left[0, 350 - \text{Fty}\right] * 0.002276 \end{aligned}$$

$$+ \text{maximum}\big[0,\ \text{Mortar strength} - 40\big] * 0.07677].$$

$$\theta_{\text{cracking}}(\text{rad/m}) = [0.0053610664 + \max\ (0,\ \text{Fly} - 350) * 0.0000008816 + \max\ (1.62926 - \text{Afl})\ * 0.0002050043 - (40 - \text{mortar strength}) * 0.0000723768].$$

$$\theta_{\text{ultimate}}(\text{rad/m}) = [0.03558 - \text{maximum}[0, \text{Fly} - 350] * 0.0003376 - \text{maximum}[0, 350 - * \text{Fly}] * 0.00008786 + \text{maximum}\big[0, \text{spacing of stirrup}\big] * 0.00102665].$$

2.3.2 WASPAS method

The method of weighted aggregated sum product assessment (WASPAS) for solution of Multi-criteria decision-making (MCDM) problems was suggested by Zavadskas et al. in [12]. This method is a combination of two methods. One is weighted sum method (WSM) and other one is weighted product method (WPM). The various steps of WASPAS procedure are taken from Madic et al. [13, 14] and [12]. The process continues in the following manner.

(A) Matrix Initialization.
(B) Apply normalization to matrix using maximization and minimization criteria Matrix Initialization.

$$\overline{x}_{ij} = x_{ij}/\text{max}_i x_{ij} \tag{1}$$

$$\overline{x}_{ij} = \text{min}_i x_{ij} \tag{2}$$

where xij, assessment value of the ith alternative with respect to the jth criterion.

(C) Calculation of total relative importance of ith alternative,

$$Q_i^{(1)} = \sum\nolimits_{j=1}^{n} \overline{x}_{ij}.w_j \tag{3}$$

(D) Application of WPM method next to find

$$Q_i^{(2)} = \prod\nolimits_{j}^{n} \overline{x}_{ij}^{w_j} \tag{4}$$

(E) Calculation of optimum value

$$Q_i = \lambda.Q_i^{(1)} + (1 - \lambda).Q_i^2 \quad (5)$$

Consider the values of λ from 0 to 1.0. In general, it takes a value of 0.5.

The experimental and predicted values of torque and twist at cracking and ultimate are presented in Table 2.

3 Interpretation of Test Results

Experimental and predicted results of torsional parameters were analysed, errors in predicted values were estimated to note the appropriateness of methods.

3.1 *General Behaviour of High Strength Beams*

All the beams were of same cross section, variation was in the strength of core concrete, number of mesh layers in outer 'U' wrap and in mortar strength. It was observed that the initiation of first crack in unwrapped face for all the beams as the tensile strength of concrete on unwrapped face is lower than the tensile strength of mortar in wrapped face. The inclination of the crack was found to be 450.

3.2 *Torsional Capacity of Specimens*

Under reinforced beams with three, four and five layers of mesh were cast with M35 grade concrete and M40 grade mortar for ferrocement 'U' wrap. There was no such difference observed in cracking torque. The cracking torque depends on mortar grade, concrete grade and amount of longitudinal reinforcement. This may be due to the fact that all three parameters mortar grade, concrete grade and amount of longitudinal reinforcement were same for three under reinforced beams (U3N, U4N and U5N). The same behaviour is observed for completely over reinforced beams (Co3N, Co4N and Co5N) with normal strength concrete of grade M35. Average cracking torque of normal strength under reinforced beams was found to be 5.59 kNm while the same was found to be 5.83 kNm for completely over reinforced beams. The average increase in cracking torque of completely over reinforced beams was found to be 4.34% over it's under reinforced beams. For high strength under reinforced beam U4H and completely over reinforced beam Co4H, cracking torque was observed as 6.42 and 6.72 kNm, respectively. The percentage of increase in cracking torque of completely over reinforced beam Co4H against its under reinforced section is found to be 4.54%. The percentage of increase in cracking torque of high strength concrete beams for completely over reinforced beam over its under reinforced beam was found to be same for normal strength concrete beams also. Increase in cracking

Table 2 Experimental and predicted torque and twist at cracking

Beams	Experimental		Analytical		Mars		Waspas		Regression analysis	
	Cracking	Ultimate	Cracking	Ultimate	Cracking	Ultimate	Cracking	Ultimate	Cracking	Ultimate
Torque (kNm) of specimens tested at cracking and ultimate										
U3N	5.53	5.82	5.55	5.56	5.65	6.46	5.58	5.50	5.58	5.73
U4N	5.61	6.01	5.55	5.89	5.65	6.46	5.62	5.64	5.60	5.88
U5N	5.62	6.01	5.55	5.90	5.65	6.46	5.67	5.79	5.62	6.02
Co3N	5.82	9.11	5.55	9.12	5.79	9.11	5.78	8.98	5.84	8.32
Co4N	5.82	9.43	5.55	9.37	5.79	9.43	5.83	9.13	5.86	8.46
Co5N	5.85	9.62	5.55	9.59	5.79	9.62	5.88	9.28	5.88	8.60
U4H	6.42	7.68	6.52	7.73	6.42	7.90	6.55	7.47	6.43	7.59
Co4H	6.72	12.91	6.52	12.98	6.78	12.91	6.74	11.25	6.95	10.70
Twist (rad/m) of specimens tested at cracking and ultimate										
U3N	0.00530	0.15500	0.00522	0.16540	0.00544	0.13820	—		0.00536	0.15539
U4N	0.00530	0.14000	0.00517	0.13000	0.00536	0.13820	—		0.00531	0.14600
U5N	0.00530	0.12000	0.00513	0.08970	0.00536	0.13825			0.00527	0.13660
Co3N	0.00550	0.12500	0.00497	0.11310	0.00552	0.10786			0.00559	0.09853
Co4N	0.00540	0.11000	0.00492	0.10299	0.00544	0.10780			0.00555	0.08914
Co5N	0.00540	0.09800	0.00489	0.09589	0.00544	0.10786			0.00550	0.07974
U4H	0.00541	0.13050	0.00548	0.13736	0.00536	0.10745			0.00523	0.11063
Co4H	0.00541	0.07540	0.00505	0.07361	0.00544	0.07706			0.00583	0.05431

torque for a particular grade of concrete over its state of torsion is due to variation of reinforcement.

Maximum error in predicted cracking torque for all of these six beams U3N, U4N, U5N, Co3N, Co4N, Co5N, U4H and Co4H was found to be −5.16% for Co5N beam in analytical method. The same was found to be −0.97% for MARS method in beam Co5N. In WASPAS method maximum error was noticed as 2.0% U4H and in regression analysis method the maximum error was noticed 3.47% for beam Co4H. Cracking torque of all beams by different methods was presented in Fig. 2 along with the percentage of error in predicted values.

The ultimate torques of beams U3N, U4N, U5N, Co3N, Co4N, Co5N, U4H and Co4H were found to be 5.82 kNm, 6.01 kNm, 6.01 kNm, 9.11 kNm, 9.43 kNm, 9.62 kNm, 7.68 kNm and 12.91 kNm, respectively. The ultimate torque of beams of a particular state of torsion was found to be increasing very less over higher number of mesh layers. Ultimate torque of U4N was found to be 3.34% higher than U3N and the same was nil for U5N over U4N. For completely over reinforced beams enhancement of Co4N over Co3N and Co5N over Co4N was found to be 3.52% and 2.06%, respectively. There was higher enhancement of ultimate torque for different state of torsion as noticed for completely over reinforced beams over under reinforced beams. Ultimate torque of Co3N over U3N, Co4N over U4N and Co5N over U5N was found to be 56.55%, 56.84% and 60.07%, respectively.

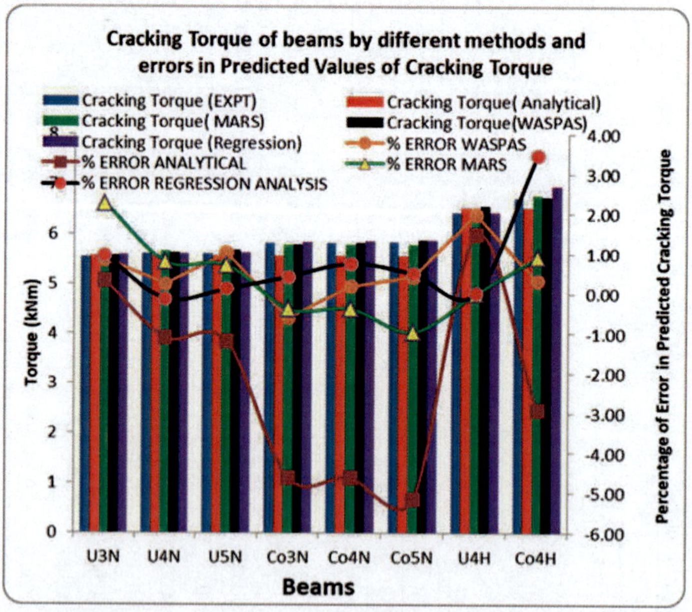

Fig. 2 Cracking torque of beams by different methods and errors in predicted values of cracking torque

Maximum error in predicted ultimate torque for all of these six beams U3N, U4N, U5N, Co3N, Co4N, Co5N, U4H and Co4H was found to be −0.72% in U4N, 11.04% in U3N, −21.48% Co4H and −17.16% in Co4H for analytical, MARS, WASPAS and regression analysis method respectively. From this, it is concluded that analytical method best predicts the ultimate torque in comparison to other three methods MARS, WASPAS and regression analysis. The same was presented in Fig. 3. Ultimate torques of these beams U3N, U4N, U5N, Co3N, Co4N, Co5N, U4H and Co4H were found to increasing 5.17%, 7.13%, 7.03%, 56.55%, 62.07%, 64.44%, 19.54% and 92.22% over their cracking torques respectively. It was observed that the ultimate torque found to be increasing more in completely over reinforced beams rather than under reinforced beams due to availability of more reinforcement. When grade of ferrocement and core concrete is higher, the increase in ultimate torque was higher. There is maximum increase in torque in Co4H. Ultimate torque of Co4H was found to be 1.68 times than U4H, the same for Co4N and U4N was calculated as 1.57.

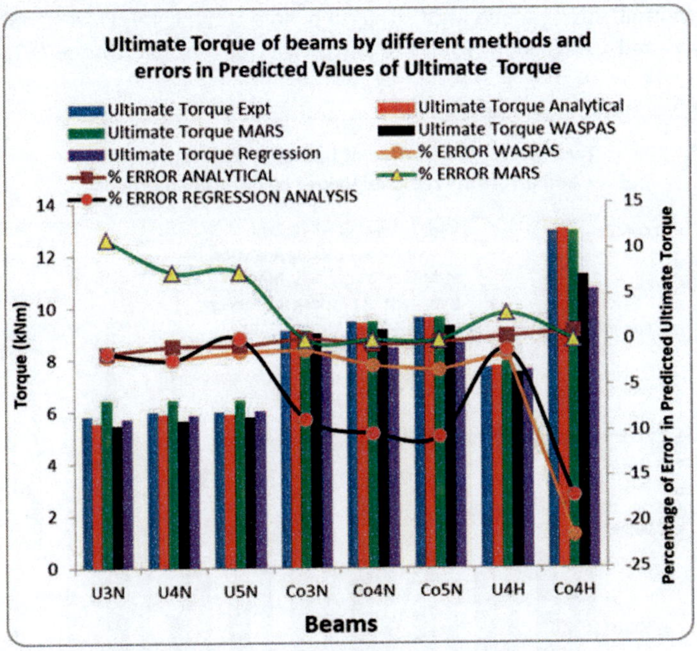

Fig. 3 Ultimate torque of beams by different methods and errors in predicted values of ultimate torque

3.3 Twist

The effect of reinforcement on twist at cracking torque of high strength and normal strength ferrocement 'U' wrap beams was discussed in this section. The reinforcement gets activated after cracking stage. All the normal strength beams U3N, U4N, U5N, Co3N, Co4N and Co5N were found to have cracking twist 0.0053 rad/m, 0.0053 rad/m, 0.0053 rad/m, 0.0055 rad/m, 0.0054 rad/m and 0.0054 rad/m respectively. The twist at cracking torque for beam U4H and Co4H was reported as 0.00541 rad/m for both the beams. There is no such variation in the cracking twist for normal and high strength beams. This may be due to the fact that the ratio of cracking torque to stiffness was found to be same for normal and high strength beams.

The maximum percentage of error in predicted value by analytical, MARS and regression analysis was found to be −9.73% for Co3N, 2.73% for U3N and 7.78% for U4H respectively. MARS method well predicts the twist at cracking torque among all three methods. The twist at cracking torque along with percentage of error in different predicted methods is presented in Fig. 4.

All the normal strength and high strength beams U3N, U4N, U5N, Co3N, Co4N, Co5N, U4H and Co4H were found to have twist at ultimate torque 0.155 rad/m,

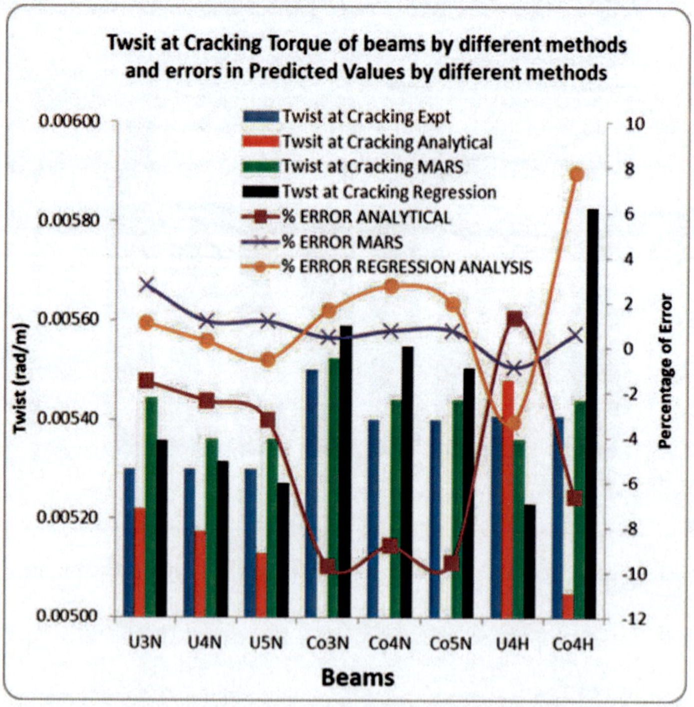

Fig. 4 Twist (rad/m) at cracking torque of beams by different methods and errors in predicted values of twist at cracking torque

0.14 rad/m, 0.12 rad/m, 0.125 rad/m, 0.11 rad/m and 0.98 rad/m, 0.13 rad/m and 0.0754 rad/m respectively. The ratio of twist at ultimate torque to twist at cracking was found to be 29.25, 26.42, 22.64, 22.73, 18.15, 24.13 and 13.94 for beams U3N, U4N, U5N, Co3N, Co4N, Co5N, U4H and Co4H, respectively. The beams with under reinforced section exhibit more twist at ultimate torque due to less stiffness than the completely over reinforced section. For the same reason, higher-grade material beams exhibit less twist than lower strength material. The maximum percentage of error in predicted value by analytical, MARS and regression analysis was found to be −25.25 for U5N, −17.66for Co5N and −27.98 U4H respectively. All these three methods failed to predict the twist at ultimate torque, hence should not be used for the prediction of twist at the ultimate torque. Twist at ultimate torque of normal strength under reinforced beam U4N was found to be increasing 7.28% over its high strength beam U4H and the same for completely over reinforced beam Co4N was reported as 45.89% over the beam Co4H. The twist at ultimate torque along with percentage of error in different predicted methods is presented in Fig. 5

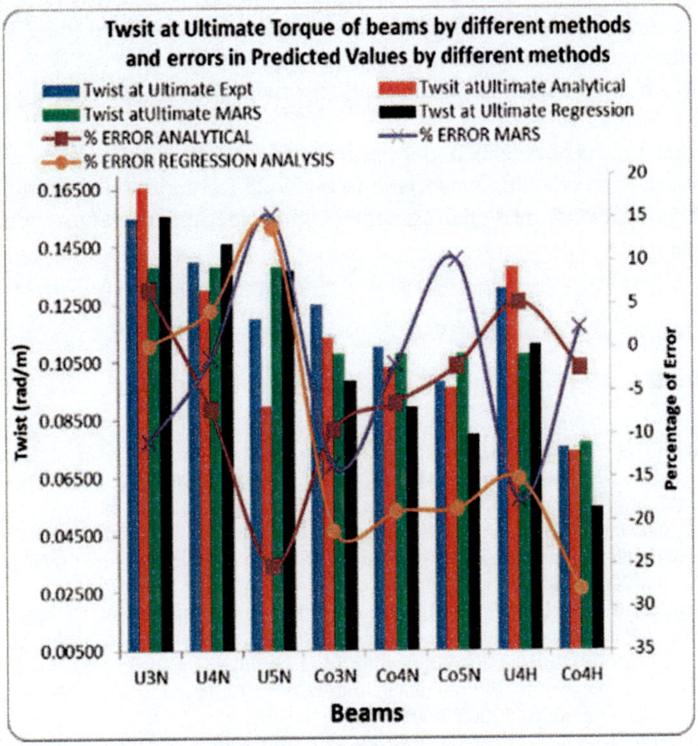

Fig. 5 Twist (rad/m) at ultimate torque of beams by different methods and errors in predicted values of twist at ultimate torque

4 Conclusions

Prediction of torque and twist at cracking and ultimate stage of ferrocement 'U' wrapped under reinforced and completely over reinforced beams with different grades of concrete and mortar were presented here by five methods. From above results, the following conclusions were drawn.

The increase in cracking torque due to longitudinal reinforcement is marginal. Enhancement of torque at cracking and ultimate stage for a particular state of torsion over different numbers of ferrocement layers is very less. The maximum increase is 3.52%. Maximum torque at cracking and ultimate stage is noticed for completely over reinforced beams.

- Increase in cracking torque of completely over reinforced beams is noticed to be 4.54% over their respective under reinforced beams.
- There is noticeable increase in ultimate torque over their cracking torque. There is 92.22% increase in ultimate torque in comparison to the cracking torque for beam Co4H.
- Ultimate torque of completely over reinforced section is found to be 1.68 times than under reinforced section.
- There is no such variation in the cracking twist for normal and high strength beams.
- Maximum twist is noticed in under reinforced beams. The ratio of twist at ultimate torque to twist at cracking was found to be 29.25 for under reinforced beam.
- The predicted results are well in agreement with experimental values for prediction of torque.

References

1. Karayannis CG, Chalioris CE, Sirkelis GM (2008) Local retrofit of exterior RC beam-column joints using thin RC jackets—An experimental study. Earthquake Eng Struct Dyn 37–5:727 746 (2008). https://doi.org/10.1002/eqe.783
2. Salom P, Greeley J, Young D (2004) Torsional strengthening of spandrel beams with fiber-reinforced polymer laminates. J Compos Constr ASCE 157–162. https://doi.org/10.1061/(ASCE)1090-0268(2004)8:2(157)
3. Panchacharam S, Belarbi A (2002) Torsional behavior of reinforced concrete beams strengthened with FRP composites. In: Proceedings of first FIB congress. Osaka, Japan, pp 1–11
4. Chalioris CE (2008) Torsional strengthening of rectangular and flanged beams using carbon fibre-reinforced-polymers—Experimental study. Constr Build Mater 22–1:21–29. https://doi.org/10.1016/j.conbuildmat.2006.09.003
5. Behera GC (2018) A model to predict the torsional stiffness of 'U-wrapped' reinforced concrete beams. Struct Buildings, Proc Inst Civ Eng 171(9):676–687. https://doi.org/10.1680/JSTBU.16.00226
6. Behera GC, Rao TDG, Rao CBK (2008) Torsional capacity of high strength concrete beams jacketed with ferrocement U wraps. Asian J Civ Eng 9–4:411–422 (2008). Corpus id: 136770610

7. Behera GC, Rao TDG, Rao CBK (2014a) Analytical model for torsional response of RC beams strengthened with ferrocement U-Wraps. Struct Eng Int 4:509–520. https://doi.org/10.2749/101686614X13854694314847
8. ACI Committee 549 (1979) Ferrocement-materials and applications. In: ACI symposium proceedings SP-61: American Concrete Institute 549, Farmington Hills, Michigan (1979)
9. Shannag MJ, Mourad SM (2012) Flowable high strength cementations matrices for ferrocement applications. Constr Build Mater 36:933–939. https://doi.org/10.1016/j.conbuildmat.2012.06.051
10. Behera GC, Rao TDG, Rao CBK (2014b) A study on post cracking torsional behaviour of high strength reinforced concrete beams with ferrocement "U" wraps. Slovak J Civ Eng 1–12. https://doi.org/10.2478/sjce-2014-0012
11. Friedman J (1991) Multivariate adaptive regression splines. Ann Stat 19:1–141
12. Zavadskas EK, Turskis Z, Antucheviciene J, Zakarevicius A (2012) Optimization of weighted aggregated sum product assessment. Elektron. Elektrotech 122:3–6. https://doi.org/10.5755/j01.eee.122.6.1810
13. Madi´c M, Gecevska V, Radovanovi´c M, Petkovi´c D (2014) Multicriteria economic analysis of machining processes using the WASPAS method. J Prod Eng 17:1–6
14. Madic M, Antucheviciene J, Radovanovic M, Petkovic D (2016) Determination of manufacturing process conditions by using MCDM methods: application in laser cutting. Eng Econ 27:144–150. https://doi.org/10.5755/j01.ee.27.2.13428

A Comparative Study on Bond Behavior of Ternary Blended Geopolymer Concrete and Conventional Concrete

Padakanti Rakesh, S. Venkateswara Rao, Rathish Kumar Pancharathi, and S. Rakesh

1 Introduction

High demand for the infrastructure over the world led to increase in concrete production which demanded a high production of cement every year [1]. That laid a negative impact on environment causing increased CO_2 emission and high-energy consumption for the production of raw material. For every one ton of production of cement directly and indirectly responsible for the emission of approximately one ton of cement [2]. Over 1.6 million ton of CO_2 emitted from cement industry in the year 2021 [3, 4]. In retrospect, synthesis of geopolymer concrete (GPC) gained interest for its low carbon emission and energy consumption. Prof Davidovits introduced the GPC in the year 1987 which encapsulates the industrial waste's with the help of alkaline activator solution [5]. GPC utilizes industrial byproducts as binder material in higher volumes which ultimately reduces another impact of environment for landfills. Apart from the sustainability aspects it also possess a better structural and durability behavior which made the most suited alternate material for the replacement of traditional cement concrete [6–9]. But the application of GPC in real time was scarce since for not having proper guidelines for structural performance and synthesis of GPC. This led the researchers to understand its structural and microstructural behavior.

One of the most significant aspect negated the application of GPC in real-time application was requirement of oven temperature for polymerization in case of flyash based GPC [10]. This constraint over heeled by the utilization of GGBS and other calcareous materials for high early age strength. The works on the slag-based GPC proven that initial compressive strength was considerably increased at ambient

P. Rakesh (✉) · S. V. Rao · R. K. Pancharathi
National Institute of Technology Warangal, Hanamkonda, India
e-mail: padakantirakesh@gmail.com

S. Rakesh
VNR Vignana Jyothi Institute of Engineering and Technology Hyderabad, Hyderabad, India

R. K. Pancharathi et al. (eds.), *Low Carbon Materials and Technologies for a Sustainable and Resilient Infrastructure*, Lecture Notes in Civil Engineering 440,
https://doi.org/10.1007/978-981-99-7464-1_34

temperature [2]. Along with slag, microbinder materials such as silica fume also improve the early age strength properties among which Mustakim et al. 2021 studied the effect of nano and micro silica on the mechanical and microstructural performance of ambient cured GPC [11]. But these calcareous has variability in chemical composition from source to source. For this, it is important to understand the structural aspects of GPC. To this extent, there are limited research work have been conducted on bond, shear and torsional strength few of them are [8, 12–14]. Few works on the evaluation of flexural and modulus of elasticity are also conducted and proven promising behavior to replace OPC [15–17]. Pradip Nath and Prabir kumer 2017, determined the flexural strength and elastic modulus of ambient cured slag based fly ash geopolymer concrete and stated that flexural strength of GPC was higher and modulus of elasticity was lower than the OPC concrete [18]. There is need to establish a correlation between the structural aspects for its mass utilization in construction industry. Among all the structural aspects, bond strength is considered as a significant parameter since it is responsible for composite action over all reinforced concrete members. The strain compatibility maintained with the bond interaction between the concrete and steel. Basically bond strength is the shear force resisted by the contact area between the steel and concrete. It mainly depends on structural characteristics, bar properties and concrete properties. Few researches determined the bond behavior of GPC and observed that the bond strength was equal to and above the bond strength offered by the OPC. The available models from the codal provisioning and empirical equations provides a conservative results for GPC. The present work focuses to understand the bond behavior of GPC in comparison with the conventional concrete through pullout bond test.

2 Research Significance

The present work focusses on comparison of bond behavior of GPC with the conventional concrete of grade 60 MPa. The significant parameters affecting the bond strength, bar diameter (12 and 16 mm), embedment length (2.5D, 5D and Full depth) are considered for the research. Pullout test according to IS: 2770:1967 was considered for experimental investigation of bond behavior for both OPC and GPC. The bond strength and corresponding slip curves were generated and compared between OPC and GPC. The modes of failure for GPC and OPC are compared.

3 Experimental Study

Pullout test according to BIS 2770- Part 1 [19] was conducted on the 18 GPC pullout specimens to evaluate the effect of bar diameter and embedment length. For 12 mm bar diameter 100 × 100 mm and for 16 mm bar 150 × 150 mm cubes were casted.

For the comparison of bond stress with conventional concrete, pullout test was also conducted on 18 OPC Pullout specimens.

3.1 Materials and Mix Designs

The class F flyash used in the study was acquired from locally available National Thermal Power Plant confirmed to IS: 3812–1981 [20]. GGBS from steel industry according to 12089–1987 [21] and silica fume confirming to IS: 15388–2003 [22]. The chemical composition from the X-Ray fluorescent (XRF) analysis for the binder system is given in Table 1. The mix design adopted for the preparation of OPC and GPC are given in Table 2. The physical and mechanical properties of bar for pullout test are given in Table 3. Alkaline activation was provided by sodium hydroxide (NaOH) (SH) and sodium silicate (Na_2SiO_3) (SS) solution. The molar ratio of SiO_2/Na_2O for the Na_2SiO_3 was maintained as 2.19. NaOH was prepared with slakes of 98% purity and density of 2.13. For improving the mixing ability of GPC, sulfonated naphthalene-based super plasticizer of pH 4 was used. Based on the literature the optimized fine and coarse aggregates ratio 45/55 was adopted for the gradation of aggregates. The coarse aggregates of size 12.5 mm maximum size of aggregate was acquired from locally available industry. For the fine aggregate natural river sand confirming to Zone II of IS: 383:2016 [23] was used. The physical properties of coarse and fine aggregates are given in Table 4. For the GPC mix, the concentration of NaOH was maintained at 14M and activator to binder ratio of 0.38. The ratio of SS/SH was considered as 2.5. SH solution was prepared 24 h prior to the casting and 2–3 h before casting SS solution was added to SH solution. Traditional mixing procedure was adopted. After casting of pullout specimens, GPC specimens are allowed to cure at ambient temperature and OPC specimens are cured by water atmospheric temperature for 28 days.

3.2 Mechanism of Bond and Testing Procedure

The composite action in case of reinforced concrete structural elements mainly depends on the bond between the reinforcement and the concrete. The flexural resistance and stiffness offered by the reinforced structural members are due to bond or other mechanical connection. Without any bond interaction between steel and concrete strain compatibility cannot be achieved in reinforced structural members. The bond is offered by chemical adhesion of C–S–H gel in OPC and three types of polymeric gels in GPC, and frictional resistance offered by the reinforcement, and mechanical interlock of ribs as shown in Fig. 1. The mechanism of bond is influenced by mainly three categories like structural characteristics, bar properties and concrete properties. The structural characteristics include the concrete cover and spacing of bars, embedment length of bar, the transverse reinforcement, bar casting position and

Table 1 Chemical composition of binder materials from XRF analysis

Material	Specific surface area (m^2/kg)	Al_2O_3 (%)	CaO (%)	K_2O (%)	MgO (%)	SO_3 (%)	Fe_2O_3 (%)	Na_2O (%)	SiO_2 (%)	L.O.I (%)
Fly ash	322.27	24.35	4.87	0.0031	1.16	0.26	4.54	0.12	63.45	1.97
GBFS	410.26	14.29	39.59	0.79	9.01	0.33	0.52	0.71	34.01	0.25
Silica fume	21,000	5.99	0.001	0.001	0.002	0.22	2.11	0.5	90.24	0.1
OPC	299.25	3.44	64.58	0.76	1.81	3.41	4.31	0.07	21.01	1.41

Table 2 Mix design of OPC and GPC

Material	OPC (kg/m^3)	GPC (kg/m^3)
Fly ash	100	213.05
GGBS	100	273.92
Silica fume	-	121.74
OPC	400	-
Fine aggregate	808	702
Coarse aggregate	817	858
Super plasticizer	2.5	12.2
Water	180	-

Table 3 Physical and mechanical characteristics of steel bar

Property	Diameter of the bar	
	12 (mm)	16 (mm)
Rib spacing (mm)	5.06	6.12
Rib width (mm)	2.04	3.06
Rib height (mm)	2.18	3.24
Yield stress (mm)	512	515
Ultimate stress (mm)	569	571

Table 4 Physical properties of fine and coarse aggregate

Aggregate	Specific gravity	Fineness modulus	Void ratio	Water absorption (%)	Porosity	Bulk density (kg/m^3)
Fine aggregate (FA)	2.68	2.67	0.67	0.71	0.46	1619
Coarse aggregate (CA)	2.73	6.25	0.83	0.63	0.57	1456

the use of non-contact lap splice. The bar properties include bar diameter and geometry, stress in steel and yield strength, tensile strength and fracture energy, type and quantity of aggregate, workability of concrete, fiber reinforcement and the degree of consolidation.

The shear stress generated between the reinforcement and concrete is generally characterized by bond stress, which could be determined by pullout test and beam end tests. From literature study, it was observed that most of the bond studies are conducted on pullout test. The stress distribution was shown Fig. 1. As in case of pullout specimen, the bond stress can be determined from the equilibrium of concrete

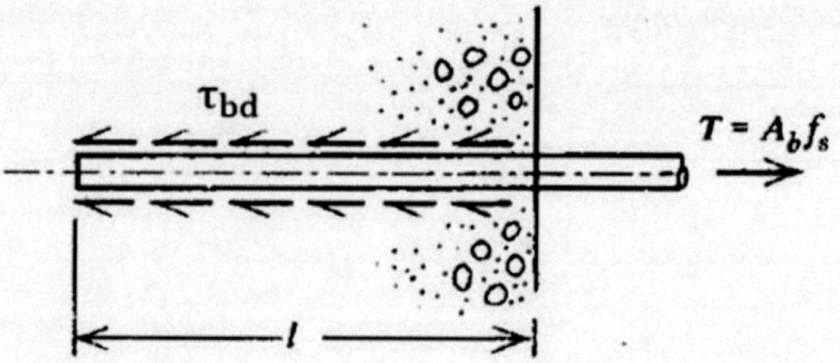

Fig. 1 Anchorage bond

and bar forces. Equation 1 can be used for determination of bond stress in case of straight rebar anchored in concrete similar to the case of pullout test.

$$T = A_b f_s = \tau_{bd} A_{\text{contact}} = \tau_{bd} \Pi d l$$
$$\tau_{bd} = \frac{f_s d_b}{4 l} \tag{1}$$

where, τ_{bd} = bond stress f_s = stress in steel, l = embedment length d_b = bar diameter

According to the IS 2770:1967 Part 1, the pullout specimens are shown in Fig. 2. A 20-mm-thick steel plate with 20-mm wide slit was arranged over the specimen for uniform distribution of stress. The load was applied at a rate of 2250 kg/min. The bond stress is calculated as the load per unit contact area between the steel and concrete. For calculation of bond stress, the Eq. (2) was used.

$$\tau_{\max} = \frac{P_{\max}}{\pi D l} \tag{2}$$

$P_{\max}$ Is the maximum load taken by the specimen i.e., load corresponding to failure of the specimen, kN

D Diameter of the bar, mm

l Embedment length, mm

4 Results and Discussion

The effect of embedment length, diameter of bar on the bond strength of geopolymer concrete are evaluated based on pullout test and compared with OPC concrete.

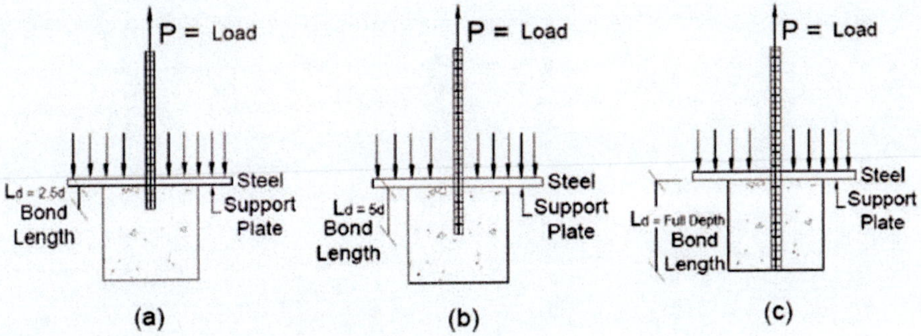

Fig. 2 Force distribution in pullout specimens **a** 2.5D **b** 5D **c** Full depth

4.1 Influence of Embedment Length of Bar on Bond Strength of GPC and OPC

The bond stress-slip curves with respect to embedment length for GPC and OPC are shown in Figs. 3 and 4. It can be observed that, increasing in embedment length of bar shows decrease in bond strength of both OPC and GPC, which indicates the effect of embedment length on GPC is similar to OPC. The maximum bond strength corresponding to embedment length for OPC and GPC are given in Table 5. The maximum bond strength of GPC is higher in all embedment lengths compared to OPC. The average bond strength of GPC is more about 4.25% compared to OPC. In GPC, the bond strength was decreased by 29.89 with increase in embedment length from 2.5D to 5D and for 5D to FD it was 24.46 in 12 mm bar diameter. The average bond strength decreased in OPC and GPC are almost equal indicating the similar response to change in embedment length. From the bond stress vs slip, the initial stiffness for the OPC is higher compared to GPC. With increase in embedment length, the slip corresponding to bond strength was increased, whereas in GPC the slip was decreased.

4.2 Influence of bar Diameter on Bond Strength of GPC and OPC

The bond strength versus slip curves with respect to bar diameter for OPC and GPC are shown in Figs. 5 and 6. It can be observed that with increase in bar diameter there is decrease in bond strength in both OPC and GPC. The maximum bond strength corresponding to bar diameter for OPC and GPC are given in Table 6. The maximum bond strength in 12 and 16 mm bar diameter in GPC is higher compared to OPC. The bond strength with respect to bar diameter in GPC is increased by 5.41% compared to OPC. In GPC, the bond strength is decreased with increase bar diameter. The percentage decrease in bond stress is high in case of full embedment length in OPC

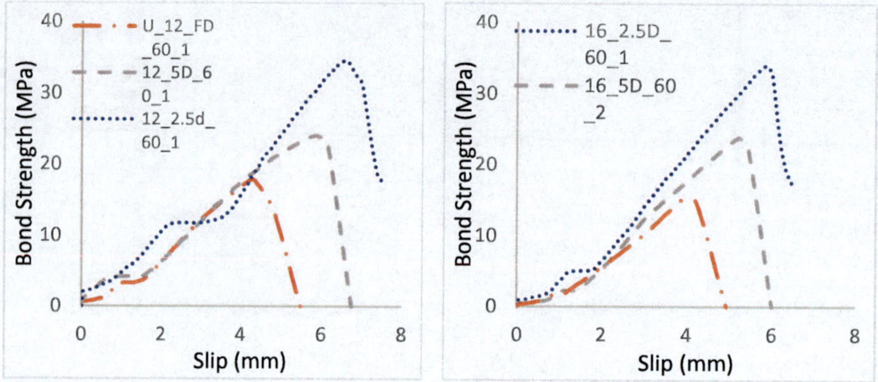

Fig. 3 Bond stress versus slip with respect to embedment length of GPC

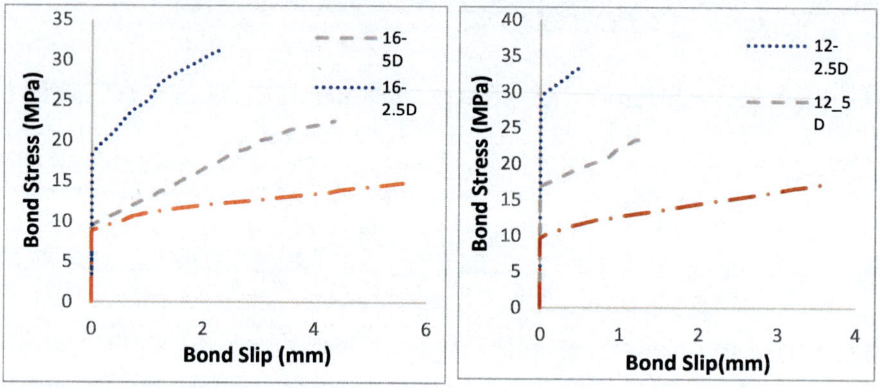

Fig. 4 Bond stress versus slip with respect to embedment length of OPC

Table 5 Maximum bond strength values with respect to embedment length of GPC and OPC

D (mm) –L (mm)	Maximum bond strength			% decrease in OPC	% decrease in GPC
	OPC	GPC	% increase		
12–30	33.42	34.66	3.72	–	–
12–60	23.43	24.06	2.67	29.89	30.60
12–100	17.70	17.87	0.97	24.46	25.71
16–40	31.50	33.93	7.70	–	–
16–80	22.46	23.85	6.21	28.71	29.70
16–150	14.80	15.03	1.52	34.09	37.00
Average	23.88	24.90	4.25	29.29	30.75

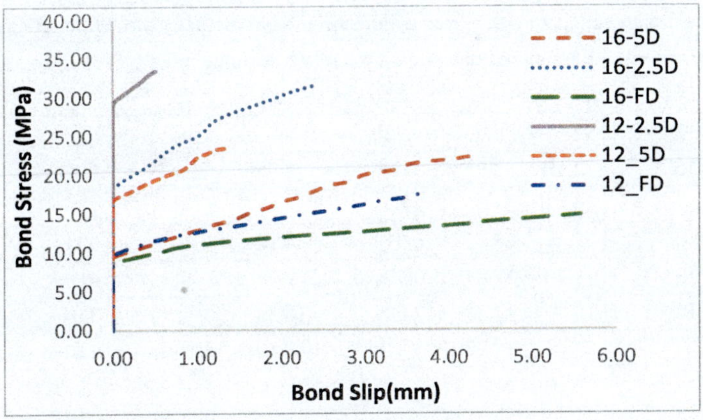

Fig. 5 Bond stress versus slip curves with respect to bar diameter of OPC

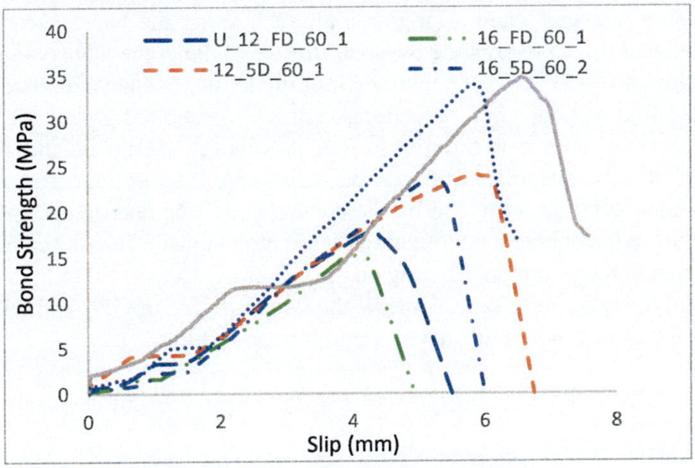

Fig. 6 Bond stress versus slip curves with respect to bar diameter of GPC

and GPC. The average percentage decrease in OPC was 8.16% and GPC was 2.01%. With increase in bar diameter the slip corresponding to bond strength was increases whereas in GPC the slip was decreased.

4.3 Failure Modes of GPC and OPC

The pullout specimens have a tendency to fail in three categories. (i) bar failure, (ii) pullout failure (iii) splitting failure. Bar failure occurs when pullout load acting on

Table 6 Maximum bond strength values with respect to bar diameter of OPC and GPC

L (mm)	D (mm)	Maximum bond stress (MPa)			% decrease in OPC	% decrease in GPC
		OPC	GPC	% increase		
2.5 D	12	33.42	34.66	3.72	–	–
	16	32.10	33.93	5.69	3.95	2.13
5D	12	23.43	24.06	2.67	–	–
	16	22.46	23.85	6.21	4.16	0.86
FD	12	17.70	17.87	0.97	–	–
	16	14.80	17.33	17.08	16.38	3.04
average		23.98	25.28	5.41	8.16	2.01

the specimens exceeds the ultimate load of the bar. This type of failure of specimens the maximum load can't be considered for maximum bond stress failure since before reaching maximum capacity of bond stress ultimate stress of bar is reached. Pullout failure happens when a longitudinal crack along the bar generated by the pullout load and bar tends to slide over the surface of concrete. This type of failure provides ultimate bond stress of concrete. Splitting failure of specimens occurs when pullout load applied on the bar, radial tensile stresses generated around the bar. This radial tensile stresses exceeds the splitting tensile strength of concrete the splitting of concrete occurs and specimen fails abruptly. Since splitting of concrete occurs, the maximum bond strength provided by the concrete can't be determined in this type of failure. Since the concrete no longer resist the radial tensile stresses caused by the pullout load, the maximum load acting on the specimen considered as the maximum bond load of the specimen. Table 7 shows the failure modes of GPC and OPC. It can be observed that GPC and OPC have a similar bond failure.

Table 7 Failure modes of OPC and GPC

D (mm)-L (mm)	Failure mode	
	OPC	GPC
12–30	Pullout of bar	Pullout of bar
12–60	Splitting of concrete	Splitting of concrete
12–100	Splitting of concrete	Splitting of concrete
16–40	Pullout of bar	Pullout of bar
16–80	Splitting of concrete	Splitting of concrete
16–150	Splitting of concrete	Splitting of concrete

5 Conclusions

In this study, the bond behavior of GPC, OPC with same bar dia and embedment length of bar was investigated by using IS 10262 pullout specimens. The bond strength influencing parameters such as embedment length and bar diameters on OPC and GPC are determined. The bond stress vs slip response for OPC and GPC were investigated. Based on this study, the following conclusions are drawn.

- The effect of embedment length on the bond behavior of OPC and GPC are similar and with increasing the embedment length the bond strength was decreased. Since with the increase in embedment length the effect of non-uniform stress distribution along the embedded length of the bar increased and causes the bond stress to decrease. The average bond strength with respect to change in embedment length in GPC was 4.25% which is higher than the OPC.
- With increasing embedment length the slip corresponding to bond strength was decreased in GPC. The magnitude of decreased bond strength with increase in embedment length in GPC and OPC are almost equal, indicating that GPC is a promising alternative material to OPC
- The effect of bar diameter on the bond behavior of GPC and OPC are similar and with increasing the bar diameter the bond strength was decreased. The average bond strength with respect to bar diameter in GPC was 5.41% higher than the OPC.
- Similar to the embedment length the slip correspond to bond strength was decreased in GPC. The average decrease in bond strength with increase in bar diameter is higher than OPC. The percentage decrease in bond strength was higher in full embedment lengths of OPC and GPC.
- The failure modes of GPC are similar to the OPC indicating that the bond behavior of GPC is almost similar to the OPC.

References

1. Alexander AE, Shashikala AP (2022) Studies on the microstructure and durability characteristics of ambient cured FA-GGBS based geopolymer mortar. Constr Build Mater 347. https://doi.org/10.1016/j.conbuildmat.2022.128538
2. Nagajothi S, Elavenil S (2021) Effect of GGBS addition on reactivity and microstructure properties of ambient cured fly ash based geopolymer concrete. https://doi.org/10.1007/s12633-020-00470-w/Published
3. Ou Z, Feng R, Mao T, Li N (2022) Influence of mixture design parameters on the static and dynamic compressive properties of slag-based geopolymer concrete. J Build Eng 53. https://doi.org/10.1016/j.jobe.2022.104564
4. Trabacchin G, Sebastian W, Zhang M (2022) Experimental and analytical study of bond between basalt FRP bars and geopolymer concrete. Constr Build Mater 315:125461. https://doi.org/10.1016/j.conbuildmat.2021.125461
5. Davidovits J (1993) Geopolymer cements to minimize carbon dioxide greenhouse warming. Ceram Trans 37:165–182

6. Verma M, Dev N (2021) Effect of liquid to binder ratio and curing temperature on the engineering properties of the geopolymer concrete. SILICON. https://doi.org/10.1007/s12633-021-00985-w
7. Amani K, Reddy YR (2018) Comparative study on bond strengths of reinforcing bars embedded in conventional and geopolymer concretes. Int J Eng Tech 4:1303–2395
8. Cong X, Zhou W, Elchalakani M (2020) Experimental study on the engineering properties of alkali-activated GGBFS/FA concrete and constitutive models for performance prediction. Constr Build Mater 240. https://doi.org/10.1016/j.conbuildmat.2019.117977
9. Shehab HK, Eisa AS, Wahba AM (2016) Mechanical properties of fly ash based geopolymer concrete with full and partial cement replacement. Constr Build Mater 126:560–565. https://doi.org/10.1016/j.conbuildmat.2016.09.059
10. Diaz-Loya EI, Allouche EN, Vaidya S (2011) Mechanical properties of fly-ash-based geopolymer concrete. ACI Mater J 108:300
11. Mustakim SM, Das SK, Mishra J, Aftab A, Thamer, Alomayri S, Hasan, Assaedi S, Cyriaque, Kaze R (2021) Improvement in fresh, mechanical and microstructural properties of fly ash-blast furnace slag based geopolymer concrete by addition of nano and micro silica. https://doi.org/10.1007/s12633-020-00593-0/Published
12. Nguyen H-A, Chang T-P, Shih J-Y (2018) Engineering properties and bonding behavior of self-compacting concrete made with no-cement binder. J Mater Civ Eng 30:04017294. https://doi.org/10.1061/(asce)mt.1943-5533.0002136
13. Saranya P, Nagarajan P, Shashikala AP (2021) Experimental investigation on bond strength properties of geopolymer concrete. Lect Notes Civil Eng 83:731–740. https://doi.org/10.1007/978-981-15-5644-9_57
14. Kaze RC, Beleuk à Moungam LM, Cannio M, Rosa R, Kamseu E, Melo UC, Leonelli C (2018) Microstructure and engineering properties of Fe_2O_3(FeO)-Al_2O_3-SiO_2 based geopolymer composites. J Clean Prod 199:849–859. https://doi.org/10.1016/J.JCLEPRO.2018.07.171
15. Ding Y, Dai JG, Shi CJ (2016) Mechanical properties of alkali-activated concrete: a state-of-the-art review. Constr Build Mater 127:68–79. https://doi.org/10.1016/J.CONBUILDMAT.2016.09.121
16. Lee NK, Lee HK (2013) Setting and mechanical properties of alkali-activated fly ash/slag concrete manufactured at room temperature. Constr Build Mater 47:1201–1209. https://doi.org/10.1016/J.CONBUILDMAT.2013.05.107
17. Ling Y, Wang K, Li W, Shi G, Lu P (2019) Effect of slag on the mechanical properties and bond strength of fly ash-based engineered geopolymer composites. Compos B Eng 164:747–757. https://doi.org/10.1016/J.COMPOSITESB.2019.01.092
18. Nath P, Sarker PK (2017) Flexural strength and elastic modulus of ambient-cured blended low-calcium fly ash geopolymer concrete. Constr Build Mater 130:22–31. https://doi.org/10.1016/j.conbuildmat.2016.11.034
19. BM (1967) Methodes of testing bond in reinforced concrete. Is : 2770
20. BIS:3812-1981 (1992) Specification for fly ash for use as pozzolana and admixture. Bureau of Indian Standards, New Delhi, India 1981
21. IS:12089-1987 (1987) Specification for granulated slag for the manufacture of Portland slag cement. Bureau of Indian Standard, New Delhi, pp 1–14
22. IS:15388 (2003) Silica fume-specification. Bureau of Indian Standards, New Delhi, India, pp 1–13
23. IS:383-2016 (2016) Indian standard coarse and fine aggregate for concrete- specification. Bureau of Indian Standards, New Delhi, India, pp 1–21

A Study on Pull-Out Strength of H-end Rebar in Normal Strength Concrete

Adla Saraswathi, C. B. K. Rao, D. Rama Seshu, and Aagya Dahal

1 Introduction

Reinforced concrete is a composite material made up of two components, concrete and steel, with unequal behaviour and physical features. Satisfactory operation of reinforced concrete as construction material depends on combined action of concrete and steel. For the composite action of reinforced concrete, the bond between steel and concrete is very important and essential for no slip condition in a loaded structure. The load is transferred from the concrete to the bar by means of bond/anchorage and develops at the interface of the steel and concrete [1–4]. Generally, a reinforcement bar in concrete fails in bond either by direct pull-out or by splitting failure. In direct pull-out failure the bond between concrete and reinforcement fails when the bar is subjected to a tensile force. Where as in splitting failure the deformations or lugs present on the bar develops a wedging action and exerts a lateral tension on concrete, results in development of a crack. The full bond between rebar and concrete will develop by providing a development length that is minimum length of the bar need to embed in concrete to develop full stress in the bar at that section. However, the actual available length will be not sufficient for full development of bond, to develop full bond stress anchorages needs to be provide, like hooks [5], bends and mechanical anchorages [6] conforming to standards [7, 8].

Conventional anchorage system like hooked bars poses certain drawbacks such as steel congestion, honeycombing in concrete, difficulties in compaction, detailing

A. Saraswathi (✉) · C. B. K. Rao · D. R. Seshu · A. Dahal
Department of Civil Engineering, National Institute of Technology, Warangal 506004, India
e-mail: asaraswathi@student.nitw.ac.in

C. B. K. Rao
e-mail: raocbk@nitw.ac.in

D. R. Seshu
e-mail: drseshu@nitw.ac.in

R. K. Pancharathi et al. (eds.), *Low Carbon Materials and Technologies for a Sustainable and Resilient Infrastructure*, Lecture Notes in Civil Engineering 440,
https://doi.org/10.1007/978-981-99-7464-1_35

problems and problems in joint strength. This leads to the development of mechanical anchorage system, i.e. headed bars which consist of heads of different head shapes, square, rectangular, circular or oval and different head profiles [9], attached at bar end by friction or forged welding or by threading technique [10]. They provide ease of installation, outnumbering the disadvantages of conventional anchorages without affecting the structural performance. As in beam-column joints, there may be the formation of plastic hinges. Use of headed bars helps in their relocation, avoiding the requirement of increased depth of beam and longitudinal bar [11–14]. Contribution to anchorage/bond in case of headed bars is provided by bearing and bond promoting the development of diagonal compression mechanism within the joint. The couplers create least congestion compared to hooks and headed bars. However, couplers have its difficulty in threading, aligning the bar and installation at the site.

Some of the investigations observed that the bond stress decreases with increase in diameter and length of rebar in normal concrete and high strength concrete. The bond strength is more in case of high strength concrete i.e., it is nearly 15–28% more compared to normal strength concrete [15]. In high strength concrete the brittleness will be more. The brittleness of high strength concrete can be reduced by inclusion of fibres in concrete mix. These fibres increase bond strength by arresting longitudinal splitting cracks but in pull-out failure the bond performance unaltered with addition of fibres because splitting cracks are not formed in this failure [16].

Fibres of different geometry hooked, crimped, twisted etc. are used in fibre reinforced concrete. Fibres with spiral geometry are one kind. A study was carried out with three embedment lengths, coil diameter, pitch length and pull-out rates with steel fibre of 0.6 mm diameter. Test results showed that the increase of embedment length resulted in rupture of fibre instead of pull-out, large diameter coil had less pull-out resistance and large pitch resulted in smaller displacement [17–20]. Spiral geometry was found to be superior in terms of the frictional bond, ductility, crack control, dynamic strengths and energy absorption [21].

2 Research Objective

It is well known that any amount of bend at the end of rebar provide a proportionate extent of anchorage accounting for equivalent development length, but standard bends and hooks lead to some congestion. Rebars are connected by couplers to avoid overlapping and for the continuation of bars. Another alternative is to provide, head at the end of the rebar. In fact, even overlap, headed bar or coupler may give rise to some amount of congestion. The Authors explored alternatives to meet the need of development length/anchorage of rebar. Pull out tests are conducted on concrete specimen with embedded H-end rebars to study the anchorage/bond.

3 Experimental Program

An experimental programme was designed to conduct the pull-out of helical end reinforcement embedded in a concrete cylinder of M35 grade. Rebar of 6 8 and 10 mm diameter (*d*) of yield strength 650 ultimate strength 700 N/mm^2 was used in the present study. Typical helical end rebar is shown in Fig. 1. The parameters studied are the rebar diameter and number of turns of the helix. Three different diameter of the rebar, i.e., 6, 8 and 10 mm and three variations in the number of turns (coils), i.e., one, one and half turns, and two turns are used as parameters for H-end rebars. There are nine parameters to study the influence of helical end on the anchorage effect. There are three specimens to arrive at average for each type. The specimens are designated with diameter-number of turns. Hence the specimen designated as A-1 means the diameter of rebar 6 mm and one turn. Here, A, B and C represents 6, 8 and 10 mm diameter rebar and 1, 1.5, 2 represents number of turns of helix. The helical end is embedded at a depth of 5d (30 mm for 6 mm rebar 40 and 50 mm for 8 mm and 10 mm rebar respectively) from top of the specimen.

3.1 *Materials*

Ordinary Portland Cement (OPC) of 53 grade (IS 12269-2013) was used [22]. Fine aggregate passing through 4.75 mm sieve and retained on 150-micron sieve (IS 383–2016) of Zone II gradation was used. The coarse aggregate consists of 60% passing through 12.5 mm sieve and retained on 10 mm sieve and 40% passing through 10 mm sieve and retained on 4.75 mm sieve (IS 383-2016) was used [23]. Mix proportions conforming to M35 are arrived at by trial and error, as given in Table 1, obtained as per IS: 10262-2009 [24]. Three standard cubes of size 150 × 150 × 150 mm were

Fig. 1 Helical end rebar (H-end rebar)

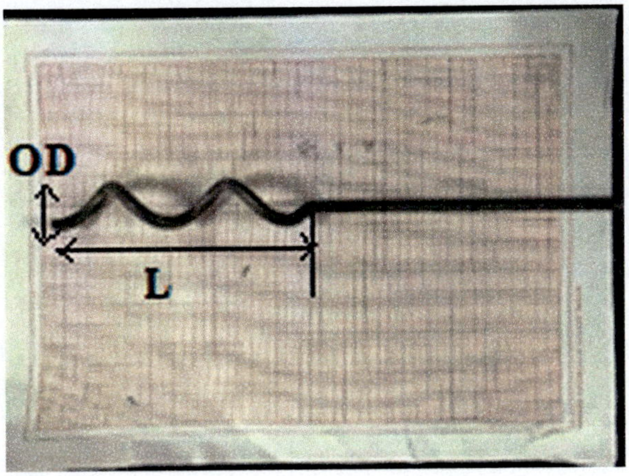

Table 1 Concrete mix composition details

Concrete grade	Mix proportion	Quantity of concrete materials per Cum				
		Cement (kg)	Fine aggregate (kg)	Coarse aggregate (kg)	Water (kg)	SP (kg)
M35	1:1.67:2.9 w/c 0.42	405	678	1198	170	3.4

SP-superplasticizer (conplast SP430)

used to determine the compressive strength as control specimens for each batch of casting and obtained the average compressive strength of concrete is 37 N/mm^2.

Helical end Rebar (H-end Rebars). Rebar of 6, 8 mm and 10 mm nominal diameter is wound on a spring making machine. Then the bar was bent to 90 degrees so that the axis of the rebar is parallel to the axis of the helical end, as shown in Fig. 1. All the specimens are wound to five turns in the factory and cut to required turns in the laboratory. Internal diameter and pitch are measured before casting each specimen.

3.2 Casting of Specimens

Standard Cylindrical moulds, diameter 150 mm and height 300 mm are used to cast the specimens. H-end rebars were placed vertically such that the helix would lie precisely at the centre of the mould and 5d from top of the mould. Concrete was filled in the moulds with proper hand compaction without disturbing the position of rebar as shown in Fig. 2.

3.3 Test Rig

A test rig was fabricated to hold the specimen and to perform the pull-out test on the concrete cylinder with embedded H-end rebars. The specimen with Test Rig shown in Fig. 3.

3.4 Testing of Concrete Specimens

Specimens were cured for 28 days. The specimen was set up in the test rig. Then this entire setup was fixed in the UTM, as shown in Fig. 4 to conduct the pull-out test as per IS:2770 (part 1) [25]. Dail guage (0.001 mm least count and 50 mm maximum travel capacity) was attached to the bar to record the displacement of helical end

Fig. 2 Specimens cast in moulds

Fig. 3 Specimen with test rig

Fig. 4 Test setup

rebar. The load was applied till either the concrete failed or the bar failed. Real-time load and displacement readings are captured through data acqasition system.

4 Results and Discussion

4.1 Behaviour Under Load

Concrete cylinders of grade M35 were used to test H-end rebars embedded in it to study their pull-out behaviour. The tested specimens are shown in Fig. 5, for 6, 8 and 10 mm. An examination of these specimens' shows that 6 mm specimens and 8 mm specimens have not shown any cracks and the reinforcing bar has failed under the load. In case of specimens with 10 mm concrete cracked but the bars were not pulled out and results were tabulated in Table 2. Further examination of the specimens revealed that the 10 mm bars showed a tendency to unwind and to attain the normal straight bar shape. In this process, the reinforcing bar has exerted an outside pushing force creating tension in the concrete. Thus, the specimens have developed cracks only from the top of the specimens up to the helix and the helix was found to be well embedded in the concrete. So, this clearly shows that helical form of reinforcement is providing sufficient anchorage.

Fig. 5 Specimen failure pattern of **a** 6 mm **b** 8 mm and **c** 10 mm diameter rebars

Table 2 Results of the tested specimens

0	*D* mm (2)	*P* Mm (3)	N	Peak load (kN) (4)	Displacement at peak load (mm) (5)	K (6)	σ_u (N/mm^2) (7)	Comments (8)
A-1	18	60	1	17	5.6	0.3	3.42	Bar failed
A-1.5			1.5	18	1.92	0.45	2.49	Bar failed
A-2			2	19.3	1.7	0.6	2.03	Bar failed
B-1	24	80	1	30	5.7	0.3	3.39	Bar failed
B-1.5			1.5	32	4.14	0.45	2.49	Bar failed
B-2			2	33	4.5	0.6	1.95	Bar failed
C-1	30	100	1	39	7.9	0.3	2.82	Concrete failed
C-1.5			1.5	37	1.48	0.45	1.84	Concrete failed
C-2			2	38	2.5	0.6	1.42	Concrete failed

D-diameter of the *H*-end rebar, *P*-pitch of the *H*-end rebar, *N*-number of turns of the *H*-end rebar, *K*-helix coefficient, and σ_u—bond stress at ultimate

4.2 Load–Displacement Curves

A dial gauge was attached to the reinforcing bar to measure the relative displacement of the bar with respect to concrete. These load displacement curves for 6 and 8, 10 mm are given in Fig. 6. An examination of these load displacement curves shows that almost up to 80% of the peak load there is no displacement of the bars. Beyond this, the relative displacement was observed to increase slowly and then displacements increased at a faster rate when the load reached the stage of failure of bar or concrete.

So, this clearly shows that the helical form for reinforcing bars have given very good anchorage and allowed the reinforcing bar to develop its full stress. With increase in the diameter of rebar and number of turn load-carrying capacity increases shown in Fig. 7.

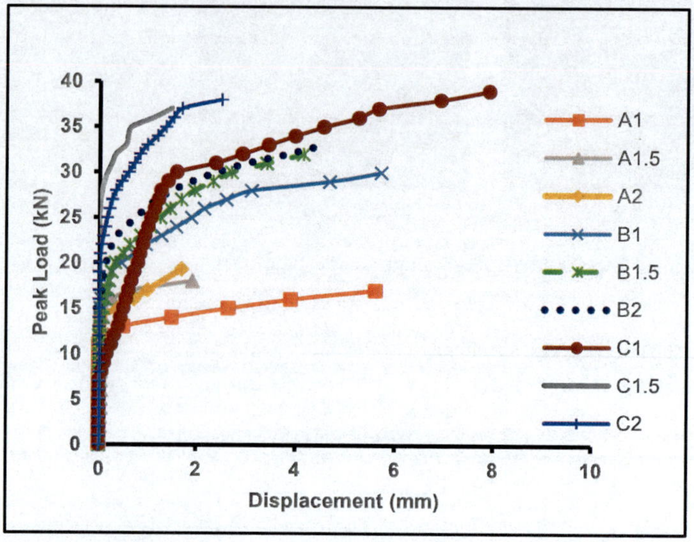

Fig. 6 Load–Displacement curves

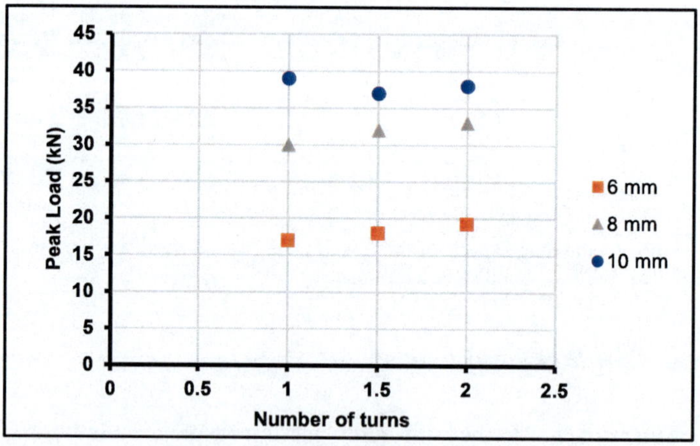

Fig. 7 Relation between peak load and number of turns

4.3 Effect of Helix Coefficient on Bond Stress

Coil diameter, pitch and number of turns effect anchorage/bond, these three are combined, and taken as helix coefficient (K) [26].

$$K = [(D/P) * N] \quad (1)$$

where, D-diameter of H-end rebar, P-pitch of H-end rebar and N-number of turns of H-end rebar

Observation of tested specimen vide Fig. 5, shows that all helical turns are intact in concrete at ultimate, offering good anchorage, except for first half turn. Hence the helical end of rebar is visualised as a solid cylinder with outer diameter (OD) of helix as diameter of cylinder and depth of embedment as length of cylinder (L) (Fig. 1). Bond stress was calculated using the outer diameter of the coil and depth of embedment of the helical end. Bond stress at ultimate (σ_u) is calculated (for 6 mm, 8 mm, 10 mm diameter rebars) and is given in column number 7 of Table 2. Points are plotted for the specimens with helix coefficient (K) as absica and bond stress at ultimate as ordinate. There is good correlation between helix coefficient (K) and bond stress (σ_u) with R^2 value 0.99 as shown in Fig. 8. Bond stress between H-end rebar and concrete decreased with increase in helix coefficient. Concrete cylindrical specimens with helically wound rebar diameter 6 and 8 mm are shown similar correlation between helix cofficient and ultimate bond stress, where as for 10 mm diameter rebar has correlation between helix coefficient and ultimate bond stress is different, in this full bond stress not developed bacause of failure of concrete.

5 Conclusions

Helical end rebars have shown good anchorage capacity. However, at the kink of the helix, the concrete need to be provided with necessary hoops to resist the tensile forces developed in concrete due to opening of the kink. The load-carrying capacity increases with increase in the diameter of rebar and number of turns. Parameter influencing the behaviour of Helical end rebar is found to be diameter of helix, pitch and number of turns. All these parameters are combined into a new parameter named as Helix Coefficient (K). There is good correlation between helix coefficient (K) and bond stress (σ_u) with R^2 value 0.99 and bond stress between H-end rebar and concrete decreased with increase in helix coefficient.

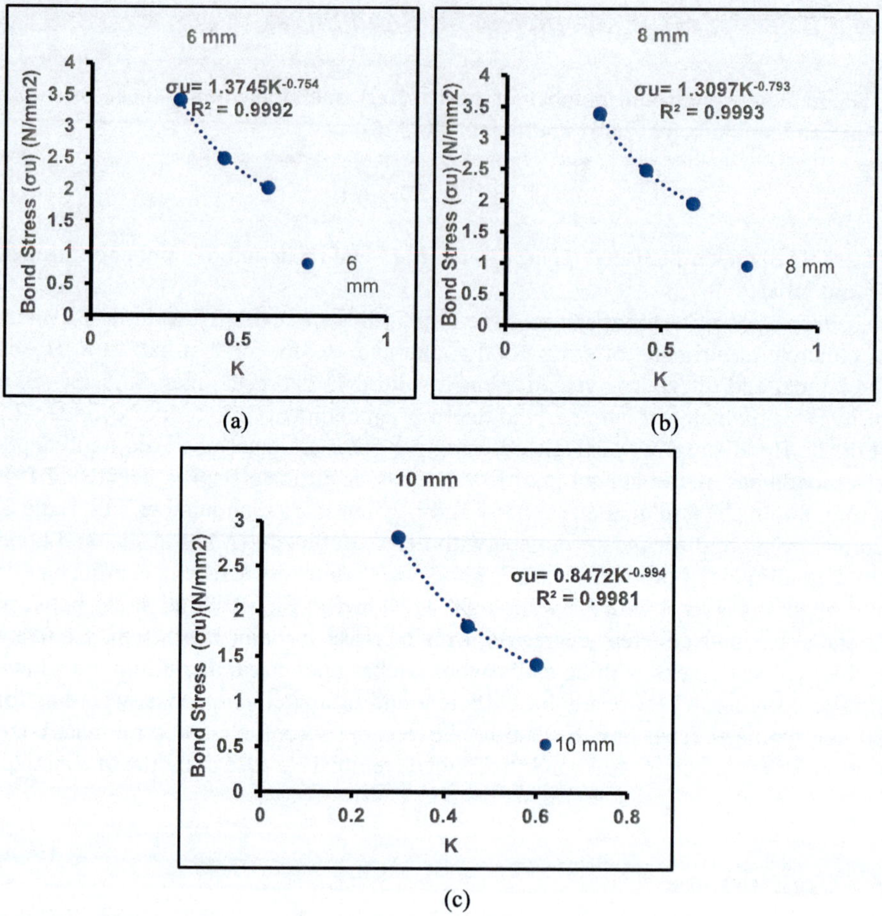

Fig. 8 Relation between bond stress at ultimate and helix coefficient for **a** 6 mm **b** 8 mm and **c** 10 mm diameter rebars

Acknowledgements The authors would like to thank the NIT Warangal, India, for providing research facilities to carry out this work.

References

1. Saleem M, Al-Kutti WA (2015) Non-destructive testing procedure to evaluate the load-carrying capacity of concrete anchors. J Constr Eng Manage 142(5)
2. Osman MH, Shukor MNMA et al (2018) Comparison of pigtail with J anchor bolt in normal concrete. MATEC Web of Conferences
3. Abrishami HH, Mitchell D (1996) Analysis of bond stress distributions in pull-out specimens. J Struct Eng 122(3)

4. Tepfers R (1973) A theory of bond applied to overlapped tensile reinforcement splices for deformed bars. Chalmers University of Technology, Sweden
5. Joj O, Goto Y (1999) Anchorage behaviour of 90-degree hooked beam bars in reinforced concrete knee joints. J Trans Built Environ 38(2)
6. Thompson MK, Jirsa JO, Breen JE, Klingner RE (2002) Anchorage behavior of headed reinforcement : a literature review. Res Rep 1855-1, 7(22):1–114
7. IS 456-2000 Plain and reinforced concrete-code of practice. Bureau of Indian Standards. New Delhi, India
8. ACI 318-14 (2019) Building code requirements for structural concrete and commentary. Reported by American Concrete Institute Committee
9. Singhal S, Chourasia A, Parashar J (2020) Anchorage behaviour of headed bars as connection system for precast reinforced concrete structural components. J Struct
10. Chourasia A, Gupta S (2019) Influential parameters for headed bars in RC beam-column joint. Curr Sci
11. Eom TS, Park HG, Hwang HJ, Kang SM (2016) Plastic hinge relocation methods for emulative PC beam-column connections. J Struct Eng (United States)
12. Kang THK, Shin M, Mitra N, Bonacci JF (2009) Seismic design of reinforced concrete beam-column joints with headed bars. J ACI Struct
13. Ghimire KP, Shao Y, Darwin D, O'Reilly M (2019) Conventional and high-strength headed bars-Part 1: anchorage tests. J ACI Struct J
14. Abed HS, Abed JM, Al-Rawe HS (2021) Anchorage behavior of headed bars in reinforced concrete beams. J Struct 29:1954–1959
15. Bashir MT (2019) Pull-out behavior of conventional steel reinforcement in normal and high strength concrete. Int J Sci Eng Sci
16. Ahmed K, Mahmood (2014) Effect of cover on bond behaviour of fiber reinforced high strength concrete. Int J Sci Eng Res
17. Hao Y, Hao H (2017) Pull-out behaviour of spiral-shaped steel fibres from normal-strength concrete matrix. J Constr Build Mater
18. Hao Y, Hao H (2013) Dynamic compressive behaviour of spiral steel fibre reinforced concrete in split Hopkinson pressure bar tests. J Constr Build Mater
19. Xu Z, Hao H, Li HN (2012) Dynamic tensile behaviour of fibre reinforced concrete with spiral fibres. J Mater Des
20. Hao Y, Hao H, Chen G (2016) Experimental investigation of the behaviour of spiral steel fibre reinforced concrete beams subjected to drop-weight impact loads. J Mater Struct Constr
21. Hajsadeghi M, Chin CS (2019) Geometrical and material optimisation of deformed steel fibres: spirally deformed fibres. J Eng Struct
22. IS:12269-2013 Ordinary Portland cement, 53 grade specification (First Revision). Bureau of Indian Stand New Delhi, India
23. IS 383-2016 Coarse and fine aggregate for concrete—Specification. Bureau of Indian Satandards
24. IS: 10262-2009 Concrete mix proportioning—Guidelines (First Revision). Bureau of Indian Satandards
25. IS: 2770-2007 (part-1)-2007 Methods of testing bond in reinforced concrete, part 1: pull-out test. Bureau of Indian Standards
26. Saraswathi A, Rao CBK, Seshu DR (2021) Pull-out tests to study anchorage/bond on concrete specimen with embedded Helical End Rebar (HER). J Struct Eng 48(4):265–273

Fatigue Analysis of Beam Column Joint

T. Bhavani Chowdary, A. Raghu, A. Bharat, and Charan

1 Introduction

Beam-column junction performance has a considerable impact on how Reinforced Concrete (RC) structures react to seismic occurrences. For nonseismically developed (NSD) structures, where transverse reinforcement is incredibly uncommon at beam-column joints, this is undoubtedly relevant. Previous seismic activities have proven that the failure of beam-column junctions has a negative effect on the general performance of RC buildings and can potentially cause their collapse.

A key factor in how reinforced (RC) structures react to seismic activity is the functionality of the beam column joints. Nonseismically designed (NSD) constructions refer to buildings that were not specifically designed to resist earthquakes. These buildings often lack the necessary features, such as transverse reinforcement, to withstand seismic forces. Transverse reinforcement is typically used in beam-column connections to provide additional strength and ductility to the structure. During an earthquake, the beam and column experience horizontal forces that can cause the beam to shift or rotate at the connection. Without proper reinforcement, this can lead to the failure of the connection and collapse of the structure. The most of the time, the behavior of joints is assessed at the level of the subassembly, where other sides of the column are hinge-supported and the contraflexure sites in a construction are typically

T. Bhavani Chowdary (✉) · A. Raghu · A. Bharat · Charan
Department of Civil Engineering, Vardhaman College of Engineering, Hyderabad, India
e-mail: Bhavanitalasila971@vardhaman.org

A. Raghu
e-mail: atheliraghuvamshidhar20ce@vardhaman.org

A. Bharat
e-mail: akulabharatraj19ce@vardhaman.org

Charan
e-mail: vallapucharan20ce@vardhaman.org

R. K. Pancharathi et al. (eds.), *Low Carbon Materials and Technologies for a Sustainable and Resilient Infrastructure*, Lecture Notes in Civil Engineering 440,
https://doi.org/10.1007/978-981-99-7464-1_36

found near the midpoint of the columns and beams. The vertical shear load applied to the beam tip occurs periodically. The first report of such tests was made by Hanson and Connor [1], who showed that joints without hoops can cause the subassembly to behave poorly under load displacement. The diagonal cracks that finally developed in the junction eventually resulted in a brittle joint shear collapse that left the transverse reinforcing bars of the beam unyielding. The things found out of this development spurred a frenzy of investigation into the empirical and numerical aspects of the dynamic response of beam column joints, results supported the early appearance of the JS failure. Being brittle and absorbing significantly less energy than a flexural failure characterized by steel yielding, this failure mode is understandably undesired in the seismic design philosophy. The literature has looked into the sub-assemblies' susceptibility to non-seismic detailing that has been used. They include lap splices in columns directly above the joint, joint cores without insufficient transverse or lateral reinforcement, and inadequate anchorage of the beam reinforcing reinforcement in the junction, to name a few. It is possible to conclude that the joint shear strength of beam-column connections decreases with an increasing aspect ratio by analyzing the joint shear strength values of these connections and their aspect ratios, as described in literature [2–4].

Buildings also feature longitudinal beams and integrally cast slabs, but much study has concentrated on two dimensional sub-assemblies. In order to effectively represent the joint shear behavior, it is essential to consider 3D phenomena such slab contributing and restriction of the longitudinal beams to the joint. This study examines how the aspect ratio affects the connections between 3D NSD external beams and columns that have transverse beams and slabs. The examination is carried out as part of a numerical analysis using the University of Stuttgart's MASA FE code.

2 Influence of Joint Aspect Ratio

The average notion is typically employed to assess the joint shear capacity. Simple technique for strain and stress According to design rules for new structures, the average active horizontal joint's shear stress must be kept under permissible bounds for minimizing the joint shear failure. FEMA 356 [5] contains suggested parameters for the necessary joint shear stress for analysis of the beam column joints for current RC structures that weren't built seismically. The principal tensile stress (pt) technique is a different model based on the same methodology. Acting joint stresses at the horizontal and vertical axes are converted into principal stresses using the Mohr circle theory. Priestley [6] provided initial joint cracking values of $p_t = 0.29f'c^{0.5}$ and final joint strength values of $p_t = 0.42f'c^{0.5}$ for 2-dimensional external joints with rebar bent in the intersection with a 90° angle hook. The CEB240 [7] recommends these values.

The joint shear strength's critical values should be determined while taking into consideration the following, though the aforementioned design rules did not take aspect ratio (alpha) into account. In his study of 2-Dimensional joints with the three

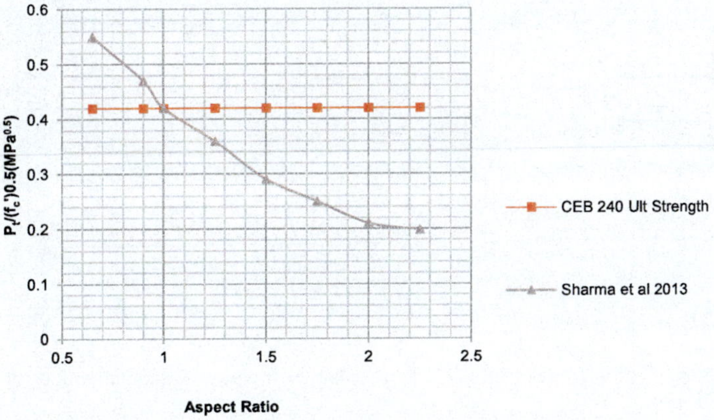

Fig. 1 Database containing information on beam-column connections with longitudinal beam bars bent to form exterior joints

distinct aspect ratios (1, 1.5, and 2), According to Wong [8], the joint shear strength rises with the ratio of the depths of the beam to column. Similar findings have been observed regarding corner joints between a longitudinal beam and slab, Park [2] and Hassan [3] both agreed. A data of 50 2D outer beam-column connections with bending beam transverse reinforcement has been constructed in order to better understand the influence of joint aspect ratio.

Figure 1 shows how a higher aspect ratio affects the joint shear strength. Internal twisting stresses within the beam and within the column combined to produce a transverse or lateral strut (S) in the joint wall itself when a descending shear force will be applied to tip of the beam [9].

The joint shear force V_{jh} which operate in the horizontal direction and joint shear force V_{jv}, which operate in the vertical direction are illustrated in Fig. 2c. The lateral joint shear force reaches horizontal stability at the middle of the joint panel and adjusts at this point the differential in between shear force with in column and the force applied in the transverse beam bars. Because the lateral force of diagonal strut S resists V_{jh}, the amount of compressive stresses in somewhat slant struts must increase to oppose the same V_{jh}. Figure 2 visually depicts it. Figures 2a and 2b additionally, it could be recorded that once diagonal fracturing in joint begins Situated between the steel reinforcing bars, the truss stabilizes.

The slab's presence deserves additional emphasis. The torsion of the transverse beam is believed to impart joint shear [10], which is said to provide a mechanism for resisting the tensile pressures in the slab bars. To calculate V_{jh} (as shown in Fig. 2c), the tensile force of the longitudinal beam bars (T_{sb}) and the force applied by the slab frames (T_{ss}) are combined. When calculating pt in this study, this information is taken into consideration.

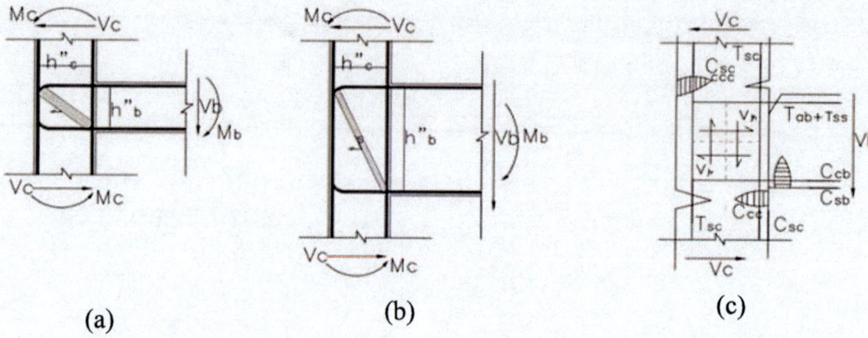

Fig. 2 Formation of diagonal strut for **a** α = 1, **b** α = 2 and **c** horizontal and vertical shear of exterior BCJ

3 Numerical Studies

3.1 FE Model

ANSYS v.22 is used to run the numerical simulations. The relaxed kinematic constraint micro plane model is the constitutive law for concrete [9, 11, 12]. Using 8-node hexahedral components with mesh widths ranging from 20 to 30 mm, we simulate concrete. Modeling of re-steel involves the use of 2-node elements of truss using a tri-linear uniaxial stress strain law. Lettow's bond connection components with zero width are used to illustrate the bonding between reinforcing steel and concrete.

3.2 Explored Connections (Beam-Column Joints)

In this work, the examined beam-column junctions are depicted in Fig. 3. The purpose of the study is to assess how the principal tensile stress pt has changed over time in relation to various aspect ratios. It is specifically looked at how the existence of transverse beams and slab affects how the pt trend for 3D joints differs from that for 2D joints. It should be noted that the behavior of the joints is typically analyzed under the assumption that the beam is uniformly pulled downwards, as would occur when the slab is in tension.

The sub-assembly of the beam-column joint and the example with a 400 mm beam level is shown in Fig. 4 along with its dimensions and reinforcing features. The 400 × 400 mm column cross section is consistently preserved, and aspect ratio is consistently altered by merely adjusting beam height. When the concrete coverings are 50 and 50 mm from the bar diameter and bar core., aspect ratio in Fig. 4 is equal

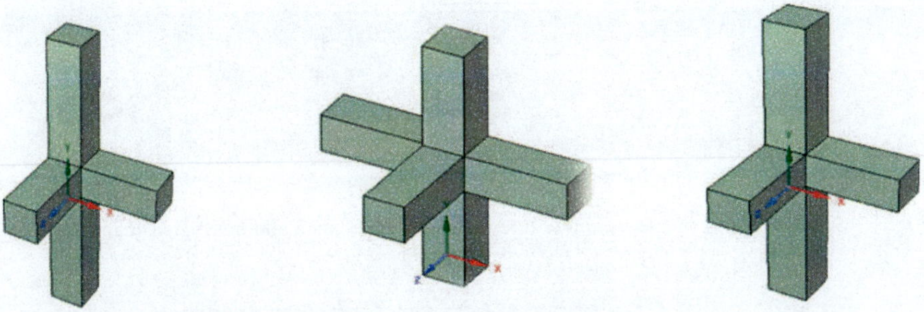

Fig. 3 Exterior BCJ; **a** 3D joint **b** Corner joint and **c** Edge joint

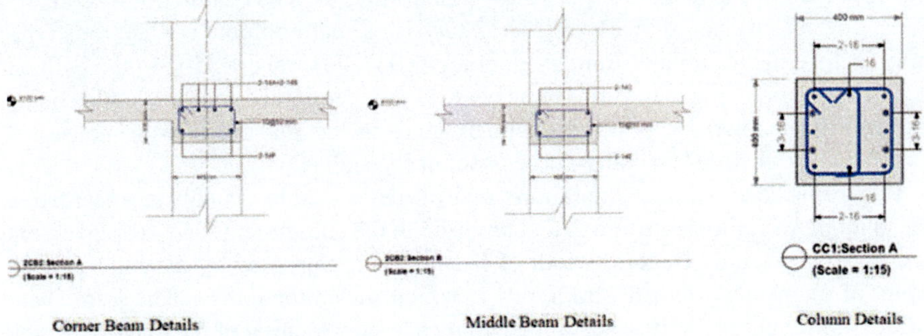

Fig. 4 Structural configuration and intricate features of beam-column joints

to 1.3. For every simulation, concrete cylinder with a 30 MPa compressive strength is used.

The longitudinal beam's transverse beam bars were similarly fastened with a 90° angle hook in the corner joint case. In view of an edge junction, the bars penetrate exactly through it. Bottom slab reinforcement bars are fixed straight, but the top slab reinforcement bars end together in 90° hook. The sub-width, assembly's which consists of the column and width of slab, is 1200 mm and for edge joint and 2050 mm for edge junction. The statistical simulations are summarized in Table 1.

Table 1 Outline of virtual simulations

Joint type	Aspect ratio			
3D	0.625	0.90	1.13	2.0
Corner	0.625	0.90	1.13	2.0
Edge	0.625	0.90	1.13	2.0

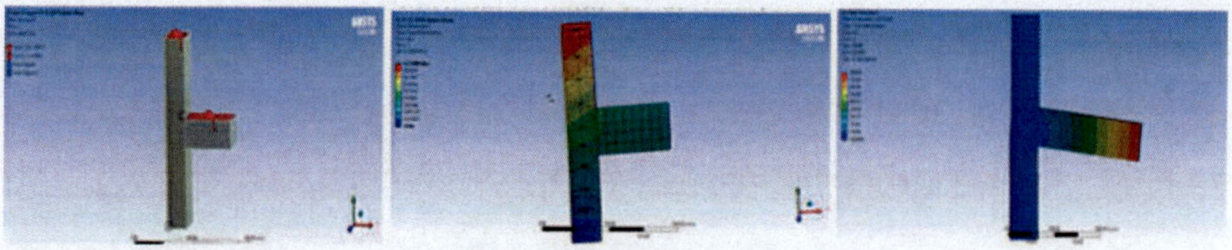

Fig. 5 Modelled Ansys members and failure pattern of edge beam and transverse beam

3.3 Formation of Cracks in 3D Beam-Column Joints

The simulated load–displacement curves illustrate the different stages of cracking, aiding in the comprehension of the behavior of the investigated beam-column connections. The various stages involved in the analysis of beam-column connections typically include the following: Joint shear cracking (J), flexural cracking in the column (C), torsional cracking in the transverse beam (T), flexural fracture in the loaded beam for two-dimensional joints (B), flexural fracture in the slab for three-dimensional joints (S), and flexural cracking at the onset of the slab (O) (T).

Regarding corner joint at the failure, many types of cracks visually represented in Fig. 5 along with the descriptions that are used in the subsequent load–displacement curves. The essential fracture width of 0.3 mm is represented by the red color on a plot of the primary tensile strain, or 11, which shows the cracks. The joint shear failure of the subassembly is accompanied by a large widening of the diagonal crack in the core of the joint. Due to the joint's horizontal expansion upon breakdown, vertical cracks progress on the back part of the joint. Furthermore, the longitudinal beam is torsionally stressed by the slab as evidenced by the helical fissures that run down its backside.

3.4 Without a Transverse Beam and Slab, a Beam-Column Joint (2D)

Figure 6 illustrates load–displacement patterns of 2D beam-column connections with different aspect ratios. The various cracking phases are also included to the curves. It is clearly seen that the curve loses stiffness with each new crack that forms. The earliest cracks in the beam are flexural ones, which affect all sub-assemblies. The curves demonstrate that as aspect ratios are increased, the stiffness of the beam rises and the number of cracks decreases. The increased moment of inertia of the beam is related to both outcomes. Additionally, the higher shear loads at the ends of the column are enhanced due to the increased shear strain on the beam, which results in flexural cracking in the column.

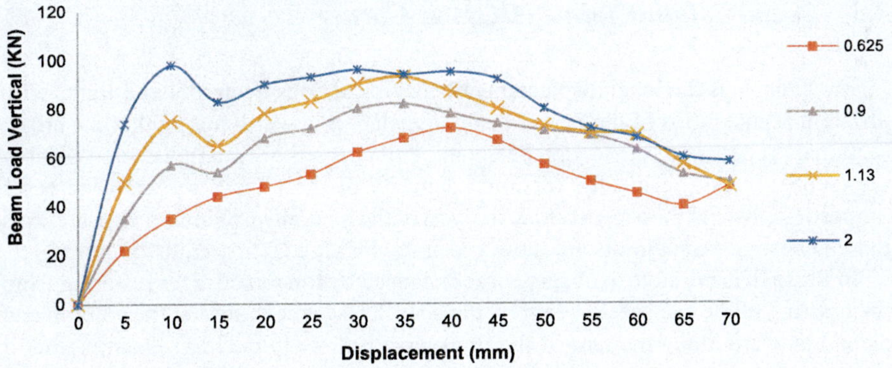

Fig. 6 The load–displacement (l-d) behavior of the 2D geopolymer beam-column (GBC) joint is illustrated through curves

The major tensile stresses over the aspect ratio are represented by the numbers in Fig. 7. The estimated values' prospective trend is also shown. They clearly show that Sharma's model fits them well [13].

As the aspect ratios increase, the primary tensile stress values (pt) for initial joint cracking and ultimate joint strength exhibit a similar trend to the vertical beam load. The examined 2D joints all experience shear failure.

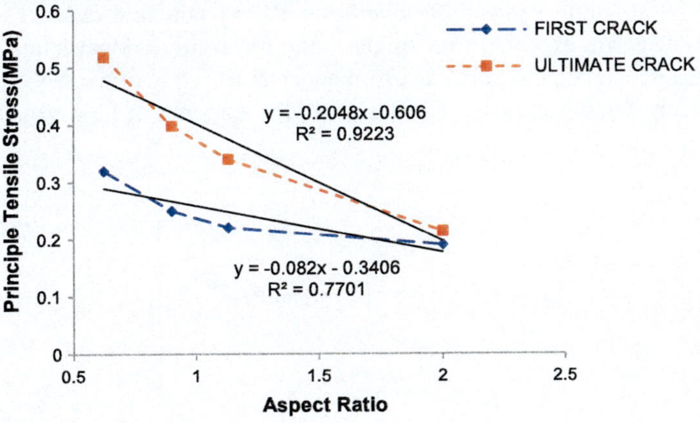

Fig. 7 Principle tensile stress to aspect ratio 2D GBC joint

3.5 Beam-Column Joints (BCJ) @ Corner

Figure 8 displays the load–displacement behavior of edge beam-column joints with different aspect ratios of the lateral beam and slab. It is worth noting that the cross-sectional properties of the loaded beam and the transverse beam in the subassembly are identical. As a result, the torsional capacity of the transverse beam increases proportionally with the aspect ratio. Analysis of the load–displacement curves reveals that 3D beam-column joints are more prone to cracking compared to 2D joints.

In Fig. 5, it is possible to observe cracks caused by torsional forces resulting from the twisting of the transverse beam by the slab. These cracks start at the column and extend outward along the rear of the transverse beam. Upon closer examination, it has been found that as the aspect ratio increases, the torsional cracking appears at lower displacements and higher loads at the tip of the beam.

Nevertheless, the loads in the corner joint exceed those in the 2D joint, leading to excessive bending or flexural cracks in both the slab and column.

The load-carrying capacity of the slab and the restrictive effect of the transverse beam on the joint can account for the higher load values. It is also interesting to observe that, in comparison to 2D joints, the rigidity does not recover as quickly following the first joint crack. Particularly, for the aspect ratios of 1.7 and 2.3, this is true. The reason for this outcome could be attributed to the fact that tensile stresses in the slab bars decrease the ability of the struts to remain stable by increasing the tensile pressures that the joint shear must resist.

The values of the initial tensile stresses, maximum joint shear strength, and initial joint fracture are shown in Fig. 9, together with any potential changes. The critical values of pt are expected to be higher than for 2-dimensional joints due to 3-dimensional phenomena like slab involvement and restriction of the transverse beam and transverse slab toward joint. The decline of the tendency is, however, essentially the same.

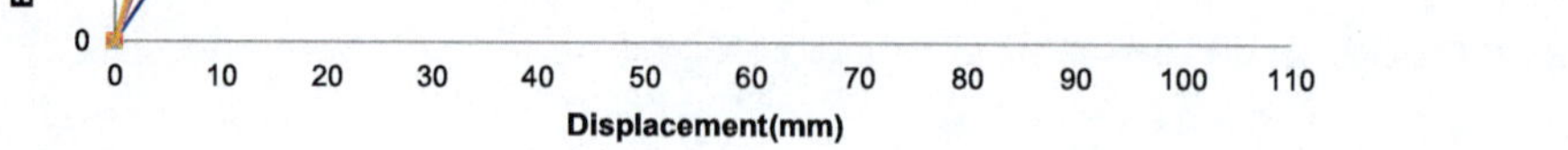

Fig. 8 Load–displacement curves for the corner joint of a 2D geopolymer beam-column (GBC) connection

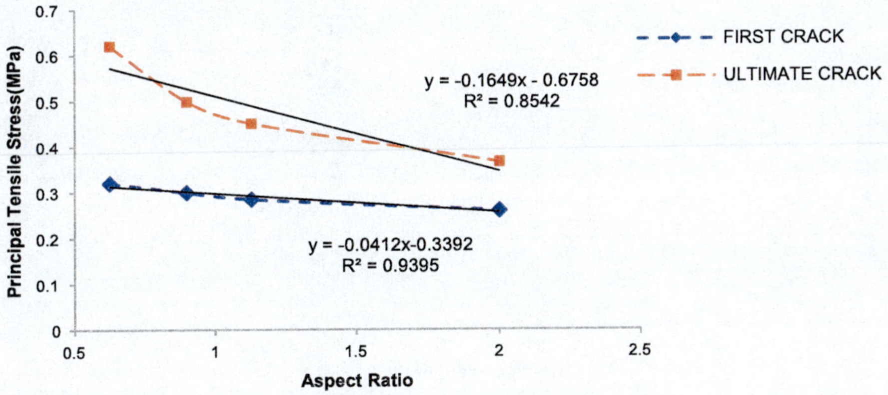

Fig. 9 Principle tensile stress to aspect ratio 2D corner GBC joint

3.6 *Beam-Column Joints@ Edge*

Load–displacement behaviors of the end connections of beam-column joints are depicted in Fig. 10. Like the corner joint, which failed due to torsion of the transverse beam, the edge joint with an alpha value of 0.92 also undergoes failure. The torsional crack appeared to originate prior to the joint cracking, according to detailed examination. This proves that transverse beam's earlier failure prevented the joint's ultimate shear strength from being reached. However, the edge joint has a higher peak load than the corner joint, which also indicates that it has a higher ultimate shear strength. Like with the other aspect ratios, a joint shear failure results in a loss in the sub-ability assembly's capacity to carry loads. That the very first joint crack also happens first within those cases. Due to the edge joints' increased slab contribution, which causes a greater disturbance of the strut stabilization, the stiffness recovers after the initial joint crack less rapidly than at corner joints.

Figure 11 displays the computed primary tensile stress values (pt) for the analyzed edge joints. Only the pt at initial joint fracture is taken into consideration in the trends for the scenario with alpha identical to 0.92 due to the transverse beam's earlier torsional failure. The findings indicate that the ultimate shear capacity of corner and 2D joints is less prone to degradation.

According to the available data, its transverse beam and corresponding slab possess the largest an impact on how edge joints behave during joint shear. The slab contribution has increased and a transverse beam is now projecting into the joint from two sides. For example, the primary tensile stress threshold that leads to the initial fracture of the joint and Priestley's suggested aspect ratio of 1 is 0.29 fck 0.5, however the value rises to 0.46 fck 0.5 and 0.6429 fck 0.5 for the corner and edge joint, respectively.

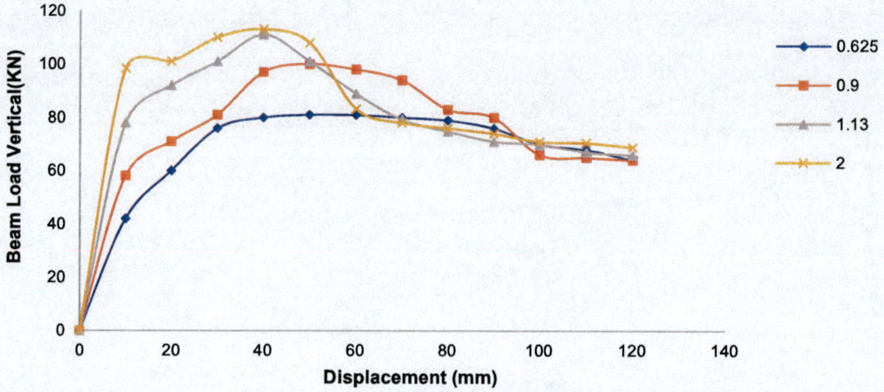

Fig. 10 Load displacement (ld) curves for 2D geopolymer beam-column (GBC) edge beam joint

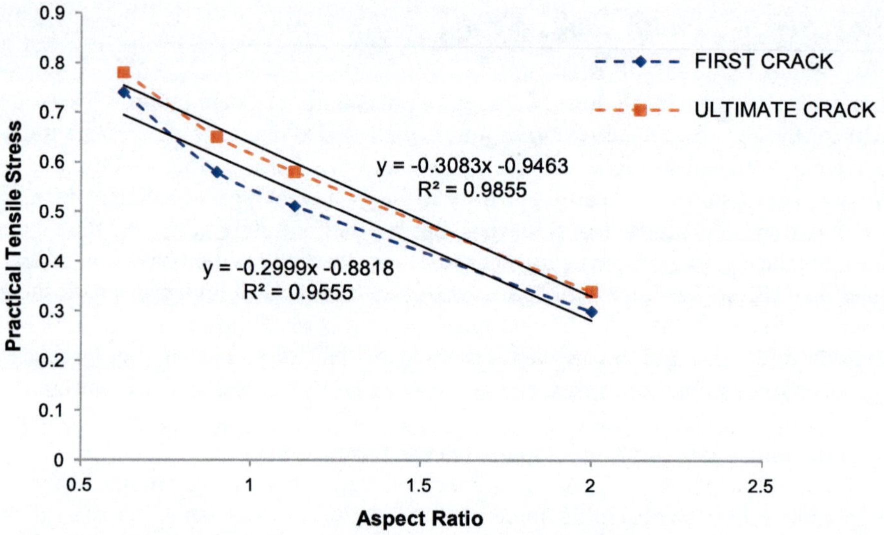

Fig. 11 Practical tensile stress to aspect ratio of 2D Edge GBC joint

4 Conclusions

A 3D beam-column joint's shear strength was evaluated by utilizing finite-element analysis to investigate the impact of the joint aspect ratio. It is well known that the primary tensile stresses for 2D joints and 3D joints have a variety of critical values suggested by the literature. This has happened as a result of the slab's involvement and the joint's limitation by the longitudinal beam and the slab. As the slab bars experience increasing tensile stresses, they transfer these stresses to the joint through the transverse beam's twisting action, thereby intensifying the shear resistance demands on the joint. According to an example [4], the maximum joint shear resistance may not be reached when the transverse beam has relatively low torsional resistance, which was frequently the issue for low aspect ratios, because of prior torsional collapse of the lateral or transverse beam. Similar to 2D joints, the investigated 3D joints also experience a decrease in shear resistance as the aspect ratio increases. Upon comparing the reduction in ultimate joint shear capacity of 2D joints and corner joints to the acquired pt values trend line, it was observed that the decline in edge joints is slightly more significant.

References

1. Hanson NW, Connor HW (1967) Seismic resistance of reinforced concrete beam-column joints. J Struct Div Proc Am Soc Civ Eng 93(5). https://doi.org/10.1061/JSDEAG.0001785
2. Priestley MJN (1997) Displacement based seismic assessment of reinforced concrete buildings. J Earthq Eng 1(1):157–172. https://doi.org/10.1142/S1363246997000088
3. Committee Euro-International Du Beton (1998) Seismic design of reinforced concrete structures for controlled inelastic response. Thomas Telford Ltd.
4. Federal Emergency Management Agency (2000) Pre standard and commentary for the seismic rehabilitation of buildings. FEMA, Washington, D.C.
5. Pantazopoulou S, French C (2001) Slab participation in practical design of R.C. frames. ACI Struct J 98(4):479–489
6. Ozbolt J, Li Y, Kozar I (2001) Microplane model for concrete with relaxed kinematic constraint. Int J Solids Struct 38(16):2683–2711. https://doi.org/10.1016/S0020-7683(00)00177-3
7. Lettow S (2007) Ein Verbundelementfürnichtlineare Finite ElementeAnalysen—Anwendung auf Übergreifungsstöße. PhD Thesis. University of Stuttgart
8. Tsonos AG (2007) Cyclic load behavior of reinforced concrete beam-column sub assemblages of modern structures. ACI Struct J 104(4):468–478. https://doi.org/10.2495/eres050421
9. Sharma A (2013) Seismic behavior and retrofitting of RC frame structures with emphasis on beam-column joints—Experiments and numerical modeling. Ph.D. Thesis. University of Stuttgart
10. Hassan WM (2011) Analytical and experimental assessment of seismic vulnerability of beam-column joints without transverse reinforcement in concrete buildings. Ph.D. Thesis. University of California, Berkeley
11. Wong HF, Kuang JS (2008) Effects of beam-column depth ratio on joint seismic behavior. Struct Build 161(2):91–101. https://doi.org/10.1680/stbu.2008.161.2.91
12. Sharma A, Hofmann J (2016) Modeling parameters for beam-column joints in seismic performance assessment of structures—a new proposal. Fib Symp Capetown
13. Park S (2010) Experimental and analytical studies on old reinforced concrete buildings with seismically vulnerable beam-column joints. Ph.D. Thesis. University of California, Berkeley

Strength Characteristics and Impact Resistance of Fiber-Reinforced Geopolymer Concrete Elements

Sambaiah Rayapudi and T. Chandra Sekhar Rao

1 Introduction

Alkali-activated compounds called Geopolymer are used in a variety of industries. Fly ash and GGBS are the two types of industrial wastes which are produced in India at a rate of roughly 300 and 11 million metric tonnes per annum, respectively. The key benefit of GP concrete over OPC concrete is that it should not require water for curing. By doing this, curing – which enable Geopolymer concrete to attain the required strength. To create geopolymer, silica, alumina –rich raw materials like FA and GGBS react with an alkaline liquid (NaOH and Na_2SiO_3) [1].

Because the greater silica content in Fine Aggregate requires heat curing to create strength, the combination of GGBS and Fine Aggregate based Geopolymer concrete helps it to acquire strength at ambient temperature [2, 3]. The prior studies also confirmed that the alkaline activators solution's NaOH content can speed up the reaction and impair the gel's fineness. GPC is usually made with 8 Molarity, 10 Molarity, and 12 Molarity of NaOH, but 12 Molarity of NaOH enhances its mechanical qualities better than the other two [4]. The mechanical characteristics which are compressive strength, tensile strength and flexural strengths are augmented by addition of steel fibers. Steel fiber enhanced compressive strength by 34% with an addition of 1.5%. Also, there was an raise in the tensile and flexural strengths of 62.8% and 50.7%, respectively, in normal grade cement concrete [5]. Slag-based geopolymer concrete's punching shear and impact resistance are impacted by adding of steel fibers at 0.5%, 1%, and 1.5%. Greater energy absorption was demonstrated by concrete that includes 1.0% steel fiber-integrated geopolymer with toughness

S. Rayapudi (✉)
Acharya Nagarjuna University, Guntur, India
e-mail: samba1057@gmail.com

T. C. S. Rao
Bapatla Engineering College, Bapatla, India

R. K. Pancharathi et al. (eds.), *Low Carbon Materials and Technologies for a Sustainable and Resilient Infrastructure*, Lecture Notes in Civil Engineering 440,
https://doi.org/10.1007/978-981-99-7464-1_37

values of 4123.88 Nm for ultimate failure and 3774.40 Nm for first crack toughness. A steel-fiber-reinforced slag-based geopolymer having a punching shear of 224 KN performs better [6]. SIFCON slabs with a 12% fiber volume fraction outperform RCC and PCC slabs of the same size and reinforcing ratio in terms of strength and energy absorption. A regression model is developed to predict the impact energy absorbed by SIFCON slab specimens for different % fibers [7]. Resistance of slabs against impact load can be improved by using fiber reinforced concrete [8]. Short columns of blended geopolymer concrete (FA and GGBS) with hybrid fibers are tested for uniaxial loading. Test results of the GPC columns with hybrid fibers show similar modes of failure and responses on par with the conventional RCC columns with the same volume fraction of hybrid fibers [9].

The objective of this research is to create sustainable GPC employing binders like FA and GGBS along with traditional fine and coarse aggregates. The impact resistance of Fine Aggregate-GGBS based GPC was increased by the addition of steel fibers, and the GPC's impact resistance was assessed using the drop hammer method.

2 Experimental Program

The experimental program is aimed at developing M45-grade GP concrete using trial mixes to investigate the mechanical properties and response to impact load. In the first phase of the study, after being cast and ambientally cured for 28 days, the compressive, split tensile and flexural strengths of geopolymer concrete are determined. The cube, cylindrical and beam specimens of sizes 150 × 150 × 150 mm, 150 × 300 mm, and 500 × 100 × 100 mm, respectively, are used for the above study. All cube and cylindrical specimens are tested in a digital CTM with a 2000 kN capacity and beam specimens in a UTM with a 100 T capacity.

In the 2nd phase, five slab specimens of 800 mm × 600 mm × 50 mm were cast with 0% (plain), 0.5%, 1%, 1.5%, and 2% steel fibers. In each type, two slabs are cast and tested under impact loading and the average value from the two test slabs is reported as the test result in each case. All slab elements are simply supported on all four edges and arranged in an iron box filled with sand on the sides to avoid lateral movements during testing. A weight of 20 kg is used as the falling weight, with a drop height of 1 m (Fig. 1).

2.1 *Materials*

2.1.1 Fly Ash and GGBS

The binding ingredients for the preparation of Geopolymer concrete are Class F fly ash and GGBS. Fly ash and GGBS results meet the standards of ASTM C 618 F

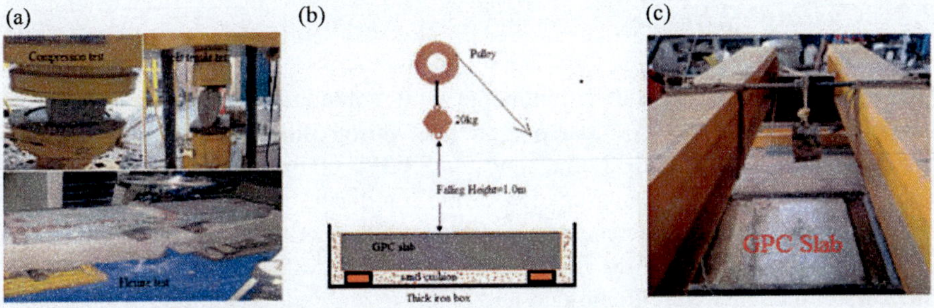

Fig. 1 **a** GPC specimens, **b** and **c** Testing of Slab elements under impact

and ASTM C 1697-16, respectively, in order to their both chemical and physical composition.

2.1.2 Aggregates

As a fine aggregate, sand available at river bank was employed. Geopolymer concrete was made using coarse particles of 20 and 12 mm. According to IS: 2386, tests were carried out to ascertain the aggregates' physical characteristics. The findings indicated that the aggregates met the IS 383 standards.

2.1.3 Alkaline Activator Solution

12M concentration of sodium hydroxide (NaOH) was prepared using pellets with 97–98 percentage of purity. The sodium silicate solution (Na_2SiO_3) was purchased from nearby vendors. The chemical formula for the sodium silicate solution is Na_2O = 14.7%, SiO_2 = 29.4%, and water = 55.9% by mass. By mass, the ratio of NaOH to Na_2SiO_3 is 2.5.

2.1.4 Fiber Reinforcement

At doses of 0.5%, 1.0%, 1.5%, and 2%, Geopolymer concrete was reinforced with glued steel fibers of 35 mm length, 0.5 mm diameter, and aspect ratio 70. Fibers are obtained from Fiber-Zone India in Ahmadabad, Gujarat.

2.2 *Casting of Slabs*

Using wooden molds, the slab components were molded to the required size. Before being applied to a mound of coarse aggregate, various fiber ratios are blended in a dry mixture of fly ash, GGBS, and sand for FRGPC slabs. To ensure uniform fiber dispersion and avoid fiber segregation or balling during mixing, hand mixing was performed after adding the necessary amount of alkaline liquid. Table vibration was used for slab specimens of the FRGPC and plain GPC. After exact one day of time, the test components were demolded and cured in ambient condition for the period of 28 days.

2.3 *Testing of Slab Elements*

The impact test was performed utilizing an in- house set up constructed in the loading frame with a pulley rigidly fixed to the shaft. The shaft is firmly fixed on the cross-girders of the loading frame. Figure 1b depicts in details the test setup that was utilized to conduct impact testing on slab elements.

3 Test Results and Discussions

3.1 *Compressive, Split Tensile and Flexural Strengths*

Table 1 shows the outcomes of the compression, split tensile and flexural strengths. According to the results, more fiber volume increases the compression, split tensile and flexural strengths of GPC. The GPC specimens evaluated for this study had compressive strengths ranging between 53 and 69 MPa, as well as tensile and flexural strengths of 3.6–4.92 MPa and 5.1–9.48 MPa, respectively, for different volume fractions of fibers. At 2% volume fraction of fiber, maximum compression, split tensile, and flexural strengths were achieved (Fig. 2).

For a 0.5%, 1.0%, 1.5%, and 2.0% volume fraction of fiber, the improvements in compressive strength at 28 days were 9.91%, 19.81%, 25.94%, and 30.19%, respectively. For the same volume fraction of fibers, the percentage increases for split tensile are 10%, 18.89%, 30.56%, and 36.67%, while for flexural strengths, they are 15.69%, 28.63%, 69.41%, and 75.88%. The findings demonstrate that the concrete strength increases then volume fraction of fiber incorporation also increased Fig. 2.

The workability value for geopolymer mixes without steel fibers (GPCS0) is 100 mm. The workability values for the geopolymer mixture including 0.5%, 1%, 1.5%, and 2% steel fibers (GPCS 0.5, 1, 1.5 and 2) are 82 mm, 70 mm, 62 mm, and 54 mm, respectively. This slump values satisfy the specifications of beam and slab

Table 1 Mix proportion

Materials used		Quantity kg/m^3
Coarse aggregate (1293)	20 mm 12 mm	776 517
Fine aggregate	–	605
Fly ash	–	197.15
GGBS	–	197.15
Sodium hydroxide (12 M)	–	41
Sodium silicate (Na_2SiO_3)	–	102.0
Water added	–	8%
Fibers (steel)	Diameter	0.5 mm
	Length	30 mm
	Aspect ratio	60
	Specific gravity	7.85
	Tensile strength	1.45 N/mm^2

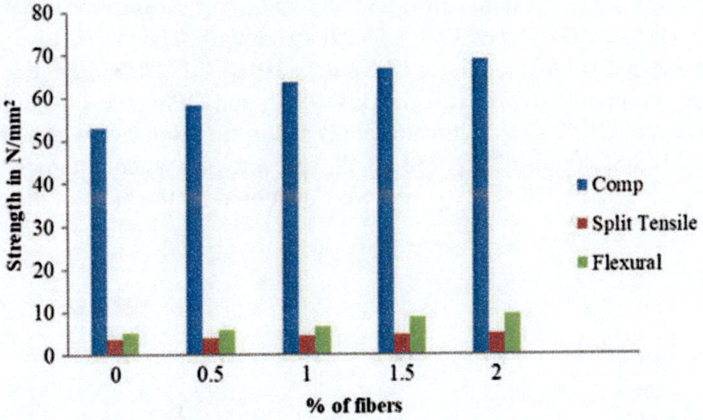

Fig. 2 Variation of strength characteristics with % fibers

elements in casting fresh state. From GPCS 0 to GPCS 2, the workability values rapidly decline as the amount of steel fibers increases.

3.2 *Number of Blows for Initial Crack and Ultimate Failures*

The impact load was applied by repeatedly dropping a 20 kg mass onto the center of the slab's top surface from a height of 1.0 m. The impact loading requirements for each slab elements considered in this study are shown in Table 1, along with the number of hits required to cause the initial crack. On each structural slab element, the number of strokes required to intimate the first crack was recorded. The amount of blows necessary for the first crack to appear is known as the first crack impact strength. The impact test was continued by dropping the hammer after the first crack appeared until the final failure stage was reached.

3.3 Energy Absorption by the Slabs Under Impact Load

Equation 1 was used to determine the energy absorption capability of each structural element utilized in this test.

$$\text{Energy absorption} = \text{Weight of hammer} \times \text{height of fall} \times \text{Number of blows} \quad (1)$$

Weight of the hammer (20 × 9.81 N) and height of the fall (1000 mm) are kept constant in the above equation throughout the experiment. At both the first crack and ultimate phases, the GPS2 slabs shows the excellent energy absorption behaviour than the GPC0, GPS0.5, GPS1 and GPC1.5 slab specimens. The % increase in energy absorption capacity at first crack for GPS2 slabs is 10%, 29.81%, 50%, and 60.02%, respectively, compared to GPS1.5, GPS1, GPS0.5 and GPS0 slabs. A similar trend was observed for GPS2 slabs' ultimate energy absorption capacity, with a percentage increase of 14.45, 40.84, 63.35, and 80.49, respectively, when compared to other constituent GPC slabs. The GPS0 slabs were shattered by the blow of the impact of hammer (Fig. 3).

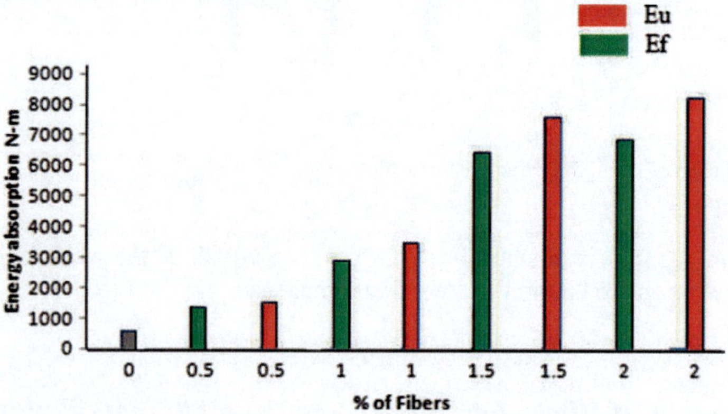

Fig. 3 Influence of % fibers on energy absorption in fiber-reinforced slabs

3.4 Regression Model for Energy-Absorption Capacity of GPC Slabs

Based on the findings of the present experiments, a simple regression model was developed to predict the ability of GPC slabs with and without fibers to absorb energy. The method of linear regression has been used to develop the model for energy absorption capacity.

$Y = K_1 + K_2X$ is a linear regression form, Where X and Y are the independent and dependent variable, K_1 and K_2 are the regression coefficients.

The energy- absorption capabilities up to the Initial crack (E_f) and ultimate phases are linked to the fiber volume fraction (f_v) and the square of the 28- days cube compressive strength (fck^2) (E_u). To get a reduced standard deviation and higher correlation, the $f_{ck}{}^2$ term was applied. So, the following equations are suggested for energy-absorption capabilities.

$$E_f = 0.064 f_{ck}^2 * f_v + 0.312 f_{ck}^2 - 122.8 \tag{2}$$

$$E_u = 0.158\, f_{ck}^2 *\, f_v + 0.75\, f_{ck}^2 - 1395.75 \tag{3}$$

As shown in Fig. 4 and Table 2, the suggested regression model is capable of accurately predicting energy absorption at both the initial crack and ultimate impact strengths.

Table 3 displays the number of impacts required to induce the first apparent crack and the final fracture of both plain and fiber-reinforced concrete. It is clear

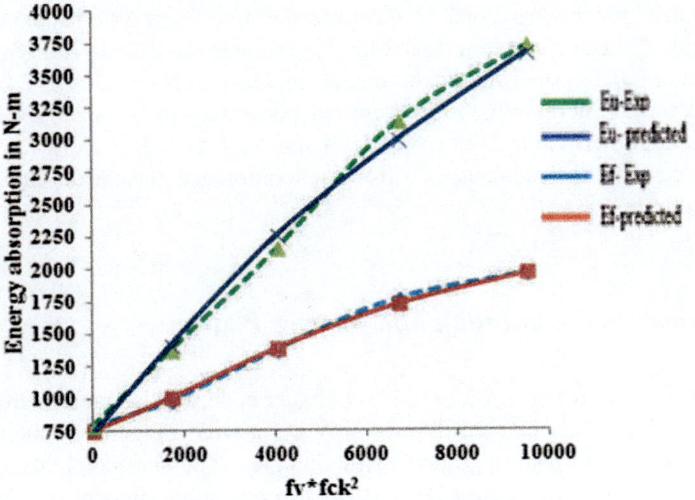

Fig. 4 Variation of experimental and predicted energy absorption capacity of slabs with respect to $V_f * f_{ck}$

Table 2 Test results of strength characteristics of GPC mixes

Designation	f_{ck} at 28 days in N/mm^2	f_t specimen at 28days in N/mm^2	f_{cr} at 28 days in N/mm^2	The no. of blows required to cause initial crack Impact strength (N_1)	The no. of blows at ultimate impact strength (N_2)
GPCS0	53	3.6	5.1	4	4
GPCS0.5	58.25	3.96	5.9	5	7
GPCS1	63.5	4.28	6.56	6	11
GPCS1.5	66.75	4.7	8.64	9	16
GPCS2	69	4.92	9.48	10	19

Note Compressive strength (f_{ck}), Split tensile Strength (f_t), Flexural Strength (f_{cr})

Table 3 Experimental and predicted values of energy absorption of slab elements at first and ultimate loads

Designation	N1	Actual value	Predicted value	% Difference	N2	Actual value	Predicted value	% Difference
	E_f in (N-m)				E_u in (N-m)			
GPCS0	4	784.4	750.80	4.28	4	784.4	725.05	7.57
GPCS0.5	5	981	1037.63	−5.77	7	1373.4	1417.10	−3.18
GPCS1	6	1377.2	1389.29	−0.88	11	2158.2	2245.53	−4.05
GPCS1.5	9	1765.8	1690.61	4.26	16	3139.2	3001.89	4.37
GPCS2	10	1962	1967.28	−0.27	19	3728	3679.48	1.30

that the presence of steel fibers in the mix significantly increased the number of blows required for the first crack to form and the final fracture to occur. When the fiber volume fractions were increased to 1%, N2 grew by N1 of 2.5, whereas the mix containing 0.5% steel fiber exhibited an increase in N2 by N1 of 2.0 times at 28 days. A further increase in the fiber volume percentage to 1.5% and 2% revealed, respectively, an increase in N2 by N1 of 1.4 and 1.5 times. This shows the amount of reserve strength the elements had from the commencement of the first crack to complete failure.

3.5 Numerical Simulation and Failure Pattern

In ABACUS/Explicit Version 6.7 (2017) computer code, five FE models are created: The ABACUS/Explicit element library provides a wide range of element formulations based on various integration methodologies. The slab component and the impactor are discretized using a 3-D- stress 8- node solid element (C3D8R) with reduced integration. This element is commonly used in structural member impact simulations. Free falling velocity of 4.43 m/sec is applied to drop weight to create

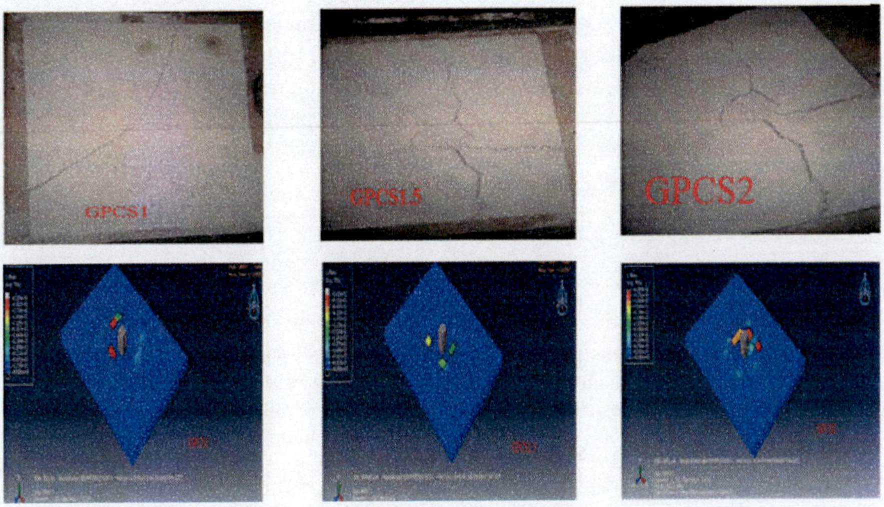

Fig. 5 A comparison of experimental and numerical damage of fiber reinforced GPC slabs

impact on the slab with a step time of 5 s. The total energy absorbed by the specimen before failure and stress contours are extracted from the visualization module. It is evident from Fig. 5 that the failure pattern shown by the numerical analysis performed on the plain and fibrous GPC slabs is consistent with that shown in the experimental investigation. As a consequence, it was established that the experimental and numerical results were in good accord. When contrasted to the experimental failure of the elements, the deformed plot of every specimen (Fig. 5) exhibits a comparable crack pattern.

The incorporation of fiber into concrete changed the failure pattern from a single large crack to a group of narrow cracks, demonstrating the beneficial effects of fiber-reinforced concrete when subjected to impact loading, which is consistent with previous studies [10, 11]

4 Conclusions

1. As the percentage of fiber increased, the compressive, tensile, and flexural strengths of GPC improved. The compressive, tensile, and flexural strengths were increased by a maximum of 30.19%, 36.67%, and 75.88%, respectively, with the addition of 2% steel fiber in comparison to plain GPC.
2. All the fiber-reinforced GPC showed higher first- crack impact energy than the plain GPC due to the fiber's ability to bridge micro fissures.
3. When compared to plain GPC and other fiber-reinforced slabs, the GPCS2 slab specimens withstood high impact loads before failure and exhibited smaller crack widths.

4. Based on the current experimental data, regression models were presented to predict the impact energy at the initial crack and ultimate phases. Theoretical hypothesis match well with the experimental results reported here.
5. Numerical analysis is consistent with the experimental investigations in predicting failure patterns.

References

1. Kashyap AMN (2018) Prediction of setting and strength characteristic of binary blended geopolymer matrix. I-manager's J Struct Eng 6(4):16–22
2. Mo KH (2014) Impact resistance of hybrid fiber-reinforced oil palm shell concrete. Constr Build Mater 50(15):499–507. https://doi.org/10.1016/j.conbuildmat.2013.10.016
3. Mindness S, Banthia N (1986) The behavior of concrete under impact loading-experimental procedures and method of analysis. Mater Constr 19:371–378
4. Sudarsana Rao H, Ramana NV, Gnaneswar K (2008) Behaviour of steel reinforced slurry infiltrated fibrous concrete (SIFCON) two way slabs in punching shear. Indian J Eng Mater Sci 334–342
5. Murali G, Santhi AS, Mohan Ganesh G (2013) Empirical relationship between the impact energy and compressive strength for fiber reinforced concrete. J Sci Ind Res 73:469–473
6. Deb PS, Nath P, Sarker PK (2014) The effects of ground granulated blast-furnace slag blending with fly ash and activator content on the workability and strength properties of geopolymer concrete cured at ambient temperature. Mater Des 62:32–39
7. Sudarsana Rao H (2010) Response of SIFCON two-way slabs under impact loading. Int J Impact Eng 37:452–458. https://doi.org/10.1016/j.ijimpeng
8. IS 456 (2000) Plain and reinforced concrete code for practice. Bureau of Indian Standards, New Delhi
9. Verma M, Dev N (2021) Sodium hydroxide effect on the mechanical properties of flyash-slag based geopolymer concrete. Struct Concr 22:368–379. https://doi.org/10.1002/suco.202000068
10. Atef B, Ashraf F, Andrew K (2006) Statistical variations in impact resistance of polypropylene fibre-reinforced concrete. Int J Impact Eng 32:1907–1920
11. Chen X, Ding Y-N, Azevedo C (2011) Combined effect of steel fibres and steel rebars on impact resistance of high performance concrete. J Cntrl South Univ Technol 18:1677–1684

Evaluation of Influence of Diaphragm Flexibility on the Seismic Response of RCC Buildings with Slab Openings

R. Pooja and B. Kavitha

1 Introduction

Due to architectural requirements, asymmetric structures become inevitable but they suffer more seismic damage when compared with symmetric structures due to plan irregularity. Shear walls are suggested in such asymmetrical structures to decrease the lateral load in the columns [1]. The suitable seismic analysis for asymmetrical structures is studied to support new guidelines. Static analysis and dynamic analysis were carried out for L and W shape. It is found that static analysis generates deflection values that are within the limit; however, time history analysis generates deflection values that are significantly higher than the permissible limit. Response spectrum analysis also does not satisfy the deflection criteria but it yields deflection values less than the time history analysis. Staad Pro, Etabs, and SAP 2000 are used for seismic analysis to study the effects of seismic parameters in irregular buildings. It is suggested to adopt dynamic analysis for irregular structures which are more than height of 40 m [2]. Both linear and non-linear analysis were carried out to study the dynamic behavior of irregular structures. It is noticed that equivalent static analysis ignores the irregularity effects yielding abnormal results while pushover analysis yields detrimental results for vertical distribution of lateral force [3].

Diaphragm discontinuities, like slab openings affects the structure's stiffness Baby and Sreeja [4]. In recent studies with mass irregularity, force is directly proportional to the displacement. Therefore, since displacement is high for irregular models, shear force is high which causes an increase in time period compared to regular models thereby reducing the performance of the structure in earthquakes [5]. The

R. Pooja (✉) · B. Kavitha
National Institute of Technology, Warangal, India
e-mail: prce21135@student.nitw.ac.in

B. Kavitha
e-mail: kavithab@nitw.ac.in

R. K. Pancharathi et al. (eds.), *Low Carbon Materials and Technologies for a Sustainable and Resilient Infrastructure*, Lecture Notes in Civil Engineering 440,
https://doi.org/10.1007/978-981-99-7464-1_38

effect of slab openings at different locations with different column shapes including rectangular, square and square are investigated using response spectrum analysis. Base shear is observed to increase with increase in the area of openings. From the base shear, displacement and drift point of view, slab opening at center position is found to be more suitable under seismic behavior [4]. Equivalent static analysis and pushover analysis are done for structure with slab openings at center, corner and periphery. It is observed that slab opening at the periphery case is effective.

The diaphragm is a horizontal load resisting element, which ties the structural components together maintaining the structural homogeneity. In general, diaphragms are thought of being rigid, indicating no relative displacement between the member's nodes. Assumption of no relative displacement in the diaphragm makes the computational process simpler by reducing the degree of freedom. Therefore, the concept of a flexible diaphragm is primarily considered in structures like composite structures. But in many practical cases of RCC buildings, such as plan asymmetric buildings and buildings with slab openings, rigid diaphragm consideration is invalid due to relative displacement.

From various nation codes which describe diaphragm flexibility in terms of qualitative criteria and quantitative criteria. It is concluded that quantitative criteria are not accurate enough and an error formula is given to reform those criteria. It is also inferred from previous work that diaphragm flexibility affects the time period and mode shape of the structure [6]. In this paper, asymmetric buildings were studied for seismic behavior with flexible diaphragms. It is inferred from the results that diaphragm flexibility leads to a high fundamental time period and these results can be generalized for low rise buildings [7]. In this paper RCC building with varying levels of flexibility along with plan irregularity is analyzed by linear time history analysis using Etabs software. The results indicate that seismic demand parameters, beam forces and column forces were affected by diaphragm flexibility [8]. In this study, the definition of Center of Rigidity has been extended to flexible diaphragm. The torsional provisions building with flexible floor as specified by the code is illustrated by superposition-based analysis procedure.

1.1 Classification of Diaphragm

The classification of diaphragm is different in various nation codes which is briefly furnished in Table1. As per IS 1893-2016, a diaphragm is rigid when displacement at its midpoint is less than 1.2 times the average of end displacements and hence vice-versa for flexible diaphragm which is presented in Fig. 2. Rigid diaphragm is capable of rotational or torsional behavior whereas flexible diaphragm is not capable. In rigid diaphragms, the load is distributed as per the relative stiffness of the elements and in the flexible diaphragms, the load is distributed according to the tributary area which is pictorially represented in Fig. 1.

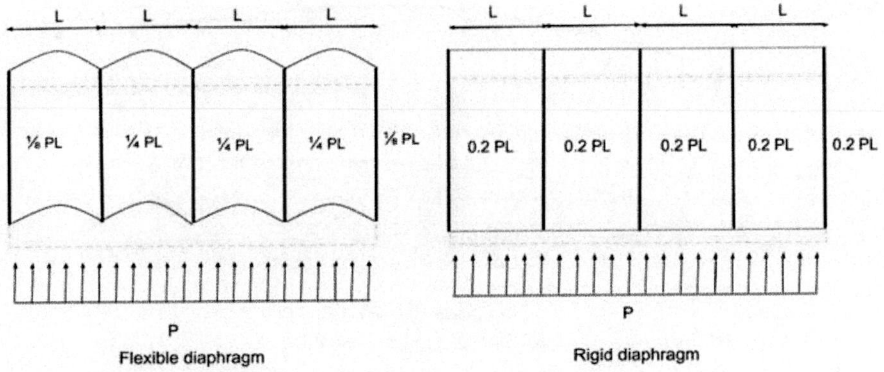

Fig. 1 Pictorial representation of rigid and flexible diaphragm

Table 1 Definition of flexible diaphragm as per various codes

Code No	Code nation	Year	Description
EC8	Eurocode 8	1994	If horizontal displacement of a diaphragm exceed the results by more than 10% of the rigid diaphragm assumption, then it is considered as flexible diaphragm
UBC	Uniform building code	1994	If the max. lateral displacement is more than 2 times the average story drift of the diaphragm, then it is termed as flexible diaphragm
IS 1893	Indian standard code	2016	If a diaphragm deforms in such a way that the maximum lateral displacement measured from the chord of the deformed shape at any point of the diaphragm is more than 1.2 times the average displacement at ends
ASCE 7	American society of civil engineers	2022	Under lateral load, if the maximum deformation is more than 2 times the average drift of the lateral force-resisting system, then it is flexible diaphragm

2 Objectives

(1) To provide openings in diaphragm at different places and study the seismic response of the structure.
(2) To incorporate the different levels of flexibility in diaphragm under the seismic action.

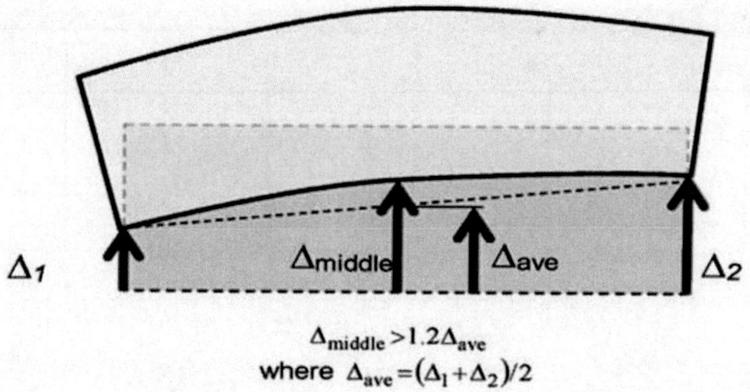

Fig. 2 Flexible floor diaphragm as per IS 1893:2016

(3) To assess the seismic reaction parameters utilizing response spectrum analysis in the Etabs program, such as base shear, displacement, and inter-story drift.

3 Building Properties and Model

The properties of the building model considered and loading conditions given were presented in Table 2. The 3D representation of the model is shown in Fig. 3. The plan dimensions, bay width, no. of bays and percentage of openings were discussed in Table 3.

Table 2 Properties and loading conditions

No of story	G + 5
Story height	3.5 m
Number of bays and width	6 bays in both directions of 6 m × 3.5 m
Grade of concrete	M 30
Grade of steel	Fe 415
Beam size and column size	200 mm × 400 mm and 400 mm × 400 mm
Live load	3 kN/m^2
Roof live load	1.5 kN/m^2
Floor finish	1 kN/m^2
Seismic zone	III
Response reduction factor	5
Importance factor	1
Soil type	Medium

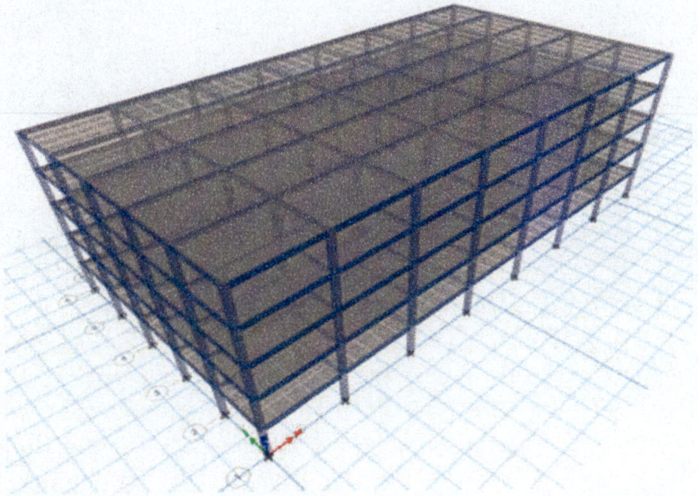

Fig. 3. 3D Model of the structure

Table 3 Model details

Model name	Plan dimensions	No. of bays in the direction of		Bay width in direction of		Size of opening	% of opening
		X	Y	X	Y		
No opening	42 m × 22.5 m	7	5	6 m	4.5 m	–	–
Center opening	42 m × 22.5 m	7	5	6 m	4.5 m	1 no. of 3 m × 3 m	1%
Corner opening	42 m × 22.5 m	7	5	6 m	4.5 m	4 nos. of 2 m × 2 m	7%

4 Results and Discussions

The base shear parameter is represented graphically in Fig. 4a and b as Base shear in X and Y direction respectively. Similarly, maximum story displacement in X and Y direction is represented in Fig. 5a and b, respectively. The story displacement is plotted in Fig. 6a and b for X and Y direction, respectively. The axial force, shear force and bending moment were graphically presented in Figs. 7, 8 and 9, respectively.

4.1 Base Shear

From the above graph, under each opening case, it is observed that base shear decreases with increase in diaphragm flexibility. The base shear is maximum at no

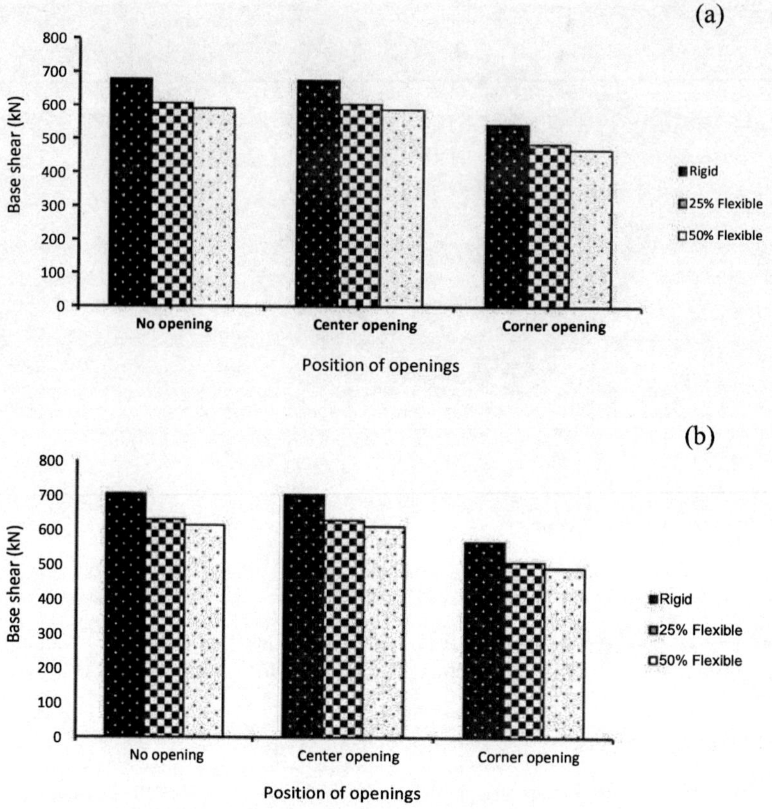

Fig. 4 **a** Base shear in X direction, **b** Base shear in Y direction

opening case, decreasing in the order of center opening followed by corner opening. Base shear value is found to be higher in Y direction compared to X direction.

4.2 Maximum Story Displacement

In both X and Y direction, maximum displacement increases with increase in flexibility in all opening cases. Displacement is found to be higher in Y direction compared to X direction. Corner opening has maximum displacement compared to other cases.

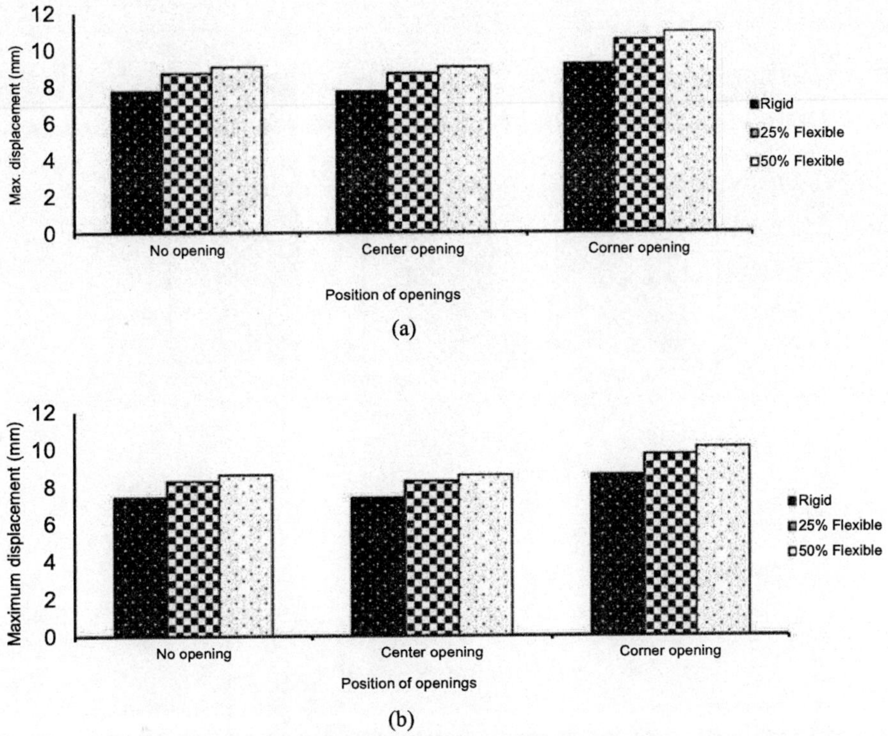

Fig. 5 **a** Max. story displacement in X direction, **b** Max. story displacement in Y direction

4.3 Maximum Story Drift

Story drift is constant for no opening and center opening case in both directions. It increases drastically for the corner opening case.

4.4 Column Forces

As the flexibility increases, axial load carrying capacity of the column increases. This demonstrates that there is a force transfer from lower to higher stiffer materials. In comparison to other cases, the corner opening case has a much lower load carrying capacity.

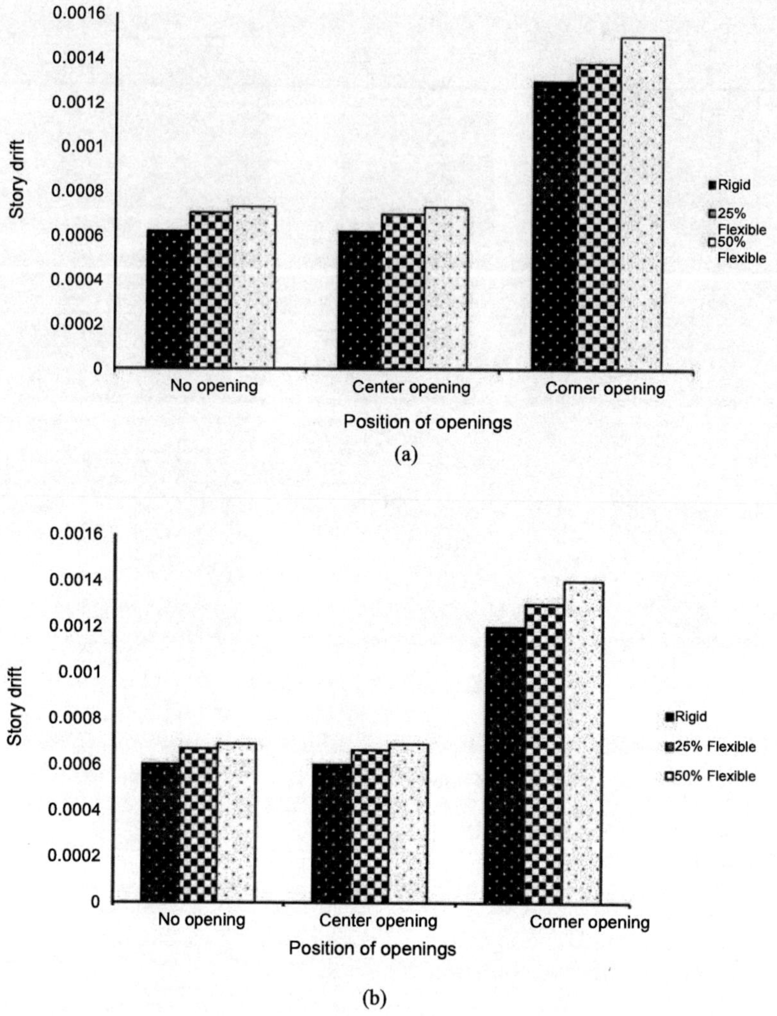

Fig. 6 **a** Story drift in X direction, **b** Story drift in Y direction

4.5 Beam Forces

As the flexibility % rises, the shear force and bending moments decrease. The corner opening instance has the lowest beam forces.

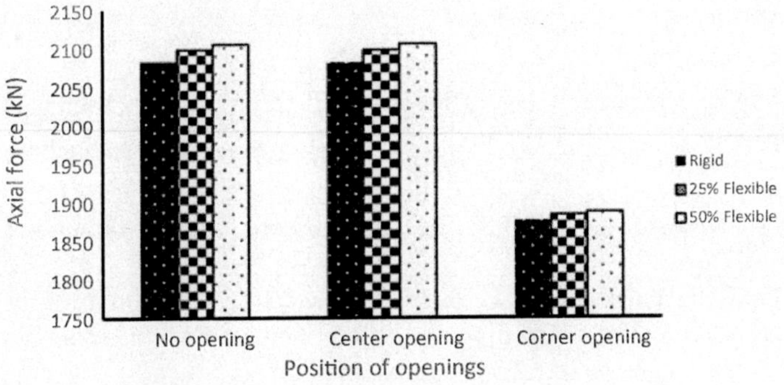

Fig. 7 Axial force

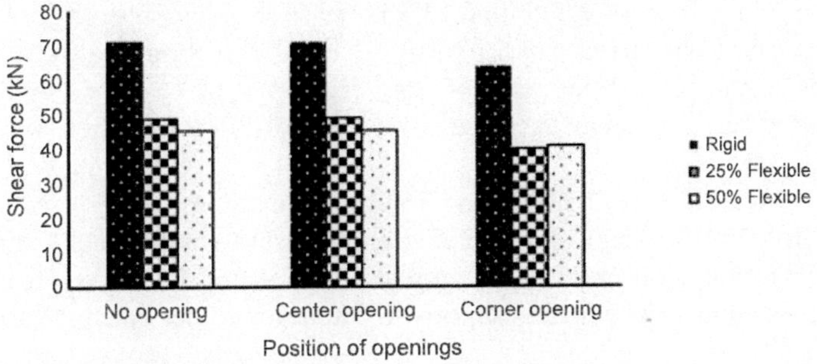

Fig. 8 Shear force

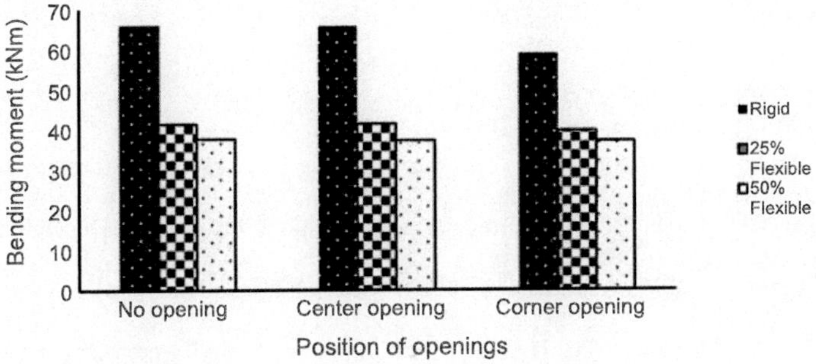

Fig. 9 Bending moment

5 Conclusion

Seismic response parameters were investigated for rigid diaphragm and flexible diaphragm with 25% and 50% flexibility no opening, a center opening, a corner opening cases. The results are quantitatively expressed as follows:

(1) Base shear decreases by 10.5% and 13% in both X and Y directions under all opening cases.
(2) In both X and Y direction, increase of 11% and 14.5% maximum story displacement is observed for no opening and center opening. For corner opening case, in X direction, increase of 12.5% and 15.75% displacement is observed whereas in Y direction, 11.2% and 14.65% increase is.
(3) An increase of 12% and 15% story drift is observed under no opening and center opening case in X direction. In Y direction, an increase of 14% and 13% in no opening case whereas 9.77% and 13% increase for center opening case. Corner opening case has increase of 5.8% and 13% in X direction whereas Y direction yields 7.69% and 20% increase of story drift.
(4) Under axial force, slight increase is seen as flexibility increases, which is not significant.
(5) Shear force decreases by 31% and 36% in X direction and 37% and 43% in Y direction for no opening and center opening cases in both X and Y directions.
(6) Bending moment is observed to decrease by 37% and 43% for no opening and center opening case which is slightly different for corner opening case as 33% and 37%.

References

1. Pokharel S (2019) Seismic performance of symmetric and asymmetric multi-storeyed buildings. Int J Recent Technol Eng 8(1S3):364–369
2. Firoj M, Singh SK (2018) Response spectrum analysis for irregular multi-storey structure in seismic zone V. 16th Symp Earthq Eng 300
3. Ravikumar CM (2012) Effect of irregular configurations on seismic vulnerability of RC buildings. Archit Res 2(3):20–26
4. Baby BE, Sreeja S (2013) Analysis of buildings with slab discontinuity. Int J Sci Res 5(9):999–1003
5. Manmathan AV, Aiswarya S (2017) Analysis of buildings with varying percentages of diaphragm openings. IJERT 6(6):461–466
6. Eivani H (2017) Effects of diaphragm flexibility on seismic response of asymmetric-plan buildings. Gradevinar 70(18):965–974. https://doi.org/10.14256/JCE.2146.2017
7. Nagpure A, Sanghai SS (2018) Effect of diaphragm flexibility on the seismic response of RCC framed building considering diaphragm discontinuity. Int J Innovations Eng Sci 3(5):100–108
8. Basu D, Jain SK (2004) Seismic analysis of asymmetric buildings with flexible floor diaphragms. J Struct Eng (JoSE) 130(8):1169–1176